Combinatorial Theory and Statistical Design

Probability and Mathematical Statistics (Continued)

PURI, VILAPLANA, and WERTZ • New Perspectives in Theoretical and Applied Statistics
RANDLES and WOLFE • Introduction to the Theory of Nonparametric Statistics
RAO • Linear Statistical Inference and Its Applications, *Second Edition*
RAO • Real and Stochastic Analysis
RAO and SEDRANSK • W.G. Cochran's Impact on Statistics
RAO • Asymptotic Theory of Statistical Inference
ROHATGI • An Introduction to Probability Theory and Mathematical Statistics
ROHATGI • Statistical Inference
ROSS • Stochastic Processes
RUBINSTEIN • Simulation and The Monte Carlo Method
SCHEFFE • The Analysis of Variance
SEBER • Linear Regression Analysis
SEBER • Multivariate Observations
SEN • Sequential Nonparametrics: Invariance Principles and Statistical Inference
SERFLING • Approximation Theorems of Mathematical Statistics
SHORACK and WELLNER • Empirical Processes with Applications to Statistics
TJUR • Probability Based on Radon Measures

Applied Probability and Statistics

ABRAHAM and LEDOLTER • Statistical Methods for Forecasting
AGRESTI • Analysis of Ordinal Categorical Data
AICKIN • Linear Statistical Analysis of Discrete Data
ANDERSON, AUQUIER, HAUCK, OAKES, VANDAELE, and WEISBERG • Statistical Methods for Comparative Studies
ARTHANARI and DODGE • Mathematical Programming in Statistics
ASMUSSEN • Applied Probability and Queues
BAILEY • The Elements of Stochastic Processes with Applications to the Natural Sciences
BAILEY • Mathematics, Statistics and Systems for Health
BARNETT • Interpreting Multivariate Data
BARNETT and LEWIS • Outliers in Statistical Data, *Second Edition*
BARTHOLOMEW • Stochastic Models for Social Processes, *Third Edition*
BARTHOLOMEW and FORBES • Statistical Techniques for Manpower Planning
BECK and ARNOLD • Parameter Estimation in Engineering and Science
BELSLEY, KUH, and WELSCH • Regression Diagnostics: Identifying Influential Data and Sources of Collinearity
BHAT • Elements of Applied Stochastic Processes, *Second Edition*
BLOOMFIELD • Fourier Analysis of Time Series: An Introduction
BOX • R. A. Fisher, The Life of a Scientist
BOX and DRAPER • Empirical Model-Building and Response Surfaces
BOX and DRAPER • Evolutionary Operation: A Statistical Method for Process Improvement
BOX, HUNTER, and HUNTER • Statistics for Experimenters: An Introduction to Design, Data Analysis, and Model Building
BROWN and HOLLANDER • Statistics: A Biomedical Introduction
BUNKE and BUNKE • Statistical Inference in Linear Models, Volume I
CHAMBERS • Computational Methods for Data Analysis
CHATTERJEE and PRICE • Regression Analysis by Example
CHOW • Econometric Analysis by Control Methods
CLARKE and DISNEY • Probability and Random Processes: A First Course with Applications, *Second Edition*
COCHRAN • Sampling Techniques, *Third Edition*
COCHRAN and COX • Experimental Designs, *Second Edition*
CONOVER • Practical Nonparametric Statistics, *Second Edition*
CONOVER and IMAN • Introduction to Modern Business Statistics
CORNELL • Experiments with Mixtures: Designs, Models and The Analysis of Mixture Data

Applied Probability and Statistics (Continued)

COX • Planning of Experiments
COX • A Handbook of Introductory Statistical Methods
DANIEL • Biostatistics: A Foundation for Analysis in the Health Sciences, *Fourth Edition*
DANIEL • Applications of Statistics to Industrial Experimentation
DANIEL and WOOD • Fitting Equations to Data: Computer Analysis of Multifactor Data, *Second Edition*
DAVID • Order Statistics, *Second Edition*
DAVISON • Multidimensional Scaling
DEGROOT, FIENBERG and KADANE • Statistics and the Law
DEMING • Sample Design in Business Research
DILLON and GOLDSTEIN • Multivariate Analysis: Methods and Applications
DODGE • Analysis of Experiments with Missing Data
DODGE and ROMIG • Sampling Inspection Tables, *Second Edition*
DOWDY and WEARDEN • Statistics for Research
DRAPER and SMITH • Applied Regression Analysis, *Second Edition*
DUNN • Basic Statistics: A Primer for the Biomedical Sciences, *Second Edition*
DUNN and CLARK • Applied Statistics: Analysis of Variance and Regression, *Second Edition*
ELANDT-JOHNSON and JOHNSON • Survival Models and Data Analysis
FLEISS • Statistical Methods for Rates and Proportions, *Second Edition*
FLEISS • The Design and Analysis of Clinical Experiments
FOX • Linear Statistical Models and Related Methods
FRANKEN, KÖNIG, ARNDT, and SCHMIDT • Queues and Point Processes
GALLANT • Nonlinear Statistical Models
GIBBONS, OLKIN, and SOBEL • Selecting and Ordering Populations: A New Statistical Methodology
GNANADESIKAN • Methods for Statistical Data Analysis of Multivariate Observations
GREENBERG and WEBSTER • Advanced Econometrics: A Bridge to the Literature
GROSS and HARRIS • Fundamentals of Queueing Theory, *Second Edition*
GUPTA and PANCHAPAKESAN • Multiple Decision Procedures: Theory and Methodology of Selecting and Ranking Populations
GUTTMAN, WILKS, and HUNTER • Introductory Engineering Statistics, *Third Edition*
HAHN and SHAPIRO • Statistical Models in Engineering
HALD • Statistical Tables and Formulas
HALD • Statistical Theory with Engineering Applications
HAND • Discrimination and Classification
HOAGLIN, MOSTELLER and TUKEY • Exploring Data Tables, Trends and Shapes
HOAGLIN, MOSTELLER, and TUKEY • Understanding Robust and Exploratory Data Analysis
HOEL • Elementary Statistics, *Fourth Edition*
HOEL and JESSEN • Basic Statistics for Business and Economics, *Third Edition*
HOGG and KLUGMAN • Loss Distributions
HOLLANDER and WOLFE • Nonparametric Statistical Methods
IMAN and CONOVER • Modern Business Statistics
JAGERS • Branching Processes with Biological Applications
JESSEN • Statistical Survey Techniques
JOHNSON • Multivariate Statistical Simulation
JOHNSON and KOTZ • Distributions in Statistics
Discrete Distributions
Continuous Univariate Distributions—1
Continuous Univariate Distributions—2
Continuous Multivariate Distributions

(*continued on back*)

Combinatorial Theory and Statistical Design

GREGORY M. CONSTANTINE

JOHN WILEY & SONS

New York • Chichester • Brisbane • Toronto • Singapore

Library of Congress Cataloging in Publication Data:

Constantine, Gregory M.
Combinatorial theory and statistical design.

(Wiley series in probability and mathematical statistics. Applied probability and statistics, ISSN 0271-6356)
Includes bibliographies and index.
1. Combinatorial analysis. 2. Experimental design.

I. Title. II. Series.
QA164.C688 1987 511′.6 86-26782
ISBN 0-471-84097-1

Printed in the United States of America

10 9 8 7 6 5 4 3 2 1

To my brother
Gregory Magda
In memoriam

Preface

This book is addressed to those who are interested in understanding the fundamental techniques of discrete mathematics as applied to statistical design. It provides a link between the many available texts on elementary combinatorics and specialized treatises on combinatorial theory, design of experiments, coding theory, and finite geometries. Part of the material will be useful to experimenters (engineers and managers) who are concerned with the actual planning of statistical studies.

Looking at the material as a whole, four major themes predominate: the enumerative aspect, represented in Chapters 1 through 3 (along with Chapter 6); graphs and networks, the subjects of Chapters 4 and 5; the statistical and combinatorial designs studied in Chapters 7 and 8; and last, the material on partially ordered sets with special emphasis on Möbius inversion.

Many of the concepts that are studied, such as t-designs, partial designs, partial geometries, and latin squares, arise out of considerations of maximizing the precision of certain statistical estimates under a variety of models. Seeking graphs with a maximal number of spanning trees, among all graphs with a specified number of vertices and edges, is another general problem that stems out of such considerations. It is discussed in Section 3 of Chapter 4. Understanding how a response of interest is affected by several factors leads to the concept of factorial designs and the related combinatorial structures with which they are identified: orthogonal arrays, balanced arrays, and orthogonal latin squares. This material is part of Chapter 8 on statistical design. While writing that chapter I tried to show *why and how* such structures arise from statistical concerns. Once identified, these structures may be studied on purely geometric or combinatorial grounds. Awareness of their origin and applicability, however, should help a research worker discern issues of potential use from other issues.

As to the style of presentation, I have chosen the subsections indicated by double numbers (e.g., Section 7.12) as the basic structural unit. Certain ideas and technical details often require examination from several angles in order to be clearly understood. It is for this reason that a correct overall picture usually emerges only upon reading an entire subsection. I thus suggest that these subsections be viewed as the fundamental unit for assimilating information. The theory is often presented in a conversational way, gearing the reader toward the important ideas, showing the necessity and benefits of introducing new techniques. An appropriate "atmosphere" is thus created, the results of interest being then within easy reach. The basic attitude is to share and rediscover the important results, rather than present them in a formal, detached way. References through the actual text are kept to a minimum. Explanatory notes are included, however, at the end of each chapter, to inform on the original contributions.

I have been privileged to work with Professor Ching-Shui Cheng of the Department of Statistics at Berkeley and Professor Ravi S. Kulkarni of the Max Planck Institute in Bonn. Many a time they helped me assimilate those finer points on which my power of perception proved somewhat weak. I am pleased to thank them both. During the late stages of writing the book, I visited the University of Wisconsin. Comments from and conversations with Professor George Box proved to be most helpful. Attending the informal Monday night seminars at his home benefited and influenced me in more ways than he will ever know. To Professor Jeff Wu, Professor Bob Wardrop, and Professor G. Bhattacharyya, all of the University of Wisconsin, I wish to express my thanks.

In its initial state the text consisted of notes prepared for a graduate course taught in the spring of 1983 jointly with Professor George Minty. The course gradually evolved into a seminar that continued through one of the summer sessions, and my notes got thicker. George lectured on the subject of network flows, exposing us to a generous supply of examples, sprinkled with a good deal of humor. Sharing information and ideas with the members of the class has been a joyous experience. Among the participants from whose presence I benefited in many ways were Professor Maria Wonenburger, Professor John Duncan, and Professor Richard Bradley. I thank them for their interest, and the many spirited discussions. Of the students actively involved I wish to mention Bill D'Amico, who offered some nice insights into the enumeration of graphs.

The writing of the book has been indirectly but decisively influenced by the numerous conversations with Grahame Bennett and Vinay Deodhar. Portions of the manuscript were read by R. N. Bhattacharya. I am thankful to all three.

Personally and professionally I owe much to Carolyn, my special friend. During the two years of writing many friends have in fact helped in many ways. For the encouraging pat on the back that I received from most, I say thank you.

For those that help in times of hardship a lasting fondness will remain. It is with such thoughts that I shall long remember Herbert J. Curtis of the University of Illinois in Chicago. To my teachers Sam Hedayat, Noboru Ito, and David Foulser I remain grateful: they instilled the fascination for the subject on which I now write.

I kindly thank Miss Medlock for her patience with my organizational skills, and for the excellent typing.

To Beatrice Shube, the statistics editor at Wiley, I owe a good deal; foremost for the professional way in which she handled my manuscript, as well as the many constructive conversations and suggestions.

ON THE POSSIBLE SELECTIONS OF TOPICS

Of the nine chapters, the first three may well be easiest to read. There is, in fact, a unifying thread among these, which makes me think of them as one whole. Each of the remaining six is rather more self-contained. Cross-references rarely happen, and when they do I almost always discuss shared conceptual aspects rather than specific technical points.

The less experienced reader, such as a student with no exposure to discrete mathematics, is invited to first take a glance at Chapters 1 and 2 or at the chapter on graphs. She or he will then be in a position to select (out of the remaining ones) those subjects which his or her background would assimilate best. A more experienced reader, one who has had an upper undergraduate course in combinatorial theory, may wish to begin with Chapter 3 and proceed (onward). Generally a certain ease in handling groups and finite fields proves helpful throughout, but especially when studying the latter part of the book. These two notions are of fundamental importance indeed. They occur naturally in the construction of block designs, when one works with Möbius functions, or when counting configurations under the action of a group.

For use as a textbook, a typical upper undergraduate selection of topics (for a one-year course) would include: the bulk of Chapters 1 and 2, Sections 2 and 3 of Chapter 3, the first three sections from Chapter 4 on graphs, the first four from Chapter 5 on network flows, along with the first three sections of Chapters 6 and 7.

A one-year graduate course would normally cover the first five chapters during the first semester, and the remaining four during the second.

A WORD ON PREREQUISITES

Having assimilated this material in bits and pieces (as the need arose), I find musing over a description of formal prerequisites an amusing circumstance. Luckily, I can think of something that would easily suffice.

The commodity in question cannot be gazed at, squeezed, or identified through any of the more usual senses, other than the intellectual. It is, in other words, an abstraction. I can't quite recall a name by which it goes other than curiosity, followed by the fascination of discovery and the peaceful feeling of conceptual understanding. Even in small doses the combination proves powerful.

The first four chapters do not rely on formal prior knowledge, and large portions of the remaining five assume only rudimentary information on finite groups and vector spaces. What I give in Appendices 1 and 2 is more than enough, in terms of algebra and geometry, to have the reader sail comfortably through the whole text. References to the sources of these results are listed at the end of each appendix. My research interests prompted me to study the statistical and combinatorial works of the eminent Raj Chandra Bose (a man I never met). And I find his recent book with Manvel [4] well suited as a prerequisite at the undergraduate level.

Many of the motivational ideas on block designs and extreme spectral behavior may be traced back to Sir R. A. Fisher's contributions to what we broadly call controlled scientific experimentation [1]. At least three major areas of research subsequently emerged: the effective planning and analysis of statistical experiments (led by G. E. P. Box, D. R. Cox, and the British school), the intrinsically geometric approach of R. C. Bose and his students, and the spectral aspect pursued by J. C. Kiefer, Jaya Srivastava, and C.-S. Cheng. This text accentuates the last of the three. Pertinent material on the statistical analysis of designs may be found in Box, Hunter, and Hunter [2], or in the text by Cochran and Cox [3].

Among the more specialized writings that offer a deeper understanding of certain portions of my text, I am glad to mention Cameron and van Lint [5] and Lander [6], both on block designs, and the fine work of Andrews [7] on partitions. I have been inspired and impressed by (portions I could understand of) MacMahon's classic treatise *Combinatory Analysis* [8]. The reader should take a peek at it as well.

A last remark on the exercises. Some of them are routine and designed to illustrate a general result by a special case. In range of difficulty they vary a lot, from easy applications to proving uniqueness of certain planes, or proving the fundamental theorem of projective geometry. A reader who is not disposed to give proofs could, I guess, at least enjoy the statements. Suggestions and answers are found at the end of the book.

REFERENCES

[1] R. A. Fisher, *The Design of Experiments*, Oliver and Boyd, Edinburgh, 1949.

[2] G. E. P. Box, W. G. Hunter, and J. S. Hunter, *Statistics for Experimenters*, Wiley, New York, 1978.

[3] W. G. Cochran and G. M. Cox, *Experimental Designs*, Wiley, New York, 1957.

[4] R. C. Bose and B. Manvel, *Introduction to Combinatorial Theory*, Wiley, New York, 1984.

[5] P. J. Cameron and J. H. van Lint, *Graphs, Codes, and Designs*, Cambridge Univ. Press, Cambridge, 1980.

[6] E. S. Lander, *Symmetric Designs: An Algebraic Approach*, Cambridge Univ. Press, Cambridge, 1983.

[7] G. E. Andrews, *The Theory of Partitions*, Addison-Wesley, Reading, MA, 1976.

[8] P. A. MacMahon, *Combinatory Analysis*, *Vols. 1 and 2*, Cambridge Univ. Press, London and New York, 1915–1916 (reprinted by Chelsea, New York, 1960).

GREGORY M. CONSTANTINE

Northbrook, Illinois
December 1986

Contents

Though this be madness, yet there is method in't.

Hamlet, W. SHAKESPEARE

Combinatorial Theory
and Statistical Design

CHAPTER 1

Ways to Choose

O hell! to choose by another's eye.
A Midsummer Night's Dream, WILLIAM SHAKESPEARE

Aside from considerations of habit and fate, much of what we do appears to rest on deliberate and (one should hope) not infrequently intelligent choice. It is to this latter kind of choice that we devote our attention in these introductory pages.

The discussion we are about to undertake is intended primarily as a refresher. This is not to say that our level of presentation rests upon any formal prerequisites of much sophistication; it, rather, springs out of the usual dose of common wisdom that most of us share. The art of counting may rest primarily on the innate ability to discern patterns or systematic formations. When placed in the presence of well-developed mathematical technique this ability is usually strengthened, especially if the methods tend to assimilate well with the intuitive.

Discerning choice entails accurate assessment of chance which, in turn, requires good counting abilities. We shall be glad if the handful of techniques presented here enhance these abilities, even if only in marginal ways.

1 THE ESSENTIALS OF COUNTING

1.1

We begin with fundamental definitions and customary notation. A *set* is an assemblage of elements listed in any order without repetitions. The notation $\{3,1,2\}$ describes a set with 1, 2, and 3 as elements; on the other hand $\{1,2,2,3\}$ is not a set. The number of elements in a set could be finite or infinite. *All the sets in this book are finite*, unless specific mention to the contrary is made. By $|A|$ we denote the number of elements (or *cardinality*, or size) of the set A. The notation $a \in A$ indicates the fact that element a belongs to the set A.

Given two sets A and B, the symbol of *inclusion* $A \subseteq B$ intimates the fact that all the elements of A could be found among those of B; we say that A is a *subset* of B. The *intersection* $A \cap B$ consists of elements common to both A and B. By $A \cup B$, the *union* of the two sets, we denote the elements that are in either A or B (or in the intersection). When A is part of a larger set, $\bar{A}$ denotes the elements not in A (but in that larger set); $\bar{A}$ is called the *complement* of A. The symbols introduced so far are related by the rules $\overline{A \cap B} = \bar{A} \cup \bar{B}$ and "dually" $\overline{A \cup B} = \bar{A} \cap \bar{B}$. Further, $B - A$ denotes the elements of B that are not in A. One may think of $B - A$ as the complement of A in B. The symbol $\varnothing$ is reserved for the *empty set*, that is, the set with no elements whatsoever. Due to its being vacuous, the empty set is somewhat illusory and thus difficult to grasp. Luckily its usefulness rests often in writing $A \cap B = \varnothing$, which merely indicates that sets A and B have no common elements (we call such A and B *disjoint*). By $\mathscr{P}(A)$ we denote the set of all subsets of A (including $\varnothing$ and the set A itself). Two sets are *equal* if they consist of the same elements.

The *Cartesian product* of sets A and B is the set of ordered pairs defined as follows:

$$A \times B = \{(a, b): a \in A \text{ and } b \in B\}.$$

And lastly, by N, Q, and R we indicate the (infinite) sets of natural, rational, and real numbers, respectively. (The natural numbers are sometimes called positive integers.)

1.2

A *function* from set A to set B is a rule by which we associate to each element of A a single element of B. Let $f: A \to B$ be a function. The most important implication in dealing with functions is that for two disjoint subsets B_1 and B_2 of B (i.e., $B_1 \cap B_2 = \varnothing$) we have disjoint preimages [i.e., $f^{-1}(B_1) \cap f^{-1}(B_2) = \varnothing$, where $f^{-1}(C)$ denotes the subset $\{a \in A: f(a) \in C\}$].

We call f *injective* (or one to one) if $i \neq j$ implies $f(i) \neq f(j)$.

We call f *surjective* (or onto) if for every b in B there exists a in A such that $f(a) = b$.

The function f is said to be a *bijection* if it is both injective and surjective. (Given an injection from A to B it is often helpful to effectively identify A with its image through the injection. A surjection spreads, in a sense, A over B, possibly several times.)

Though clear, it is important enough to state explicitly that

If f is injective, then $|A| \leq |B|$.
If f is surjective, then $|A| \geq |B|$.
If f is bijective, then $|A| = |B|$.

1.3

With regard to Cartesian products we have

$$|A \times B| = |A|\,|B|.$$

This is easy to see, since we count the number of ordered pairs with $|A|$ possibilities for the first entry and $|B|$ possibilities for the second:

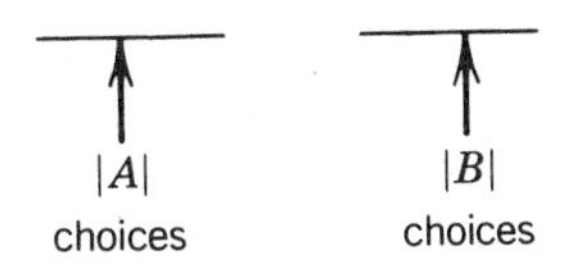

(a totality of $|A|\,|B|$ choices).

As an example, suppose your mate prepares for lunch three kinds of soup, five kinds of main course, six types of dessert, and five brands of drinks. How many choices for lunch do you have? The answer is $3 \cdot 5 \cdot 6 \cdot 5$, the cardinality of the Cartesian product of the sets of soup, main course, dessert, and drinks. (A lunch is understood to consist of precisely one choice out of each of the four available courses.)

1.4

A set with n elements has 2^n subsets. [Or, in terms of symbols, $|\mathscr{P}(A)| = 2^{|A|}$.]

The simplest way to see this is to list the n elements of A as

$$(a_1, a_2, a_3, a_4, \ldots, a_{n-1}, a_n)$$

and then associate a vector of length n to each subset of A by placing a 1 in the position of the elements that occur in that subset and 0 in the remaining positions. The question now is how many vectors of length n with 0 or 1 as entries are there? Well, there are n positions to fill, with two choices (0 or 1) for each entry, so there are 2^n such vectors. We thus conclude that A ($|A| = n$) has 2^n subsets.

[Representing subsets as vectors is a nice little trick that we shall use again. Just to make sure the reader understands it we give a small example. If $A = \{a_1, a_2, a_3, a_4, a_6, a_6\}$ is our set, then we associate as follows:

$$\{a_1, a_2, a_4, a_6\}$$
$$\updownarrow$$
$$(1, 1, 0, 1, 0, 1).]$$

1.5

Let x be a real number or an indeterminate. Define $[x]_n$ to be $x(x-1)(x-2)\cdots(x-n+1)$. We set for convenience $[x]_0 = 1$. With $n \le m$ natural numbers, the number $[m]_n = m(m-1)(m-2)\cdots(m-n+1)$ has combinatorial meaning. Specifically:

∗ *The number of ordered n-tuples with no repeated entries from a set of m symbols equals* $[m]_n$.

∗ *The number of sequences of length n with no repeated letters formed from m available (distinct) letters is* $[m]_n$.

∗ *The number of injections from A into B equals* $[m]_n$, *if* $|A| = n$ *and* $|B| = m$.

In all these cases we have m choices at step 1, $m-1$ choices at step 2, $m-2$ choices at step 3, ..., and $m-n+1$ choices at step n. This gives a total of $m(m-1)(m-2)\cdots(m-n+1) = [m]_n$ possibilities:

$$\begin{array}{ccccc} [\;\underline{\qquad} & \underline{\qquad} & \underline{\qquad} & \cdots & \underline{\qquad}\;]. \\ \uparrow & \uparrow & \uparrow & & \uparrow \\ m & m-1 & m-2 & & m-n+1 \\ \text{choices} & \text{choices} & \text{choices} & & \text{choices} \end{array}$$

When $m = n$ we denote $[n]_n$ by $n!$ (n *factorial*). In addition $0!$ is by convention set to 1. Small values for factorials are: $1! = 1$, $2! = 2$, $3! = 6$, $4! = 24$, $5! = 120$, $6! = 720$.

Example. The number of sequences of length 2 made with the elements of the set $\{a, b, c, d\}$ is $[4]_2 = 4 \cdot 3 = 12$.

Indeed, they are:

$$\begin{array}{cccc} ab & ba & ca & da \\ ac & bc & cb & db \\ ad & bd & cd & dc. \end{array}$$

1.6 Binomial Numbers

For $0 \le n \le m$ we denote $[m]_n/n!$ by $\binom{m}{n}$ (to be read "m choose n"); when $n > m$ we define $\binom{m}{n}$ to be 0.

The numbers $\binom{m}{n}$ are called *binomial numbers* and they have several possible combinatorial interpretations:

∗ *The number of subsets with n elements of a set with m elements is* $\binom{m}{n}$.

∗ *The number of sequences of length m with precisely n ones and* $(m-n)$ *zeros is* $\binom{m}{n}$. (Allowing 0 and 1 as sole possibilities explains the term "binomial" as nomenclature.)

∗ *The number of nondecreasing paths of length m from* $(0, 0)$ *to* $(n, m-n)$ *on the planar lattice of integral points equals* $\binom{m}{n}$.

Example.

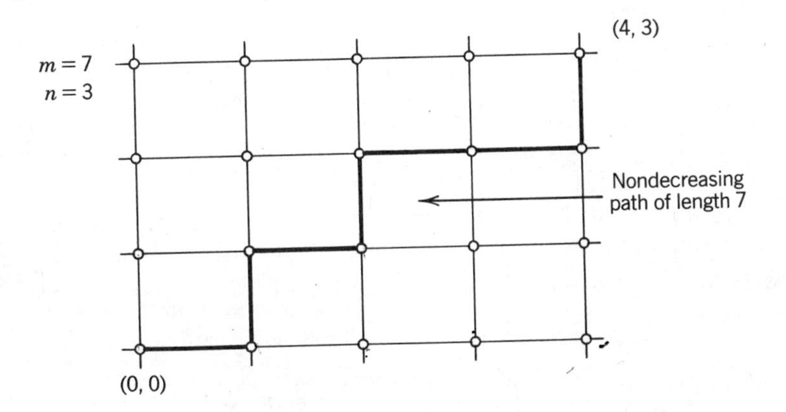

A path is called *nondecreasing* if the sequence of coordinates of the points we successively select is nondecreasing in each of the coordinates.

Example. The number of subsets of a set with four elements is $[4]_2/2! = (4 \cdot 3)/(2 \cdot 1) = 6$. Indeed, the six subsets are:

$$\{a, b\}$$

$$\{a, c\}\{b, c\}$$

$$\{a, d\}\{b, d\}\ \{c, d\}.$$

Let us prove the first assertion. Pick a subset of n elements ($n \leq m$). One can make $n!$ sequences with the elements of this fixed subset. There are $[m]_n$ sequences of length n in all. The quotient $[m]_n/n!$ counts therefore the number of subsets with n elements.

The second assertion is the same as the first upon identifying the 1's in a sequence of length m with the subset of positions in which they occur.

The third statement is the same as the second if we attach a sequence to a path by writing a 1 whenever we move horizontally, and a 0 whenever we move vertically.

A short list of small values of $\binom{m}{n}$ may be helpful: $\binom{m}{0} = 1$, $\binom{m}{m} = 1$, $\binom{m}{1} = \binom{m}{m-1} = m$, $\binom{7}{2} = 21$, $\binom{9}{3} = 84$.

Properties of the Binomial Numbers

(a) $\binom{m}{n} = \binom{m-1}{n} + \binom{m-1}{n-1}$.

This property is known as Pascal's triangle:

$$\begin{array}{c} 1 \\ 1\ 1 \\ 1\ 2\ 1 \\ 1\ 3\ 3\ 1 \\ 1\ 4\ 6\ 4\ 1 \\ 1\ 5\ 10\ 10\ 5\ 1 \\ \cdots \end{array} \qquad \left(10 = \binom{5}{3} = \binom{4}{3} + \binom{4}{2} = 6 + 4\right).$$

This property can be proved any number of ways. The number $\binom{m}{n}$ of nondecreasing paths between $(0,0)$ and $(n, m-n)$ equals the number of such paths between $(0,0)$ and $(n, m-1-n)$ [i.e., $\binom{m-1}{n}$] plus those between $(0,0)$ and $(n-1, m-n)$ [i.e., $\binom{m-1}{n-1}$]. This is one possible proof.

(b) $\binom{m}{n} = \binom{m}{m-n}$.

One can see this by realizing that each time we pick a subset of size n we uniquely identify a subset of size $m-n$, namely its complement.

(c) $\binom{m}{n-1} \leq \binom{m}{n}$, *for* $1 \leq n \leq (m+1)/2$.

Indeed,

$$\binom{m}{n-1} = \frac{[m]_{n-1}}{(n-1)!} \leq \frac{[m]_n}{n!} = \binom{m}{n}$$

if and only if $n \leq m - n + 1$ (or $n \leq (m+1)/2$).

(d) ***The Vandermonde Convolution***: $\binom{m+n}{k} = \sum_{j=0}^{m} \binom{m}{j}\binom{n}{k-j}$.

One can see the proof at once from the following figure:

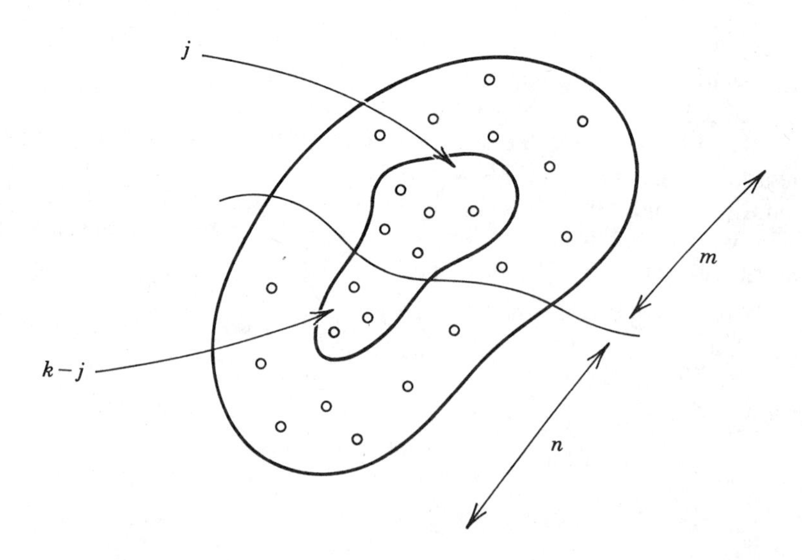

By coloring m of the objects red and n blue, a choice of k objects out of the $m+n$ would contain j red objects and $k-j$ blue ones. Sorting out by the possible values of j leads to the formula stated above.

Liebnitz's Formula

A natural way in which the binomial numbers arise, apart from those already mentioned, is when computing higher order derivatives of a product of two functions.

Let f and g be functions differentiable any number of times. Denote by $D^n f$ the nth derivative of f. Conveniently write f_n for $D^n f$ and understand also that $f_0 = f$ and $g_0 = g$. Applying the product rule we obtain

$$Dfg = f_1 g_0 + f_0 g_1,$$

$$D^2 fg = f_2 g_0 + f_1 g_1 + f_1 g_1 + f_0 g_2 = f_2 g_0 + \binom{2}{1} f_1 g_1 + f_0 g_2,$$

$$D^3 fg = f_3 g_0 + \binom{3}{1} f_2 g_1 + \binom{3}{2} f_1 g_2 + f_0 g_3,$$

and so on. Assume that $D^n fg = \sum_{k=0}^{n} \binom{n}{k} f_k g_{n-k}$. Then

$$D^{n+1} fg = \sum_{k=0}^{n} \binom{n}{k} D f_k g_{n-k} = \sum_{k=0}^{n} \binom{n}{k} (f_{k+1} g_{n-k} + f_k g_{n-k+1})$$

$$= \sum_{k=0}^{n} \left(\binom{n}{k} + \binom{n}{k-1} \right) f_k g_{n-k+1} + \binom{n}{n} f_{n+1} g_0$$

$$= \sum_{k=0}^{n+1} \binom{n+1}{k} f_k g_{n+1-k},$$

which completes our inductive proof.

We thus proved Liebnitz's formula:

$$D^n fg = \sum_{k=0}^{n} \binom{n}{k} f_k g_{n-k}.$$

1.7 Stirling Numbers of the Second Kind

Subsets $A_1, A_2, \ldots, A_m$ of set A form a *partition* of the set A if $A_i \neq \varnothing$, for all $1 \leq i \leq m$,

$$A_i \cap A_j = \varnothing \qquad \text{for } i \neq j,$$

and

$$A_1 \cup A_2 \cup \cdots \cup A_m = A.$$

The subsets A_i are called the *classes* of the partition.

A partition of a set with n elements is said to be of type $1^{\lambda_1}2^{\lambda_2}\cdots n^{\lambda_n}$ ($\lambda_i \geq 0$) if it contains:

$$\lambda_1 \text{ classes of cardinality } 1$$

$$\lambda_2 \text{ classes of cardinality } 2$$

$$\vdots$$

$$\lambda_n \text{ classes of cardinality } n.$$

We first observe that:

(a) *The number of partitions of type* $1^{\lambda_1}2^{\lambda_2}\cdots n^{\lambda_n}$ *of a set with n elements* (where $n = \sum_{i=1}^{n} i\lambda_i$) *is*

$$\frac{n!}{(1!)^{\lambda_1}(2!)^{\lambda_2}\cdots(n!)^{\lambda_n}(\lambda_1!)(\lambda_2!)\cdots(\lambda_n!)}.$$

Proof. Keeping in mind at all times the type of partition we seek (i.e., $1^{\lambda_1}2^{\lambda_2}\cdots n^{\lambda_n}$), we initially list all the $n!$ sequences we can possibly make with the n elements of our set. The order of elements within any class being of no importance, we should divide $n!$ by $(1!)^{\lambda_1}(2!)^{\lambda_2}\cdots(n!)^{\lambda_n}$. But classes of the same size can also be permuted among themselves in any way whatever without changing the partition; we should, therefore, also divide by $(\lambda_1!)(\lambda_2!)\cdots(\lambda_n!)$. This ends our proof, for any other permutation of elements would in fact change the partition.

Example. Let $n = 14$ and the type be $1^2 2^3 3^2 4^0 \cdots 14^0$. That is:

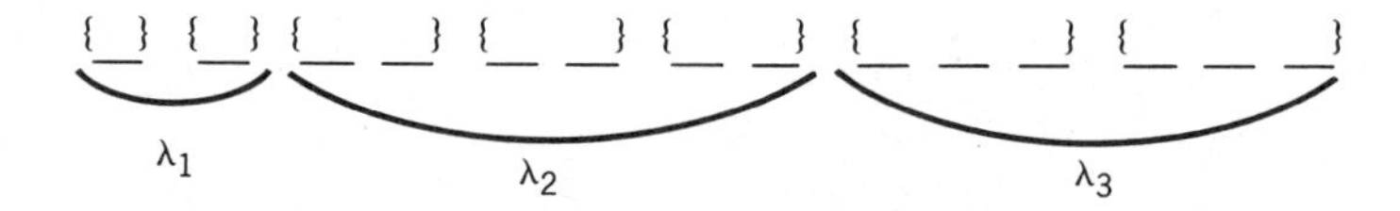

We have

$$\frac{14!}{(1!)^2(2!)^3(3!)^2(2!)(3!)(2!)}$$

partitions of this type.

The number of partitions of a set of n objects into exactly m classes is called the *Stirling number of the second kind*. We denote this number by S_n^m.

In terms of the calculation performed in (a), we may write

$$S_n^m = \sum \frac{n!}{(1!)^{\lambda_1}(2!)^{\lambda_2}\cdots(n!)^{\lambda_n}(\lambda_1!)(\lambda_2!)\cdots(\lambda_n!)}$$

the sum extending over all vectors $(\lambda_1, \lambda_2, \ldots, \lambda_n)$, with λ_i nonnegative integers satisfying $\sum_{i=1}^{n}\lambda_i = m$ and $\sum_{i=1}^{n} i\lambda_i = n$.

(b) *The numbers S_n^m satisfy the following recurrence relations*:

$$S_n^1 = S_n^n = 1,$$

and

$$S_{n+1}^m = S_n^{m-1} + mS_n^m, \qquad \textit{for } 1 < m \leq n.$$

(We define for convenience $S_n^0 = 0$, and $S_n^m = 0$ for values of m exceeding n.)

Proof. Consider the list of all partitions of $n + 1$ objects into m classes. There are S_{n+1}^m such partitions, by definition. Let w be an object among the $n + 1$. Separate the list of such partitions into two disjoint parts: those in which w is the sole element of a class and those in which any class containing w has cardinality 2 or more. There are S_n^{m-1} of the former kind and mS_n^m of the latter. Thus $S_{n+1}^m = S_n^{m-1} + mS_n^m$.

To summarize graphically, denote by A the set of the $n + 1$ elements:

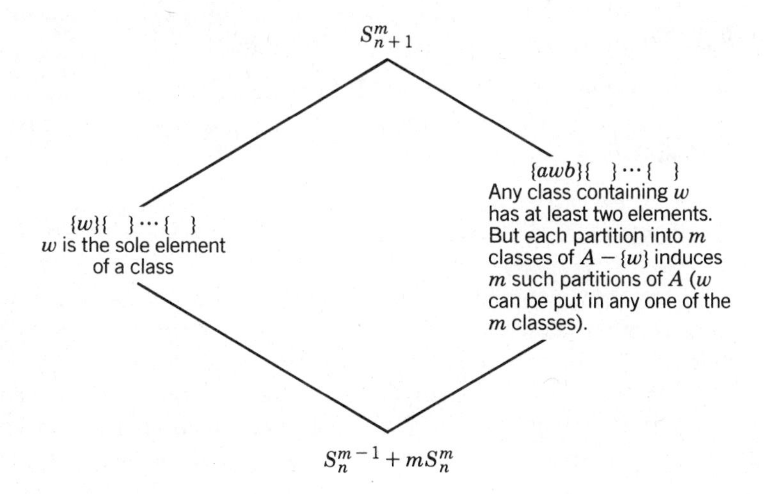

This ends the proof.

The recurrence relations established in (b) allow us to write the small values of S_n^m:

n \ m	1	2	3	4	5	6
1	1	0	0	0	0	0
2	1	1	0	0	0	0
3	1	3	1	0	0	0
4	1	7	6	1	0	0
5	1	15	25	10	1	0
6	1	31	90	65	15	1

Example. We have $S_4^2 = 7$ partitions with two classes of a set with four elements. Let the set be $\{a, b, c, d\}$. The seven partitions are:

$$\begin{array}{ll} \{a\}\{b,c,d\}, & \\ \{b\}\{a,c,d\}, & \{a,b\}\{c,d\}, \\ \{c\}\{a,b,d\}, & \{a,c\}\{b,d\}, \\ \{d\}\{a,b,c\}, & \{a,d\}\{b,c\}. \end{array}$$

For a real number x we denote the product $x(x-1)(x-2)\cdots(x-k+1)$ by $[x]_k$. The powers of x are related to the $[x]_k$'s via the Stirling numbers S_n^k. Specifically, we now prove that:

(c) $x^n = \sum_{k=0}^n S_n^k [x]_k$.

Proof. The proof involves counting the number of functions from set A to set B in two different ways. Let $|A| = n$ and $|B| = m$.

First Way of Counting. Each element of A can be mapped into any one of the m elements of B. Hence there are $m \cdot m \cdot \cdots \cdot m = m^n$ functions in all between A and B.

Second Way of Counting. Fix a subset C of B and count all the functions from A with precisely C as image (then sum over all nonempty subsets C of B).

If $|C| = k$, then a function from A onto C can be identified with a partition of A into precisely k classes (a class being the preimage of a point in C). Conversely, and more importantly, a partition of A into k classes gives rise to $k!$ functions from A onto C (the possible ways of mapping the k classes onto the elements of C). The number of functions from A onto C equals therefore $k!$ times the number of partitions of A into k classes, that is, it equals $k!S_n^k$.

Since there are $\binom{m}{k}$ subsets of size k in B, we conclude that the number of all functions from A to B is $\sum_{k=1}^m \binom{m}{k} k!S_n^k = \sum_{k=1}^m S_n^k [m]_k$.

By the two ways of counting we conclude that

$$m^n = \sum_{k=1}^{m} S_n^k [m]_k. \tag{1.1}$$

Let us now look at the polynomial

$$P(x) = x^n - \sum_{k=1}^{n} S_n^k [x]_k.$$

By (1.1) it is clear that the numbers $1, 2, \ldots, n$ are all roots of P. But 0 is also a root of P. We just finished exhibiting $n+1$ distinct roots of P, and since P is a polynomial of degree n we conclude that P must necessarily be the zero polynomial. That is,

$$x^n = \sum_{k=1}^{n} S_n^k [x]_k$$

which (considering that S_n^0 is 0) is the content of (c). This ends our proof. (The technique employed to derive this result is a disguised form of Möbius inversion, which we study in detail in Chapter 9.)

Let us conclude Section 1.7 by proving the following formula:

(d) $S_{n+1}^m = \sum_{k=0}^{n} \binom{n}{k} S_k^{m-1}$ (when reading this formula the reader should recall that by definition $S_k^p = 0$ for $k < p$).

One may establish the above formula as follows. Consider the list of all partitions with m classes of a set with $n + 1$ elements. (There are S_{n+1}^m such partitions.)

Let us count these partitions in a different way. Fix an element, say w, of the $n + 1$ available elements. In each of the partitions eliminate the class containing w. We thus obtain partitions with precisely $m - 1$ classes, on k elements, where k is at most n. Sorting out the partitions so obtained by the values of k we obtain the formula stated in (d).

[To pay attention to detail, if w belongs to a class of size j, by eliminating the class containing w, we obtain a partition with $m - 1$ classes of a set with $n - j + 1$ elements. Each choice of (a class of) j elements leads to S_{n-j+1}^{m-1} partitions of the remaining $n - j + 1$ elements. The element w being always among the j elements selected, we have $\binom{n}{j-1}$ possibilities to select a class with j elements that contains w, a totality of $\binom{n}{j-1} S_{n-j+1}^{m-1}$ choices. Observe, in addition, that the largest value j may take is $n + 2 - m$ (since we must have m classes in the original partition). Summing up over j we obtain

$$S_{n+1}^m = \sum_{j=1}^{n+2-m} \binom{n}{j-1} S_{n-j+1}^{m-1} = \sum_{k=0}^{n} \binom{n}{k} S_k^{m-1}.]$$

1.8 Stirling Numbers of the First Kind

The set of bijections from a set A to itself is the object of study in this section. Let us formally denote

$$\text{Sym } A = \left\{ \sigma \colon A \overset{\sigma}{\to} A, \sigma \text{ bijection} \right\}.$$

An element of Sym A is called a *permutation*. If A has n elements, then the cardinality of Sym A is $n!$.

To fix ideas, say $A = \{1, 2, 3, 4, 5\}$. Line up the elements of A in some order, say 1 2 3 4 5, and keep this order as reference at all times. The writing (or sequence) 3 4 5 2 1 indicates the bijection σ:

$$\begin{matrix} 12345 \\ \downarrow\sigma \\ 34521 \end{matrix} \qquad \left(\text{i.e.,} \quad \begin{matrix} \sigma(1) & \sigma(2) & \sigma(3) & \sigma(4) & \sigma(5) \\ \| & \| & \| & \| & \| \\ 3 & 4 & 5 & 2 & 1 \end{matrix} \right).$$

(The reader now understands why $|\text{Sym}\, A| = n!$, if $|A| = n$; this is so simply because $|\text{Sym}\, A|$ counts the number of sequences on n symbols with no repetitions, i.e., $|\text{Sym}\, A| = [n]_n = n!$.)

Although this sequential notation for a permutation comes in handy quite often, it is the decomposition of a permutation into disjoint cycles that we want to emphasize. The decomposition of a permutation σ into disjoint cycles is carried out as follows: pick a symbol, say 1, and list the symbols obtained by repeated applications of σ, that is, $(1\sigma(1)\sigma(\sigma(1))\sigma(\sigma(\sigma(1)))\cdots)$. Close the cycle when we get back to 1 again. If any symbols are left, repeat the process until none are left.

For example, the permutation

$$\begin{matrix} 12345 \\ \downarrow\sigma \\ 34521 \end{matrix}$$

has cycle decomposition $(1\,3\,5)\,(2\,4) = \sigma$.

More Examples.

$$\begin{matrix} 123456 \\ \downarrow\sigma \\ 635421 \end{matrix}$$

corresponds to $(1\,6)\,(2\,3\,5)\,(4)$ and

$$\begin{matrix} 12345678 \\ \downarrow\sigma \\ 58374612 \end{matrix}$$

corresponds to $(1\,5\,4\,7)\,(2\,8)\,(3)\,(6)$. (The reader should write the cyclic decompositions of the permutations 2 7 3 6 9 1 4 5 8 and 6 3 4 7 2 5 1.)

It is self-evident that *each permutation has a unique decomposition into disjoint cycles* (up to a rearrangement of the cycles).

A cycle decomposition of a permutation on n symbols is said to be of type $1^{\lambda_1}2^{\lambda_2}\cdots n^{\lambda_n}$ (just notation), if it has

λ_1 cycles of length 1

λ_2 cycles of length 2

$\vdots$

λ_n cycles of length n

(clearly here $\sum_{i=1}^{n} i\lambda_i = n$).

We shall count the number of permutations on n symbols with precisely m cycles. One should conveniently think of a permutation simply as a collection of disjoint cycles in a formal way. (If necessary we can interpret this symbolic writing as a bijection on a finite set, but this is seldom necessary.) It should, however, be clear to the reader that a same cycle of length n can be written out in n different ways: for example, $(3\,2\,4\,1) = (1\,3\,2\,4) = (4\,1\,3\,2) = (2\,4\,1\,3)$. In other words the cyclic motion within a cycle gives the same cycle. And this cyclic motion is the only change within a cycle that preserves it.

Let us observe that:

(a) *The number of permutations of type* $1^{\lambda_1}2^{\lambda_2}\cdots n^{\lambda_n}$ *of a set with n elements* (or symbols) *is*

$$\frac{n!}{1^{\lambda_1}2^{\lambda_2}\cdots n^{\lambda_n}(\lambda_1!)(\lambda_2!)\cdots(\lambda_n!)}$$

(The proof is almost the same as that of statement (a) in Section 1.7; replace the word "class" with "cycle" and recall that, apart from the cyclic order, the order of the elements within a cycle is relevant.)

Let $(-1)^{n-m}s_n^m$ be the number of permutations with exactly m (disjoint) cycles on a set of n elements. The numbers s_n^k are called *Stirling numbers of the first kind*.

By (a) above,

$$(-1)^{n-m}s_n^m = \sum\frac{n!}{1^{\lambda_1}2^{\lambda_2}\cdots n^{\lambda_n}(\lambda_1!)(\lambda_2!)\cdots(\lambda_n!)}$$

where the sum extends over all vectors $(\lambda_1, \lambda_2, \ldots, \lambda_n)$, with λ_i positive integers satisfying $\Sigma\lambda_i = m$ and $\Sigma i\lambda_i = n$.

Comparing this last expression with its counterpart for S_n^m in Section 1.7 one observes straightaway that $S_n^m \le |s_n^m|$, simply because in the former we divide to $k!$ while in the latter we divide only to k.

Another helpful thing to notice is that:

(b) *The numbers* s_n^m *satisfy the following recurrence relations*:

$$s_n^0 = 0, \qquad s_n^n = 1,$$

and

$$s_{n+1}^m = s_n^{m-1} - ns_n^m, \qquad \textit{for } 1 \le m \le n$$

(for convenience we set $s_n^m = 0$ for m larger than n).

Proof. The chain of arguments parallels that of the proof of (b) in Section 1.7. Consider the list of the $(-1)^{n+1-m}s_{n+1}^m$ permutations on the $n+1$ elements of set A with m cycles. Fix an element w and split this list into two parts as follows:

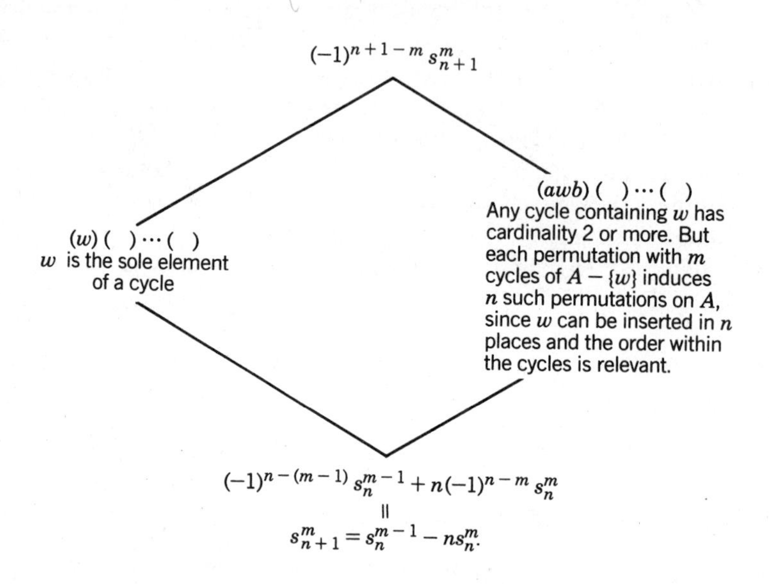

This concludes the proof.

A list of small values for s_n^m is given below:

n \ m	1	2	3	4	5	6
1	1	0	0	0	0	0
2	−1	1	0	0	0	0
3	2	−3	1	0	0	0
4	−6	11	−6	1	0	0
5	24	−50	35	−10	1	0
6	−120	274	−225	85	−15	1

Example. There are $(-1)^{4-2}s_4^2 = 11$ permutations on four symbols with exactly two cycles. These are:

(1) (2 3 4) (3) (1 2 4) (1 2) (3 4)
(1) (4 3 2) (3) (4 2 1) (1 3) (2 4)
(2) (1 3 4) (4) (1 2 3) (1 4) (2 3).
(2) (4 3 1) (4) (3 2 1)

We now prove the following identity:

(c) $[x]_n = \sum_{k=0}^n s_n^k x^k$.

Proof. Let

$$[x]_n = x(x-1)\cdots(x-n+1) = c_n^0 + c_n^1 x + c_n^2 x^2 + \cdots + c_n^n x^n$$

(the symbol i in c_n^i being an *index* and *not a power*). Then

$$\begin{aligned}\cdots + c_{n+1}^m x^m + \cdots &= [x]_{n+1} = [x]_n(x-n)\\ &= \left(\cdots + c_n^{m-1}x^{m-1} + c_n^m x^m + \cdots\right)(x-n)\end{aligned}$$

and we see that the numbers c_n^m satisfy

$$c_n^0 = 0, \qquad c_n^n = 1$$

and

$$c_{n+1}^m = c_n^{m-1} - nc_n^m.$$

But we showed in (b) that the Stirling numbers s_n^m satisfy these recurrence relations and initial conditions. Thus $s_n^m = c_n^m$, and this ends our proof.

In the same way we proved the formula in (d), Section 1.7, we can prove the following:

(d) $|s_{n+1}^m| = \sum_{k=0}^n [n]_{n-k} |s_k^{m-1}|$

(e.g., $225 = |s_6^3| = [5]_3|s_2^2| + [5]_2|s_3^2| + [5]_1|s_4^2| + [5]_0|s_5^2|$.)

1.9 Bell Numbers

Let B_n denote the number of partitions of a set with n elements. The B_n's are called *Bell numbers*. By the definition of B_n and the definition of S_n^m, the Stirling numbers of the second kind, we have

$$B_n = \sum_{m=1}^{n} S_n^m.$$

As an example, let $A = \{a, b, c, d\}$. Then the list of all the partitions of A is

$\{a,b,c,d\}$, $\{a\}\{b,c,d\}$, $\{a\}\{b\}\{c,d\}$, $\{a\}\{b\}\{c\}\{d\}$,
$\{b\}\{a,c,d\}$, $\{a\}\{c\}\{b,d\}$,
$\{c\}\{a,b,d\}$, $\{a\}\{d\}\{b,c\}$,
$\{d\}\{a,b,c\}$, $\{b\}\{c\}\{a,d\}$,
$\{a,b\}\{c,d\}$, $\{b\}\{d\}\{a,c\}$,
$\{a,c\}\{b,d\}$, $\{c\}\{d\}\{a,b\}$,
$\{a,d\}\{b,c\}$,

(15 partitions in all).

Observe that

$$\begin{aligned}B_{n+1} &= \sum_{m=1}^{n+1} S_{n+1}^m = \sum_{m=1}^{n+1}\sum_{k=0}^{n}\binom{n}{k}S_k^{m-1}\\ &= \sum_{k=0}^{n}\binom{n}{k}\sum_{m=1}^{n+1}S_k^{m-1} = \sum_{k=0}^{n}\binom{n}{k}B_k.\end{aligned}$$

The second equality sign is explained by (d), Section 1.7.

We proved the following:

(a) $B_{n+1} = \sum_{k=0}^{n}\binom{n}{k}B_k$.

A formula that is appropriate to state here, though we postpone its proof until introducing generating functions in the next chapter (see specifically, Section 2.7, formula 3), is the following:
(b) $B_n = e^{-1}\sum_{m=0}^{\infty} m^n/m!$. (This is *Dobinski's formula*, where e is the Eulerian constant $e \cong 2.71828\ldots$).

Observe, by the way, that the analog of the Bell number B_n for permutations is $n!$, that is, $n! = \sum_{m=1}^{n}|s_n^m|$. And since we know that $S_n^m \le |s_n^m|$, this tells us that $B_n \le n!$. A more refined perception of the magnitude of B_n can be cultivated upon reading Section 2.7.

2 OCCUPANCY

1.10

As need arises, we allow A to be an assemblage of elements *distinguishable* from one another (in which case A is a set) or have it consist of a number of *indistinguishable* elements, such as $A = \{1,1,1,1\}$. The number of elements of A, distinguishable or not, is indicated by $|A|$.

Without specifying whether the elements of A and B are distinguishable or indistinguishable, by a *function* from A to B we understand a rule by which we assign to each element of A a single element of B.

Compiled below are results regarding the numbers of functions, injections, surjections, and bijections from A to B ($|A| = n$, $|B| = m$).

	Elements of A	Elements of B	Functions	($n \le m$) Injections	($m \le n$) Surjections	($m = n$) Bijections
α	Distinguishable	Distinguishable	m^n	$[m]_n$	$m!S_n^m$	$n!$
β	Indistinguishable	Distinguishable	$\binom{n+m-1}{n}$	$\binom{m}{n}$	$\binom{n-1}{m-1}$	1
γ	Distinguishable	Indistinguishable	$\sum_{k=1}^{m} S_n^k$	1	S_n^m	1
δ	Indistinguishable	Indistinguishable	$\sum_{k=1}^{m} P_k(n)$	1	$P_m(n)$	1

In this table the entry S_n^m, for example, signifies the number of surjections from a set with n distinguishable elements to a set with m indistinguishable elements. (S_n^m is the Stirling number of the second kind.)

1.11

We devote the remainder of Section 2 to explaining the notation and proving the statements contained in this table.

(α) {*Distinguishable*} $\xrightarrow{f}$ {*Distinguishable*}

Functions. ($A \xrightarrow{f} B$; $|A| = n$, $|B| = m$.) For each element in A we have m choices to map it to. A total of m^n choices.

Injections. ($n \leq m$.) When defining an injection, we have m choices for the first element of A, $m - 1$ choices for the second, $m - 2$ for the third, ..., $m - n + 1$ for the nth. In all $[m]_n$ possibilities.

Surjections. ($m \leq n$.) For each element b in B, consider the subset $A_b = \{a \in A: f(a) = b\}$. Since f is a surjective function, the subsets $\{A_b: b \in B\}$ form a partition of A with m classes. Conversely, to each partition of A with m classes, there correspond $m!$ surjections from A to B. There are S_n^m partitions with m classes of A, and this gives a total of $m!S_n^m$ surjections.

Bijections. ($n = m$.) There are $[n]_n = n!$ of these.

(β) {*Indistinguishable*} $\xrightarrow{f}$ {*Distinguishable*}

Bijections. ($n = m$.) There is only one bijection.

Injections. ($n \leq m$.) Think of injecting the m (indistinguishable) elements of A into B. This corresponds to a choice of n elements out of the m available in B. The order in which we identify these n elements of B is of no consequence since the elements of A are indistinguishable. This shows that there are $\binom{m}{n}$ injections (one associated to each subset of size n in B).

Functions. The sort of counting we perform here is quite relevant throughout the upcoming chapter. There is a little trick by which we do the counting, actually, which we hope the reader will appreciate. An entire section is now devoted to this.

1.12

We assert that the following four statements are effectively the same, and we substantiate this assertion by exhibiting explicit bijections between them:

(a) *The number of ways to distribute n indistinguishable balls into m distinguishable boxes is* $\binom{n+m-1}{n}$.

(*b*) *The number of vectors* $(n_1, n_2, \ldots, n_m)$ *with nonnegative integer entries satisfying*

$$n_1 + n_2 + \cdots + n_m = n$$

is $\binom{n+m-1}{n}$.

(c) *The number of ways to select n objects with repetitions from m different types of objects is* $\binom{n+m-1}{n}$. (We assume that we have an unlimited supply of objects of each type and that the order of selection of the n objects is irrelevant.)
(d) *The number of functions from n indistinguishable elements to m distinguishable elements is* $\binom{n+m-1}{n}$.

When we assert that these statements are the same we mean that there are bijective correspondences between them. In other words, it's like meeting the same person on four different occasions, each time wearing a different attire. When appearances are ignored you do recognize the same person.

The disguise between (a) and (b) is pretty thin, actually. When placing the n balls within the m boxes we place n_1 in box 1, n_2 in box $2, \ldots, n_m$ in box m ($\sum_{i=1}^{m} n_i = n$). To this process we naturally attach the vector $(n_1, n_2, \ldots, n_m)$. Counting the number of such vectors is the same thing as counting the ways of placing the n indistinguishable balls into the m distinguishable boxes. [We prove shortly that the answer to this counting problem is $\binom{n+m-1}{n}$.]

With regard to (a) and (c): Selecting n objects with repetitions from m distinct types of objects involves a selection of n_1 objects of type 1, n_2 of type $2, \ldots, n_m$ of type m. This is the same as placing n_1 balls in box 1, n_2 in box $2, \ldots, n_m$ in box m.

Statement (d) is surely the same as (a) upon thinking of each distinguishable element as a box and each indistinguishable element as a ball. A function is merely a way of assigning the indistinguishable balls to the distinguishable boxes.

(That all four statements are the same follows now by transitivity.)

The numerical answer shared by statements (a) through (d) is $\binom{n+m-1}{n}$. To see this we first solve the following problem:

∗ *The number of ways of placing n distinguishable balls into m distinguishable boxes, paying attention to the order in which the balls are placed within the boxes, is equal to* $[n+m-1]_n = m(m+1)\cdots(m+n-1)$.

Proof. The *trick* is to visualize an assignment of balls to boxes as a sequence of n balls and $m-1$ /'s, as displayed below:

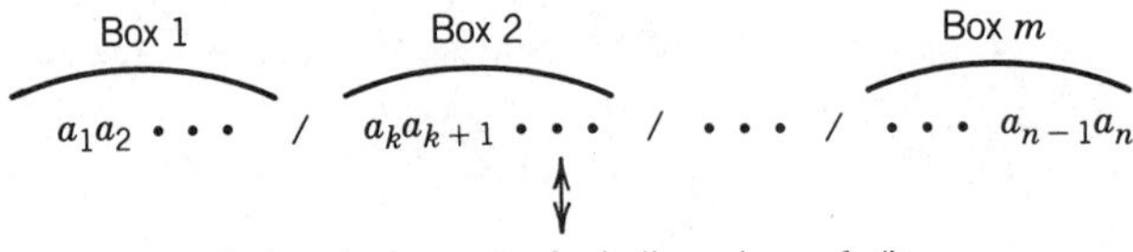

We now wish to count the number of distinct sequences of $n+m-1$ objects consisting of the n distinguishable balls and the $m - 1$ indistinguishable /'s. If the /'s were also distinguishable, we would have had $(n + m - 1)!$ sequences in all. Since they are not, we may permute the $m - 1$ available /'s among themselves in whichever positions they occur [and this can be done in $(m - 1)!$ ways] without changing the sequence. The number of distinct sequences is therefore $(n + m - 1)!/(m - 1)!$ or $[n + m - 1]_n$. This ends the proof.

When the n objects are indistinguishable, we obtain the following consequence: The number of ways of placing n indistinguishable objects into m distinguishable boxes is

$$\frac{[n + m - 1]_n}{n!} = \binom{n + m - 1}{n}.$$

This, however, is statement (a).

Example. With two distinguishable balls z and w, and three distinguishable boxes, the $[2 + 3 - 1]_2$ possibilities are:

$$\begin{array}{cccccc} zw//, & /zw/, & //zw, & z/w/, & z//w, & /z/w, \\ wz//, & /wz/, & //wz, & w/z/, & w//z, & /w/z. \end{array}$$

If z and w are indistinguishable we have

$$\frac{[2 + 3 - 1]_2}{2!} = \binom{2 + 3 - 1}{2}$$

possibilities:

$$\cdot\cdot//, \quad /\cdot\cdot/, \quad //\cdot\cdot, \quad \cdot/\cdot/, \quad \cdot//\cdot, \quad /\cdot/\cdot.$$

Initially we wanted to count the number of functions from n indistinguishable elements to m distinguishable elements. By the counting we just did, and the equivalence of statements (a) and (d), we conclude that there are $\binom{n+m-1}{n}$ functions.

Surjections. ($m \leq n$.) Think of the elements of A being n indistinguishable balls and the elements of B being m distinguishable boxes. We then ask for the number of ways of placing the n balls into the m boxes with no box left empty.

We count as follows: place initially one ball within each box. There are $n - m$ indistinguishable balls remaining, which we can now place into the m boxes without restrictions. The number of surjections from n indistinguishable elements to m distinguishable ones equals therefore the number of functions from the remaining $n - m$ indistinguishable elements to the m distinguishable elements. We just finished counting the number of such functions. There are

$$\binom{n - m + m - 1}{n - m} = \binom{n - 1}{n - m} = \binom{n - 1}{m - 1}$$

of them.

1.13

Let us proceed with explaining the remaining entries in our table.

(γ) {*Distinguishable*} $\xrightarrow{f}$ {*Indistinguishable*}

Bijections. ($n = m$.) There is only one bijection.

Injections. ($n \leq m$.) There is only one injection.

Surjections. ($n \geq m$.) Given a surjection from A to B, the preimages of the indistinguishable elements of B form a partition of A with m classes. Conversely, given a partition with m classes of A, one canonically constructs a surjection by mapping all the elements of a class to an (indistinguishable) element of B. The number of surjections is therefore S_n^m, the number of partitions of a set of n elements into m classes (a Stirling number).

Functions. We have $\sum_{k=1}^{m} S_n^k$ functions from n distinguishable elements to m indistinguishable ones. This is apparent from the above way in which we counted the surjections, since any function is a surjection onto its image.

(δ) {*Indistinguishable*} $\xrightarrow{f}$ {*Indistinguishable*}

We have one bijection and one injection. Counting surjections is difficult. No closed formula is known. Let us denote by $P_m(n)$ the number of surjections from n indistinguishable objects to m indistinguishable objects ($n \geq m$). With this notation, the number of functions we seek is $\sum_{k=1}^{m} P_k(n)$, again, because any function is a surjection on its image.

The task of specifying a surjection from a collection of n indistinguishable objects to another of m indistinguishable ones is equivalent to that of writing n as a sum of precisely m positive integers. The positive integers whose sum is n are commonly called the *parts* of m. Hence $P_m(n)$ can be interpreted as the number of ways of writing n as a sum of m (positive) parts. The order in which we list the parts is, of course, irrelevant here. We may as well then list the parts nonincreasingly, and specify $P_m(n)$ as

$$P_m(n) = \left\{ (\alpha_1, \alpha_2, \dots, \alpha_m) : \alpha_1 \geq \alpha_2 \geq \cdots \geq \alpha_m, \right.$$

$$\left. \sum_{i=1}^{m} \alpha_i = n, \text{ and } \alpha_i\text{'s are positive integers} \right\}.$$

We refer the reader to Section 7 of Chapter 2 for more information on the numbers $P_m(n)$, which we call the number of partitions of n with m parts.

3 MORE ON COUNTING

1.14 Multinomial Numbers

For n and n_i $(1 \leq i \leq m)$ positive integers satisfying $\sum_{i=1}^{m} n_i = n$, the numbers $n!/(n_1!n_2! \cdots n_m!)$ have combinatorial meaning. [Observe that for $m = 2$, $n_1 + n_2 = n$, we obtain the binomial numbers $\binom{n}{n_1}$.] We call $n!/(n_1!n_2! \cdots n_m!)$ *multinomial numbers*. We prove the following three statements:

* *Given n distinguishable objects and m distinguishable boxes* (labeled $1, 2, \ldots, m$), *the number of ways of placing n_1 objects in box 1, n_2 objects in box 2 $, \ldots,$ n_m objects in box m is* $n!/(n_1!n_2! \cdots n_m!)$ (the order in which the n_i objects are placed in box i is irrelevant) $(\sum_{i=1}^{m} n_i = n)$.

* *Given*

n_1 *indistinguishable objects of color* 1

n_2 *indistinguishable objects of color* 2

$\vdots$

n_m *indistinguishable objects of color m*

the number of distinct sequences of length n possible to make with these objects is $n!/(n_1!n_2! \cdots n_m!)$. (How many sequences of length 11 can there be made with the letters of the word Mississippi?)

* *The number of nondecreasing paths from* $(0, 0, \ldots, 0)$ *to* $(n_1, n_2, \ldots, n_m)$ (*with* $\sum_{i=1}^{m} n_i = n$) *on the m-dimensional lattice of integral points is* $n!/(n_1!n_2! \cdots n_m!)$. (A path is understood to be nondecreasing if the coordinates of the points successively selected form nondecreasing sequences in each of the coordinates.)

Let us prove these three statements. The first assertion is proved as follows: Box 1 can be filled in $\binom{n}{n_1}$ ways; with box 1 full, box 2 can be filled in $\binom{n - n_1}{n_2}$ ways; with boxes 1 and 2 full, box 3 can be filled in $\binom{n - n_1 - n_2}{n_3}$ ways, and so on. The total number of ways is therefore

$$\binom{n}{n_1}\binom{n - n_1}{n_2}\binom{n - n_1 - n_2}{n_3} \cdots \binom{n_m}{n_m} = \frac{n!}{n_1!n_2! \cdots n_m!}.$$

Statement two is also quite easy to prove. If the n objects were all distinguishable we would have had $n!$ sequences. Objects of color 1, however, can be permuted among themselves in $n_1!$ ways (in whichever positions they occur). The same is true for the other colors. We are thus left with $n!/(n_1!n_2! \cdots n_m!)$ distinct sequences.

One can easily see that the last statement is effectively the same as the second. Indeed, color coordinate i with color i. Then attach a sequence of colors to a nondecreasing path by marking down the color of the coordinate that changes at each step of the way. (Through this process the notions of a nondecreasing path and a sequence of colors become identical.) By statement two we conclude that there are $n!/(n_1!n_2!\cdots n_m!)$ such paths.

The last of the three combinatorial interpretations of the multinomial numbers appears to be the richest and most convenient to rely on, chiefly because of its geometrical appeal. One good example of this is the proof of the *multinomial Vandermonde convolution*:

$$\frac{(p+q)!}{n_1!n_2!\cdots n_m!} = \sum \frac{p!}{k_1!k_2!\cdots k_m!} \frac{q!}{(n_1-k_1)!(n_2-k_2)!\cdots(n_m-k_m)!},$$

where the sum is over all points $(k_1, k_2, \ldots, k_m)$, *with* $0 \le k_i \le n_i$, k_i *integral, and* $\sum_{i=1}^{m} k_i = p$.

Indeed, $(p+q)!/(n_1!\cdots n_m!)$ counts the number of paths of length $p+q$ between $(0,\ldots,0)$ and $(n_1,\ldots,n_m)$. We now count these paths in a different way. Any path of length $p+q$ ending at $(n_1,\ldots,n_m)$ is composed of a path of length p ending, say, at $(k_1,\ldots,k_m)$, followed by a path of length q from $(k_1,\ldots,k_m)$ to $(n_1,\ldots,n_m)$. By placing the origin at $(k_1,\ldots,k_m)$ momentarily, the latter path of length q can be envisioned as a path between $(0,\ldots,0)$ and $(n_1-k_1,\ldots,n_m-k_m)$. There are $p!/(k_1!\cdots k_m!)$ choices for the initial path of length p and $q!/((n_1-k_1)!\cdots(n_m-k_m)!)$ for its follow-up path of length q. With the point $(k_1,\ldots,k_m)$ fixed we thus have

$$\frac{p!}{k_1!\cdots k_m!} \frac{q!}{(n_1-k_1)!\cdots(n_m-k_m)!}$$

paths passing through it. Summing up over all intermediate points $(k_1,\ldots,k_m)$ p steps away from the origin, we obtain the multinomial Vandermonde convolution. (Throughout the proof of this formula we understand by a path a nondecreasing path, of course.)

1.15 Lah Numbers

The numbers $(n!/m!)\binom{n-1}{m-1}$ have combinatorial meaning, specifically as follows:

* *The number of ways of placing n distinguishable objects into m indistinguishable boxes with no box left empty, paying attention to the order in which they are put inside the boxes, is equal to* $(n!/m!)\binom{n-1}{m-1}$.

To see this, think initially of the boxes being distinguishable and symbolized by the slashes, /.

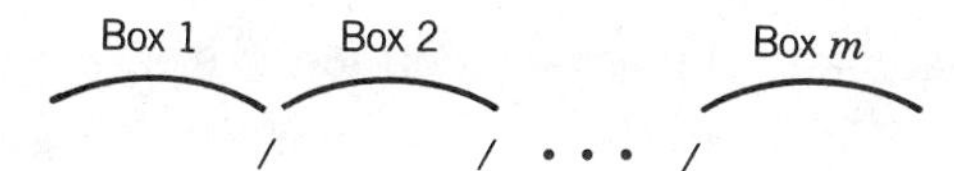

(We have $m - 1$ slashes in all.)

Take any sequence formed by the n objects

$$a_1 {}_\wedge a_2 {}_\wedge a_3 {}_\wedge a_4 {}_\wedge a_5 {}_\wedge \cdots {}_\wedge a_{n-1} {}_\wedge a_n.$$

The $n - 1$ hats "$\wedge$" indicate the possible places where slashes can be inserted. No consecutive slashes are permitted. We have to insert $m - 1$ slashes in $n - 1$ places (indicated by the hats), a task that can be accomplished in $\binom{n-1}{m-1}$ ways. Since there are $n!$ sequences possible to make with the n distinguishable objects, we obtain a total of $n!\binom{n-1}{m-1}$ ways of assigning the n distinguishable objects to the m distinguishable boxes with no box left empty. But the m boxes are in fact indistinguishable, thus leaving us with $(n!/m!)\binom{n-1}{m-1}$ ways.

We define numbers L_n^m by setting

$$L_n^m = (-1)^n \frac{n!}{m!}\binom{n-1}{m-1}$$

and call them *Lah numbers*. They have many nice properties, some of which are described in Section 2.9 of Chapter 2.

Example. $n = 4$, $m = 3$. The $(4!/3!)\binom{4-1}{3-1} = 12$ ways are indicated below:

1/2/34	1/3/24	1/4/23
1/2/43	1/3/42	1/4/32
2/3/14	2/4/13	3/4/12
2/3/41	2/4/31	3/4/21.

EXERCISES

1. Is it more likely to obtain a sum of 9 when rolling two fair dice or when rolling three?

2. Are there as many subsets of even cardinality as there are of odd cardinality? (Exhibit a "constructive" bijection, if possible.)

3. How many sequences of length five with 0 or 1 as entries contain at least two consecutive 0's? (Attempt to generalize and answer this question for sequences of length n that contain runs of k or more 0's.)

4. A basket contains four red, five yellow, and seven green apples. Pick six apples at random. What is the chance that two are red, three yellow, and one green?

5. How many subsets of three (distinct) integers between 1 and 90 are there whose sum is: (a) even, (b) divisible by 3, (c) divisible by 9?

6. Give a combinatorial interpretation to the inequality

$$\binom{m}{n-1} \le \binom{m}{n}, \qquad \text{for } 1 \le n \le \frac{m+1}{2}.$$

[*Hint*: Construct a bijection between two Cartesian products.]

7. Show that:

(a) $\sum_{k=m}^{n} \binom{k}{m} = \binom{n+1}{m+1}$.

(b) $\sum_{k=0}^{n} \binom{n}{k}^2 = \binom{2n}{n}$.

(c) $\sum_{k=0}^{p} \binom{p}{k}\binom{q}{k}\binom{n+k}{p+q} = \binom{n}{p}\binom{n}{q}$.

[*Hint*: Use nondecreasing lattice paths.]

8. What is the probability that a random hand of eight cards (out of a usual deck of 52 cards) has:

(a) Two cards of each suit?

(b) The same number of hearts and spades?

(c) Two hearts and at least three spades, and the values of the spades are all strictly greater than the values of the hearts?

9. What is the chance that two or more people have the same birthdate (i.e., same month and day) among a random group of n people?

10. How many ways are there to invite one of three friends over for dinner on six successive evenings such that no friend is invited more than three times?

11. If you flip a fair coin 20 times and get 12 heads and 8 tails, what is the chance that there are no consecutive tails?

12. How many sequences formed with the letters in the word "Mississippi" have no consecutive i's and no consecutive s's?

13. Let p be a prime. For $1 \le k \le p-2$ show that p divides $\binom{p}{k}$. Show also that $\binom{2p}{p} = 2$ (modulo p). Prove that $\binom{pn}{pm} = \binom{n}{m}$ (modulo p), where m and n are nonnegative integers.

14. How many triangles can be drawn all of whose vertices are among those of a given n-gon and all of whose sides are diagonals (but not sides) of the n-gon?

15. Like the binomial numbers, the Stirling numbers $S_n^0, S_n^1, \ldots, S_n^k, \ldots, S_n^n$ (and $|s_n^0|, |s_n^1|, \ldots, |s_n^n|$) start off by increasing up to a certain value of k and then strictly decrease. The maximum, however, is not achieved at the middle terms, as is the case with the binomial numbers. Prove this.

16. Show that the (absolute value of the) Stirling number of the first kind, $|s_n^k|$, is the sum of all products of $n - k$ different integers taken from $\{1, 2, \ldots, n - 1\}$. For example,

$$s_5^3 = 1 \cdot 2 + 1 \cdot 3 + 1 \cdot 4 + 2 \cdot 3 + 2 \cdot 4 + 3 \cdot 5.$$

Show that $|s_n^k| < (n - k)!\binom{n-1}{k-1}^2$.

17. Show that the Stirling number of the second kind, S_n^k, is the sum of all products of $n - k$ not necessarily distinct integers taken from $\{1, 2, \ldots, k\}$. For example,

$$S_5^3 = 1 \cdot 1 + 1 \cdot 2 + 1 \cdot 3 + 2 \cdot 2 + 2 \cdot 3 + 3 \cdot 3.$$

Prove also that $S_n^k < k^{n-k}\binom{n-1}{k-1}$.

18. Prove that $\sum_{k=m}^{n}\binom{k}{m}|s_n^k| = |s_{n+1}^{m+1}|$.

19. Prove that $S_n^m = (1/m!)\sum_{k=0}^{n-1}(-1)^k\binom{m}{k}(m - k)^n$.

20. Prove that the number of sequences of integers $x_1 x_2 \cdots x_r$ $(1 \le x_i \le n)$ containing less than i entries less than or equal to i (for $i = 1, 2, \ldots, n$) is $(n - r)n^{r-1}$; $1 \le r \le n$.

21. Persons A and B independently proofread a book. A finds a errors, B finds b errors, while c errors were spotted by both A and B. Estimate the number of remaining errors in the book.

22. In how many ways can r rooks be placed on a chessboard with m rows and n columns such that no rook can attack another?

23. Show that

$$\sum |A_1 \cup A_2 \cup \cdots \cup A_k| = n(2^k - 1)^{k(n-1)},$$

where the sum is taken over all ordered selections of subsets $A_1, A_2, \ldots, A_k$ of a set with n elements.

24. What is the largest number of subsets noncomparable with respect to inclusion of a set with n elements? (Noncomparable with respect to inclusion means $A_i \not\subseteq A_j$, for $i \ne j$.)

25. Prove that among a set of $n + 1$ positive integers, none of which exceed $2n$, at least one divides another.

26. Select the minimal element in each of the $\binom{n}{r}$ subsets of $\{1, 2, \ldots, n\}$. Show that the average of the numbers so obtained is $(n + 1)(r + 1)^{-1}$.

27. Show that the number of permutations σ of the set $\{1, 2, \ldots, n\}$ that have the property that there exist exactly k elements j for which $\sigma(j) > \sigma(i)$, for every $i < j$, is equal to $|s_n^k|$.

HISTORICAL NOTE

Fascination with producing configurations with remarkable properties, or counting the exact number of available choices goes back in history a rather long way. The square

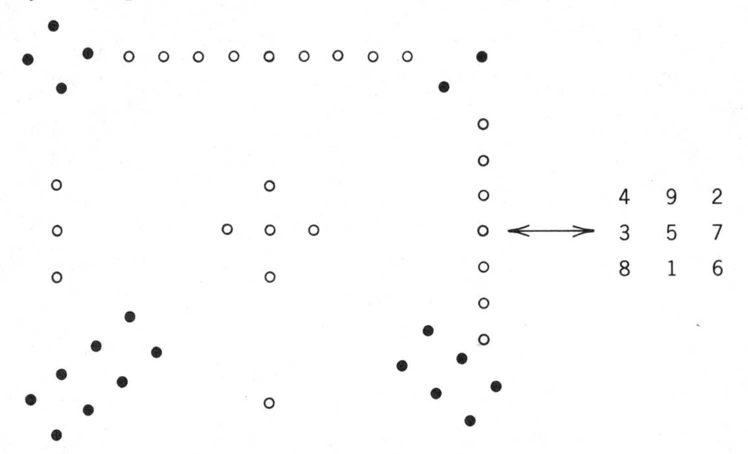

has the "magical" property that the sum of dots along each of the four edges and the two diagonals is always 15. Legend tells [1] that it was observed on the back of a divine tortoise that emerged from the River Lo (in China) some 4000 years ago.

River Ho was known to shelter a no less remarkably decorated tortoise, displaying a magic figure which in present day notation takes the following form:

6
1
10 10
9 4 5 3 8
10 10
2
7

The reader will observe the symmetry displayed with respect to the center, for example 5 + 4 = 9; 4 + 10 + 1 = 9 + 6; 2 + 10 + 4 = 9 + 7.

Written documentation on binomial numbers dates back to the twelfth century's Indian school of arithmetic led by Bhaskra. The binomial recurrence that we call "Pascal's triangle" was actually known to Nasir-Ad-Din [2], a Persian philosopher of the thirteenth century. In connection with games of chance the binomial numbers were rediscovered by Pascal and Fermat about the middle of the seventeenth century, to become commonplace in the works of Euler and Laplace.

James Stirling is responsible for bringing to attention the two species of numbers that now bear his name. Descendant of one of the oldest landed families of Scotland, he was educated at Oxford and was in correspondence with most of the elite mathematicians of his day, such as Euler, Newton,

Bernoulli, and MacLaurin (see [4]). The numbers in question appear in his major work [3], along with the more famous Stirling formula that approximates factorials. Taking not quite the form we recognize today, his original formula was an expansion of the logarithm of the factorial. A few generations earlier, Napier, the inventor of the logarithms (and a neighboring landowner), had married Elizabeth, a Stirling. Intermarriage between the two families has in fact occurred more than once.

We recommend [5] and [6] for more information on the topics covered in this chapter.

REFERENCES

[1] F. M. Müller, Ed., *Sacred Books of the East*, Vol. XVI ("The Yi-King"). Oxford University Clarendon Press, London, 1882.

[2] Nasir-Ad-Din At-Tusi, "Handbook of arithmetic using board and dust," *Istor. Math. Issled.*, **15**, 431–444 (1963); *Math. Rev.*, **31**, 5776 (1966).

[3] J. Stirling, *Methodus Differentialis, Sive Tractus de Summatione et Interpolatione Serierum Infinitarum*, Oxford University, London, 1730.

[4] C. Tweedie, *James Stirling*, Oxford Clarendon Press, London, 1922.

[5] C. Berge, *Principles of Combinatorics*, Academic Press, New York, 1971.

[6] J. Riordan, *An Introduction to Combinatorial Analysis*, Wiley, New York, 1958.

CHAPTER 2

Generating Functions

Do not pray for tasks equal to your powers. Pray for powers equal to your tasks.

Twenty Sermons, PHILLIPS BROOKS

Generating functions provide an algebraic machinery for solving combinatorial problems. The usual algebraic operations (convolution, especially) facilitate considerably not only the computational aspects but also the thinking processes involved in finding satisfactory solutions. More often than not we remain blissfully unaware of this disinterested service, until trying to reproduce the same by direct calculations (a task usually accompanied by no insignificant mental strain). The main reason for introducing formal power series is the ability to translate key combinatorial operations into algebraic ones that are, in turn, easily and routinely performed within a set (usually an algebra) of generating functions. Generally this is much easier said than done, for it takes great skill to establish such a happy interplay. Yet notable examples exist, and we examine a couple of better known ones in considerable detail.

We begin by introducing the ordinary and exponential generating functions. Upon closely investigating the combinatorial meaning of the operation of convolution in these two well-known cases, we turn to specific generating functions associated with the Stirling and Lah numbers. The latter part of the chapter touches briefly upon the uses of formal power series to recurrence relations and introduces the Bell polynomials, in connection with Faa DiBruno's formula, for explicitly computing the higher order derivatives of a composition of two functions. In ending the chapter we dote upon subjects such as Kirchhoff's tree generating matrix (along with applications to statistical design), partitions of an integer, and a generating function for solutions to Diophantine systems of linear equations in nonnegative integers.

1 THE FORMAL POWER SERIES

2.1

The *generating function* of the sequence (a_n) is the (formal) power series $A(x) = \Sigma_n a_n x^n = a_0 + a_1 x + a_2 x^2 + \cdots + a_n x^n + \cdots$. The summation sign always starts at 0 and extends to infinity in steps of one. By x we understand an indeterminate.

Most of the time we view generating functions as formal power series. Occasionally, however, questions of convergence may arise and the analytic techniques would then come to play an important role. We recall for convenience that two formal power series are equal if (and only if) the coefficients of the corresponding powers of x are equal.

By writing $(a_n) \leftrightarrow A(x)$ we indicate the bijective association between the sequence (a_n) and its generating function $A(x)$. In terms of this association we observe that if $(a_n) \leftrightarrow A(x)$, $(b_n) \leftrightarrow B(x)$, and c is a constant, then

$$(a_n + b_n) \leftrightarrow A(x) + B(x)$$

$$(ca_n) \leftrightarrow cA(x)$$

and, most importantly, multiplication by convolution

$$\left(\sum_{i=0}^{n} a_i b_{n-i} \right) \leftrightarrow A(x)B(x).$$

(The set of generating functions endowed with these operations is said to form an algebra.)

Generating functions A and B are said to be *inverses* of each other if $A(x)B(x) = 1 = B(x)A(x)$. This last relation we sometimes write as $B = A^{-1}$, $B = 1/A$, $A = B^{-1}$, or $A = 1/B$. Note, for example, that $A(x) = 1 - x$ and $B(x) = \Sigma_n x^n$ are a pair of inverses.

An important operation with power series is that of *composition* (or substitution). By $A \circ B$ we understand the series defined as follows: $(A \circ B)(x) = A(B(x))$. More explicitly still, if $A(x) = \Sigma_n a_n x^n$ and $B(x) = \Sigma_n b_n x^n$, then $(A \circ B)(x) = A(B(x)) = \Sigma_n a_n (B(x))^n$. In order that $A(B(x))$ be a well-defined power series, the original series A and B need be such that the coefficient of each power of x in $A(B(x))$ is obtained as a sum of finitely many terms. [Thus if $A(x) = \Sigma_n x^n$ and $B(x) = x + x^2$, $A(B(x))$ is well defined, but if $B(x) = 1 + x$, then $A(B(x))$ is not well defined. In the latter case the constant term of $A(B(x))$ involves the summation of infinitely many 1's.] We can see therefore that $A(B(x))$ makes sense essentially under two conditions: when $A(x)$ has infinitely many nonzero coefficients then the constant term in $B(x)$ must be 0, and if $A(x)$ has only finitely many nonzero coefficients [i.e., if $A(x)$ is a polynomial], then $B(x)$ can be arbitrary.

Whenever well defined, the series $A \circ B$ is called the composition of A with B (or the substitution of B into A).

We also let the linear operator D (of *formal differentiation*) act upon a generating function A as follows:

$$DA(x) = D\left(\sum_n a_n x^n\right) \underset{\text{def.}}{=} \sum_n (n+1) a_{n+1} x^n.$$

As an example, let $A(x) = 2 - 5x + 3x^2$ and $B(x) = \Sigma_n (n+1)^{-1} x^n$. The reader may quickly verify that

$$A(x)B(x) = 2 - 4x + \sum_{n=2}^{\infty} (n+5) n^{-1} (n^2 - 1)^{-1} x^n.$$

Applying the differential operator D to A, B, and AB respectively, we obtain:

$$DA(x) = -5 + 3 \cdot 2x, \; DB(x) = \sum_n (n+1)(n+2)^{-1} x^n$$

and

$$D(A(x)B(x)) = -4 + \sum_{n=2}^{\infty} (n+5)(n^2-1)^{-1} x^{n-1}.$$

In closing, let us mention that the operator of formal differentiation satisfies the familiar rules of differentiation:

$$D(AB) = (DA)B + A(DB)$$

$$DA^{-1} = -A^{-2} DA,$$

and most importantly, the "chain rule,"

$$D(A \circ B) = ((DA) \circ B) DB.$$

2.2

The *exponential generating function* of the sequence (a_n) is the (formal) power series

$$E(x) = \sum_n a_n \frac{x^n}{n!} = a_0 + a_1 \frac{x}{1!} + a_2 \frac{x^2}{2!} + \cdots + a_n \frac{x^n}{n!} + \cdots.$$

In as much as the exponential generating functions are concerned, if $(a_n) \leftrightarrow E(x)$, $(b_n) \leftrightarrow F(x)$, and c is a constant, then

$$(a_n + b_n) \leftrightarrow E(x) + F(x)$$

$$(ca_n) \leftrightarrow cE(x)$$

and

$$\left(\sum_{i=0}^{n} \binom{n}{i} a_i b_{n-i}\right) \leftrightarrow E(x)F(x).$$

In this case we say that the multiplication of two exponential generating functions corresponds to the binomial convolution of sequences.

As before, we call E and F inverses if $E(x)F(x) = 1 = F(x)E(x)$.

The operator D of formal differentiation acts here as follows:

$$DE(x) = D\left(\sum_n a_n \frac{x^n}{n!}\right) \underset{\text{def.}}{=} \sum_n a_{n+1} \frac{x^n}{n!}.$$

We illustrate the multiplication of two exponential generating functions by a simple example:

$$\begin{aligned}\left(\sum_n 3^n \frac{x^n}{n!}\right)\left(\sum_n \frac{1}{2^n} \frac{x^n}{n!}\right) &= \sum_n \left(\sum_{i=0}^n \binom{n}{i} 3^i \frac{1}{2^{n-i}}\right)\frac{x^n}{n!} \\ &= \sum_n \left(\sum_{i=0}^n \binom{n}{i} 3^i \left(\frac{1}{2}\right)^{n-i}\right)\frac{x^n}{n!} = \sum_n \left(3 + \frac{1}{2}\right)^n \frac{x^n}{n!} \\ &= \sum_n \left(\frac{7}{2}\right)^n \frac{x^n}{n!}.\end{aligned}$$

The next to the last equality sign is explained by the fact that $\sum_{i=0}^n \binom{n}{i} a^i b^{n-i} = (a+b)^n$, where a and b are two entities that commute (such as 3 and $\frac{1}{2}$).

With regard to differentiation,

$$D\left(\sum_n 3^n \frac{x^n}{n!}\right) = \sum_n 3^{n+1} \frac{x^n}{n!}.$$

2.3

The vector space of sequences can be made into an algebra by defining a multiplication of two sequences. We require the rule of multiplication to be "compatible" with the rules of addition and scalar multiplication. Two such rules of multiplication have been described in Sections 2.1 and 2.2. Other rules could be conceived, but one wonders of how much use in combinatorial counting they would be. One well-known multiplication, of interest to number theorists, is as follows:

$$(a_n)(b_n) = \left(\sum_{\substack{d \\ dm=n}} a_d b_m\right)$$

and is called the *Dirichlet convolution*. In this case we attach the formal Dirichlet series $\sum_n (a_n/n^x)$ to the sequence (a_n).

Eulerian generating functions are known to be helpful in enumeration problems over finite vector spaces and with inversion problems in sequences.

The *Eulerian series* of the sequence (a_n) is defined as

$$E_q(x) = \sum_n \frac{a_n x^n}{(1-q)(1-q^2)\cdots(1-q^n)}.$$

We briefly discuss these series in Chapter 3.

Let us now make ourselves more aware of what combinatorial operations the generating functions and the exponential generating functions perform for us.

2 THE COMBINATORIAL MEANING OF CONVOLUTION

2.4

In Section 1.2 we established bijective correspondences between the three general problems listed below and showed that they all admit the same numerical solution:

(a) *The number of ways to distribute n indistinguishable balls into m distinguishable boxes is* $\binom{n+m-1}{n}$.

(b) *The number of vectors* $(n_1, n_2, \ldots, n_m)$ *with nonnegative integer entries satisfying*

$$n_1 + n_2 + \cdots + n_m = n$$

is $\binom{n+m-1}{n}$.

(c) *The number of ways to select n objects with repetition from m different types of objects is* $\binom{n+m-1}{n}$. (We assume that we have an unlimited supply of objects of each type and that the order of selection of the n objects is irrelevant.)

The three problems just mentioned consociate well to the operation of convolution with generating functions. Specifically, let us explain how we attach combinatorial meaning to the multiplication by convolution of several generating functions with coefficients 0 or 1:

1. *The number of ways of placing n indistinguishable balls into m distinguishable boxes is the coefficient of* x^n *in*

$$(1 + x + x^2 + \cdots)^m = \left(\sum_k x^k\right)^m = (1-x)^{-m}.$$

Indeed, we can describe the possible contents of our boxes as follows:

Box 1	Box 2	Box 3	$\cdots$	Box m
1	1	1		1
x	x	x		x
x^2	x^2	x^2		x^2
x^3	x^3	x^3		x^3
$\vdots$	$\vdots$	$\vdots$		$\vdots$

(*)

The symbol x^i beneath box j indicates the fact that we may place i balls in box j. Think of m (the number of boxes) being fixed, but keep n unspecified. With this in mind we can assume that the columns beneath the boxes are of infinite length. How do we then obtain the coefficient of x^n in the product $(1 + x + x^2 + x^3 + \cdots)^m$? We select x^{n_1} from column 1 of (*), x^{n_2} from column $2, \ldots,$ x^{n_m} from column m such that $x^{n_1}x^{n_2} \cdots x^{n_m} = x^n$, and do this in all possible ways. The number of such ways clearly equals the number of vectors $(n_1, n_2, \ldots, n_m)$ satisfying

$$\sum_{i=1}^{m} n_i = n,$$

with $0 \le n_i$, n_i integers; $1 \le i \le m$. By (b) above we conclude that there are precisely $\binom{n+m-1}{n}$ solutions, which is also in agreement with (a), thus proving our statement.

In terms of generating functions, this shows that

$$\left(\sum_k x^k\right)^m = \sum_n \binom{n+m-1}{n} x^n. \tag{2.1}$$

By observing that $(1-x)^{-1} = \Sigma_n x^n$ we can rewrite relation (2.1) as follows:

$$(1-x)^{-m} = \sum_n \binom{n+m-1}{n} x^n. \tag{2.2}$$

2. *The number of ways of placing n indistinguishable objects into m distinguishable boxes with at most r_i objects in box i is the coefficient of x^n in*

$$\prod_{i=1}^{m} (1 + x + x^2 + \cdots + x^{r_i}).$$

The contents of our m boxes is now as follows:

Box 1	Box 2	Box 3	$\cdots$	Box m
1	1	1		1
x	x	x		x
x^2	x^2	x^2		x^2
$\vdots$	$\vdots$	$\vdots$		$\vdots$
x^{r_1}	x^{r_2}	x^{r_3}		x^{r_m}

Again, the coefficient of x^n is the number of selections of powers of x (one from each column) such that the sum of these powers is n. To be more precise, the coefficient of x^n is the number of all vectors $(n_1, n_2, \ldots, n_m)$ satisfying

$$\sum_{i=1}^{m} n_i = n,$$

with $0 \le n_i \le r_i$, n_i integers; $1 \le i \le m$.

Example. At suppertime Mrs. Jones rewards her children, Lorie, Mike, Tammie, and Johnny, for causing only a limited amount of damage to each other during the day. She decides to give them a total of ten identical candies. According to their respective good behavior she chooses to give at most three candies to Lorie, at most four to Mike, at most four to Tammie, and at most one to Johnny. In how many ways can she distribute the candies to the children?

In this problem we make the abstractions as follows:

children ↔ distinguishable boxes

candies ↔ indistinguishable balls

The possibilities of assignment are described by

Lorie	Mike	Tammie	Johnny
1	1	1	1
x	x	x	x
x^2	x^2	x^2	
x^3	x^3	x^3	
	x^4	x^4	

The generating function in question is

$$(1 + x + x^2 + x^3)(1 + x + x^2 + x^3 + x^4)^2(1 + x)$$

and the numerical answer we seek will be found in the coefficient of x^{10}. As it seems simple enough to write a computer program that multiplies two formal power series (and, in particular, two polynomials), calculating the coefficient of a power of x can be done expeditiously. Indeed, all it takes to program multiplication by convolution is a DO loop. [The coefficient in question equals, as we saw, the number of solutions (n_1, n_2, n_3, n_4) to

$$\begin{aligned} &n_1 + n_2 + n_3 + n_4 = 10 \\ &0 \le n_1 \le 3 \\ &0 \le n_2, n_3 \le 4 \\ &0 \le n_4 \le 1, \qquad n_i \text{ integers.} \end{aligned}$$

There are precisely nine such vectors, which we actually list below:

Lorie	Mike	Tammie	Johnny
1	4	4	1
2	3	4	1
2	4	3	1
2	4	4	0
3	2	4	1
3	3	3	1
3	3	4	0
3	4	2	1
3	4	3	0 .]

3. *The number of ways of assigning n indistinguishable balls to m distinguishable boxes such that box j contains at least* s_j *balls is the coefficient of* x^n *in* $\prod_{j=1}^{m}(x^{s_j}(1 + x + x^2 + \cdots)) = x^{\Sigma s_j}(1 - x)^{-m} = \Sigma_n\binom{n - \Sigma s_j + m - 1}{m - 1}x^n$.

The composition of the m boxes is, in this case,

Box 1	Box 2	$\cdots$	Box m
x^{s_1}	x^{s_2}		x^{s_m}
x^{s_1+1}	x^{s_2+1}		x^{s_m+1}
x^{s_1+2}	x^{s_2+2}		x^{s_m+2}
$\vdots$	$\vdots$		$\vdots$

Taking in common factor x^{s_j} from column j we obtain the generating function written above. The process of extracting x^{s_j} in common factor from column j and the writing down of the generating function by multiplying all factors together parallels the combinatorial argument of solving this problem by first leaving s_j balls in box j and then distributing the remaining $n - \Sigma_{j=1}^{m} s_j$ balls without restrictions to the m boxes. This shows in fact that the coefficient of x^n in the generating function written above is

$$\binom{n - \Sigma s_j + m - 1}{n - \Sigma s_j} = \binom{n - \Sigma s_j + m - 1}{m - 1}.$$

4. *The number of ways to distribute n indistinguishable balls into m distinguishable boxes with box i having the capacity to hold either* s_{i1}, *or* $s_{i2}, \ldots,$ *or* s_{ir_i} (and no other number of) *balls equals the coefficient of* x^n *in* $\prod_{i=1}^{m}(x^{s_{i1}} + x^{s_{i2}} + \cdots + x^{s_{ir_i}})$.

The composition of the boxes is

Box 1	Box 2	$\cdots$	Box m
$x^{s_{11}}$	$x^{s_{21}}$		$x^{s_{m1}}$
$x^{s_{12}}$	$x^{s_{22}}$		$x^{s_{m2}}$
$\vdots$	$\vdots$		$\vdots$
$x^{s_{1r_1}}$	$x^{s_{2r_2}}$		$x^{s_{mr_m}}$

Placing s_{ij} balls in box j corresponds to selecting the power $x^{s_{ij}}$ in the jth column. Distributing a total of n balls to the m boxes amounts to selecting a vector of powers of x (one from each column), say $(s_{1n_1}, s_{2n_2}, \ldots, s_{mn_m})$, such that $\Sigma_{i=1}^{m} s_{in_i} = n$. The number of all such distributions of n balls is therefore the coefficient of x^n in the generating function given above. It also equals the number of integer solutions to

$$\sum_{i=1}^{n} s_{in_i} = n$$

with s_{in_i} restricted to belong to $\{s_{i1}, s_{i2}, \ldots, s_{ir_i}\}$, $1 \le i \le m$.

One can geometrically visualize the solutions to these constraints as the points $(y_1, y_2, \ldots, y_m)$, with y_i belonging to the finite set $\{s_{i1}, s_{i2}, \ldots, s_{ir_i}\}$, which are also on the hyperplane $\sum_{i=1}^m y_i = n$.

5. *We conclude this long section with a revision of several useful relations among generating functions. These are*:

(i) $(1-x)^{-1} = \sum_n x^n$.

(ii) $(1-x^{n+1})(1-x)^{-1} = 1 + x + x^2 + \cdots + x^n$.

(iii) $(1-x)^{-m} = \sum_n \binom{n+m-1}{n} x^n$; *m positive integer.*

(iv) $(1+x)^m = \sum_{n=0}^m \binom{m}{n} x^n$; *m positive integer.*

(v) $(x_1 + x_2 + \cdots + x_r)^m$
$$= \sum_{\substack{(n_1,\ldots,n_r) \\ \sum n_i = m}} m!/(n_1!n_2!\cdots n_r!)x_1^{n_1}x_2^{n_2}\cdots x_r^{n_r}.$$

(vi) $\prod_{i=1}^m(\sum_j a_{ij}x^j) = \sum_n(\sum_{\substack{(j_1,\ldots,j_m) \\ \sum_k j_k = n}} a_{1j_1}a_{2j_2}\cdots a_{mj_m})x^n$

The contents of (i) and (ii) can be straightforwardly verified by multiplying out. The statement made in (iii) has been established in (2.2).

To understand (iv), write $(1+x)^m$ as

$$\begin{matrix} 1 & 1 & 1 & \ldots & 1 \\ x & x & x & & x \end{matrix} \quad (m \text{ columns}).$$

A formal term in the expansion of $(1+x)^m$ involves the choice of one entry from each of the m columns. A term containing exactly n x's is obtained by picking x from a subset of n columns, and 1's from the remaining $m-n$ columns. The number of such terms equals the number of subsets with n elements out of the set of m columns, that is, it equals $\binom{m}{n}$. This explains (iv).

The proof of (v) is similar. Write out m columns

$$\begin{matrix} x_1 & x_1 & \cdots & x_1 \\ x_2 & x_2 & & x_2 \\ \vdots & \vdots & & \vdots \\ x_r & x_r & & x_r \end{matrix}$$

A formal product is obtained by picking an x_i from each column. The coefficient of $x_1^{n_1}x_2^{n_2}\cdots x_r^{n_r}$ is the number of formal products of length m containing n_1 x_1's, n_2 x_2's, $\ldots$, n_r x_r's. There are $m!/(n_1!n_2!\cdots n_r!)$ such products (see also Section 1.14). This establishes (v).

To realize that (vi) is true, line up m columns of infinite length:

$$\begin{matrix} a_{10} & a_{20} & \cdots & a_{m0} \\ a_{11}x & a_{21}x & & a_{m1}x \\ a_{12}x^2 & a_{22}x^2 & & a_{m2}x^2 \\ a_{13}x^3 & a_{23}x^3 & & a_{m3}x^3 \\ \vdots & \vdots & & \vdots \end{matrix}.$$

A term involving x^n is obtained by picking $a_{kj_k}x^{j_k}$ from column k $(1 \leq k \leq m)$ and making the product $\prod_{k=1}^{m} a_{kj_k}x^{j_k}$, with the exponents of x satisfying $\sum_{k=1}^{m} j_k = n$. The totality of such terms equals

$$\sum_{\substack{(j_1,\ldots,j_m)\\ \sum_{k=1}^{m} j_k = n}} \prod_{k=1}^{m} a_{kj_k}x^{j_k},$$

thus explaining the coefficient of x^n.

2.5

We turn our attention now to exponential generating functions. These generating functions are helpful when counting the number of sequences (or words) of length n that can be made with m (possibly repeated) letters and with specified restrictions on the number of occurrences of each letter; such as the number of distinct sequences of length four that can be made with the (distinguishable) letters a, b, c, d, e in which b occurs twice, c at least once, e at most three times, and with no restrictions on the occurrences of a and d.

For convenience we denote the exponential generating function $\sum_n x^n/n!$ by e^x. We invite the reader to observe at once that $(e^x)^m = e^{mx}$. Indeed, the coefficient of $x^n/n!$ in e^{mx} is m^n, while the coefficient of $x^n/n!$ in $(e^x)^m$ is

$$\sum_{\substack{(n_1,\ldots,n_m)\\ \sum n_i = m\\ 0 \leq n_i,\ \text{integers}}} \frac{n!}{n_1!n_2!\cdots n_m!}.$$

These two expressions count the same thing, however, namely the number of sequences of length n that can be made with m distinguishable letters and with no restrictions on the number of occurrences of each letter. (To be specific, we have n spots to fill with m choices for each spot, and this gives us m^n choices; on the other hand we can sort out the set of sequences by the number of occurrences of each letter, thus obtaining the second expression.)

The mechanism of using exponential generating functions to solve problems in counting is similar to that described in Section 2.4. We present an example that captures all the relevant features of a general case.

Assume at all times that we have available an abundant (and if necessary infinite) supply of replicas of the letters a, b, c, d, e. We want to count *the number of distinct sequences of length four containing two* b*'s, at least one* c*, at most three* e*'s, and with no restrictions on the occurrences of* a *and* d.

The recipe that leads to the solution is the following: With each distinct letter attach a column in which the powers of x indicate the number of times that letter is allowed to appear in a sequence. Such powers of x are divided by the respective factorials. In this case we have

a	b	c	d	e
1			1	1
$\frac{x}{1!}$		$\frac{x}{1!}$	$\frac{x}{1!}$	$\frac{x}{1!}$
$\frac{x^2}{2!}$	$\frac{x^2}{2!}$	$\frac{x^2}{2!}$	$\frac{x^2}{2!}$	$\frac{x^2}{2!}$
$\frac{x^3}{3!}$		$\frac{x^3}{3!}$	$\frac{x^3}{3!}$	$\frac{x^3}{3!}$
$\frac{x^4}{4!}$		$\frac{x^4}{4!}$	$\frac{x^4}{4!}$	
$\vdots$		$\vdots$	$\vdots$	

The exponential generating function we attach to this problem is (as before) the product of the columns, that is,

$$\left(\sum_k \frac{x^k}{k!}\right)\frac{x^2}{2!}\left(\sum_{k=1}^{\infty}\frac{x^k}{k!}\right)\left(\sum_k \frac{x^k}{k!}\right)\left(1+\frac{x}{1!}+\frac{x^2}{2!}+\frac{x^3}{3!}\right)$$

$$= e^x\left(\frac{x^2}{2!}\right)(e^x-1)e^x\left(1+\frac{x}{1!}+\frac{x^2}{2!}+\frac{x^3}{3!}\right)$$

$$= e^{2x}(e^x-1)\frac{x^2}{2!}\left(1+\frac{x}{1!}+\frac{x^2}{2!}+\frac{x^3}{3!}\right).$$

The numerical answer we seek is simply the coefficient of $x^4/4!$. If, with the same restrictions, we become interested in the number of sequences of length n, the answer is the coefficient of $x^n/n!$ in the above exponential generating function.

To see that the coefficient of $x^4/4!$ is indeed the answer to our problem one has to observe the following. The act of picking $x^{n_i}/n_i!$ from column i ($1 \le i \le 5$) corresponds to looking at sequences consisting of precisely n_1 a's, n_2 b's, n_3 c's, n_4 d's, and n_5 e's. Taking the product

$$\prod_i \frac{x^{n_i}}{n_i!} = \frac{(\Sigma_i n_i)!}{\Pi_i n_i!}\,\frac{x^{\Sigma_i n_i}}{(\Sigma_i n_i)!}$$

(with $\Sigma_i n_i = 4$) produces a coefficient of

$$\frac{(\Sigma_i n_i)!}{\Pi_i n_i!} = \frac{4!}{n_1!n_2!n_3!n_4!n_5!}$$

for $x^4/4!$, which equals the number of sequences with precisely n_i copies of each letter. The totality of such pickings, with values of n_i restricted to the exponents of x that appear in column i, leads to the coefficient of $x^4/4!$, which equals, therefore, the number of sequences with occurrences restricted as specified.

Specifically, we have

$$1 \cdot \frac{x^2}{2!} \cdot \frac{x}{1!} \cdot 1 \cdot \frac{x}{1!} + 1 \cdot \frac{x^2}{2!} \cdot \frac{x}{1!} \cdot \frac{x}{1!} \cdot 1$$

$$+1 \cdot \frac{x^2}{2!} \cdot \frac{x^2}{2!} \cdot 1 \cdot 1 + \frac{x}{1!} \cdot \frac{x^2}{2!} \cdot \frac{x}{1!} \cdot 1 \cdot 1$$

$$= \left(\frac{1}{2} + \frac{1}{2} + \frac{1}{4} + \frac{1}{2}\right)x^4 = 4!\left(\frac{1}{2} + \frac{1}{2} + \frac{1}{4} + \frac{1}{2}\right)\frac{x^4}{4!}.$$

We thus conclude that there are 42 such sequences.

Let us look at some examples of a more general nature.

Example 1. Find the number of (distinct) sequences of length n formed with m letters ($m \geq n$), with no letter repeated.

The m columns, one for each letter, are:

$$\begin{matrix} 1 & 1 & 1 & \ldots & 1 \\ x & x & x & & x \end{matrix}.$$

This gives the exponential generating function $(1 + x)^m$. We thus seek the coefficient of $x^n/n!$. And since $(1 + x)^m = \sum_{n=0}^{m}\binom{m}{n}x^n$ this shows that $x^n/n!$ has coefficient $m!/(m - n)!$ ($= [m]_n$), as expected.

Example 2. Find the number of sequences of length n, formed with m letters ($m \leq n$), in which each letter appears at least once.

The m columns are all the same, namely $x/1!$, $x^2/2!$, $x^3/3!, \ldots$. Hence the exponential generating function is $(\sum_{n=1}^{\infty}x^n/n!)^m = (e^x - 1)^m$. The coefficient of $x^n/n!$ turns out to be $m!\, S_n^m$, where S_n^m is the Stirling number, as we shall see in Section 3.

Example 3. How many sequences of length n can be made with the digits $1, 2, 3, \ldots, m$ such that digit i is not allowed to appear n_{i1} or n_{i2} or $\cdots$ or n_{ir_i} times (these being the only restrictions)?

The ith column in this case consists of the terms of e^x with precisely $x^{n_{ij}}/n_{ij}!$ missing ($1 \leq j \leq r_i$). We conclude therefore that the exponential generating function in question is

$$\prod_{i=1}^{m}\left(e^x - \sum_{j=1}^{r_i}\frac{x^{n_{ij}}}{n_{ij}!}\right).$$

The numerical answer we seek is the coefficient of $x^n/n!$.

2.6

Having thus shown the computational power of generating functions we address a problem that involves the permutations of the ordered set $1 < 2 < \cdots < m$. If $i < j$ and $\sigma(i) > \sigma(j)$ we say that the permutation σ has an *inversion* at the pair (i, j). Denote by a_{mk} the number of permutations on $\{1, 2, \ldots, m\}$ with precisely k inversions; $0 \le k \le \binom{m}{2}$. *We seek the generating function for* a_{mk}.

For a permutation σ and an integer j $(1 \le j \le m)$ denote by $\bar{\sigma}(j)$ the cardinality of the set $\{i\colon 1 \le i < j \text{ and } \sigma(i) > \sigma(j)\}$. The number of inversions of σ can now be written as $\bar{\sigma}(1) + \bar{\sigma}(2) + \cdots + \bar{\sigma}(m)$. (Note that $\bar{\sigma}(1) = 0$.)

Thus the number of permutations with exactly k inversions is the number of solutions in nonnegative integers to

$$n_1 + n_2 + \cdots + n_m = k$$

with restrictions $0 \le n_i \le i - 1$. (For a fixed permutation σ, n_i corresponds to $\bar{\sigma}(i)$.) We know how to interpret the set of such solutions (cf. Section 2.4). Think of m distinguishable boxes (as columns), with column i consisting of $1, x, x^2, \ldots, x^{i-1}$. The generating function that we associate is $\prod_{i=0}^{m} (1 + x + \cdots + x^{i-1})$ and then a_{mk}, being the same as the number of solutions to the constraints mentioned above, equals the coefficient of x^k in this generating function. We conclude, therefore, that

$$\sum_{k=0}^{\binom{m}{2}} a_{mk} x^k = \prod_{i=0}^{m} (1 + x + \cdots + x^{i-1}) = \prod_{i=0}^{m} \left(\frac{1 - x^i}{1 - x} \right).$$

EXERCISES

1. How many ways are there to get a sum of 14 when 4 (distinguishable) dice are rolled?

2. Find the generating function for the number of ways a sum of n can occur when rolling a die an infinite (or at least n) number of times.

3. How many ways are there to collect \$12 from 16 people if each of the first 15 people can give at most \$2 and the last person can give either \$0 or \$1 or \$4?

4. How many ways are there to distribute 20 jelly beans to Mary (G), Larry (B), Sherry (G), Terri (G), and Jerry (B) such that a boy (indicated by B) is given an odd number of jelly beans and a girl is given an even number (0 counts as even).

5. Find the coefficient of x^n in $(1 + x + x^2 + x^3)^m (1 + x)^m$.

6. Find the generating function for the sequence (a_n) if (a) $a_n = n^2$, (b) $a_n = n^3$, (c) $a_n = \binom{n}{2}$, and (d) $a_n = \binom{n}{3}$.

7. In how many ways can ten salespersons be assigned so that two are assigned to district A, three to district B, and five to district C? If five of the salespersons are men and five are women, what is the chance that a random assignment of two salespersons to district A, three to B and five to C will result in segregation of the salespersons by sex? What is the probability that

a random assignment will result in at least one female salesperson being assigned to each of the three districts?

8. How many distinct formal words can be made with the letters in the word "abracadabra"?

9. Show that $\Sigma_k(-1)^k\binom{n}{k}((1+kx)/(1+nk)^k) = 0$, for all x and all positive integers n. What do we obtain by taking $x = 0$, or $x = 1$? [Hint: Write

$$0 = \left(1 - \frac{1}{1+nx}\right)^n - \left(1 - \frac{1}{1+nx}\right)^n$$

$$= \left(1 - \frac{1}{1+nx}\right)^n - \frac{nx}{1+nx}\left(1 - \frac{1}{1+nx}\right)^{n-1},$$

expand using the binomial expansion and sort out by $\binom{n}{k}$.]

3 GENERATING FUNCTIONS for STIRLING NUMBERS

Let x and y be indeterminates and denote $\Sigma_n x^n/n!$ by e^x, $\Sigma_n(-1)^n x^{n+1}/(n+1)$ by $\ln(1+x)$, and $e^{x \ln y}$ by y^x. We occasionally yield to the temptation of looking at these formal power series as series expansions of analytic functions. While this contemplative attitude is in itself harmless enough, the effective act of assigning numerical values to x and y becomes an unmistakable cause of concern. Questions of convergence immediately arise and they are of crucial importance. It can be shown that both e^x and $\ln(1+x)$ converge for positive values of x. The relations $e^{\ln x} = x = \ln e^x$ are also known to hold and are used freely in what follows. The formal expansion

$$(1+y)^x = \sum_k [x]_k \frac{y^k}{k!}$$

is needed as well; it holds for $|x| < 1$ [here $[x]_k = x(x-1)\cdots(x-k+1)$]. The reader can find these series expansions in most calculus books. We take them for granted here.

2.7

Taking advantage of the new tools just introduced, let us take another look at the Stirling and Bell numbers:

∗ *Compiled beneath are several generating functions for these numbers* (expanding the right-hand side and equating like powers yields many identities):

1. $\Sigma_n S_n^k y^n/n! = (1/k!)(e^y - 1)^k$.
2. $\Sigma_n s_n^k y^n/n! = (1/k!)(\ln(1+y))^k$.
3. $\Sigma_k S_n^k x^k = e^{-x}\Sigma_m m^n x^m/m!$.

4. $\Sigma_n \Sigma_k S_n^k x^k y^n / n! = e^{x(e^y - 1)}$.

5. $\Sigma_n B_n y^n / n! = e^{e^y - 1}$.

6. $\Sigma_n S_n^k x^{n-k} = (1-x)^{-1}(1-2x)^{-1} \cdots (1-kx)^{-1}$.

7. *The Bell numbers* B_n *satisfy*

$$\lim_{n\to\infty} \frac{n^{-\frac{1}{2}}(\lambda(n))^{n+\frac{1}{2}} e^{\lambda(n)-n-1}}{B_n} = 1,$$

where $\lambda(n)$ *is defined by* $\lambda(n) \ln \lambda(n) = n$. (We recall the usual conventions with indices: $S_n^k = 0$ for all $k \geq n$, and $S_n^0 = 0$ for all n.)

Proof. 1. The proof relies on Stirling's formula

$$x^n = \sum_k S_n^k [x]_k,$$

which we proved in (c) of Section 1.7. We proceed as follows:

$$\sum_k \sum_n S_n^k \frac{y^n}{n!}[x]_k = \sum_n \sum_k S_n^k [x]_k \frac{y^n}{n!} = \sum_n x^n \frac{y^n}{n!}$$

$$= \sum_n \frac{(xy)^n}{n!} = e^{xy} = (e^y)^x = (1 + (e^y - 1))^x$$

$$= \sum_k \frac{1}{k!}(e^y - 1)^k [x]_k.$$

Identifying the coefficients of $[x]_k$ gives

$$\sum_n S_n^k \frac{y^n}{n!} = \frac{1}{k!}(e^y - 1)^k.$$

2. Start out with $[x]_n = \Sigma_k s_n^k x^k$, a formula that we proved in (c) of Section 1.8. Multiply both sides by $y^n/n!$, sum over n, and use known series expansions to obtain:

$$\sum_k \sum_n s_n^k \frac{y^n}{n!} x^k = \sum_n [x]_n \frac{y^n}{n!} = (1+y)^x = e^{x\ln(1+y)}$$

$$= \sum_k \frac{1}{k!}(\ln(1+y))^k x^k.$$

Identifying the coefficients of x^k yields the result.

3. Observe first that $x^k e^x = \Sigma_i x^{i+k}/i! = \Sigma_m [m]_k x^m/m!$, since $[m]_k = 0$ in the first $k-1$ terms. By Stirling's formula, recalling also that $m^n = \Sigma_k S_n^k [m]_k$, we have

$$e^x \sum_k S_n^k x^k = \sum_k S_n^k x^k e^x = \sum_k S_n^k \sum_m [m]_k \frac{x^m}{m!}$$

$$= \sum_m \frac{x^m}{m!} \sum_k S_n^k [m]_k = \sum_m \frac{m^n x^m}{m!}.$$

If we set $x = 1$, we obtain Dobinski's formula

$$B_n = e^{-1} \sum_m \frac{m^n}{m!}.$$

4. Start with the formula established in **3**, multiply it by $y^n/n!$, and sum. What results is

$$\sum_n \sum_k S_n^k x^k \frac{y^n}{n!} = e^{-x} \sum_m \sum_n \frac{m^n x^m}{m!} \frac{y^n}{n!} = e^{-x} \sum_m \frac{x^m}{m!} \sum_n \frac{(my)^n}{n!}$$

$$= e^{-x} \sum_m \frac{x^m}{m!} e^{my} = e^{-x} \sum_m \frac{(xe^y)^m}{m!} = e^{-x} e^{xe^y} = e^{x(e^y - 1)}.$$

5. Recall that $\Sigma_k S_n^k = B_n$. Set $x = 1$ in **4** to obtain **5**.

6. (Induction on k.) The relation is true for $k = 1$ since it reduces to $1 + x + x^2 + \cdots = 1/(1 - x)$. Assume that it holds for $k - 1$ and show that it holds for k. Let $f(x) = \Sigma_{n,\, n \geq k} S_n^k x^{n-k}$. Then

$$\begin{aligned}
f(x) &= \sum_{\substack{n \\ n \geq k}} S_n^k x^{n-k} = \left\{\text{by the recurrence } S_n^k = S_{n-1}^{k-1} + kS_{n-1}^k\right\} \\
&= \sum_{\substack{n \\ n \geq k}} \left(S_{n-1}^{k-1} + kS_{n-1}^k\right) x^{n-k} \\
&= \sum_{\substack{n \\ n-1 \geq k-1}} S_{n-1}^{k-1} x^{(n-1)-(k-1)} + k \sum_{\substack{n \\ n \geq k}} S_{n-1}^k x^{n-k} \\
&= \{\text{by induction}\} = \prod_{m=1}^{k-1} (1 - mx)^{-1} + k \sum_{\substack{n \\ n \geq k}} S_{n-1}^k x^{n-k} \\
&= \prod_{m=1}^{k-1} (1 - mx)^{-1} + k\left(S_{k-1}^k x^0 + S_k^k x + S_{k+1}^k x^2 + S_{k+2}^k x^3 + \cdots\right) \\
&= \prod_{m=1}^{k-1} (1 - mx)^{-1} + k \sum_{\substack{n \\ n \geq k}} S_n^k x^{n-k+1} \\
&= \prod_{m=1}^{k-1} (1 - mx)^{-1} + kx \sum_{\substack{n \\ n \geq k}} S_n^k x^{n-k} \\
&= \prod_{m=1}^{k-1} (1 - mx)^{-1} + kxf(x).
\end{aligned}$$

We can now solve for $f(x)$ and thus obtain the formula we want.

7. The proof of this asymptotic result is somewhat analytic in nature and we omit it to preserve continuity. See reference [10].

2.8

The Stirling numbers occur when relating moments to lower factorial moments. Call $\mathbf{M}_n(f) = \Sigma_x f(x) x^n$ the nth *moment* of f and $\mathbf{m}_n(f) = \Sigma_x f(x)[x]_n$ the

nth *lower factorial moment* of f. (The sum over x could be an integral as well. The variable x is understood to belong to some subset of the real line.) Stirling's formulas give us immediately

$$\mathbf{M}_n = \sum_k S_n^k \mathbf{m}_k \quad \text{and} \quad \mathbf{m}_n = \sum_k s_n^k \mathbf{M}_k.$$

We now describe another situation in which the Stirling numbers pop up.

∗ *Let D be the operator of differentiation (i.e., $D = d/dx$) and let $\theta = xD$. Then*

$$\theta^n = \sum_{k=0}^{n} S_n^k x^k D^k \qquad \left(\text{and} \quad x^n D^n = \sum_{k=0}^{n} s_n^k \theta^k \right).$$

Proof. Proceed as follows:

$$\theta = xD = S_1^1 xD$$

$$\theta^2 = xD(\theta) = xD(xD) = x(D + xD^2) = xD + x^2D^2 = S_2^1 xD + S_2^2 x^2 D^2$$

$$\vdots$$

$$\theta^n = \sum_{k=0}^{n} S_n^k x^k D^k \text{ (assume this)}.$$

Then

$$\begin{aligned}
\theta^{n+1} &= xD(\theta^n) = xD\left(\sum_{k=0}^{n} S_n^k x^k D^k \right) \\
&= x\left(\sum_{k=0}^{n} S_n^k (kx^{k-1}D^k + x^k D^{k+1}) \right) \\
&= \sum_{k=0}^{n} S_n^k k x^k D^k + \sum_{k=0}^{n} S_n^k x^{k+1} D^{k+1} \\
&= \sum_{k=0}^{n} S_n^k k x^k D^k + \sum_{k=1}^{n+1} S_n^{k-1} x^k D^k \\
&= \sum_{k=1}^{n+1} (S_n^{k-1} + kS_n^k) x^k D^k \\
&= \sum_{k=0}^{n+1} S_{n+1}^k x^k D^k.
\end{aligned}$$

This ends the proof, by induction.

The second formula, written in parentheses in the statement above, is equivalent to the first through a process of inversion. This process is presented in detail in Chapter 3.

2.9

We discuss here several properties of the Lah numbers L_n^k. A combinatorial interpretation of these numbers was given in Section 1.15, where we labeled

$$L_n^k = (-1)^n \frac{n!}{k!}\binom{n-1}{k-1}.$$

For small values of n and k we have the following table for L_n^k:

n \ k	1	2	3	4	5
1	−1				
2	2	1		0	
3	−6	−6	−1		
4	24	36	12	1	
5	−120	−240	−120	−20	−1

Define now numbers L_n^k (we show that these are the same as the Lah L_n^k above) by

$$[-x]_n = \sum_{k=1}^{n} L_n^k [x]_k; \qquad L_n^k = 0, \qquad \text{for } k > n.$$

* *We prove the following*:

1. $[-x]_n = \Sigma_{k=1}^n L_n^k [x]_k$ *if and only if* $[x]_n = \Sigma_{k=1}^n L_n^k [-x]_k$.
2. $L_{n+1}^k = -L_n^{k-1} - (n+k)L_n^k$.
3. $\Sigma_n L_n^k t^n/n! = (1/k!)(-t/(1+t))$.
4. $L_n^k = (-1)^n (n!/k!)\binom{n-1}{k-1}$.
5. $\Sigma_k \Sigma_n L_n^k x^k t^n/n! = \exp(-xt/(1+t))$.
6. $L_n^k = \Sigma_{j=k}^n (-1)^j s_n^j S_j^k$.

Proof. 1. Interchange x and $-x$.

2.

$$\sum_{k=1}^{n+1} L_{n+1}^k [x]_k = [-x]_{n+1} = (-x-n)[-x]_n$$

$$= (-x-n)\sum_{k=1}^{n} L_n^k [x]_k = \sum_{k=1}^{n} L_n^k (-x-n)[x]_k$$

$$= \sum_{k=1}^{n} L_n^k (-(x-k) - (n+k))[x]_k$$

$$= \sum_{k=1}^{n} \left(-L_n^k [x]_{k+1} - (n+k) L_n^k [x]_k\right).$$

Identifying coefficients of $[x]_k$ gives **2**.

3. Start with $\Sigma_k L_n^k [x]_k = [-x]_n$. Multiply by $t^n/n!$ and sum:

$$\sum_k [x]_k \sum_n L_n^k \frac{t^n}{n!} = \sum_n \sum_k L_n^k [x]_k \frac{t^n}{n!} = \sum_n [-x]_n \frac{t^n}{n!}$$

$$= (1+t)^{-x} = \left(\frac{1}{1+t}\right)^x = \left(1 - \frac{t}{1+t}\right)^x$$

$$= \sum_k \frac{[x]_k (-t/(1+t))^k}{k!},$$

yielding **3**.

4.

$$\sum_n L_n^k \frac{t^n}{n!} = \frac{(-t/(1+t))^k}{k!} = \frac{1}{k!}\left(-t^k(1 - t + t^2 - t^3 + \cdots)^k\right)$$

$$= \sum_n (-1)^n \frac{n!}{k!}\binom{n-1}{k-1}\frac{t^n}{n!}.$$

See (2.2) for an explanation of the last equality sign.

5. Start with $\Sigma_n L_n^k t^n/n! = (-t/(1+t))^k/k!$, multiply by x^k, and sum

$$\sum_k \sum_n L_n^k \frac{t^n}{n!} x^k = \sum_k \frac{(-tx/(1+t))^k}{k!} = \exp\left(\frac{-xt}{1+t}\right).$$

6.

$$\sum_{k=1}^{n} L_n^k [x]_k = [-x]_n = \sum_{j=1}^{n} (-1)^j s_n^j x^j$$

$$= \sum_{j=1}^{n} (-1)^j s_n^j \sum_{k=1}^{j} S_j^k [x]_k$$

$$= \sum_{k=1}^{n} \sum_{j=1}^{n} (-1)^j s_n^j S_j^k [x]_k.$$

∗ *The Lah numbers occur when expressing the upper factorial moments, defined by* $\overline{\mathbf{m}}_n(f) = \Sigma_x f(x)[x]^n$, *in terms of the lower factorial moments* $\mathbf{m}_n(f)$, *which we defined earlier in this section.* (*Here* $[x]^n = x(x+1)\cdots(x+n-1)$.) *Specifically,*

$$\overline{\mathbf{m}}_n = \sum_k (-1)^n L_n^k \mathbf{m}_k \quad \textit{and} \quad \mathbf{m}_n = \sum_k (-1)^k L_n^k \overline{\mathbf{m}}_k.$$

Further, in terms of differential operators, if $\theta = x^2 D$ (where D stands for d/dx), *then*

$$\theta^n = \sum_{k=1}^{n} (-1)^n L_n^k x^{n+k} D^k$$

(or, equivalently, $D^n = \Sigma_{k=1}^n (-1)^k L_n^k x^{-n-k} \theta^k$).

The proof is similar to the case of $\theta = xD$ involving Stirling numbers.

4 BELL POLYNOMIALS

The object of this section is to bring to attention *an explicit formula by which the higher derivatives of a composition of two functions can be computed.* Partitions of a set, and thus Bell numbers, will enter these calculations in a natural way.

Let $h = f \circ g$ be the composition of f with g, that is, $h(t) = f(g(t))$ where t is an argument. We assume that the functions f, g, and h have derivatives of all orders.

Denote by D_y the operator d/dy of differentiation with respect to y. By D_y^n we indicate the n-fold application of D_y, that is, the nth derivative with respect to y.

We denote as follows:

$$h_n = D_t^n h, \qquad f_n = D_g^n f, \qquad g_n = D_t^n g.$$

Our aim is to find an explicit formula for h_n in terms of the f_k's and g_k's.

To begin with, let us look at the first few expressions for h_n:

$$\begin{aligned}
h_1 &= f_1 g_1 \\
h_2 &= f_2 g_1^2 + f_1 g_2 \\
h_3 &= f_3 g_1^3 + f_2(2g_1g_2) + f_2 g_1 g_2 + f_1 g_3 \\
&= f_3 g_1^3 + f_2(3g_1g_2) + f_1 g_3 \\
h_4 &= f_4 g_1^4 + f_3(6g_2g_1^2) + f_2(4g_3g_1 + 3g_2^2) + f_1 g_4 \\
&\vdots
\end{aligned}$$

Write in general $h_n = \sum_{k=1}^n f_k \alpha_{nk}$. Here the α_{nk}'s are polynomials in g_i's that do not depend upon the choice of f. The h_n's are called *Bell polynomials*. (As is plain to see, these polynomials are linear in the f_k's but highly nonlinear in the g_k's.)

We proceed in establishing the explicit form of the α_{nk}'s and do so in "steps." To this end, define polynomials $\overline{B}_n$ by

$$\overline{B}_n = \sum_{k=1}^{n} \alpha_{nk}.$$

Step 1. $\overline{B}_n = e^{-g}(D_t^n e^{-g})$.

Indeed, let $f(z) = e^z$ be the exponential series. Then $h = e^g$ and $h_n = D_t^n e^g = \sum_{k=1}^n f_k \alpha_{nk} = \sum_{k=1}^n e^g \alpha_{nk} = e^g \sum_{k=1}^n \alpha_{nk} = e^g \overline{B}_n$.

Step 2. $\overline{B}_{n+1} = \sum_{k=0}^{n} \binom{n}{k} g_{k+1} \overline{B}_{n-k}$.

Recall that if α_0 and β_0 are functions of t, differentiable any number of times, then

$$
\begin{aligned}
D_t^0 \alpha_0 \beta_0 &= \alpha_0 \beta_0 \qquad \text{(ordinary multiplication)} \\
D_t^1 \alpha_0 \beta_0 &= \alpha_1 \beta_0 + \alpha_0 \beta_1 \\
D_t^2 \alpha_0 \beta_0 &= \alpha_2 \beta_0 + \alpha_1 \beta_1 + \alpha_1 \beta_1 + \alpha_0 \beta_2 \\
&= \alpha_2 \beta_0 + 2\alpha_1 \beta_1 + \alpha_0 \beta_2 \\
D_t^3 \alpha_0 \beta_0 &= \alpha_3 \beta_0 + 3\alpha_2 \beta_1 + 3\alpha_1 \beta_2 + \alpha_0 \beta_3 \\
&\vdots \\
D_t^n \alpha_0 \beta_0 &= \sum_{k=0}^{n} \binom{n}{k} \alpha_k \beta_{n-k}. \qquad \text{(Leibnitz's formula)}.
\end{aligned}
$$

This formula is not hard to prove, and it was derived at the end of Section 1.6.

With this at hand,

$$
\begin{aligned}
\overline{B}_{n+1} &= e^{-g}\left(D_t^{n+1} e^g\right) = e^{-g} D_t^n \left(g_1 e^g\right) \\
&= \{\text{let } \alpha_0 = g_1 \quad \text{and} \quad \beta_0 = e^g\} \\
&= e^{-g} \sum_{k=0}^{n} \binom{n}{k} g_{k+1} D_t^{n-k} e^g \\
&= \sum_{k=0}^{n} \binom{n}{k} g_{k+1} e^{-g}\left(D_t^{n-k} e^g\right) \\
&= \sum_{k=0}^{n} \binom{n}{k} g_{k+1} \overline{B}_{n-k}.
\end{aligned}
$$

Step 3. $\ln(\sum_n \overline{B}_n x^n/n!) = \sum_n g_{n+1} x^{n+1}/(n+1)!$.

Indeed (formally) differentiating both sides with respect to x we obtain

$$
\left(\sum_n \overline{B}_n \frac{x^n}{n!}\right)^{-1} \left(\sum_n \overline{B}_{n+1} \frac{x^n}{n!}\right) = \sum_n g_{n+1} \frac{x^n}{n!}
$$

if and only if

$$
\sum_n \overline{B}_{n+1} \frac{x^n}{n!} = \left(\sum_n \overline{B}_n \frac{x^n}{n!}\right)\left(\sum_n g_{n+1} \frac{x^n}{n!}\right)
$$

if and only if

$$
\overline{B}_{n+1} = \sum_{k=0}^{n} \binom{n}{k} g_{k+1} \overline{B}_{n-k}
$$

(which is true by Step 2, above). This shows that the two series in Step 3 are equal, up to a constant term. But the constant term is clearly zero on both sides. This completes the proof of Step 3.

[We used here the nontrivial but familiar fact that the formal derivative of $\ln y$ (where y is a formal power series in x) equals y^{-1} times the formal derivative of y with respect to x. While true, the verification of this statement is omitted, to preserve continuity. In passing we remind the reader that $\ln(1+y) = \Sigma_n(-1)^n y^{n+1}/(n+1)$.]

Step 4.

$$\overline{B}_n = \sum_{k=1}^{n} \sum_{\substack{\lambda_i \geq 0 \\ \Sigma_{i=1}^k \lambda_i = k \\ \Sigma_{i=1}^k i\lambda_i = n}} \frac{n!}{(1!)^{\lambda_1} \cdots (k!)^{\lambda_k}(\lambda_1!) \cdots (\lambda_k!)} g_1^{\lambda_1} g_2^{\lambda_2} \cdots g_k^{\lambda_k}$$

(The inner sum is over all partitions of $\{1, 2, \ldots, n\}$ with exactly k classes;

λ_1 classes of size 1

λ_2 classes of size 2

$\vdots$

λ_k classes of size k).

Indeed, exponentiating both sides of Step 3 we obtain

$$\sum_n \overline{B}_n \frac{x^n}{n!} = \exp\left[\sum_n g_{n+1} \frac{x^{n+1}}{(n+1)!}\right] = \prod_{n=1}^{\infty} \exp\left(g_n \frac{x^n}{n!}\right)$$

$$= \prod_{n=1}^{\infty}\left(\sum_{k=0}^{\infty} \frac{(g_n x^n/n!)^k}{k!}\right) = \prod_{n=1}^{\infty}\left(\sum_{k=0}^{\infty} \frac{1}{k!}\left(\frac{g_n}{n!}\right)^k x^{nk}\right)$$

$$= \left\{\text{write } a_{nk} \text{ (double index) for } \frac{1}{k!}\left(\frac{g_n}{n!}\right)^k\right\}$$

$$= \prod_{n=1}^{\infty}\left(\sum_{k=0}^{\infty} a_{nk} x^{nk}\right)$$

$$= \left(a_{10} + a_{11}x^{1\cdot 1} + a_{12}x^{1\cdot 2} + \cdots\right)\left(a_{20} + a_{21}x^{2\cdot 1} + a_{22}x^{2\cdot 2} + \cdots\right)$$
$$\cdot\left(a_{30} + a_{31}x^{3\cdot 1} + a_{32}x^{3\cdot 2} + \cdots\right)\cdots$$

$$= \sum_{n=0}^{\infty}\left(\sum_{\Sigma_{i=1}^n i\lambda_i = n} a_{1\lambda_1} a_{2\lambda_2} \cdots a_{n\lambda_n}\right) x^n$$

$$= \sum_{n=0}^{\infty} \sum_{\substack{\lambda_i \geq 0 \\ \Sigma i\lambda_i = n}} \frac{1}{(\lambda_1!) \cdots (\lambda_n!)}\left(\frac{g_1}{1!}\right)^{\lambda_1} \cdots \left(\frac{g_n}{n!}\right)^{\lambda_n} x^n$$

$$= \sum_{n=0}^{\infty}\left[\sum_{k=1}^{n} \sum_{\substack{\Sigma\lambda_i = k \\ \Sigma i\lambda_i = n}} \frac{1}{(1!)^{\lambda_1} \cdots (k!)^{\lambda_k}(\lambda_1!) \cdots (\lambda_k!)} g_1^{\lambda_1} g_2^{\lambda_2} \cdots g_k^{\lambda_k}\right] x^n.$$

Equating the coefficients of x^n on both sides explains Step 4.

[*Aside*: The polynomial $\bar{B}_n$ evaluated at $g_1 = 1$, $g_2 = 1, \ldots,$ $g_n = 1$ becomes the Bell number B_n (this follows immediately from Step 4).]

Step 5.

$$h_n = \sum_{k=1}^{n} f_k \left(\sum_{\substack{\lambda_i \geq 0 \\ \Sigma\lambda_i = k \\ \Sigma i\lambda_i = n}} \frac{1}{(1!)^{\lambda_1} \cdots (k!)^{\lambda_k} (\lambda_1!) \cdots (\lambda_k!)} g_1^{\lambda_1} g_2^{\lambda_2} \cdots g_k^{\lambda_k} \right)$$

(i.e., α_{nk} is the inner sum in Step 5). This is *Faa DiBruno's formula*. To prove this formula denote the inner sum in Step 5 by α_{nk}^*, for convenience. Recall that α_{nk} has been defined by $h_n = \Sigma_{k=1}^n f_k \alpha_{nk}$ and that the content of Step 4 is (in this notation) $\Sigma_{k=1}^n \alpha_{nk} = \Sigma_{k=1}^n \alpha_{nk}^* (= \bar{B}_n)$. Our aim is to prove that $\alpha_{nk} = \alpha_{nk}^*$.

We have the following chain of implications:

$$\sum_{k=1}^{n} \alpha_{nk} = \sum_{k=1}^{n} \alpha_{nk}^* \Rightarrow \sum_{k=1}^{n} (\alpha_{nk} - \alpha_{nk}^*) = 0$$

$$\Rightarrow \alpha_{nk} - \alpha_{nk}^* = 0 \Rightarrow \alpha_{nk} = \alpha_{nk}^*.$$

The first implication is just rewriting. Let us study the second implication: It is clear that the α_{nk}^*'s are homogeneous polynomials of degree k in the g_i's. We now show that the α_{nk}'s are also homogeneous of degree k. It is easy to verify this statement for small values of n and k. Assume it is so for the α_{nk}'s, for all $1 \leq k \leq n$, and show, by induction, that the $\alpha_{n+1,k}$'s are homogeneous of degree k, $1 \leq k \leq n+1$. Recall that $h_n = \Sigma_{k=1}^n f_k \alpha_{nk}$. The coefficient of f_k in the expression of h_{n+1}, that is, $\alpha_{n+1,k}$, is obtained by differentiating $f_{k-1}\alpha_{n,k-1} + f_k \alpha_{nk}$. That is, $D_t(f_{k-1}\alpha_{n,k-1} + f_k\alpha_{nk}) = f_k g_1 \alpha_{n,k-1} + f_{k-1} D_t \alpha_{n,k-1} + f_{k+1} g_1 \alpha_{nk} + f_k D_t \alpha_{nk}$. We hence have

$$\alpha_{n+1,k} = g_1 \alpha_{n,k-1} + D_t \alpha_{nk}.$$

The right-hand side in this relation has both terms homogeneous of degree k, the first by the inductive assumption, the second using the product rule and induction (on k). Hence the α_{nk}'s are homogeneous polynomials of degree k. The second implication now follows by equating to zero all the *homogeneous components* of the sum. The third implication follows because the monomials of degree k in the g_i's are linearly independent (since the g_i's themselves are, in general). This completes Step 5 and ends the proof of Faa DiBruno's formula.

Bell Polynomials

$$h_1 = f_1 g_1$$

$$h_2 = f_1 g_2 + f_2 g_1^2$$

$$h_3 = f_1 g_3 + f_2(3 g_2 g_1) + f_3 g_1^3$$

$$h_4 = f_1 g_4 + f_2(4g_3 g_1 + 3g_2^2) + f_3(6g_2 g_1^2) + f_4 g_1^4$$

$$h_5 = f_1 g_5 + f_2(5g_4 g_1 + 10g_3 g_2) + f_3(10g_3 g_1^2 + 15g_2^2 g_1) + f_4(10g_2 g_1^3) + f_5 g_1^5$$

$$h_6 = f_1 g_6 + f_2(6g_5 g_1 + 15g_4 g_2 + 10g_3^2) + f_3(15g_4 g_1^2 + 60g_3 g_2 g_1 + 15g_2^3) + f_4(20g_3 g_1^3 + 45g_2^2 g_1^2) + f_5(15g_2 g_1^4) + f_6 g_1^6$$

$$h_7 = f_1 g_7 + f_2(7g_6 g_1 + 21g_5 g_2 + 35g_4 g_3) + f_3(21g_5 g_1^2 + 105g_4 g_2 g_1 + 70g_3^2 g_1 + 105g_3 g_2^2) + f_4(35g_4 g_1^3 + 210g_3 g_2 g_1^2 + 105g_2^3 g_1) + f_5(35g_3 g_1^4 + 105g_2^2 g_1^3) + f_6(21g_2 g_1^5) + f_7 g_1^7$$

$$h_8 = f_1 g_8 + f_2(8g_7 g_1 + 28g_6 g_2 + 56g_5 g_3 + 35g_4^2) + f_3(28g_6 g_1^2 + 168g_5 g_2 g_1 + 280g_4 g_3 g_1 + 210g_4 g_2^2 + 280g_3^2 g_2) + f_4(56g_5 g_1^3 + 420g_4 g_2 g_1^2 + 280g_3^2 g_1^2 + 840g_3 g_2^2 g_1 + 105g_2^4) + f_5(70g_4 g_1^4 + 560g_3 g_2 g_1^3 + 420g_2^3 g_1^2) + f_6(56g_3 g_1^5 + 210g_2^2 g_1^4) + f_7(28g_2 g_1^6) + f_8 g_1^8$$

$$h_9 = f_1 g_9 + f_2(9g_8 g_1 + 36g_7 g_2 + 84g_6 g_3 + 126g_5 g_4) + f_3(36g_7 g_1^2 + 252g_6 g_2 g_1 + 504g_5 g_3 g_1 + 378g_5 g_2^2 + 315g_4^2 g_1 + 1260g_4 g_3 g_2 + 280g_3^3) + f_4(84g_6 g_1^3 + 756g_5 g_2 g_1^2 + 1260g_4 g_3 g_1^2 + 1890g_4 g_2^2 g_1 + 2520g_3^2 g_2 g_1 + 1260g_3 g_2^3) + f_5(126g_5 g_1^4 + 1260g_4 g_2 g_1^3 + 840g_3^2 g_1^3 + 3780g_3 g_2^2 g_1^2 + 945g_2^4 g_1) + f_6(126g_4 g_1^5 + 1260g_3 g_2 g_1^4 + 1260g_2^3 g_1^3) + f_7(84g_3 g_1^6 + 378g_2^2 g_1^5) + f_8(36g_2 g_1^7) + f_9 g_1^9$$

$$
\begin{aligned}
h_{10} = {} & f_1 g_{10} + f_2\bigl(10 g_9 g_1 + 45 g_8 g_2 + 120 g_7 g_3 + 210 g_6 g_4 + 126 g_5^2\bigr) \\
& + f_3\bigl(45 g_8 g_1^2 + 360 g_7 g_2 g_1 + 840 g_6 g_3 g_1 + 630 g_6 g_2^2 \\
& + 1260 g_5 g_4 g_1 + 2520 g_5 g_3 g_2 + 1575 g_4^2 g_2 + 2100 g_4 g_3^2\bigr) \\
& + f_4\bigl(120 g_7 g_1^3 + 1260 g_6 g_2 g_1^2 + 2520 g_5 g_3 g_1^2 \\
& + 3780 g_5 g_2^2 g_1 + 1575 g_4^2 g_1^2 + 12600 g_4 g_3 g_2 g_1 \\
& + 3150 g_4 g_2^3 + 2800 g_3^3 g_1 + 6300 g_3^2 g_2^2\bigr) \\
& + f_5\bigl(210 g_6 g_1^4 + 2520 g_5 g_2 g_1^3 + 4200 g_4 g_3 g_1^3 \\
& + 9450 g_4 g_2^2 g_1^2 + 12600 g_3^2 g_2 g_1^2 + 12600 g_3 g_2^3 g_1 + 945 g_2^5\bigr) \\
& + f_6\bigl(252 g_5 g_1^5 + 3150 g_4 g_2 g_1^4 + 2100 g_3^2 g_1^4 + 12600 g_3 g_2^2 g_1^3 + 4725 g_2^4 g_1^2\bigr) \\
& + f_7\bigl(210 g_4 g_1^6 + 2520 g_3 g_2 g_1^5 + 3150 g_2^3 g_1^4\bigr) \\
& + f_8\bigl(120 g_3 g_1^7 + 630 g_2^2 g_1^6\bigr) + f_9\bigl(45 g_2 g_1^8\bigr) + f_{10} g_1^{10}.
\end{aligned}
$$

5 RECURRENCE RELATIONS

The general question that we address here is as follows: *From a rule of recurrence among the elements of a sequence (a_n) determine explicitly that sequence.*

Examples are many. If the recurrence is $a_n = a_{n-1} + n$ with $a_0 = 1$ $(n = 1, 2, 3, \dots)$, then it easily follows that $a_n = 1 + \binom{n+1}{2}$. On the other hand, if $a_0 = 1$, $a_1 = 1$, and the recurrence relation is $a_n = \sum_{k=1}^{n-1} a_k^2 a_{n-k}$ $(n \geq 2)$, then it is not so easy to determine a_n as a function of n. Indeed, more often than not one will not be able to find a_n explicitly.

Generating functions provide, nonetheless, a powerful technique that leads to complete solutions in many situations. Let us illustrate this by a classical example.

2.10

Mr. Fibonacci just bought a pair of baby rabbits (one of each sex) possessing some remarkable, and perhaps enviable, properties:

They take a month to mature.

When mature, a pair gives birth each month to precisely one new pair (again one of each sex), and with the same remarkable properties.

The mating takes place only between the members of a pair born from the same parents.

They live forever!

(Excepting these particulars, the rabbits do resemble in all other respects their more usual mortal counterparts.)

How many pairs of rabbits will Fibonacci have at the beginning of the nth month?

The picture below shows the beginning values of the sequence a_n = the number of pairs of rabbits at the beginning of the nth month ($n \geq 0$). By ----- we indicate the month to mature, and _____ indicates the month of pregnancy. We see from above that $a_0 = 1$, $a_1 = 1$, $a_2 = 2$, $a_3 = 3$, $a_4 = 5$, $a_5 = 8, \ldots$.

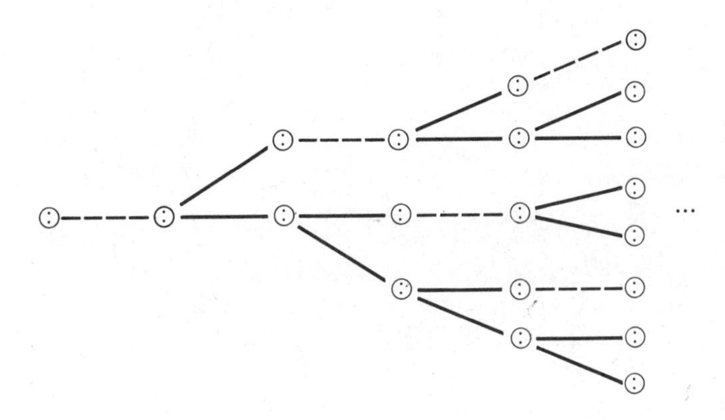

The sequence (a_n) satisfies in fact the recurrence relation

$$a_{n+2} = a_{n+1} + a_n; \qquad n \geq 0.$$

(To see this observe that at stage $n + 2$ we have all the a_{n+1} pairs that we had at stage $n + 1$ plus the a_n children or grandchildren of the pairs we had at stage n, that is, $a_{n+2} = a_{n+1} + a_n$.)

To find a_n as a function of n only we proceed as follows. Denote by $A(x)$ the generating function of (a_n), that is, $A(x) = \Sigma_n a_n x^n$. Then

$$a_{n+2} = a_{n+1} + a_n$$

implies

$$a_{n+2}x^{n+2} = a_{n+1}x^{n+2} + a_n x^{n+2}$$

implies

$$\sum_n a_{n+2}x^{n+2} = \sum_n a_{n+1}x^{n+2} + \sum_n a_n x^{n+2}$$

implies

$$A(x) - a_1 x - a_0 x = x(A(x) - a_0) + x^2 A(x)$$

implies

$$A(x) - x - 1 = xA(x) - x + x^2 A(x),$$

which leads to

$$A(x) = \frac{1}{1 - x - x^2}.$$

We use this closed form expression of $A(x)$ to find an explicit power series expansion for $A(x)$. Observe first that $1 - x - x^2 = -(a - x)(b - x)$, where $a = \frac{1}{2}(-1 - \sqrt{5})$ and $b = \frac{1}{2}(-1 + \sqrt{5})$. Now

$$\begin{aligned} A(x) &= \frac{1}{1 - x - x^2} = \frac{-1}{(a - x)(b - x)} \\ &= (a - b)^{-1}\left((a - x)^{-1} - (b - x)^{-1}\right) \\ &= (a - b)^{-1}\left(a^{-1}\left(1 - \frac{x}{a}\right)^{-1} - b^{-1}\left(1 - \frac{x}{b}\right)^{-1}\right) \\ &= (a - b)^{-1}\left(a^{-1}\sum_n \left(\frac{x}{a}\right)^n - b^{-1}\sum_n \left(\frac{x}{b}\right)^n\right) \\ &= \sum_n \left[(a - b)^{-1}(a^{-n-1} - b^{-n-1})\right] x^n. \end{aligned}$$

Hence $a_n = (a - b)^{-1}(a^{-n-1} - b^{-n-1})$, or

$$a_n = \frac{1}{\sqrt{5}}\left[\left(\frac{2}{-1 + \sqrt{5}}\right)^{n+1} - \left(\frac{2}{-1 - \sqrt{5}}\right)^{n+1}\right], \qquad n \geq 0.$$

In general an explicit expression for a_n in terms of n only (although not always desirable) usually gives a more accurate idea of the magnitude of a_n, a fact that the recurrence might not immediately convey. We have thus found how many pairs of rabbits Fibonacci will have at the beginning of the nth month.

Note: If we expand the generating function $A(x) = (1 - (x + x^2))^{-1}$ as the power series $1 + (x + x^2) + (x + x^2)^2 + (x + x^2)^3 + \cdots$ what expression for the Fibonacci numbers do we obtain?

2.11

The case of the Fibonacci sequence, which we just described, is part of a more general class of problems known as linear recurrence relations with constant coefficients.

* *Let* (a_n) *be a sequence satisfying the recurrence relation*

$$c_0 a_n + c_1 a_{n-1} + c_2 a_{n-2} + \cdots + c_k a_{n-k} = 0; \qquad (2.3)$$
$$c_0 = 1; \qquad c_k \neq 0; \qquad n \geq k$$

with c_i*'s constants* (not depending on n). *Then the generating function of* (a_n)

is of the form

$$\frac{p(x)}{q(x)} \tag{2.4}$$

where $q(x)$ is a polynomial of degree k with a nonzero constant term and $p(x)$ is a polynomial of degree less than k.

Conversely, given polynomials $p(x)$ and $q(x)$ as in (2.4), *there exists a sequence (a_n) that satisfies a recurrence relation as in* (2.3) *and whose generating function is $p(x)/q(x)$.*

Indeed, suppose (a_n) satisfies (2.3) and has initial values $a_0, a_1, \ldots, a_{k-1}$. Proceed exactly as in the case of the Fibonacci sequence treated in Section 2.10 to obtain $A(x)$, the generating function of (a_n). In fact, $A(x) = p(x)/q(x)$, where $q(x) = \Sigma_{i=0}^{k} c_i x^i$, and $p(x) = \Sigma_{j=0}^{k} c_j x^j (\Sigma_{i=0}^{k-j-1} a_i x^i)$.

Conversely, given $q(x) = b_0 + b_1 x + \cdots + b_k x^k$ with $b_0 \neq 0$, $b_k \neq 0$ and $p(x) = d_0 + d_1 x + \cdots + d_{k-1} x^{k-1}$, using partial fractions and the expansion $(1 - y)^{-1} = \Sigma_n y^n$ we can write

$$\frac{p(x)}{q(x)} = a_0 + a_1 x + a_2 x^2 + \cdots + a_n x^n + \cdots \tag{2.5}$$

Rewrite (2.5) as follows:

$$d_0 + d_1 x + \cdots + d_{k-1} x^{k-1} = \left(b_0 + b_1 x + \cdots + b_k x^k\right) \cdot \left(a_0 + a_1 x + a_2 x^2 + \cdots\right).$$

Identifying coefficients of powers of x on both sides we obtain

$$\begin{aligned} b_0 a_0 &= d_0 \\ b_0 a_1 + b_1 a_0 &= d_1 \\ &\vdots \\ b_0 a_{k-1} + b_1 a_{k-2} + \cdots + b_{k-1} a_0 &= d_{k-1} \end{aligned} \tag{2.6}$$

and

$$b_0 a_n + b_1 a_{n-1} + \cdots + b_k a_{n-k} = 0, \qquad \text{for } n \geq k.$$

Divide this last relation by b_0 and set $c_j = b_j/b_0$ to obtain the recurrence relation mentioned in (2.3). The initial values $a_0, a_1, \ldots, a_{k-1}$ can be determined from (2.6).

2.12

Merely as an exercise, *consider finding all sequences (a_n) that satisfy the recurrence relation*

$$a_{n+1} = 3a_n - 5(n+1) + 7 \cdot 2^n, \qquad n \geq 0.$$

The way we proceed is typical of how one uses generating functions to solve problems of this sort.

Let $A(x) = \Sigma_n a_n x^n$. Then

$$a_{n+1}x^n = 3a_n x^n - 5(n+1)x^n + 7 \cdot 2^n x^n$$

$$\sum_n a_{n+1}x^n = 3\sum_n a_n x^n - 5\sum_n (n+1)x^n + 7\sum_n 2^n x^n$$

$$x^{-1}(A(x) - a_0) = 3A(x) - 5(1-x)^{-2} + 7(1-2x)^{-1}$$

$$(1-3x)A(x) = a_0 - 5x(1-x)^{-2} + 7x(1-2x)^{-1}$$

$$A(x) = \frac{a_0}{1-3x} - 5x\frac{1}{(1-3x)(1-x)^2} + 7x\frac{1}{(1-3x)(1-2x)}.$$

We expand $A(x)$ in a power series again, but first we use partial fraction decompositions as follows:

$$\frac{1}{(1-3x)(1-x)^2} = \frac{A}{1-3x} + \frac{Bx+C}{(1-x)^2},$$

which upon solving for A, B, and C gives $A = \frac{9}{4}$, $B = \frac{3}{4}$, $C = -\frac{5}{4}$. Similarly

$$\frac{1}{(1-3x)(1-2x)} = \frac{3}{1-3x} - \frac{2}{1-2x}.$$

We now proceed

$$A(x) = \frac{a_0}{1-3x} - 5x\left(\frac{\frac{9}{4}}{1-3x} + \frac{\frac{3}{4}x - \frac{5}{4}}{(1-x)^2}\right) + 7x\left(\frac{3}{1-3x} - \frac{2}{1-2x}\right)$$

$$A(x) = a_0\sum_n (3x)^n - \frac{45}{4}x\sum_n (3x)^n - \frac{15}{4}x^2\sum_n (n+1)x^n$$

$$+ \frac{25}{4}x\sum_n (n+1)x^n + 21x\sum_n (3x)^n - 14x\sum_n (2x)^n.$$

Looking at the coefficient of x^n we immediately obtain

$$a_0, a_1 = 3a_0 + 2,$$

and

$$a_{n+2} = \left(3a_0 + \frac{39}{4}\right)3^{n+1} - 7 \cdot 2^{n+2} + \frac{10}{4}(n+1) + \frac{25}{4}, \qquad n \geq 0.$$

This sequence does indeed verify the original recurrence.

2.13

Let us count the number of permutations σ on the set $1 < 2 < 3 < \cdots < n$ that satisfy $\sigma(1) > \sigma(2) < \sigma(3) > \sigma(4) < \cdots$. (The signs $>$ and $<$ alternate.) Denote by a_n the number of such permutations.

To begin with, let us look at the initial values of the sequence a_n:

n:	1	2	3	4
σ:	1	21	213 312	2143 3142 3241 4132 4231
a_n:	1	1	2	5

It is well worth observing that the sequence (a_n) satisfies the recurrence

$$a_{n+1} = \sum_{\substack{k=0 \\ (k \text{ even})}} \binom{n}{k} a_k a_{n-k} \tag{2.7}$$

where, for convenience, we define $a_0 = 1$.

We explain this for $n = 5$ and the argument will carry over to any value of n. Take any permutation σ that satisfies $\sigma(1) > \sigma(2) < \sigma(3) > \sigma(4) < \sigma(5)$, say $\sigma = 32514$. Then $n + 1$, in this case 6, can be inserted in all the "even" positions in σ to produce permutations on 6 symbols with the same property, that is,

$$\uparrow\ 3\quad 2\ \uparrow\ 5\quad 1\ \uparrow\ 4$$

Possible places to insert 6.

(An "even" position is defined by the fact that there is an even number of symbols at the left of the place where $n + 1$ is inserted—note that $n + 1$ may not be inserted in "odd" positions.)

For fixed σ the insertion of $n + 1$ in position k (k even) produces precisely $\binom{n}{k} a_k a_{n-k}$ new permutations (on $n + 1$ symbols).

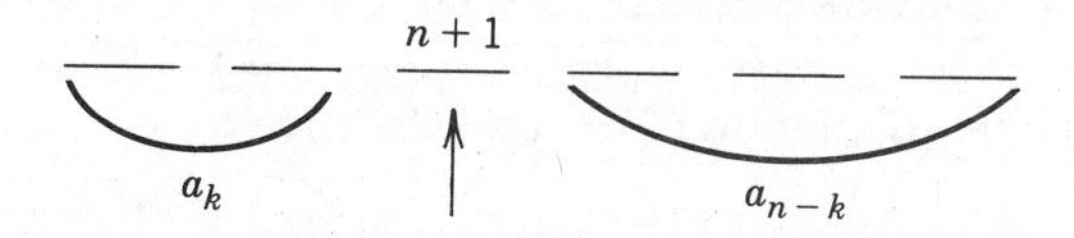

Indeed, for each selection of digits at the left of $n + 1$ there are $a_k a_{n-k}$ possible permutations. And there are $\binom{n}{k}$ possible choices for the digits left of $n + 1$.

Summing up we obtain the recurrence in (2.7).

(The case $k = 0$ requires, in fact, special attention. Note that 6 3 2 5 1 4 is not of the form we want. But by changing the digits $1 \leftrightarrow n$, $2 \leftrightarrow n-1$, etc., we produce 6 3 4 1 5 2, which is fine. This being a bijection the recurrence (2.7) remains valid as stated.)

Denote by $A(x) = \Sigma_n a_n x^n/n!$ the exponential generating function of (a_n). With the substitution $b_n = a_n/n!$ the recurrence relation (2.7) can be written as

$$(n+1)b_{n+1} = \sum_{\substack{k=0 \\ (k \text{ even})}} b_k b_{n-k}.$$

This leads to

$$(n+1)b_{n+1}x^n = \left(\sum_{\substack{k=0 \\ (k \text{ even})}} b_k b_{n-k} \right) x^n.$$

Summing, we obtain

$$\sum_n (n+1)b_{n+1}x^n = \sum_n \left(\sum_{\substack{k=0 \\ (k \text{ even})}} b_k b_{n-k} \right) x^n$$

$$= \left(b_0 + b_2x^2 + b_4x^4 + b_6x^6 + \cdots\right)\left(\sum_n b_n x^n\right)$$

$$= \tfrac{1}{2}(A(x) + A(-x))A(x).$$

Since $\Sigma_n (n+1)b_{n+1}x^n = DA(x)$ [the formal derivative of $A(x)$] we obtain

$$DA(x) = \tfrac{1}{2}(A(x) + A(-x))A(x). \tag{2.8}$$

This functional equation, along with the knowledge that the constant term is 1, force a unique solution for $A(x)$. Indeed, (2.8) and the constant term being 1, determine uniquely the coefficient of x, then that of x^2, of x^3, and so on.

If we denote $1 - (x^2/2!) + (x^4/4!) - (x^6/6!) \pm \cdots$ by $\cos x$ and $(x/1!) - (x^3/3!) + (x^5/5!) - (x^7/7!) \pm \cdots$ by $\sin x$, a solution (and therefore the solution) to (2.8) is $(\sin x/\cos x) + (1/\cos x)$. If, by analogy to the notation in trigonometry, we furter denote $(\sin x/\cos x)$ by $\tan x$ and $(1/\cos x)$ by $\sec x$, the unique solution to (2.8) can be written as

$$A(x) = \tan x + \sec x.$$

We conclude, therefore, that *the exponential generating function for the sequence of permutations* (a_n) *defined at the beginning of this paragraph is* $A(x) = \tan x + \sec x$.

[While most of us surely can appreciate a wild guess that works, the claim that $\tan x + \sec x$ is a solution to (2.8) touches undeniably upon the miraculous. Let us sketch a proof that $A'(x) = \frac{1}{2}(A(x) + A(-x))A(x)$ and $A(0) = a_0 = 1$ imply $A(x) = \sec x + \tan x$ (here the prime denotes the derivative).

Let $B(x) = \frac{1}{2}(A(x) + A(-x))$ and $C(x) = \frac{1}{2}(A(x) - A(-x))$. Note that

$$\begin{aligned} B'(x) &= \tfrac{1}{2}(A'(x) - A'(-x)) = \tfrac{1}{4}(A(x) + A(-x))(A(x) - A(-x)) \\ &= B(x)C(x) \end{aligned} \tag{2.9}$$

$$C'(x) = \tfrac{1}{2}(A'(x) + A'(-x)) = \tfrac{1}{4}(A(x) + A(-x))^2 = B(x)^2$$

Hence $(B(x)^2 - C(x)^2)' = 2B(x)B(x)C(x) - 2C(x)B(x)^2 = 0$. And since $B(0) = 1$ and $C(0) = 1$, we have

$$B(x)^2 - C(x)^2 = 1. \tag{2.10}$$

Next note that

$$\left(\frac{1}{B(x)}\right)' = -\frac{B'(x)}{B(x)^2} = -\frac{C(x)}{B(x)},$$

and

$$\begin{aligned} \left(\frac{1}{B(x)}\right)'' &= -\left(\frac{C(x)}{B(x)}\right)' = -\frac{C'B - CB'}{B^2} = \{\text{by } (2.9)\} \\ &= -\frac{B^3 - BC^2}{B^2} = -\frac{1}{B}(B^2 - C^2) = \{\text{by } (2.10)\} \\ &= -\frac{1}{B(x)}. \end{aligned}$$

Now,

$$\left(\frac{1}{B(x)}\right)'' = -\frac{1}{B(x)} \quad \text{and} \quad \frac{1}{B(0)} = 1 \quad \text{imply} \quad \frac{1}{B(x)} = \cos x.$$

Hence $B(x) = \sec x$, and by (2.10) $C(x) = \pm\tan x$. By (2.9) $C(x) = \tan x$, necessarily. This gives

$$A(x) = B(x) + C(x) = \sec x + \tan x.]$$

EXERCISES

1. Let $c_n = (n+1)^{-1}\binom{2n}{n}$.

(a) Find the number of increasing lattice paths from $(0,0)$ to (n, n) that never cross, but may touch, the main diagonal [i.e., the line joining $(0,0)$ with (n, n)].

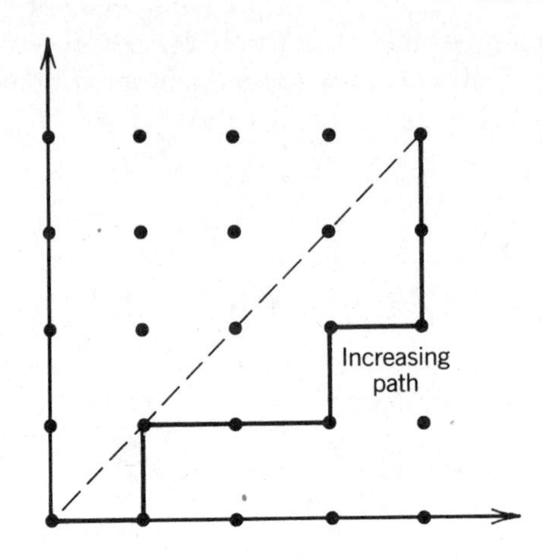

Answer: $2c_n$

(b) How many ways can the product $x_1x_2 \cdots x_n$ be parenthesized? (Note: we do not allow the order of the x's to change.)
Example: $n = 4$

$$((x_1x_2)(x_3x_4)), \quad (((x_1x_2)x_3)x_4), \quad ((x_1(x_2x_3))x_4),$$
$$(x_1((x_2x_3)x_4)), \quad (x_1(x_2(x_3x_4))).$$

Answer: c_{n-1}

(c) Let P_n be the regular n-gon on n labeled vertices. A *diagonal triangulation* of P_n is a triangulation of P_n that involves exactly $n - 3$ nonintersecting diagonals of P_n. Find the number of diagonal triangulations of P_n (Euler).
Example:

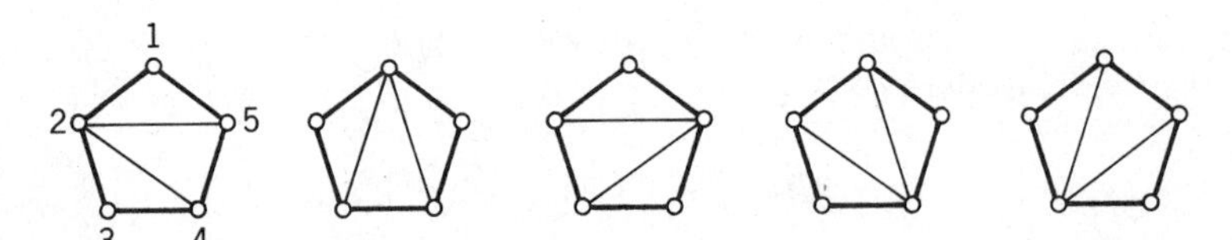

Answer: c_{n-2}

(d) Given $2n$ people of different heights, in how many ways can these $2n$ people be lined up in two rows of length n each so that everyone in the first row is taller than the corresponding person in the second row?
Answer: c_n

(e) (Application to politics.) In an election candidate A receives a votes and candidate B receives b votes ($a > b$). In how many ways can the ballots be arranged so that when they are counted, one at a time, there are always (strictly) more votes for A than B?

Answer: $((a-b)/(a+b))\binom{a+b}{a}$

(If the election ends in a tie with n votes to each, then the number of sequences in which at no time of the counting is B ahead is $2c_n$.)

(f) Show: $c_n = \sum_{k=0}^{n-1} c_k c_{n-k-1}$, $c_0 = 1$.

(g) Show: $\sum_n c_n x^n = (1 - \sqrt{1-4x})/2x$.

The (c_n)'s are called *Catalan numbers.*

$$c_n\colon\quad 1, 1, 2, 5, 14, 42, 132, 429, 1430, 4862, \ldots$$

2. Let (a_n) be a sequence satisfying the recurrence relation

$$a_n + a_{n-1} - 16a_{n-2} + 20a_{n-3} = 0, \qquad n \geq 3$$

with $a_0 = 0$, $a_1 = 1$, $a_2 = -1$. Find a_n (as a function of n).

3. Let (a_n) be the Fibonacci sequence (take $a_0 = 0$, $a_1 = 1$, $a_2 = 1$, and $a_n = a_{n-1} + a_{n-2}$, $n \geq 3$). Verify that:

(a) $a_1 + a_2 + \cdots + a_n = a_{n+2} - 1$.

(b) $a_1 + a_3 + a_5 + \cdots + a_{2n-1} = a_{2n}$.

(c) $a_2 + a_4 + a_6 + \cdots + a_{2n} = a_{2n+1} - 1$.

(d) $a_n^3 + a_{n+1}^3 - a_{n-1}^3 = a_{3n}$.

(e) $\binom{n}{0} + \binom{n-1}{1} + \binom{n-2}{2} + \cdots = a_{n+1}$.

(f) $a_{n+m} = a_m a_{n+1} + a_{m-1} a_n$. Show also that a_{mn} is a multiple of a_n.

(g) a_n is $(1/\sqrt{5})((1+\sqrt{5})/2)^n$ rounded off to the nearest integer.

(h) $a_1 a_2 + a_2 a_3 + \cdots + a_{2n-1} a_{2n} = a_{2n}^2$.

4. Place n points on the circumference of a circle and draw all possible chords through pairs of these points. Assume (at least formally) that no three chords are concurrent. Let a_n be the number of regions formed inside the circle. Find (a_n) and the generating function of (a_n).

5. Define a_0 to be 1. For $n \geq 1$, let a_n be the number of $n \times n$ symmetric matrices with entries 0 or 1 and row sums equal to 1 (i.e., symmetric permutation matrices). Show that $a_{n+1} = a_n + n a_{n-1}$ and then prove that $\sum a_n x^n / n! = \exp(x + \frac{1}{2}x^2)$.

6 THE GENERATING FUNCTION OF LABELED SPANNING TREES

Let us temporarily drift away from generating functions of sequences to present a result in graph theory: the generating function for the spanning trees of a graph.

2.14

A *graph* G is a collection of (possibly repeated) subsets of cardinality two (called *edges*) of a finite set of points (called *vertices*). Below is an example of a graph:

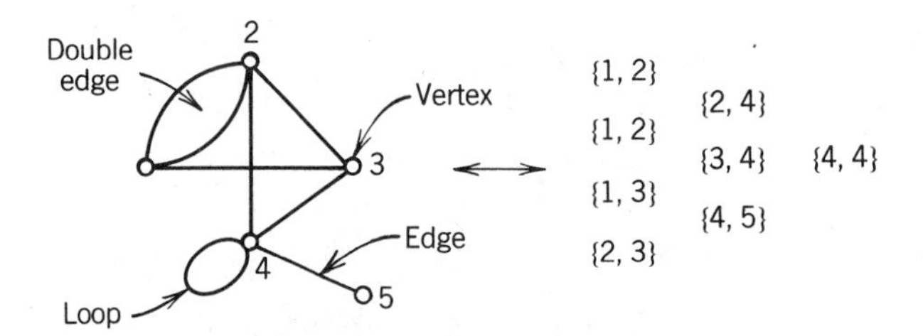

In the definition of a graph we also allow the notation $\{4, 4\}$ for an edge joining the vertex 4 to itself, which we call a *loop*. All edges, including the multiple ones, are *distinguishable* from each other. A *path* is a collection of edges like this:

(any length). A *cycle* is a collection of edges like this:

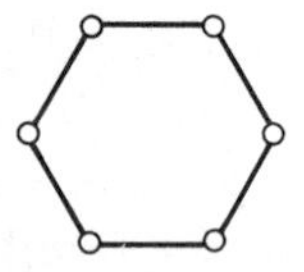

(any length). We call a graph *connected* if any two distinct vertices can be joined by a path. A *tree* is a set of edges containing no cycles. By a *spanning tree* of a graph with n vertices we understand a set of $n - 1$ edges containing no cycles. (Graphs that are not connected have no spanning trees.) A path, cycle, tree, or spanning tree is understood to contain no loops or multiple edges.

The following two pictures are spanning trees in the graph above:

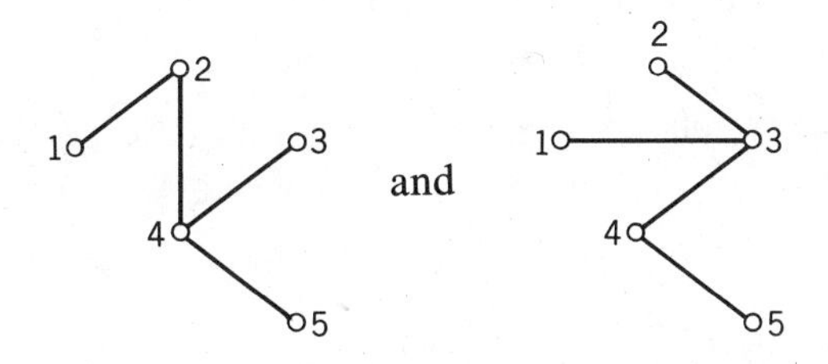

These two are not:

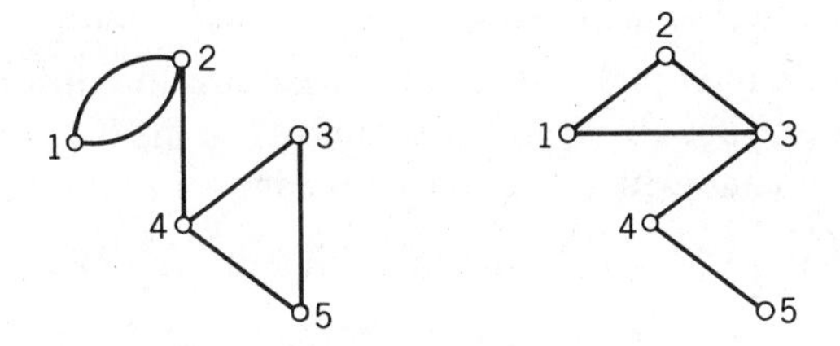

Two spanning trees are the *same* if they consist of exactly the same $n-1$ edges. (There are two spanning trees associated with our first picture of a spanning tree above, because there are two *distinguishable* edges $\{1,2\}$ in G.) Given a graph G we address two issues:

(i) *How many spanning trees does G have?*
(ii) *Generate a list of all spanning trees of G.*

2.15

To a graph G we associate its *information* (or *Kirchhoff*) matrix $C = (c_{ij})$ (both rows and columns indexed by the vertices of G in the same fixed order) as follows:

$$-c_{ij} = \text{number of edges between vertices } i \text{ and } j, \qquad i \neq j$$

$$c_{ii} = -\sum_{\substack{j \\ j\neq i}} c_{ij}.$$

Suppose G has n vertices labeled $1, 2, \ldots, n$. The matrix C is then $n \times n$. Denote by $\mathbf{1}$ the column vector with all its entries 1 (and by $\mathbf{1}'$ its transpose).

Properties of the matrix C:

1. $C\mathbf{1} = \mathbf{0}$ (i.e., $\mathbf{1} \in \ker C =$ kernel of C).
2. *If* rank $C = n - 1$, *then all cofactors of* C *are equal and nonzero.*
3. $C \geq 0$ (i.e., $x^t Cx \geq 0$, for all vectors x).
4. C *is of rank* $n - 1$ *if and only if* G *is connected.*

Proof. Statement 1 follows from the definition of C. To realize that statement 2 is true denote by C_{ij} the cofactor of c_{ij}. Then

$$(c_{ij})(C_{ij})^t = (\det C)I$$

where $\det C$ stands for the determinant of C and I is the $n \times n$ identity matrix. This equality holds for *any* square matrix. In our case $\det C = 0$ since C is singular, by statement 1. Hence all column vectors of $(C_{ij})^t$ belong to $\ker C = \langle \mathbf{1} \rangle$ (because the rank of C is $n - 1$). Thus for fixed i all C_{ij}'s are equal. Similarly (working with transposes) for fixed j, all C_{ij}'s are equal and therefore (C_{ij}) is a multiple of J, the matrix with all entries 1. This proves statement 2. For an edge $\{i, j\}$ in G denote by C^{ij} the Kirchoff matrix of the graph on n vertices and with $\{i, j\}$ the only edge (of multiplicity 1). Then C^{ij} is the $n \times n$ matrix $\begin{bmatrix} 1 & -1 \\ -1 & 1 \end{bmatrix}$ with 1's in ith and jth diagonal positions, -1 in positions (i, j) and (j, i), and 0 elsewhere. It is easy to check that $x^t C^{ij} x \geq 0$, for all x. (Note that $C^{ii} = 0$, i.e., the Kirchhoff matrix of a loop is zero.) Then

$$C = \sum_{\substack{\{i,j\} \\ \text{edge of} \\ \text{graph } G}} C^{ij} \quad \text{and} \quad x^t Cx = \sum x^t C^{ij} x \geq 0,$$

which gives statement 3. (The expression of C as a sum of C^{ij}'s is important because it shows how C gathers "information.") We now prove statement 4. If $A \geq 0$, $B \geq 0$, and $B \leq A$ (notation for $A - B \geq 0$), then the row span of B is included in the row span of A [because $\ker A \subseteq \ker B$—this is easy to check—and the column (or row) span of B is included in the column (or row) span of A as orthogonal complements of kernels]; keep this in mind. Let G be connected. Then a path γ exists between 1 and any other vertex k. Say the path is $(12 \cdots k)$ (without loss). Then the Kirchhoff $n \times n$ matrix of the path is

$$C_\gamma = \begin{bmatrix} 1 & -1 & & & & & \\ -1 & 2 & -1 & & & \mathbf{0} & \\ & -1 & 2 & -1 & & & \\ & & & \ddots & & & \\ & & & -1 & 2 & -1 & \\ & & & & -1 & 1 & \\ & & & & & & \mathbf{0} \end{bmatrix}$$

with the $\mathbf{0}$ in the bottom right-hand corner of dimension $(n-k) \times (n-k)$. Let $e_i = (0 \cdots 010 \cdots 0)$ with 1 in the ith place. The first row of C_γ is $f_1 = e_1 - e_2$, the sum of first two rows gives $f_2 = e_2 - e_3, \ldots,$ the sum of first $k-1$ rows gives $f_{k-1} = e_{k-1} - e_k$. Then $\Sigma_{i=1}^m f_i = e_1 - e_m$, $2 \le m \le k$, are in the row span of C_γ. But for any path γ, $C = C_\gamma + C_\beta$, where β is the set of edges in G but not in γ, that is, $C_\beta = \Sigma_{\{i,j\} \in \beta} C^{ij}$. Clearly $C_\gamma \le C$, and by the above remark $e_1 - e_k$ is also in the row span of C; $2 \le k \le n$. These $n-1$ vectors span a subspace of dimension $n-1$. The converse is easy. If G is not connected, then C can be written as

$$C = \begin{bmatrix} C_1 & \mathbf{0} \\ \mathbf{0} & C_2 \end{bmatrix}$$

where C_1 is the Kirchhoff matrix of a connected part (or component) of G. The vectors $(\mathbf{1}, \mathbf{0})$ and $(\mathbf{0}, \mathbf{1})$ are both in the kernel of C, showing that C can be of rank $n-2$ at the most [$\mathbf{1}$ in $(\mathbf{1}, \mathbf{0})$ has $|C_1|$ entries, and $\mathbf{1}$ in $(\mathbf{0}, \mathbf{1})$ has $|C_2|$ entries, or coordinates]. This proves statement 4.

2.16

Let G be a graph. We label by the indeterminate x_{ij} the edge between vertices i and j (if there are multiple edges between i and j we use $x_{ij}^{(1)}, x_{ij}^{(2)}, \ldots,$ etc.). To each spanning tree of G we associate a monomial of degree $n-1$, the product of all x_{ij}'s, where $\{i, j\}$'s are the $n-1$ edges of the spanning tree.

Let $C(G)$ be the (vertex versus vertex) matrix with off-diagonal (i, j)th entry $-x_{ij}$ (if multiple edges $-\Sigma_k x_{ij}^{(k)}$), 0 if there is no edge between i and j, and ith diagonal entry the negative of the sum of the off-diagonal entries in the ith row. (If G has n vertices, then $\overline{C}(G)$ is $n \times n$ with zero row and column sums.)

We now return to the issues considered at the end of Section 2.14, accomplishing (ii) and answering (i).

$*$ *Let G be a graph with matrix $\overline{C}(G)$. Delete a row and* (not necessarily same) *column of $\overline{C}(G)$. Denote the resulting matrix by K. Let* det K *be* (the formal expansion of) *the determinant of K. Then the monomials in the expansion of* det K (after cancellations) *are all square free and give a complete list of all spanning trees of G.* (Each monomial corresponds uniquely to a spanning tree.) *When setting all x_{ij}'s equal to* 1 *in* $\overline{C}(G)$ det K *equals* (up to sign) *the number of spanning trees of G.*

We call det K the generating function of the spanning trees of G.

The proof of this result may best be illustrated by an example that captures all the relevant features of a general proof:

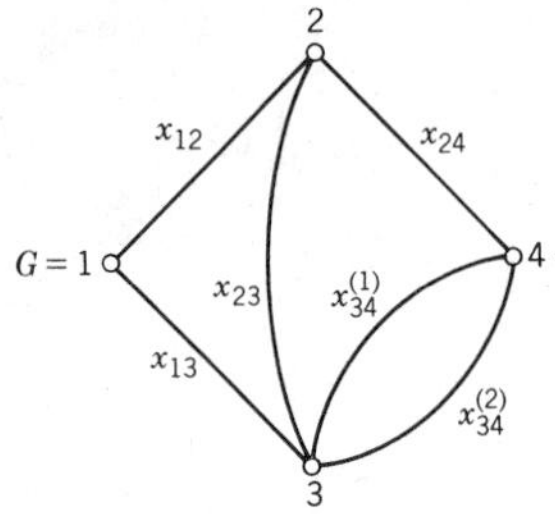

$$G \longleftrightarrow \begin{bmatrix} x_{11} & -x_{12} & -x_{13} & & 0 \\ -x_{12} & x_{22} & -x_{23} & & -x_{24} \\ -x_{13} & -x_{23} & x_{33} & -x_{34}^{(1)} & -x_{34}^{(2)} \\ 0 & -x_{24} & -x_{34}^{(1)} & -x_{34}^{(2)} & x_{44} \end{bmatrix} = \bar{C}(G).$$

[Recall that $x_{ii} = -$(sum of the off-diagonal entries in row i).]

The general idea of the proof is as follows: Select an edge of G (say $x_{34}^{(1)}$). Partition the spanning trees of G into those that do not contain the edge $x_{34}^{(1)}$ and those that do. The first class can be identified with the spanning trees of the graph $G_1 = \{G \text{ without edge } x_{34}^{(1)}\}$, while the second class consists of spanning trees (augmented with edges $x_{34}^{(1)}$) of the graph G_2, obtained from G by shrinking edge $x_{34}^{(1)}$ into a point (thus making vertices 3 and 4 the same vertex and deleting edge $x_{34}^{(1)}$). Both classes defined above involve listing spanning trees in graphs with one edge less than G (G_2 has also one vertex less) and hence we can complete the proof by induction on the number of edges of G.

Obtain K by deleting row 4 and column 4 in $\bar{C}(G)$. (The fact that $\det K$ is independent of which row or column we delete in $\bar{C}(G)$ to obtain K can be proved as property 2 of matrices C discussed in Section 2.15.) We obtain

$$\det K = \begin{vmatrix} x_{11} & -x_{12} & -x_{13} \\ -x_{12} & x_{22} & -x_{23} \\ -x_{13} & -x_{23} & x_{13} + x_{23} + x_{34}^{(2)} + x_{34}^{(1)} \end{vmatrix}$$

$$= \begin{vmatrix} x_{11} & -x_{12} & -x_{13} \\ -x_{12} & x_{22} & -x_{23} \\ -x_{13} & -x_{23} & x_{13} + x_{23} + x_{34}^{(2)} \end{vmatrix} + \begin{vmatrix} x_{11} & -x_{12} & 0 \\ -x_{12} & x_{22} & 0 \\ 0 & 0 & x_{34}^{(1)} \end{vmatrix}$$

$$\updownarrow \qquad\qquad\qquad\qquad \updownarrow$$

$$\bar{C}(G_1) = \begin{bmatrix} x_{11} & -x_{12} & -x_{13} & 0 \\ -x_{12} & x_{22} & -x_{23} & -x_{24} \\ -x_{13} & -x_{23} & x_{13} + x_{23} + x_{34}^{(2)} & -x_{34}^{(2)} \\ 0 & -x_{24} & -x_{34}^{(2)} & x_{24} + x_{34}^{(2)} \end{bmatrix} \qquad \bar{C}(G_2) = \begin{bmatrix} x_{11} & -x_{12} & -x_{13} \\ -x_{12} & x_{22} & -x_{23} - x_{24} \\ -x_{13} & -x_{23} - x_{24} & x_{13} + x_{23} + x_{24} \end{bmatrix}$$

$$\updownarrow \qquad\qquad\qquad\qquad \updownarrow$$

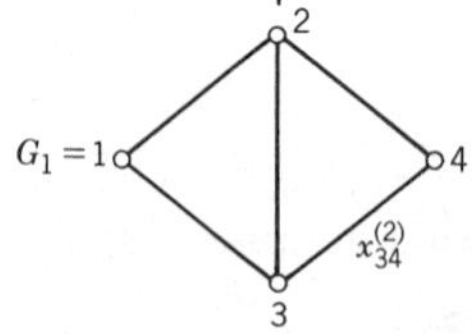

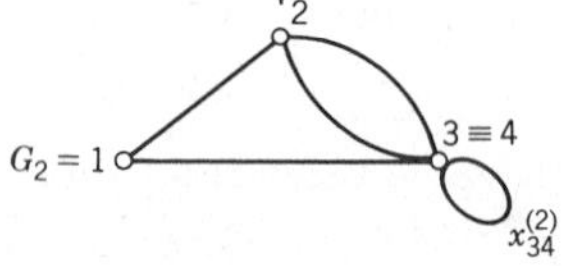

G_1 $\qquad$ G_2

The matrix $\bar{C}(G_1)$ is obtained from $\bar{C}(G)$ by setting $x_{34}^{(1)} = 0$. Add row 4 to row 3 and column 4 to column 3 in $\bar{C}(G)$, then delete row and column 4, to obtain $\bar{C}(G_2)$. [Note that by just looking at G_2 it is not clear whether x_{13} or x_{14} is an edge. But $\bar{C}(G_2)$ clears this up: x_{13} *is* an edge, x_{14} *is not*.]

The first determinant gives the list of trees not containing $x_{34}^{(1)}$ (upon expansion and cancellation). They are

$$x_{12}x_{23}x_{34}^{(2)} + x_{23}x_{34}^{(2)}x_{14} + x_{34}^{(2)}x_{14}x_{12} + x_{14}x_{12}x_{23}$$
$$+ x_{12}x_{24}x_{34}^{(2)} + x_{23}x_{24}x_{14} + x_{12}x_{24}x_{23} + x_{14}x_{24}x_{34}^{(2)} \qquad \text{(8 in all)}.$$

The second determinant gives the spanning trees containing $x_{34}^{(1)}$:

$$x_{12}x_{13}x_{34}^{(1)} + x_{12}x_{23}x_{34}^{(1)} + x_{12}x_{24}x_{34}^{(1)}$$
$$+ x_{13}x_{23}x_{34}^{(1)} + x_{13}x_{24}x_{34}^{(1)} \qquad \text{(5 in all)}.$$

Hence G contains 13 spanning trees. Indeed, when all $x_{ij} = 1$ det K becomes

$$\begin{vmatrix} 2 & -1 & -1 \\ -1 & 3 & -1 \\ -1 & -1 & 4 \end{vmatrix} = 13.$$

Application to Optimal Statistical Design

The information (or Kirchhoff) matrix C, introduced in Section 2.15, is an important representative of a class of matrices known to statisticians as Fisher information matrices (also known as C-matrices). They capture all the relevant statistical information locked into the actual planning (or design) of an experiment. Without dwelling on the general concerns that surround the planning, we wish to point out (in purely mathematical terms) a specific problem that often arises and that, as yet, has not been brought to a satisfactory solution:

Among all graphs with n vertices and m edges identify those with a maximum number of (labeled) spanning trees.

An understanding of the structure of such graphs translates directly into optimum ways of planning experiments. The resulting design will be called D-optimal by the statistician.

It might not be surprising to find that the Kirchhoff tree generating matrix plays an important part in the solution. For the necessary background in statistics we refer the reader to Chapter 8.

EXERCISES

1. How many (labeled) spanning trees does the graph displayed below have?

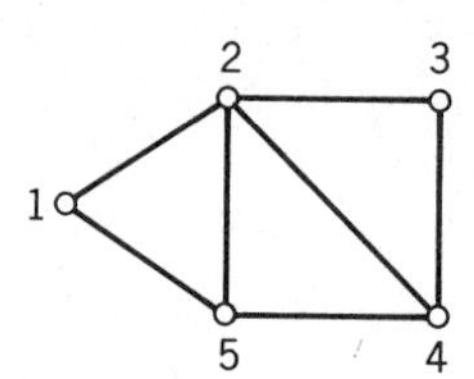

2. Let $0 = \mu_0(G) \leq \mu_1(G) \leq \cdots \leq \mu_{n-1}(G)$ be the eigenvalues of the Kirchhoff matrix $C(G)$ of a graph G on n vertices. Show that $n^{-1}\prod_{i=1}^{n-1}\mu_i(G)$ = number of labeled spanning trees of G.

3. A graph is called *simple* if between any two vertices there is at most one edge and no loops are allowed. By K_n we denote a simple graph on n vertices with an edge between any two vertices. We call K_n the *complete graph*; K_n has $\binom{n}{2}$ edges. How many labeled spanning trees does K_n have?

4. Partition $n_1 + n_2 + \cdots + n_m$ vertices into m classes, the ith class containing n_i vertices. Produce a simple graph $K(n_1, n_2, \ldots, n_m)$ by joining each vertex in class i to all vertices outside class i (and to none within class i); do this for all i. The resulting graph is called the *complete multipartite graph* $K(n_1, n_2, \ldots, n_m)$. For example $K(2, 3)$ is

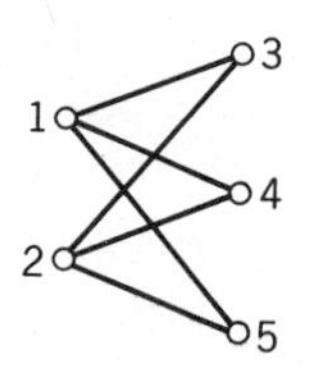

How many labeled spanning trees does $K(n_1, n_2, \ldots, n_m)$ have?

5. Place n^2 vertices into an $n \times n$ square array and join two vertices if and only if they are in the same row or same column. Call the resulting graph S_n. Compute the number of labeled spanning trees of S_n. (S_3 is drawn below.)

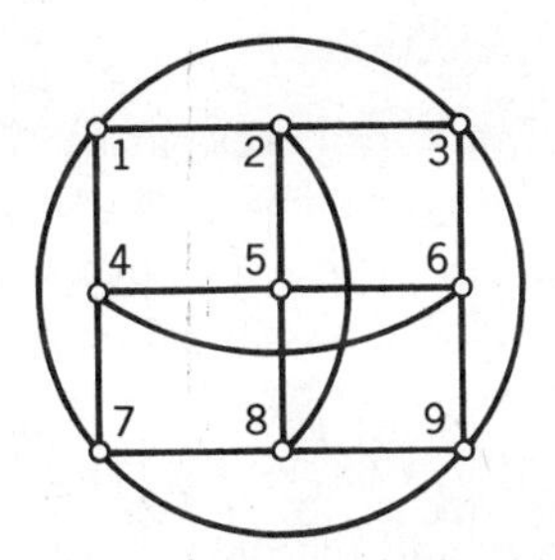

6. A graph is called *regular* if each of its vertices has the same degree. The *complementary graph* $\overline{G}$ of a simple graph G is the graph on the same set of vertices as G whose edges are precisely those that are missing in G. For G a regular and simple graph relate the eigenvalues of $C(\overline{G})$ to those of $C(G)$, and (with the help of Exercise 2) obtain a relationship between the number of labeled spanning trees in G and $\overline{G}$.

7. Show that among all graphs on n vertices and e edges (with e sufficiently large) those that have a maximal number of labeled spanning trees must have the degrees of their vertices differ by at most 1 and the number of edges between any two vertices differ by at most 1. [*Hint*: look at $\prod_{i=1}^{n-1} (\mu_i + x)$ for large values of x.]

7 PARTITIONS OF AN INTEGER

We touch only briefly here upon a rich and well-developed subject: that of partitioning an integer.

2.17

The question we raise regards *the number of* (unordered) *ways of writing the number n as the sum of exactly m positive integers*. Let us call this number $P_m(n)$. More rigorously,

$$P_m(n) = \begin{array}{l} |\{(\alpha_1, \alpha_2, \ldots, \alpha_m)\text{: each } \alpha_i \text{ is a positive} \\ \text{integer, } \alpha_1 + \alpha_2 + \cdots + \alpha_m = n, \text{ and} \\ \alpha_1 \geq \alpha_2 \geq \cdots \geq \alpha_m \geq 1\}|. \end{array}$$

The α_i's are called the *parts* of n. Clearly $m \leq n$.

The number of ways of writing n as the sum of 1 integer, as the sum of $n-1$ integers, or as the sum of n integers is unique, so that $P_1(n) = P_n(n) = P_{n-1}(n) = 1$.

We wish to find a pattern, a simple recurrence relation, for $P_m(n)$. *Our first result is the following*:

$$P_1(n) + P_2(n) + \cdots + P_k(n) = P_k(k+n), \qquad \text{for } k \leq n.$$

Proof. Let

$$P = \{\text{partitions of } n \text{ into } k \text{ or fewer parts}\}$$

$$= \left\{(\alpha_1, \alpha_2, \ldots, \alpha_m, 0, \ldots, 0) \in \{k\text{-tuples}\} \ \sum_i^m \alpha_i = n, m \le k\right\}.$$

Define a mapping on P as follows:

$$(\alpha_1, \alpha_2, \ldots, \alpha_m, 0, \ldots, 0) \to (\alpha_1 + 1, \alpha_2 + 1, \ldots, \alpha_m + 1, 1, \ldots, 1).$$

The image is a k-tuple again, and the number of single 1's is the same as the number of 0's in its preimage.

Note that the image corresponds, in fact, to a partition of $k + n$ into k parts. This mapping is injective, and for each partition of $k + n$ into k parts there is a k-tuple in P that is mapped into it, that is, the mapping is also onto the set of partitions of $n + k$ into k parts. Hence $|P| = |\text{image of } P| = P_k(n + k)$. But also $|P| = P_1(n) + P_2(n) + \cdots + P_k(n)$, from which the recurrence relation follows. This ends the proof.

For small values of m and n we have the following table for $P_m(n)$:

	m					
n	1	2	3	4	5	6
1	1	0	0	0	0	0
2	1	1	0	0	0	0
3	1	1	1	0	0	0
4	1	2	1	1	0	0
5	1	2	2	1	1	0
6	1	3	4	2	1	1

2.18 Ferrer Diagrams

We can also represent a partition by a *Ferrer diagram*, which will be very useful in visualizing many results. Given a partition we represent each part by the appropriate number of dots in a row and place the rows beneath one another. For example, the Ferrer diagram of the partition (6, 4, 3, 1) is

```
•  •  •  •  •  •   6 dots

•  •  •  •         4 dots

•  •  •            3 dots

•                  1 dot
```

Given a partition $\alpha = (\alpha_1, \alpha_2, \ldots, \alpha_m)$ we define a new partition $(\alpha_1^*, \alpha_2^*, \ldots, \alpha_k^*)$, where α_i^* is the number of parts in α that are greater than or equal to i. The new partition α^* is called the *conjugate* of α. For example, if $\alpha = (5, 3, 2)$, then $\alpha^* = (3, 3, 2, 1, 1)$. The simplest and most visual way to construct α^* is by rotating the Ferrer diagram of α about the diagonal. (It is thus clear that $\alpha^{**} = \alpha$.) For example,

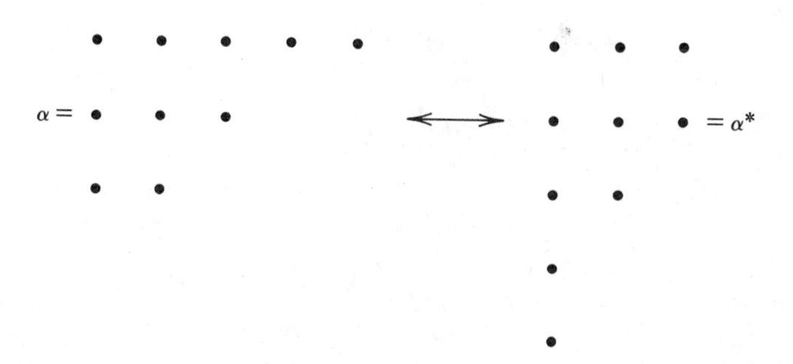

It is also clear from the way α^* is obtained on the Ferrer diagram that $\Sigma_1^m \alpha_i = \Sigma_1^k \alpha_i^*$, that is, if α is a partition of n, α^* is also a partition of n. The bijective correspondence between partitions of n and conjugate partitions suggests the following result:

* *The number of partitions of n into k parts is equal to the number of partitions of n into parts the largest of which is k.*

Proof. Let $P = \{(\alpha_1, \ldots, \alpha_k)$: partitions of n into k parts$\}$. The mapping $(\alpha_1, \ldots, \alpha_k) \to (\alpha_1^*, \alpha_2^*, \ldots)$ is a bijection, for the conjugate is obtained by a rotation of the Ferrer diagram. Also, the largest part of $(\alpha_1^*, \alpha_2^*, \ldots)$ does not exceed k.

As an easy consequence we have:

* *The number of partitions of n with at most k parts equals the number of partitions of n in which no part exceeds k.*

If $\alpha = \alpha^*$ we call α *self-conjugate*. Note that α is self-conjugate if and only if its Ferrer diagram is symmetric with respect to the diagonal. With this definition we have the following result:

* *The number of self-conjugate partitions of n is equal to the number of partitions of n with all parts unequal and odd.*

Proof. Take each (odd) part of the initial partition, bend in the middle, and reassemble as indicated below:

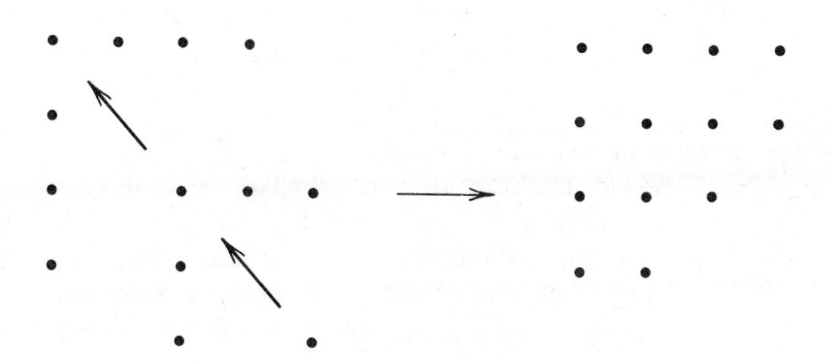

We thus obtain a self-conjugate partition. This operation produces, in fact, a (visual) bijection from partitions of n with distinct and odd parts to self-conjugate partitions of n.

Another transformation of a Ferrer diagram is used in establishing the following result.

* *The number of partitions of n into unequal parts is equal to the number of partitions of n into odd parts.*

Proof. Consider a partition of n into odd parts. Write it as $n = k_1\alpha_1 + k_2\alpha_2 + \cdots + k_m\alpha_m$ with k_i the multiplicity of α_i (where the α_i's are odd).

We produce a new partition as follows: expand each k_i in binary base, say $k_i = \varepsilon_0 2^0 + \varepsilon_1 2^1 + \cdots + \varepsilon_{r_i} 2^{r_i}$. Group together $\varepsilon_s 2^s$ rows of α_i (attached to k_i), as s ranges between 0 and r_i, and $\varepsilon_s = 0$ or 1. Form the new partition of n with parts $\varepsilon_s 2^s \alpha_i$, as s and i take values in their respective ranges.

As an example, let $\alpha = (7, 7, 7, 7, 5, 3, 3, 3, 1, 1, 1, 1, 1)$, that is, $n = 4 \cdot 7 + 1 \cdot 5 + 3 \cdot 3 + 5 \cdot 1$. Then

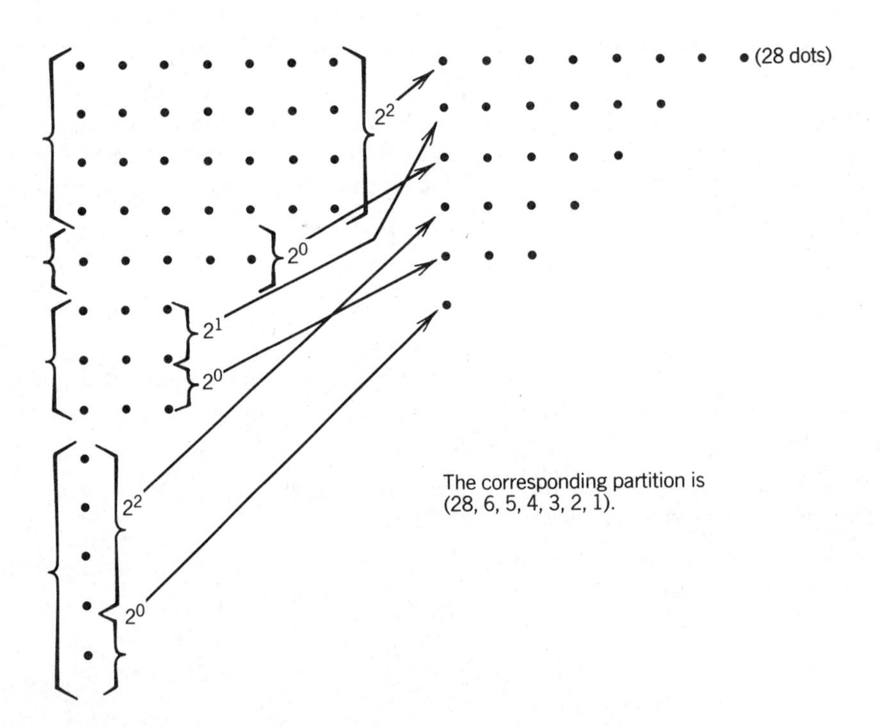

Since the α_i's are odd, $2^a\alpha_i \neq 2^b\alpha_j$ for $i \neq j$ or $a \neq b$. Hence the above transformation sends odd partitions of n to unequal partitions, and it is clear that the new partitions are of n, because the total number of dots is preserved.

This transformation can be reversed in a unique way, for given $\beta = (\beta_1, \beta_2, \ldots, \beta_m)$ with β_i's distinct, write each β_i as a product of an odd number and a power of 2. This representation of the β_i's is unique. Therefore, if $\beta_i = 2^{r_i}\alpha_i$ with α_i odd, obtain a new partition α with parts α_i of appropriate multiplicities. Clearly α is a partition of n with odd parts. We have therefore a bijection between unequal partitions of n and odd partitions of n, proving the above statement.

Our next result can be stated as follows:

* *Let $P(n; d, o)$ and $P(n; d, e)$ denote, respectively, the number of partitions of n into an odd/even number of distinct parts. Then*

$$P(n; d, e) - P(n; d, o) = \begin{cases} (-1)^m & \text{if } n = m(3m+1)/2 \\ 0 & \textit{otherwise}. \end{cases}$$

(This result is known as *Euler's pentagonal theorem*.)

Proof. We initially try to establish a bijective correspondence between the distinct partitions of n into even parts and the distinct partitions of n into odd parts.

Given a partition $\lambda = (\lambda_1, \lambda_2, \ldots, \lambda_r)$ of n into distinct parts let $s(\lambda) = \lambda_r$, that is, $s(\lambda)$ is the smallest part of λ, and let $\sigma(\lambda)$ be the number of consecutive parts of λ from λ_1 down. [More formally, $\sigma(\lambda) = \max\{j: \lambda_j = \lambda_1 - j + 1\}$.]

Examples.

$\lambda = (7, 6, 4, 3, 2)$ $\sigma(\lambda) = 2$ $s(\lambda) = 2$

$\lambda = (8, 7, 6, 3)$ $\sigma(\lambda) = 3$ $s(\lambda) = 3$

$\lambda = (4, 2)$ $\sigma(\lambda) = 1$ $s(\lambda) = 2$

We separate the proof into two cases.

Case 1. $s(\lambda) \leq \sigma(\lambda)$. Add 1 to each of the first $s(\lambda)$ parts of λ and delete the smallest part. Thus $(7, 6, 4, 3, 2) \rightarrow (8, 7, 4, 3)$.

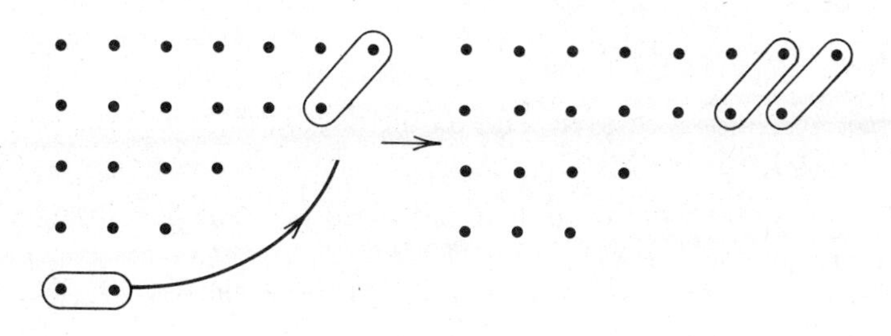

This transformation is always possible, except when the dots enumerated by $s(\lambda)$ and $\sigma(\lambda)$ meet [and $s(\lambda) = \sigma(\lambda)$], for example, if $\lambda = (5, 4, 3)$.

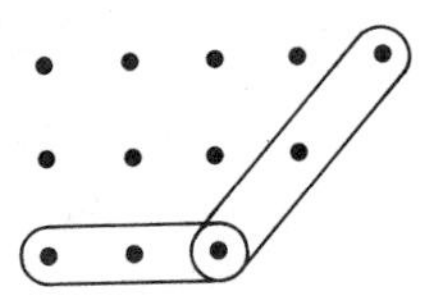

Then, if the initial partition had an odd number of distinct parts, the above transformation does not lead to a partition with an even number of parts. In all other cases, however, the above transformation establishes a bijective map between partitions into distinct, odd parts and partitions into distinct, even parts.

Case 2. $s(\lambda) > \sigma(\lambda)$. Subtract 1 from each of the $\sigma(\lambda)$ largest parts of λ and add a new smallest part of size $\sigma(\lambda)$. Thus $(8, 7, 5, 4, 3) \rightarrow (7, 6, 5, 4, 3, 2)$.

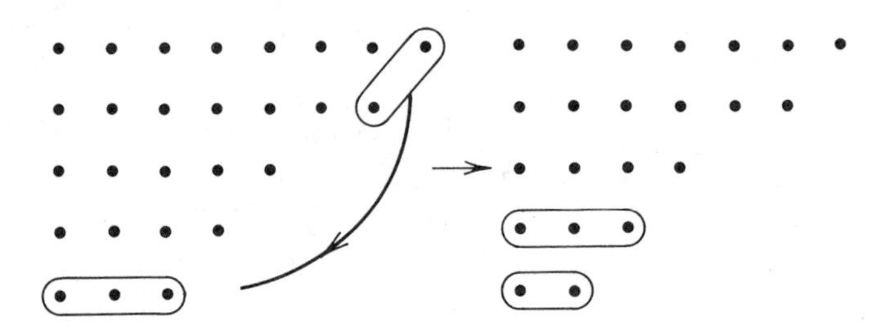

This transformation is always possible except when the dots of $\sigma(\lambda)$ and $s(\lambda)$ meet and $s(\lambda) = \sigma(\lambda) + 1$, as in $\lambda = (6, 5, 4)$.

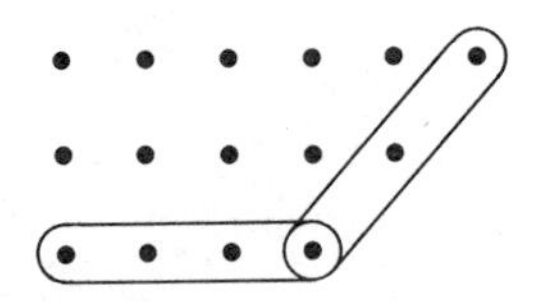

In this case the above transformation will not give a partition into distinct parts but in all other cases it will transform an odd, distinct partition into an even, distinct partition.

The two exceptional cases depend on the number n, for:

(a) If $s(\lambda) = \sigma(\lambda)$ and the dots in $s(\lambda)$ meet with the dots in $\sigma(\lambda)$, then n is divided into $\sigma(\lambda)$ parts. By writing $m = \sigma(\lambda)$ we conclude that $n = m + (m + 1) + \cdots + (m + m - 1) = m(3m - 1)/2$.

(b) If $s(\lambda) = \sigma(\lambda) + 1$ and the dots in $s(\lambda)$ meet with the dots in $\sigma(\lambda)$, then n is divided into $\sigma(\lambda)$ parts. Hence, if $m = \sigma(\lambda)$, then $n = (m + 1) + (m + 2) + \cdots + (m + 1 + m - 1) = m(3m + 1)/2$.

Therefore, if $n \neq m(3m \pm 1)/2$ for some positive integer m, then Case 1 and Case 2 establish a bijective mapping from partitions of n into an odd number of distinct parts to partitions of n into an even number of distinct parts. For such integers $P(n; d, o) = P(n; d, e)$.

We now investigate the exceptional cases (a) and (b) mentioned above. Let $n = m(3m - 1)/2$ for some odd m, $m \geq 1$. For this n only the exceptional situation described in Case (a) can occur, and this exceptional situation involves only the one partition mentioned in Case (a). For this sole partition the bijective transformation fails. The "extra" partition explains why for m odd, and $n = m(3m - 1)/2$, we have

$$P(n; d, o) = P(n; d, e) + 1.$$

Similar arguments will explain the result for even m and, in Case b, for $n = m(3m + 1)/2$. This ends our proof.

2.19

A lot of results about partitions can be obtained by means of generating functions. Let us look at some of these:

1. $F(x) = (1 - x^a)^{-1}(1 - x^b)^{-1}(1 - x^c)^{-1} \cdots$ *is the generating function of* $P(n; \{a, b, c, \ldots\})$, *the number of ways of writing n as the sum of integers from the set* $\{a, b, c, \ldots\}$ *with repetitions allowed.*

Proof. Consider the coefficient of x^n in the series expansion of $F(x)$: $(1 - x^a)^{-1}(1 - x^b)^{-1}(1 - x^c)^{-1} \cdots = (1 + x^a + x^{2a} + \cdots + x^{ka} + \cdots)$ $(1 + x^b + \cdots + x^{kb} + \cdots)(1 + x^c + \cdots) \cdots$.

If the term x^n is formed from the product of $x^{k_1 a}, x^{k_2 b}, x^{k_3 c}, \ldots$ then

$$n = \underbrace{a + \cdots + a}_{k_1 \text{ times}} + \underbrace{b + \cdots + b}_{k_2 \text{ times}} + \underbrace{c + \cdots + c}_{k_3 \text{ times}} + \cdots.$$

Hence the term x^n arises exactly as often as n can be written as the sum of a's, b's, c's,... . The coefficient of x^n is therefore $P(n; \{a, b, c, \ldots\})$.

Immediate consequences of the above observation are:

1.1. *The generating function for* $P(n)$, *the number of ways of writing n as the sum of positive integers, is*

$$F(x) = (1 - x)^{-1}(1 - x^2)^{-1}(1 - x^3)^{-1} \cdots (1 - x^k)^{-1} \cdots.$$

1.2. *The generating function for* $P(n; \{\text{odd integers}\})$ is

$$(1 - x^1)^{-1}(1 - x^3)^{-1}(1 - x^5)^{-1} \cdots (1 - x^{2k+1})^{-1} \cdots.$$

1.3. *The generating function for* $P(n; \{1, 2, \ldots, k\})$ *is*

$$(1 - x)^{-1}(1 - x^2)^{-1} \cdots (1 - x^k)^{-1}.$$

1.4. *We have*

$$\sum_n P_m(n)x^n = x^m(1-x)^{-1}(1-x^2)^{-1}\cdots(1-x^m)^{-1}.$$

Proof. We prove 1.4. As we just saw $(1-x)^{-1}(1-x^2)^{-1}\cdots(1-x^m)^{-1} = \Sigma_n P(n;\{1,2,\ldots,m\})x^n$. Multiplying by x^m we obtain $x^m(1-x)^{-1}(1-x^2)^{-1}\cdots(1-x^m)^{-1} = \Sigma_m P(n;\{1,2,\ldots,m\})x^{n+m} = \Sigma_{n=m}^{\infty} P(n-m;\{1,2,\ldots,m\})x^n = \Sigma_{n=m}^{\infty}(\Sigma_{k=1}^{m} P_k(n-m))x^n = \Sigma_{n=m}^{\infty} P_m(n)x^n$, as claimed. The last two signs of equality are explained by the first two results proved in this section. This proves 1.4.

Our next result is the following:

2. $F(x) = (1+x^a)(1+x^b)(1+x^c)\cdots$ *is the generating function of* $P(n; d, \{a,b,c,\ldots\})$, *the number of ways of writing* n *as a sum using the distinct numbers* $a, b, c, \ldots$ *at most once each.*

Proof. To form x^n we can choose either 1 or x^a from the first factor and there is no option for choosing x^a again. The same is true for $x^b, x^c, \ldots$. Hence $n = \varepsilon_a a + \varepsilon_b b + \varepsilon_c c + \cdots$, where $\varepsilon_k = 1$ or 0. We can see that x^n arises as often as n can be written in the above way; hence the coefficient of x^n is $P(n; d, \{a,b,c,\ldots\})$. This ends our proof.

Immediate consequences are:

2.1. *The generating function of* $P(n; d)$, *the number of ways of writing* n *as the sum of distinct integers, is*

$$(1+x)(1+x^2)(1+x^3)\cdots.$$

2.2. *The generating function of* $P(n; d, \{\text{odd integers}\})$ is

$$(1+x)(1+x^3)(1+x^5)\cdots.$$

2.3. *The generating function of* $P(n; d, \{2^k\colon k = 0,1,2,\ldots\})$ *is* $\prod_{k=0}^{\infty}(1+x^{2^k})$.

In Section 2.18 we proved the equality of $P(n; d)$ and $P(n; \{\text{odd integers}\})$ using Ferrer diagrams. Relying on generating functions we can prove this as follows:

$$\begin{aligned}\sum_n P(n;d)x^n &= (1+x)(1+x^2)(1+x^3)\cdots\\ &= \frac{(1-x)(1+x)(1-x^2)(1+x^2)(1-x^3)(1+x^3)\cdots}{(1-x)(1-x^2)(1-x^3)\cdots}\\ &= \frac{(1-x^2)(1-x^4)(1-x^6)(1-x^8)\cdots}{(1-x)(1-x^2)(1-x^3)(1-x^4)\cdots}\\ &= \frac{1}{(1-x)(1-x^3)(1-x^5)\cdots}\\ &= \sum_n P(n;\{\text{odd integers}\})x^n.\end{aligned}$$

REMARK. We all know that a positive integer has a unique expression in base 2. It is somewhat amusing to see how this follows from easy work with generating functions.

$$\sum_n P(n; d, \{2^k: k = 0,1,2,\dots\})x^n$$
$$= (1+x)(1+x^2)(1+x^4)(1+x^8)\cdots$$
$$= \left(\frac{1-x^2}{1-x}\right)\left(\frac{1-x^4}{1-x^2}\right)\left(\frac{1-x^8}{1-x^4}\right)\left(\frac{1-x^{16}}{1-x^8}\right)\cdots$$
$$= \frac{1}{1-x} = \sum_n x^n.$$

Hence $P(n; d, \{2^k: k = 0,1,2,\dots\}) = 1$, for all n.

Let us close this section with a series expansion version of *Euler's pentagonal theorem*:

$$\prod_{n=1}^{\infty}(1-x^n) = 1 + \sum_{m=1}^{\infty}(-1)^m\left(x^{m(3m-1)/2} + x^{m(3m+1)/2}\right).$$

Proof.

$$\sum_{m=1}^{\infty}(-1)^m\left(x^{m(3m-1)/2} + x^{m(3m+1)/2}\right)$$
$$= \sum_{m=1}^{\infty}(-1)^m x^{m(3m\pm1)/2}$$
$$= \sum_{n=1}^{\infty} x^n \begin{cases} (-1)^m & \text{if } n = m(3m\pm1)/2 \\ 0 & \text{otherwise} \end{cases}$$
$$= \{\text{by the result in Section 2.18}\}$$
$$= \sum_{n=1}^{\infty}(P(n; d, e) - P(n; d, o))x^n.$$

We need to show that

$$1 + \sum_{n=1}^{\infty}(P(n; d, e) - P(n; d, o))x^n = \prod_{n=1}^{\infty}(1-x^n).$$

Let us look at the coefficient of x^n in $\prod_{n=1}^{\infty}(1-x^n)$. Since x^n can be formed as a product of $(-x)^{k_1}(-x^2)^{k_2}\cdots(-x^n)^{k_n}$, where $k_i = 0$ or 1, we have $x^n = (-1)^{k_1+\cdots+k_n}x^{k_1+2k_2+\cdots+nk_n}$. The coefficient of x^n is therefore

$$\sum_{(k_1,\dots,k_n)}(-1)^{k_1+k_2+\cdots+k_n},$$

where each n-tuple corresponds to a partition of n into distinct integers as $n = k_1 + 2k_2 + 3k_3 + \cdots + nk_n$ ($k_i = 0$ or 1). Note that $k_1 + k_2 + \cdots + k_n$

gives us the number of parts of the partition of n. Hence $(-1)^{k_1+\cdots+k_n}$ is 1 if the partition has an even number of parts and -1 if it has an odd number of parts. This observation leads us to conclude that the coefficient of x^n in $\prod_{n=1}^{\infty}(1 - x^n)$ is

$$\sum_{(k_1,\ldots,k_n)} (-1)^{k_1+\cdots+k_n} = P(n; d, e) - P(n; d, o).$$

The constant term is clearly 1 on both sides. We have therefore proved that

$$\prod_{n=1}^{\infty}(1 - x^n) = 1 + \sum_{n=1}^{\infty}(P(n; d, e) - P(n; d, o))x^n$$

$$= 1 + \sum_{m=1}^{\infty}(-1)^m(x^{m(3m-1)/2} + x^{m(3m+1)/2}).$$

Brief Note on Terminology

The word pentagonal has been mentioned more than once in connection with Euler's result. Numbers of the form $m(3m \pm 1)/2$ are called pentagonal. These are exceptional integers for which the number of distinct even partitions does not equal the number of distinct odd ones. We call them so because each can be written as the sum of a square and a "triangular" number, thus producing the geometric effect of a pentagon (or of a house), as displayed below:

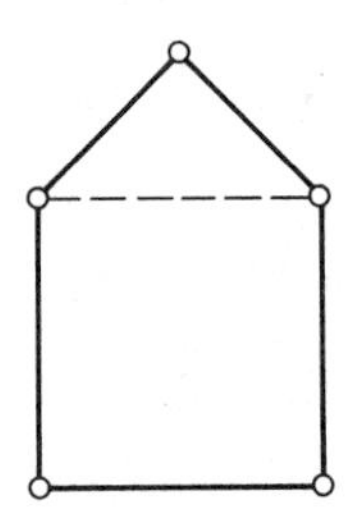

Indeed, $m(3m \pm 1)/2 = s + t$, where $s = m^2$ and $t = m(m \pm 1)/2$.

EXERCISES

1. The number of noncongruent triangles with circumference $2n$ and integer sides is equal to $P_3(n)$. Prove this.

2. A partition of the number n is called *perfect* if every integer from 1 to $(n - 1)$ can be written in a unique way as the total of a subset of the parts

of this partition. Prove that the number of perfect partitions of n is the same as the number of ways of factoring $n + 1$, where the order of the factors counts and factors of 1 are not counted. When will the trivial partition $n = 1 + 1 + \cdots + 1$ be the only solution?

3. Find a generating function for the number of integer solutions of $n = 2x + 3y + 7z$ with:

(a) $x, y, z \geq 0$.

(b) $0 \leq z \leq 2 \leq y \leq 8 \leq x$.

4. Find a generating function for the number of ways of making n cents change in pennies, nickels, dimes, and quarters.

5. Show with generating functions that every positive integer can be written as a sum of distinct powers of 10, that is, it has a unique decimal expansion.

6. Prove the identity

$$\frac{1}{1-x} = (1 + x + x^2 + \cdots + x^9)(1 + x^{10} + x^{20} + \cdots + x^{90}) \cdot (1 + x^{100} + x^{200} + \cdots + x^{900}) \cdots .$$

7. Show that the number of partitions of the integer $2r + k$ into exactly $r + k$ parts is the same for any nonnegative integer k.

8. Show that the number of partitions of n into at most two parts is $[n/2] + 1$, with $[x]$ denoting the integral part of x.

9. Prove that the number of partitions of n in which only odd parts may be repeated equals the number of partitions of n in which no part appears more than three times.

10. Prove that the number of partitions of n with unique smallest part (i.e., the smallest part occurs only once) and largest part at most twice the smallest part equals the number of partitions of n in which the largest part is odd and the smallest part is larger than half the largest part.

8 A GENERATING FUNCTION FOR SOLUTIONS OF DIOPHANTINE SYSTEMS IN NONNEGATIVE INTEGERS

The title pretty well describes our intentions with regard to the contents of this section. Consider

$$\sum_{j=1}^{n} a_{ij} x_j = b_i; \qquad i = 1, 2, \ldots, m \tag{2.11}$$

where a_{ij} and b_i are nonnegative integers. *We investigate the solutions to the Diophantine system* (2.11) *in nonnegative integers.*

Write

$$x = (x_1, \ldots, x_n), \qquad s = (s_1, \ldots, s_n)$$
$$b = (b_1, \ldots, b_m), \qquad t = (t_1, \ldots, t_m)$$
$$s^x = \prod_{j=1}^{n} s_j^{x_j}, \qquad t^b = \prod_{i=1}^{m} t_i^{b_i}.$$

The notation $x \geq 0$ or $b \geq 0$ means that the respective components are nonnegative (and, in this case, also integral).

Assume that each column of the $m \times n$ matrix (a_{ij}) has a nonzero entry. The nonnegativity of the entities involved insures then at most a finite number of solutions to system (2.11).

For $x \geq 0$ and $b \geq 0$ set

$$N_x(b) = \begin{cases} 1 & \text{if } x \text{ is a solution of } (2.11) \\ 0 & \text{otherwise} \end{cases}$$

and let $N(b)$ be the number of solutions to (2.11).

* *We assert that*

$$\sum_{\substack{x \geq 0 \\ b \geq 0}} N_x(b) s^x t^b = \prod_{j=1}^{n} \left(1 - s_j t_1^{a_{1j}} t_2^{a_{2j}} \cdots t_m^{a_{mj}}\right)^{-1}$$

and

$$\sum_{b \geq 0} N(b) t^b = \prod_{j=1}^{n} \left(1 - t_1^{a_{1j}} t_2^{a_{2j}} \cdots t_m^{a_{mj}}\right)^{-1}.$$

The proof rests upon routine expansions:

$$\sum_{\substack{x \geq 0 \\ b \geq 0}} N_x(b) s^x t^b = \sum_{x \geq 0} s_1^{x_1} \cdots s_n^{x_n} t_1^{\Sigma_j a_{1j} x_j} \cdots t_m^{\Sigma_j a_{mj} x_j}$$

$$= \prod_{j=1}^{n} \left(\sum_{x_j = 0}^{\infty} s_j^{x_j} t_1^{a_{1j} x_j} \cdots t_m^{a_{mj} x_j} \right)$$

$$= \prod_{j=1}^{n} \sum_{x_j = 0}^{\infty} \left(s_j t_1^{a_{1j}} t_2^{a_{2j}} \cdots t_m^{a_{mj}}\right)^{x_j}$$

$$= \prod_{j=1}^{n} \left(1 - s_j t_1^{a_{1j}} t_2^{a_{2j}} \cdots t_m^{a_{mj}}\right)^{-1}.$$

The second formula is explained similarly. This explains the assertion.

Further, by writing our first formula as

$$\left[\prod_{j=1}^{n} \left(1 - s_j t_1^{a_{1j}} \cdots t_m^{a_{mj}}\right) \right] \left[\sum_{\substack{x \geq 0 \\ b \geq 0}} N_x(b) s^x t^b \right] = 1$$

and equating the coefficients of $s^x t^b$ on both sides we obtain *recursive formulas* for the $N_x(b)$'s. The same can be done to the second formula to obtain recurrences for the $N(b)$'s.

In particular, the reader may wish to investigate in detail the Diophantine system

$$\left.\begin{aligned} x_0 + x_1 + x_2 + \cdots + x_n &= b_1 \\ x_1 + 2x_2 + \cdots + nx_n &= b_2 \end{aligned}\right\}.$$

It leads to the so-called *Gaussian polynomials* which we discuss in Section 6 of Chapter 3—see, in particular, Exercise 5 of that section.

HISTORICAL NOTE

What we have seen in this chapter is by and large classical material on generating functions. Much of the first two sections introduce the (formal) power series and explain the combinatorial meaning of multiplication by convolution. Of the results in Section 3 those regarding Stirling numbers rely fundamentally on Stirling's formulas, introduced in Sections 1.7(c) and 1.8(c) of Chapter 1. The Lah numbers, and their analogous behavior to those of Stirling, were only relatively recently noticed by Ivo Lah [8] of the University of Belgrade, Yugoslavia. Though less fundamental in nature than the numbers of Stirling, we meet them again in connection with inversion formulas.

Faa DiBruno observed the pattern of the higher order derivative of a composition of two functions in terms of (what we now call) Bell polynomials; this result can be found in [1]. We only briefly discussed recurrence relations and only those aspects that call for immediate use of generating functions. The contents of Section 2.13 are based upon a paper of D. André of 1879 [6].

Enumerating labeled spanning trees of a graph, as we did, was (implicitly) noted by Kirchhoff in his classic paper [5] on electrical networks of which the famous Kirchhoff laws of current form the main topic. That the Kirchhoff matrix coincides with the Fisher information matrix in the setting of statistical designs (with blocking in one direction—see [7]) is an unexpected connection with possibly interesting ramifications. We discuss these shared aspects in Chapter 8, the chapter on statistical design. The contents of Section 8 are of recent origin and appear only as part of a more substantial work on Fuchsian groups [9] by R. S. Kulkarni.

The pentagonal theorem (hereinafter written as theorem P) dates back almost to the very beginnings of the work with generating functions [4]. It had preoccupied Euler a good deal over the span of at least a decade. In 1740, while expanding $\prod_n(1 - x^n)$, Euler observed the pattern of -1's and 1's that arises in connection with the pentagonal numbers. The reader may be entertained by how Euler relates this:

> Theorem P is of such a nature that we can be assured of its truth without giving it a perfect demonstration. Nevertheless, I will present evidence for it of such a character that it might be regarded as almost equivalent to a rigorous demonstration.

We are then informed that he has compared coefficients of up to the 40th power of x and that they all follow the proposed pattern.

> I have long searched in vain for a rigorous demonstration of theorem P, and I have proposed the same question to some of my friends with whose ability in these matters I am familiar but all have agreed with me on the truth of theorem P without being able to unearth any clue of a demonstration. Thus it will be a known truth, but not yet demonstrated.... And since I must admit that I am not in a position to give it a rigorous demonstration, I will justify it by a sufficiently large number of examples.... I think these examples are sufficient to discourage anyone from imagining that it is by pure chance that my rule is in agreement with the truth.... If one still doubts that the law is precisely that one which I have indicated, I will give some examples with larger numbers.

Here he tells how he took the trouble to examine the coefficients of x^{101} and x^{301} and how they came out to be just what he had expected.

> These examples which I have just developed undoubtedly will dispel any qualms which we might have had about the truth of theorem P.

Euler did succeed in proving the pentagonal theorem in 1750. The passages above were extracted from Pólya's work mentioned also as reference [4].

Of the texts available that treat similar material we recommend [1], [2], and [3].

REFERENCES

[1] J. Riordan, *An Introduction to Combinatorial Analysis*, Wiley, New York, 1958.

[2] G. E. Andrews, *The Theory of Partitions*, Addison-Wesley, Reading, MA, 1976.

[3] C. Berge, *Graphs and Hypergraphs*, North-Holland, Amsterdam, 1973.

[4] *Leonhardi Euleri Opera Omnia*, Ser. 1, Vol. 2, 1915, pp. 241–253 (see G. Pólya, *Collected Works*, *Vol.* 4, The MIT Press, Cambridge, MA, 1984, pp. 186–187.)

[5] G. Kirchhoff, Über die Auflösung der Gleichungen, auf welche man bei der Untersuchung der linearen verteilung galvanischer Ströme gerführt wird, *Ann. Phys. Chem.*, **72**, 497–508 (1847) (English translation in *Trans. Inst. Radio Engrs.*, **CT-5**, 4–7, March, 1958).

[6] D. André, Developpement de *sec* x et de *tang* x, *C.R. Acad. Sci. Paris*, **88** (965–967 (1879).

[7] J. C. Kiefer, Optimum experimental designs, *J. Royal Stat. Soc.* (*B*), **21** 272–304 (1959).

[8] I. Lah, Eine neue Art von Zahlen, ihre Eigenschaften und Anwendung in der mathematischen Statistik, *Mitteilungsbl. Math. Statist.*, **7** 203–212 (1955).

[9] R. S. Kulkarni, An extension of a theorem of Kurosh and applications to Fuchsian groups, *Mich. Math. Journal*, **30** 259–272 (1983).

[10] L. Lovász, *Combinatorial Problems and Exercises*, North Holland, Amsterdam, 1979.

CHAPTER 3

Classical Inversion

A lie, turned topsy-turvy, can be prinkled and tinselled out, decked in plumage new and fine, till none knows its lean old carcass.

Peer Gynt, HENRIK IBSEN

The two opening sections of this chapter describe a process of inversion that explains some connections between the various counting numbers introduced in the first two chapters. Attention is first given to an inversion theorem that, although of simple theoretical content, is the cluster point of a wealth of inversion formulas. Mere awareness of this inversion principle makes one understand at once that, for example, the formulas

$$\theta^n = \sum_k S_n^k x^k D^k \qquad \text{and} \qquad x^n D^n = \sum_k s_n^k \theta^k$$

(which appear in Section 2.8) are equivalent manifestations of one and the same phenomenon. They are a pair of inverse relations. When one is aware of an underlying inversion process the proof of a formula, or an identity, can sometimes be facilitated by proving the inverse relation instead. Many of the formulas that we describe can in fact be viewed as special cases of the process of Möbius inversion studied in Chapter 9. The examples we present are some of the better known ones. They deserve special mention since they occur frequently in calculations. The proofs we give describe several general ways of producing such formulas.

Upon touching briefly on classical identities of Abel, Vandermonde, and Nörlund, attention focuses on formal Laurent series. Within the context of such series we discuss the important inversion theorem of Lagrange. Its multidimensional version is studied as well, with special emphasis on the linear case known as MacMahon's master theorem. These results are then used to extract explicit solutions from functional equations that arise when working with generating functions. Gaussian polynomials are treated in the last section.

1 INVERSION IN THE VECTOR SPACE OF POLYNOMIALS

3.1

The set V of all polynomials, with real numbers as coefficients, is a vector space with respect to usual addition and scalar multiplication. A (most) natural basis of V is the sequence $1, x, x^2, \ldots, x^n, \ldots$ of powers of x. [Indeed, any finite subset of powers of x is linearly independent, and any polynomial in V is (by definition) a finite linear combination of powers of x.] Can we eventually describe an arbitrary basis of V? It really is not hard to grasp, and quite easy to check, that

* *Any basis for the vector space of polynomials can* (upon suitably permuting its elements) *be written as* $p_0, p_1, p_2, \ldots, p_n, \ldots,$ *where* p_k *is a polynomial of degree precisely* k, *and* p_0 *is not the zero polynomial.*

We denote a basis ordered in this fashion by a column vector (p_n) of indefinite length.

Many of the counting numbers introduced in the first two chapters occur naturally when expressing the vectors of a basis in terms of another basis. The Stirling numbers S_n^k, for example, occur when we express x^n as a linear combination of the $[x]_k$'s; and the s_n^k's occur when applying the inverse transformation, that is, expressing $[x]_n$ in terms of the x^k's. To graphically summarize, we have

$$([x]_n) \underset{(s_n^k)}{\overset{(S_n^k)}{\rightleftarrows}} (x^n).$$

The lower triangular matrices (S_n^k) and (s_n^k) are therefore a pair of inverse matrices. In general we have the following result:

The Inversion Theorem. *If for two bases* (p_n) *and* (q_n) *we can write*

$$(q_n) = A(p_n), \qquad \text{and} \qquad (p_n) = B(q_n),$$

then matrices A *and* B *are inverses of each other, both being countably infinite lower triangular.*

Proof. Our assumptions imply $(BA)(p_n) = B(A(p_n)) = B(q_n) = (p_n)$, and $(AB)(q_n) = A(B(q_n)) = A(p_n) = (q_n)$, showing that $AB = I = BA$, with I the (countably infinite) identity matrix. This demonstrates that A and B are inverses of each other. And since a polynomial of degree precisely n can only be written as a linear combination of polynomials of degree at most n, A and B are necessarily lower triangular. This concludes our proof.

The relations obtained by equating $AB = I$ and $BA = I$ are called the *orthogonal relations.*

3.2

Let $A = (a_{ij})$ and $B = (b_{ij})$ be inverses of each other; each a lower triangular, countably infinite matrix. We wish to observe that two vectors of scalars, or of polynomials of degree zero, $a^t = (a_0, a_1, a_2, \ldots)$ and $b^t = (b_0, b_1, b_2, \ldots)$ verify

$$a = Ab \qquad \text{if and only if} \qquad b = Ba.$$

[By c^t we understand the transpose of the infinite vector (or sequence) c.] This relation follows immediately from knowing that A and B are a pair of inverse matrices. By equating the nth entries on both sides we may write the above relation in the form

$$a_n = \sum_{k=0}^{n} a_{nk} b_k \qquad \text{iff} \qquad b_n = \sum_{k=0}^{n} b_{nk} a_k. \tag{3.1}$$

We refer to the contents of (3.1) as a pair of *inverse relations*. Here and throughout the rest of this section "iff" abbreviates the expression "if and only if."

The inversion theorem, through the process of expressing the elements of a basis in terms of another, yields a pair of inverse matrices. These matrices can in turn be used to produce a pair of inverse relations, such as (3.1). Interesting identities can be proven in this way, often as inverses of known ones. Section 3.3 illustrates this method.

3.3

We now list an assortment of well-known inverse relations, provide proofs, and show how on occasion they can be of use.

Applications of the Inversion Theorem

(i) $a_n = \sum_{k=0}^{n} \binom{n}{k} b_k$ iff $b_n = \sum_{k=0}^{n} \binom{n}{k} (-1)^{n-k} a_k$ (binomial).

(ii) $a_n = \sum_{k=1}^{n} s_n^k b_k$ iff $b_n = \sum_{k=1}^{n} S_n^k a_k$ (Stirling).

(iii) $a_n = \sum_{k=1}^{n} L_n^k b_k$ iff $b_n = \sum_{k=1}^{n} L_n^k a_k$ (Lah).

(iv) $a_n = \sum_{k=0}^{m} \binom{n}{k} b_{n-2k}$ iff $b_n = \sum_{k=0}^{m} (-1)^k \frac{n}{n-k} \binom{n-k}{k} a_{n-2k}$
where $m = [n/2]$ (integral part) (Chebyshev).

(v) $a_n = \sum_{k=0}^{n} \binom{n+k}{2k} b_k$ iff

$$b_n = \sum_{k=0}^{n} (-1)^{n+k} \left[\binom{2n}{n-k} - \binom{2n}{n-k-1} \right] a_k \qquad \text{(Legendre).}$$

(vi) $a_n = \sum_{k=0}^{n} \binom{n}{k} x(x+n-k)^{n-k-1} b_k$ iff

$b_n = \sum_{k=0}^{n} (-1)^{n+k} \binom{n}{k} x(x-n+k)^{n-k-1} a_k$; x an indeterminate

(Abel).

(vii) $a_n = \sum_{k=0}^{n} \binom{2k}{k} b_{n-k}$ iff $b_n = a_n + \sum_{k=1}^{n} \left(-\frac{2}{k}\right) \binom{2k-2}{k-1} a_{n-k}$

(Catalan).

Let us proceed to prove these seven inverse relations.

(i) *The inverse binomial formulas.*

Take $q_n(x) = x^n$ and $p_n(x) = (x-1)^n$.

Then $q_n = x^n = ((x-1)+1)^n = \sum_{k=0}^{n} \binom{n}{k} (x-1)^k$, and $p_n = (x-1)^n = \sum_{k=0}^{n} \binom{n}{k} (-1)^{n-k} x^k$. Or, in matrix notation,

$$(q_n) = \begin{bmatrix} 1 \\ x \\ x^2 \\ \vdots \\ x^n \\ \cdot \end{bmatrix} = \begin{bmatrix} 1 & & & & 0 \\ \binom{1}{0} & \binom{1}{1} & & & \\ \binom{2}{0} & \binom{2}{1} & \binom{2}{2} & & \\ \vdots & \vdots & \vdots & & \\ \binom{n}{0} & \binom{n}{1} & \binom{n}{2} & \cdots & \binom{n}{n} \\ \cdot & \cdot & \cdot & & \cdot \end{bmatrix} \begin{bmatrix} 1 \\ x-1 \\ (x-1)^2 \\ \vdots \\ (x-1)^n \\ \cdot \end{bmatrix} = A(p_n)$$

$$(p_n) = \begin{bmatrix} 1 \\ x-1 \\ (x-1)^2 \\ \vdots \\ (x-1)^n \\ \cdot \end{bmatrix}$$

$$= \begin{bmatrix} 1 & & & & 0 \\ -\binom{1}{0} & \binom{1}{1} & & & \\ \binom{2}{0} & -\binom{2}{1} & \binom{2}{2} & & \\ \vdots & \vdots & & & \\ (-1)^n \binom{n}{0} & (-1)^{n-1} \binom{n}{1} & & \cdots & (-1)^0 \binom{n}{n} \\ \cdot & \cdot & & & \cdot \end{bmatrix} \begin{bmatrix} 1 \\ x \\ x^2 \\ \vdots \\ x^n \\ \cdot \end{bmatrix} = B(q_n).$$

By the inversion theorem B is the inverse of A. Being inverses of one another, when applied to sequences of scalars we obtain

$$a_n = \sum_{k=0}^{n} \binom{n}{k} b_k \quad \text{iff} \quad b_n = \sum_{k=0}^{n} \binom{n}{k} (-1)^{n-k} a_k.$$

Here (a_n) and (b_n) are sequences, or vectors, of scalars.

For example, let $a_n = n^p$ (with p a fixed positive integer) and $b_n = n!S_p^n$. Then we have proved, by counting two ways the number of functions from a set with p elements to another with n elements—see (1.1) for details—that

$$a_n = n^p = \sum_{k=0}^{n} \binom{n}{k} (k!S_p^k) = \sum_{k=0}^{n} \binom{n}{k} b_k.$$

By virtue of the above inversion we then also have

$$b_n = n!S_p^n = \sum_{k=0}^{n} \binom{n}{k} (-1)^{n-k} k^p = \sum_{k=0}^{n} \binom{n}{k} (-1)^{n-k} a_n,$$

or

$$S_p^n = \frac{1}{n!} \sum_{k=0}^{n} \binom{n}{k} (-1)^{n-k} k^p.$$

This last identity is known as *Stirling's formula*.

In a similar manner we can invert $2^n = \sum_{k=0}^{n} \binom{n}{k}$ to obtain $1 = \sum_{k=0}^{n} \binom{n}{k} (-1)^{n-k} 2^k$. (Here we took $a_n = 2^n$ and $b_n = 1$.)

(ii) *Stirling's inverse formulas.* We have shown in Section 1.7(c) and 1.8(c) that

$$[x]_n = \sum_{k=1}^{n} s_n^k x^k \quad \text{and} \quad x^n = \sum_{k=1}^{n} S_n^k [x]_k.$$

By the inversion theorem it now follows that the lower triangular matrices $(s_n^k)_{1 \le k \le n}$ and $(S_n^k)_{1 \le k \le n}$ of Stiring numbers of the first and second kind, respectively, are inverses of each other.

Example. Let us show how the binomial and Stirling inverse relations lead to new identities. In Section 1.7(d) we showed that

$$S_{n+1}^m = \sum_{k=0}^{n} \binom{n}{k} S_k^{m-1}.$$

In matrix notation this is rewritten as $\left(\binom{i}{j}\right)(S_j^i) = (S_{j+1}^{i+1})$. Taking inverses of both sides we obtain

$$\left(S_j^i\right)^{-1} \left(\binom{i}{j}\right)^{-1} = \left(S_{j+1}^{i+1}\right)^{-1}.$$

As we just found out, $\left(\binom{i}{j}\right)^{-1} = \left((-1)^{i+j}\binom{i}{j}\right)$ and $(S_j^i)^{-1} = (s_j^i)$, and hence we can write this last relation as follows:

$$(s_j^i)\left((-1)^{i+j}\binom{i}{j}\right) = (s_{j+1}^{i+1}).$$

We have thus established the "new" identity:

$$|s_n^m|\binom{m}{m} + |s_n^{m+1}|\binom{m+1}{m} + \cdots + |s_n^n|\binom{n}{m} = |s_{n+1}^{m+1}|.$$

(iii) *Lah inverse formulas.* The Lah numbers L_n^k are defined by the identity: $[-x]_n = \Sigma_{k=1}^n L_n^k [x]_k$. Changing x to $-x$ gives $[x]_n = \Sigma_{k=1}^n L_n^k [-x]_k$. By the inversion theorem the lower triangular matrix $(L_n^k)_{1 \le k \le n}$ is its own inverse. Hence $a_n = \Sigma_{k=1}^n L_n^k b_k$ iff $b_n = \Sigma_{k=1}^n L_n^k a_k$. [Recall that $L_n^k = (-1)^n (n!/k!)\binom{n-1}{k-1}$, and see Sections 1.15 and 2.9 for the combinatorial meaning and properties of the Lah numbers.]

We proceed in proving the remaining inverse relations. These are associated with classical orthogonal polynomials that are special instances of basic hypergeometric functions.

(iv) *The Chebyshev inverse relations.* Let $b_n(x) = \Sigma_{k=0}^m (-1)^k (n/(n-k)) \binom{n-k}{k} x^{n-2k}$, $m = [n/2]$. We show that $x^n = \Sigma_{k=0}^m \binom{n}{k} b_{n-2k}(x)$ and thus establish Chebyshev's inverse formula. [Note that $b_0(x) = 1$, $b_1(x) = x$, and $b_2(x) = x^2 - 2$.] Since $\binom{n-k}{k} + \binom{n-k-1}{k-1} = (n/(n-k))\binom{n-k}{k}$, we have

$$b_n(x) = \sum_{k=0}^{m} (-1)^k \left[\binom{n-k}{k} + \binom{n-k-1}{k-1}\right] x^{n-2k}$$

$$= \sum_{k=0}^{m} (-1)^k \left[\binom{n-k-1}{k} + \binom{n-k-2}{k-1} + \binom{n-k-1}{k-1} + \binom{n-k-2}{k-2}\right] x^{n-2k}$$

$$= x b_{n-1}(x) - b_{n-2}(x) \qquad \text{for } n \ge 3$$

[the second equality uses the basic relation $\binom{n}{m} = \binom{n-1}{m} + \binom{n-1}{m-1}$]. Verify carefully the last equality, first for n odd, then for n even. We can therefore write:

$$x b_0(x) = b_1(x), \qquad x b_1(x) = b_2(x) + 2 b_0(x) \qquad (b_{-1}(x) \equiv 0)$$

$$x b_n(x) = b_{n+1}(x) + b_{n-1}(x), \qquad n \ge 2.$$

Let $x^n = \sum_{k=0}^m a_{nk} b_{n-2k}(x)$. Then

$$x^{2m+1} = \sum_{k=0}^{m} a_{2m+1,k} b_{2m+1-2k}(x)$$

$$= x \cdot x^{2m} = \sum_{k=0}^{m} a_{2m,k} x b_{2m-2k}(x)$$

$$= \sum_{k=0}^{m} a_{2m,k} [b_{2m+1-2k}(x) + b_{2m-1-2k}(x)]$$

$$= \sum_{k=0}^{m} (a_{2m,k} + a_{2m,k-1}) b_{2m+1-2k}(x).$$

Identifying coefficients of $b_{2m+1-2k}(x)$ gives

$$a_{2m+1,k} = a_{2m,k} + a_{2m,k-1}, \qquad 0 \leq k \leq m. \tag{3.2}$$

In the same way

$$x^{2m} = \sum_{k=0}^{m} a_{2m,k} b_{2m-2k}(x) = x \cdot x^{2m-1} = \sum_{k=0}^{m-1} a_{2m-1,k} x b_{2m-1-2k}(x)$$

$$= \sum_{k=0}^{m-1} (a_{2m-1,k} + a_{2m-1,k-1}) b_{2m-2k}(x) + 2a_{2m-1,m-1} b_0(x)$$

so that

$$a_{2m,k} = a_{2m-1,k} + a_{2m-1,k-1}, \qquad 0 \leq k \leq m-1 \tag{3.3}$$

and

$$a_{2m,m} = 2a_{2m-1,m-1}.$$

The initial conditions $b_0(x) = 1$, $b_1(x) = x$ and recurrences (3.2) and (3.3) determine uniquely $a_{nk} = \binom{n}{k}$. This proves Chebyshev's inverse formula.

As an example, suppose we want to verify the identity

$$2 = \sum_{k=0}^{m} (-1)^k \frac{n}{n-k} \binom{n-k}{k} 2^{n-2k}, \qquad n = 1, 2, \ldots .$$

This looks a bit unfriendly and, as it involves Chebyshev inversion, perhaps its inverse is easier to prove.

Let $b_0 = 1$, $b_n = 2$ for $n \geq 1$ and $a_n = 2^n$. Then the inverse of the above identity is

$$2^n = a_n = \sum_{k=0}^{m} \binom{n}{k} b_{n-2k}.$$

For n odd it becomes

$$2^n = \sum_{k=0}^{m} \binom{n}{k} \cdot 2 = 2 \sum_{k=0}^{m} \binom{n}{k} = \sum_{k=0}^{n} \binom{n}{k} = (1+1)^n,$$

while for even n it leads to (recall that $b_0 = 1$)

$$2^n = \sum_{k=0}^{m-1} \binom{n}{k} \cdot 2 + \binom{n}{m} = \sum_{k=0}^{n} \binom{n}{k} = (1+1)^n.$$

We checked that the inverse identity is true and hence the original identity is also true.

(v) We derive Legendre's formulas from Chebyshev's by substitutions.

$$a_n = \sum^{m} \binom{n}{k} b_{n-2k}$$

iff $$b_n = \sum^{m} (-1)^k \frac{n}{n-k} \binom{n-k}{k} a_{n-2k}, \qquad m = \left[\frac{n}{2}\right]$$

$$\updownarrow \quad A_n = \frac{a_n}{n}, \quad B_n = \frac{b_n}{n} \qquad n = 1, 2, \ldots$$

$$nA_n = \sum^{m} \binom{n}{k} (n-2k) B_{n-2k}$$

iff $$nB_n = \sum^{m} (-1)^k \frac{n}{n-k} \binom{n-k}{k} (n-2k) A_{n-2k}$$

or

$$A_n = \sum^{m} \binom{n}{k} \frac{n-2k}{n} B_{n-2k}$$

iff $$B_n = \sum^{m} (-1)^k \binom{n-k-1}{k} A_{n-2k}$$

$$\left(\text{because } \frac{n-2k}{n-k} \binom{n-k}{k} = \binom{n-k-1}{k}\right)$$

$$\updownarrow \quad a_n = A_{n+1}, \quad b_n = B_{n+1} \qquad n = 0, 1, 2, \ldots$$

$$a_n = \sum^{m} \left[\binom{n}{k} - \binom{n}{k-1}\right] b_{n-2k}$$

iff $$b_n = \sum^{m} (-1)^k \binom{n-k}{k} a_{n-2k}$$

$$\left(\text{since } \binom{n+1}{k} \frac{n+1-2k}{n+1} = \binom{n}{k} - \binom{n}{k-1}\right)$$

$\updownarrow n \leftrightarrow 2n$

$$A_{2n} = \sum^{n}\left[\binom{2n}{k} - \binom{2n}{k-1}\right]B_{2n-2k}$$

iff $$B_{2n} = \sum^{n}(-1)^k\binom{2n-k}{k}A_{2n-2k}$$

$$\updownarrow \quad a_n = (-1)^n A_{2n}, \quad b_n = (-1)^n B_{2n}$$

$$(-1)^n a_n = \sum^{n}\left[\binom{2n}{k} - \binom{2n}{k-1}\right](-1)^{n-k}b_{n-k}$$

iff $$(-1)^n b_n = \sum^{n}(-1)^k\binom{2n-k}{k}(-1)^{n-k}a_{n-k}$$

$$\updownarrow k \leftrightarrow n-k; \qquad a_n \leftrightarrow b_n$$

$$b_n = \sum^{n}\left[\binom{2n}{n-k} - \binom{2n}{n-k-1}\right](-1)^{n+k}a_k$$

iff $$a_n = \sum^{n}\binom{n+k}{2k}b_k \qquad \text{(Legendre).}$$

In all the sums above the index of summation is k and it begins from 0. This proof illustrates how (more or less) new inverse relations can be obtained by making substitutions into known ones. An alternative proof [as in (iv)] can be given by inverting the Legendre polynomials

$$P_n(x) = 2^{-n}\sum_{k=0}^{m}(-1)^k\binom{n-k}{k}\binom{2n-2k}{n-k}x^{n-2k}, \qquad m = \left[\frac{n}{2}\right].$$

(vi) *Abel's identity.* We now prove a famous identity due to Abel:

$$x^{-1}(x+y+n)^n = \sum_{k=0}^{n}\binom{n}{k}(x+k)^{k-1}(y+n-k)^{n-k}.$$

To see this, let

$$A_n(x, y; p, q) \underset{\text{def.}}{=} \sum_{k=0}^{n}\binom{n}{k}(x+k)^{k+p}(y+n-k)^{n-k+q}.$$

Then the substitution $k \leftrightarrow n-k$ gives

$$A_n(x, y; p, q) = A_n(y, x; q, p). \tag{3.4}$$

Recalling that $\binom{n}{k} = \binom{n-1}{k} + \binom{n-1}{k-1}$ leads us to

$$A_n(x, y; p, q) = A_{n-1}(x, y+1; p, q+1) + A_{n-1}(x+1, y; p+1, q). \tag{3.5}$$

By writing $(y+n-k)^{n-k+q}$ as $(y+n-k)(y+n-k)^{n-k+q-1}$ we obtain

$$A_n(x, y; p, q) = yA_n(x, y; p, q-1) + nA_{n-1}(x, y+1; p, q). \tag{3.6}$$

[Statements (3.4), (3.5), and (3.6) are easy to check.] Now (3.5) and (3.6) lead to

$$\begin{aligned} A_n(x, y; p, q) &= y\{A_{n-1}(x, y+1; p, q) + A_{n-1}(x+1, y; p+1, q-1)\} \\ &\quad + nA_{n-1}(x, y+1; p, q) \\ &= yA_{n-1}(x+1, y; p+1, q-1) \\ &\quad + (y+n)A_{n-1}(x, y+1; p, q). \end{aligned} \tag{3.7}$$

Set $p = -1$ and $q = 0$, and denote as follows:

$$A_n(x, y; -1, 0) \underset{\text{def.}}{=} B_n(x, y) \underset{\text{def.}}{=} A_n(y, x; 0, -1).$$

Relation (3.7) above can be rewritten as

$$B_n(x, y) = yB_{n-1}(y, x+1) + (y+n)B_{n-1}(x, y+1). \tag{3.8}$$

Note that $B_0(x, y) = x^{-1}$, as required. Assume $B_{n-1}(x, y) = x^{-1}(x+y+n-1)^{n-1}$. Then (3.8) gives

$$\begin{aligned} B_n(x, y) &= yy^{-1}(x+y+n)^{n-1} + (y+n)x^{-1}(x+y+n)^{n-1} \\ &= xx^{-1}(x+y+n)^{n-1} + (y+n)x^{-1}(x+y+n)^{n-1} \\ &= x^{-1}(x+y+n)^n, \qquad \text{as desired.} \end{aligned}$$

This proof illustrates well how induction is a very handy device for checking various formulas. It unfortunately gives, comparatively speaking, very little insight into coming up with new identities. The most powerful tool, perhaps, for creating new identities is the inversion theorem of Lagrange, which we study in Section 3 of this chapter. Abel's identity will be seen to be a consequence of that method.

To prove Abel's inversion formulas, let $q_n(y) = (x+y+n)^n$, and $p_n(y) = (y+n)^n$. Then

$$(x+y+n)^n = q_n(y) = \sum_k \binom{n}{k} x(x+k)^{k-1} p_{n-k}(y)$$

(by Abel's identity)

and

$$\begin{aligned} p_n(y) &= (y+n)^n = (-x+x+y+n)^n \\ &= \sum_k \binom{n}{k}(-x)(-x+k)^{k-1}(x+y+n-k)^{n-k} \\ &\qquad\qquad \text{(by Abel's identity)} \\ &= \sum_k \binom{n}{k}(-1)^k x(x-k)^{k-1} q_{n-k}(y). \end{aligned}$$

We just accomplished a change of basis [between polynomials $q_n(y)$ and $p_n(y)$], thus establishing a pair of *inverse matrices*. When operating on sequences these matrices give

$$a_n = \sum_k \binom{n}{k} x(x+k)^{k-1} b_{n-k} \quad \text{iff} \quad b_n = \sum_k \binom{n}{k}(-1)^k x(x-k)^{k-1} a_{n-k}.$$

Switching $k \leftrightarrow n-k$ leads to (vi).

(vii) Observe that, in general, the generating function relations $A(x) = B(x)C(x)$ iff $B(x) = A(x)C^*(x)$ [with $C(x)C^*(x) = 1$] imply the inverse relations

$$a_n = \sum_{k=0}^{n} c_{n-k} b_k \quad \text{iff} \quad b_n = \sum_{k=0}^{n} c^*_{n-k} a_k$$

where $A(x) = \Sigma_n a_n x^n$, $B(x) = \Sigma_n b_n x^n$, and $C(x) = \Sigma_n c_n x^n$. If, e.g., $C(x) = \Sigma_n x^n = (1-x)^{-1}$, then $A(x) = B(x)(1-x)^{-1}$ iff $B(x) = A(x)(1-x)$ translates into $a_n = b_0 + \cdots + b_n$ iff $b_n = a_n - a_{n-1}$.

Consider the series $C(x) = \Sigma_n \binom{2n}{n} x^n$. Examining with some care the coefficient of x^n we observe that

$$\binom{2n}{n} = \frac{1 \cdot 3 \cdot 5 \cdot 7 \cdot \cdots \cdot (2n-1)}{n! 2^n} \cdot 4^n,$$

and recognize at right the constant term in the nth derivative of $(1-4x)^{-1/2}$ (divided by $n!$). This observation allows us to conclude that $C(x) = \Sigma_n \binom{2n}{n} x^n = (1-4x)^{-1/2}$. The inverse $C^*(x)$ of $C(x)$ is therefore $(1-4x)^{1/2}$. We may write

$$C^*(x) = (1-4x)^{1/2} = (1-4x)C(x) = \sum_n \left[\binom{2n}{n} - 4\binom{2n-2}{n-1} \right] x^n.$$

(Binomial coefficients with negative entries are understood to be 0.) Note, in addition, that $\binom{2n}{n} - 4\binom{2n-2}{n-1} = -(2/n)\binom{2n-2}{n-1}$, for $n \geq 1$ (a Catalan number—see Exercise 1 at the end of Section 2.13). Thus $C^*(x) = 1 + \Sigma_{n \geq 1}(-2/n)\binom{2n-2}{n-1} x^n$, and $A(x) = C(x)B(x)$ iff $B(x) = C^*(x)A(x)$ translates into the Catalan inverse relation (vii). This ends our proof.

3.4

Let us emphasize again that many inversion formulas of this nature can be obtained by inverting special functions on certain lattices using Möbius inversion. Example (4) in Section 1 of Chapter 9 is a typical example.

We can modify an inversion formula by making substitutions; example (v) was such an illustration. An inversion formula originates with (and generates) a pair of inverse matrices, A and B, say. But then A^r and B^r ($r \geq 2$) are also inverse matrices (leading to new pairs of inverse relations). So are A^t and B^t, the transposes of A and B. So is, in fact, any pair of leading principal minors of the same size.

With observations such as these, starting with the pair

$$a_n = \sum_{k=0}^{n} \binom{n}{k} b_k \quad \text{iff} \quad b_n = \sum_{k=0}^{n} \binom{n}{k} (-1)^{n-k} a_k,$$

the reader will be able to derive, for example,

$$a_n = \sum_{k=0}^{n} \binom{p-k}{p-n} b_k \quad \text{iff} \quad b_n = \sum_{k=0}^{n} \binom{p-k}{p-n} (-1)^{n-k} a_k,$$

an inverse relation differing significantly from the initial one. If interested, the reader may wish to supply the details; this is not a difficult exercise.

3.5

We end this section with a potpourri of inverse relations. On occasion they may prove helpful. For more on this subject the reader is referred to Riordan's book [2].

An Assortment of Inverse Formulas

1. $$a_n = \sum_{k=0}^{n} (-1)^k \binom{p+qk+k}{n-k} b_k \quad \text{iff}$$
$$b_n = \sum_{k=0}^{n} (-1)^k \frac{p+qk-k}{p+qn-k} \binom{p+qn-k}{n-k} a_k, \qquad p, q \in N.$$

2. $$a_n = \sum_{k=0}^{m} \binom{n+p}{k} b_{n-2k} \quad \text{iff}$$
$$b_n = \sum_{k=0}^{m} (-1)^k \frac{n+p}{n+p-k} \binom{n+p-k}{k} a_{n-2k}; \qquad m = \left[\frac{n}{2}\right].$$

3. $$a_n = \sum_{k \geq 0} \left[\binom{n}{k} + (c+1)\binom{n}{k-1}\right] b_{n+ck} \quad \text{iff}$$
$$b_n = \sum_{k \geq 0} (-1)^k \binom{n+ck+k}{k} a_{n+ck}, \qquad 0 \neq c \text{ integer}.$$

4. $$a_n = \sum_{k=0}^{n} \binom{cn+p}{n-k} b_k \quad \text{iff}$$
$$b_n = \sum_{k=0}^{n} \left[\binom{n+p-1+ck-k}{n-k} + c\binom{n+p-1+ck-k}{n-k-1}\right] (-1)^{n-k} a_k.$$

5. $$a_n = \sum_{k \geq 0} \frac{n!}{k!(n-ck)!} b_{n-ck} \quad \text{iff}$$
$$b_n = \sum_{k \geq 0} (-1)^k \frac{n!}{k!(n-ck)!} a_{n-ck} \qquad \text{(Hermite)}.$$

6. $$a_n = \sum_{k=0}^{n} \binom{n}{k} (x+2k)(x+n+k)^{n+k-1} b_k \quad \text{iff}$$
$$b_n = \sum_{k=0}^{n} (-1)^{n-k} \binom{n}{k} (x+2k)(x+n+k)^{n+k-1} a_k.$$

7. $$a_n = \sum_{k=0}^{n} \binom{2p+2k}{p+k}\binom{p+k}{k}\binom{2p}{p}^{-1} b_{n-k} \quad \text{iff}$$
$$b_n = \sum_{k=0}^{n} (-1)^k \binom{2p+1}{2k}\binom{p+k}{k}\binom{p+k}{2k}^{-1} a_{n-k}.$$

8. $$a_n = \sum_{k=0}^{n} \binom{n}{k}(k+1)^{-1} b_k \quad \text{iff} \quad b_n = \sum_{k=0}^{n} \binom{n}{k} B_k a_{n-k}.$$

9. $$a_n = \sum_{k \geq 0} \binom{n}{2k} b_{n-2k} \quad \text{iff} \quad b_n = \sum_{k \geq 0} \binom{n}{2k} E_{2k} a_{n-2k}.$$

10. $$a_{m,n} = \sum_{j,k=0} \binom{m}{j}\binom{n}{k} b_{m-2j,n-2k} \quad \text{iff}$$
$$b_{m,n} = \sum_{j,k=0} (-1)^{j+k} \frac{m}{m-j}\binom{m-j}{j}\frac{n}{n-k}\binom{n-k}{k} a_{m-2j,n-2k}.$$

The B_n's are defined by $x(e^x - 1)^{-1} = \Sigma_n (B_n/n!)x^n$ ($B_0 = 0$, $B_1 = -\frac{1}{2}$, $B_{2n+1} = 0$) and are called Bernoulli numbers. The E_n's are defined by $2(e^x + e^{-x})^{-1} = \Sigma_n (E_n/n!)x^n$ and are called Eulerian numbers.

2 TAYLOR EXPANSIONS

The Taylor expansions taught in calculus are herein extended to normal families of polynomials. This allows, in particular, for a unified treatment of identities by Vandermonde and Nörlund.

3.6

We begin with a couple of definitions. The sequence (p_n) is called a *normal family* of polynomials if $p_n(x)$ is a polynomial of degree precisely n, $p_n(0) = 0$ for all $n > 1$, and $p_0(x) = 1$.

A linear operator D (defined on the vector space of polynomials) is called a *differential operator* for the normal family (p_n) if $Dp_n(x) = np_{n-1}(x)$ for $n \geq 1$, and $Dp_0(x) = 0$.

The main result of this section is the following:

Taylor's Formula. *For each normal family of polynomials (p_n) there exists a unique differential operator D. Moreover, a polynomial $\phi(x)$ of degree n can be expressed uniquely as*

$$\phi(x) = \phi(0)p_0(x) + \frac{D\phi(0)}{1!}p_1(x) + \frac{D^2\phi(0)}{2!}p_2(x)$$
$$+ \cdots + \frac{D^n\phi(0)}{n!}p_n(x).$$

[Here $D^2\phi = D(D\phi)$, and generally, $D^k\phi = D(D^{k-1}\phi)$. Naturally, $D^k\phi(0)$ stands for the constant term of the polynomial $D^k\phi$, or (if one prefers) it equals $D^k\phi$ evaluated at 0.]

Proof. We first show that $\phi(x)$ [which we also write as $\phi_n(x)$ to emphasize its degree] can be expressed uniquely in the form

$$\phi_n(x) = \alpha_n p_n(x) + \alpha_{n-1} p_{n-1}(x) + \cdots + \alpha_0 p_0(x)$$

where $\alpha_n, \alpha_{n-1}, \ldots, \alpha_0$ are constants, that is, we show that (p_n) forms a basis.

Select α_n such that $\phi_n(x) - \alpha_n p_n(x)$ is a polynomial of degree at most $n - 1$. This gives, *uniquely*,

$$\alpha_n = \frac{\text{coeff. of } x^n \text{ in } \phi_n(x)}{\text{coeff. of } x^n \text{ in } p_n(x)}.$$

Set, by definition, $\phi_{n-1} = \phi_n(x) - \alpha_n p_n(x)$ and determine α_{n-1} such that $\phi_{n-1}(x) - \alpha_{n-1} p_{n-1}(x)$ has degree $n - 2$ or less. This determines α_{n-1} *uniquely* as

$$\alpha_{n-1} = \frac{\text{coeff. of } x^{n-1} \text{ in } \phi_{n-1}(x)}{\text{coeff. of } x^{n-1} \text{ in } p_{n-1}(x)}, \text{ etc.}$$

Now define the operator D by

$$D\phi_n(x) = n\alpha_n p_{n-1}(x) + (n-1)\alpha_{n-1} p_{n-2}(x) + \cdots + 2\alpha_2 p_1(x) + \alpha_1 p_0(x) + 0.$$

The operator D so defined is indeed a differential operator, and by the unique expression of $\phi_n(x)$ in terms of $\{p_k(x)\}$ any other differential operator must behave as D does.

To show that $\alpha_k = D^k\phi(0)/k!$ apply D k times to $\phi(x) = \alpha_n p_n(x) + \alpha_{n-1} p_{n-1}(x) + \cdots + \alpha_0 p_0(x)$ and set $x = 0$ [here we use the fact that $p_n(0) = 0$, $n \geq 1$].

3.7 Applications of Taylor's Formula

We prove the following:

(a) $(x + y)^n = \sum_{k=0}^{n} \binom{n}{k} x^k y^{n-k}$ (*binomial*).

(b) $[x + y]_n = \sum_{k=0}^{n} \binom{n}{k} [x]_k [y]_{n-k}$ (*Vandermonde*).

(c) $[x + y]^n = \sum_{k=0}^{n} \binom{n}{k} [x]^k [y]^{n-k}$ (*Nörlund*).

[Here $[z]^n \underset{\text{def.}}{=} z(z + 1) \cdots (z + n - 1)$.]

Proof. (a) Let $p_n(x) = x^n$ be our normal family; $n \geq 0$. The differential operator D is

$$D\big(\alpha_n x^n + \alpha_{n-1}x^{n-1} + \cdots + \alpha_1 x + \alpha_0\big)$$
$$= n\alpha_n x^{n-1} + (n-1)\alpha_{n-1}x^{n-2} + \cdots + \alpha_1 + 0$$
$$(\textit{ordinary}\text{ differentiation}).$$

Consider in particular the polynomial $\phi(x) = (x+y)^n$, y constant. Then (using the chain rule for ordinary differentiation) Taylor's formula becomes

$$(x+y)^n = y^n + \binom{n}{1}xy^{n-1} + \binom{n}{2}x^2y^{n-2} + \cdots + \binom{n}{n}x^n$$
$$= \sum_{k=0}^{n}\binom{n}{k}x^k y^{n-k} \text{ (the binomial formula)}.$$

(b) Let $p_n(x) = [x]_n = x(x-1)\cdots(x-n+1)$, $n \geq 0$. The differential operator D satisfies $D[x]_n = n[x]_{n-1} = (x+1)[x]_{n-1} - (x-n+1)[x]_{n-1} = [x+1]_n - [x]_n$, and by Taylor's formula for any polynomial $\phi(x)$, $D\phi(x) = \phi(x+1) - \phi(x)$. Define L_y by $L_y(\phi(x)) = \phi(x+y)$, for y a constant. Then for D as above, and any polynomial $\phi(x)$, $DL_y\phi(x) = D\phi(x+y) = \phi(x+y+1) - \phi(x+y) = L_y[\phi(x+1) - \phi(x)] = L_y D\phi(x)$, which shows that L_y and D commute. In particular, $D[x+y]_n = DL_y[x]_n = L_y D[x]_n = L_y n[x]_{n-1} = n[x+y]_{n-1}$, $D^2[x+y]_n = n(n-1)[x+y]_{n-2}$, and so on. Taylor's formula with $\phi(x) = [x+y]_n$ now yields

$$[x+y]_n = \sum_{k=1}^{n}\binom{n}{k}[x]_k[y]_{n-k}.$$

(c) Observe that $[z]^n = (-1)^n[-z]_n$. This ends our proof.

EXERCISES

1. (Multisection of series.) Let $\omega = e^{2\pi i/m}$ (a primitive mth root of unity).

(a) Show that

$$\sum_{k=0}^{m-1}\omega^{nk} = \begin{cases} m & \text{if } m \text{ divides } n \\ 0 & \text{otherwise} \end{cases}.$$

(b) Let $F(x) = \Sigma_{j\geq 0}a_j x^j$. Prove that

$$\sum_{\substack{k \\ k\equiv l (\text{mod } m)}} a_k x^k = \frac{1}{m}\sum_{j=0}^{m-1}F(x\omega^j)\omega^{-lj}.$$

2. With the help of the results in Exercise 1 show that

(a) $\displaystyle\sum_{k=0}^{n}\binom{2n+1}{2k+1}x^{2k} = \frac{1}{2x}\{(1+x)^{2n+1} - (1-x)^{2n+1}\}.$

(b) $\sum_{k=0}^{n} \binom{3n}{3k} = \frac{1}{3}(2^n + 2(-1)^n)$.

(c) $\sum_{k \equiv l \pmod m} \binom{n}{k} = \frac{2^n}{m} \sum_{j=0}^{m-1} \cos^n\left(\frac{\pi j}{m}\right) \cos \frac{(n-2l)\pi j}{m}$.

3. Derive the inverse pairs

(a) $a_n = \sum_{k=0}^{n} \frac{1}{k!} b_{n-k}$ iff $b_n = \sum_{k=0}^{n} \frac{(-1)^k}{k!} a_{n-k}$.

(b) $a_n = \sum_{k=0}^{m} \frac{1}{k!} b_{n-ck}$ iff $b_n = \sum_{k=0}^{m} \frac{(-1)^k}{k!} a_{n-ck}$.

c is a positive integer and $m = [n/c]$.

4. Prove the identities:

(a) $2^n = \sum_{k=0}^{n} (-1)^{k+n} \binom{n}{k} k^{n-k} 2(k+2)^{k-1}$.

(b) $2^n = \sum_{k=0}^{m} \binom{n}{k} \frac{(n+1-2k)^2}{n+1-k}$, where $m = \left[\frac{n}{2}\right]$.

(c) $n^n = \sum_{k=0}^{n-1} \binom{n-1}{k} n^{n-k-1}(k+1)!$.

5. Consider the sum $a_{n,m} = \sum_{k=0}^{n}(-1)^k\binom{n}{k} m/(m+k)$. Show that $a_{n,m} = a_{n-1,m} - (m/(m+1))a_{n-1,m+1}$, and thus conclude that $a_{n,m} = \binom{m+n}{n}^{-1}$. Invert to obtain

$$\frac{m}{m+n} = \sum_{k=0}^{n} (-1)^k \binom{n}{k} \binom{m+k}{k}^{-1}.$$

6. Show that

$$2^{n-m}\binom{n}{m} = \sum_{k=m}^{n} \binom{n}{k}\binom{k}{m} \quad \text{and} \quad \binom{n}{m} = \sum_{k=m}^{n} (-1)^{n+k} \binom{n}{k}\binom{k}{m} 2^{k-m}.$$

Show also that

$$\binom{n+p}{m} = \sum_{j=0} \binom{p}{j}\binom{m-j}{n}.$$

7. By selecting the normal family of polynomials $p_n(x) = x(x-n)^{n-1}$, $n \geq 0$ and identifying the corresponding differential operator, prove Abel's identity:

$$xy \sum_{k=0}^{n} (x-k)^{k-1}(y-n+k)^{n-k-1} = (x+y)(x+y-n)^{n-1}.$$

3 FORMAL LAURENT SERIES

The result before us, a theorem of Lagrange, is one of a technical nature. It is a vehicle by which we can extract explicit solutions for generating functions from functional equations that they satisfy. For example, knowledge that a generating function F satisfies the equation $F(x) = xe^{F(x)}$ allows us to conclude (by way of Lagrange's theorem) that F is unique and its explicit form is $F(x) = \Sigma_{n \geq 1}(n^{n-1}/n!)x^n$. What's more, not only can we find F, but also the explicit form of any given series of F, $G(F(x))$ say. Most problems in combinatorial enumeration lead first to functional equations involving the generating function in question. The inversion theorem of Lagrange is a mechanism by which explicit solutions are then extracted. The background for the theorem is that of formal Laurent series, which we introduce shortly.

3.8

Let us denote by $R[[x]]$ the (algebra of) *formal power series*, as described in the first section of Chapter 2. To recall, elements of $R[[x]]$ are formal series $\Sigma_n a_n x^n$, with n ranging from 0 to infinity in steps of one. The coefficients a_n are real numbers. Addition, scalar multiplication, multiplication by convolution, and composition are operations by now familiar to the reader. We wish to stress that *the composition of two series in* $R[[x]]$ *is allowed only if the coefficient of each power of* x *in that composition can be computed as the sum of finitely many monomials*. [If $\exp(x)$ indicates the exponential series, then the composition $\exp(\exp(x))$ is not allowed (since the computation of the constant term in this composition involves the infinite sum of monomials: $1 + 1/2! + 1/3! + 1/4! + \cdots$). But $\exp(\exp(x) - 1)$ is a composition that can be made.] In fact, we may compose $A(B(x))$ for any series A in $R[[x]]$, and any series B in $R[[x]]$ with constant term 0 (this being not an exhaustive description of possible compositions).

If we take a power series A of $R[[x]]$ can we easily tell whether or not A^{-1} (the inverse of A) exists? Yes, we can tell pretty easily. For *if* $A(x) = \Sigma_n a_n x^n$, *then* A^{-1} *exists in* $R[[x]]$ *if and only if the constant term* a_0 *is different from zero*. [Indeed, by writing $A^{-1}(x) = \Sigma_n b_n x^n$, and equating $A(x)A^{-1}(x) = 1$, one obtains $b_0 = a_0^{-1}$, and then recursively the rest of the b_n's; $n \geq 1$. In fact, $A^{-1}(x) = a_0^{-1}\Sigma_{n \geq 0}(1 - a_0^{-1}A(x))^n$.] It thus seems natural to partition $R[[x]]$ into $R[[x]]_1$ and $R[[x]]_0$, with $R[[x]]_1$ consisting of invertible series in $R[[x]]$, and $R[[x]]_0$ of series not invertible. Formally,

$$R[[x]]_1 = \left\{\sum_n a_n x^n \colon a_0 \neq 0\right\}$$

$$R[[x]]_0 = \left\{\sum_n a_n x^n \colon a_0 = 0\right\}.$$

(A series in $R[[x]]_0$ is therefore of the form $a_1x + a_2x^2 + \cdots + a_nx^n + \cdots$, and hence divisible by x.)

3.9 Laurent Series

In particular x itself is not invertible in $R[[x]]$ (clearly $x \in R[[x]]_0$). This is a nuisance in many situations, which we seek to remedy. In essence we visualize $R[[x]]$ as a subset (or subalgebra) of a larger set (the Laurent series algebra), this larger set having the desirable property of containing the inverse of any series in $R[[x]]$ (including those in $R[[x]]_0$, and thus of x itself, in particular).

This convenient larger algebra we denote by $R((x))$ and *let it consist of series $\Sigma_n a_n x^n$ in which finitely many negative powers of x are allowed to appear.* [The index n can thus start at any finite (possibly negative) integer and be incremented in steps of 1 indefinitely.] *We call $R((x))$ the set of Laurent series* (in honor of the French mathematician).

An example is $2x^{-6} + 3x^{-4} - 2x^{-1} + 4 - (7/5)x + \Sigma_{n \geq 2} 2^n x^n$. In particular x^{-1} is a Laurent series (for $x^{-1} = x^{-1} + \Sigma_{n \geq 0} 0x^n$). And x^{-1} is the inverse of x, since $xx^{-1} = 1 = x^{-1}x$.

We add and multiply by constants Laurent series the same way we perform such operations in $R[[x]]$. Multiplication of two Laurent series is done by convolution just as it was done in $R[[x]]$. One can thus see that $R(x)$ extends $R[[x]]$ not only in a set theoretic sense (i.e., by inclusion) but also in an operational sense (for the operations in both sets are *the same*). It seems appropriate then to call $R((x))$ the algebra of Laurent series, of which $R[[x]]$ is a subalgebra. Composition of two Laurent series is allowed only if each coefficient in the resulting series is a sum of at most finitely many monomials, and if the resulting series has at most finitely many negative powers of x.

Observe that each Laurent series, with the exception of 0, *has an inverse.* Indeed, let $A(x) = \Sigma_{n \geq m} a_n x^n$ be a Laurent series, with $a_m \neq 0$ (m a positive, zero, or negative integer). Then $x^{-m}A(x)$ is a series in $R[[x]]_1$ and thus it has an inverse. If we write $B(x) = x^{-m}A(x)$, then $B^{-1}(x)$ exists and the inverse of $A(x)$ is $A^{-1}(x) = B^{-1}(x)x^{-m}$. Let us check: $A^{-1}A = B^{-1}x^{-m}x^mB = 1 = x^mBB^{-1}x^{-m} = AA^{-1}$, as desired. In particular, any nonzero series in $R[[x]]_0$ has an inverse in $R((x))$.

Apart from inversion, understanding the existence of the logarithm of a Laurent series is very important. The reader will recall that the series $\Sigma_{n \geq 0}(-1)^n x^{n+1}/(n+1)$ is written as $\ln(1+x)$. It is not hard to verify that *the logarithm of $G(x)$ exists as a Laurent series* (in that it has at most finitely many negative powers of x) *if and only if $G(x)$ is a series in* $R[[x]]_1$. (The reader should recall the restrictions called for when composing two series.)

The formal derivative, denoted by D, of a Laurent series $\Sigma_{n \geq m} a_n x^n$ is $\Sigma_{n \geq m} n a_n x^{n-1}$, in complete analogy to the definition of the differential operator for power series in $R[[x]]$. As an example, $D(x^{-5} - 2x^{-3} + 2x^{-1} - 5 + \Sigma_{n \geq 1}\binom{2n}{2}x^n) = -5x^{-6} + 6x^{-4} - 2x^{-2} - 0 + \Sigma_{n \geq 1} n\binom{2n}{2}x^{n-1}$. It is important to observe that *there is no Laurent series whose derivative is x^{-1}.*

Finally, every Laurent series F can be written in a unique way as $F(x) = x^m G(x)$, with G in $R[[x]]_1$. The exponent m is called the *valuation* of F. (Equivalently, the valuation equals the smallest power of x with nonzero coefficient.) The valuation of $4x^{-3} - 7x^{-2} + \cdots$ is -3, for example. The valuation of $\Sigma_{n\geq 2}(n^{n+1}/(n+2)!)x^n$ is -2, the valuation of x^{-1} is -1, the valuation of any series in $R[[x]]_1$ is 0, and that of a series in $R[[x]]_0$ is a positive integer.

3.10 The Formal Residue Operator

Given a Laurent series we often need to extract the coefficient of a power of x. Specifically, if $A(x) = \Sigma_n a_n x^n$ is the series in question, the writing $[x^n]A(x)$ indicates the coefficient of x^n in $A(x)$, that is, $[x^n]A(x) = a_n$. For fixed n, $[x^n]$ is a linear operator. Of particular interest is $[x^{-1}]$, the extraction of the coefficient of x^{-1}. We call $[x^{-1}]$ the *residue operator* (and say that $[x^{-1}]A(x)$ is the *residue* of A at 0).

In what follows by series we understand Laurent series.

It is of fundamental importance that x^{-1} is not the derivative of any power of x. This fact has many noteworthy consequences. We mention two immediate ones that we need shortly. The first, and most obvious, is:

$$\text{If } F \text{ is the derivative of some series, then } [x^{-1}]F(x) = 0. \tag{3.9}$$

Occasionally we find it convenient to write G' for the formal derivative of G, normally written as DG. All the usual rules of differentiation are valid for Laurent series. In particular the product rule informs us that $(FG)' = F'G + FG'$, and hence, by (3.9),

$$0 = [x^{-1}](F(x)G(x))' = [x^{-1}]F'(x)G(x) + [x^{-1}]F(x)G'(x).$$

Or,

$$\begin{array}{l}\text{For any two series } F \text{ and } G,\\ [x^{-1}]F'(x)G(x) = -[x^{-1}]F(x)G'(x).\end{array} \tag{3.10}$$

A most important result is the change of variable lemma, which we discuss next. The lemma is stated as follows.

$*$ *If F and H are Laurent series, with H having a positive valuation m, then*

$$m[x^{-1}]F(x) = [z^{-1}]F(H(z))H'(z). \tag{3.11}$$

It shows how the residue behaves when a change of variable [namely replacing x by $H(z)$] is introduced.

We first prove the result for $F(x) = x^n$, where n is an integer.

If $n \neq -1$, then $[z^{-1}](H(z))^n H'(z) = [z^{-1}]((n+1)^{-1}(H(z))^{n+1})' = \{\text{by (3.9)}\} = 0 = m[x^{-1}]x^n$. Observe that $(H(z))^{n+1}$ is indeed a Laurent series, in that it has at most finitely many negative powers of z.

For $n = -1$ we are looking at $[z^{-1}](H(z))^{-1}H'(z)$. [Note that $(H(z))^{-1}H'(z)$ is not the derivative of $\ln H(z)$ because $H(z)$ belongs to

$R[[z]]_0$ and hence its logarithm does not exist.] We may write, however, $H(z) = z^m G(z)$, with G a series in $R[[z]]_1$. Being a series in $R[[z]]_1$, G has a logarithm. Consequently, the product and chain rule yield

$$H^{-1}H' = G^{-1}z^{-m}(mz^{m-1}G + z^m G') = mz^{-1} + G^{-1}G' = mz^{-1} + (\ln G)'.$$

Now,

$$\begin{aligned} [z^{-1}](H(z))^{-1}H'(z) &= [z^{-1}](mz^{-1} + (\ln G(z))') \\ &= [z^{-1}]mz^{-1} + [z^{-1}](\ln G(z))' = \{\text{by } (3.9)\} \\ &= m + 0 = m = m[x^{-1}]x^{-1}. \end{aligned}$$

We proved that $m[x^{-1}]x^n = [z^{-1}](H(z))^n H'(z)$, for all integers n. Let us write $F(x) = \sum_n a_n x^n$. By treating each power of x separately we obtain:

$$\begin{aligned} m[x^{-1}]F(x) &= m[x^{-1}]\sum_n a_n x^n = \sum_n a_n m[x^{-1}]x^n \\ &= \sum_n a_n [z^{-1}](H(z))^n H'(z) = [z^{-1}]\sum_n a_n (H(z))^n H'(z) \\ &= [z^{-1}]\Big(\sum_n a_n (H(z))^n\Big)H'(z) = [z^{-1}]F(H(z))H'(z), \end{aligned}$$

as desired. Observe that, since H has positive valuation, $F(H(z))$ has at most finitely many negative powers of z and is therefore a Laurent series. This proves the change of variable lemma.

3.11 The Inversion Theorem of Lagrange

With background thus developed we are in position to state and prove Lagrange's theorem. This result alone proves to be quite a powerful tool for finding solutions to functional equations, and is at the center of many combinatorial identities. We begin by enunciating and then demonstrating it, after which examples illustrating some of its uses follow.

Theorem. *If w is a series in x that satisfies the functional equation*

$$w = x\phi(w),$$

with $\phi(y)$ a series in $R[[y]]_1$, then w belongs to $R[[x]]_0$, w is the unique series that satisfies the above functional equation, and the coefficients of any Laurent series $F(w)$ are obtained as follows:

$$[x^n]F(w) = \frac{1}{n}[y^{n-1}](F'(y)\phi^n(y)),$$

for $n \neq 0$; $n \geq$ valuation of F, and

$$[x^0]F(w) = [y^0]F(y) + [y^{-1}]F'(y)\ln(\phi(y)\phi^{-1}(0)).$$

Proof. Denote $y\phi^{-1}(y)$ by $G(y)$. Since $\phi^{-1}(y)$ is a series in $R[[y]]_1$ the series $G(y)$ has valuation 1. Moreover,

$$G(w) = w\phi^{-1}(w) = x, \tag{3.12}$$

since w satisfies the functional equation $w = x\phi(w)$. Equation (3.12) forces a unique solution for w; indeed, the coefficients of w are uniquely determined (recursively) in terms of the coefficients of G. The series w is in fact the compositional inverse of G, which we denote by $\bar{G}$. [By compositional inverse we mean the series $\bar{G}$ that satisfies $G(\bar{G}(y)) = y = \bar{G}(G(y))$.] In conclusion, $w(x) = \bar{G}(x)$, which is a series in $R[[x]]_0$.

For any integer n we may write

$$[x^n]F(w) = [x^{-1}]x^{-n-1}F(w) = [w^{-1}](G(w))^{-n-1}F(w)G'(w), \quad (3.13)$$

where the first sign of equality is self-explanatory and the second follows from (3.11), the change of variable lemma [with x replaced by $G(w)$].

If $n \neq 0$ (3.13) yields

$$\begin{aligned}[x^n]F(w) &= -\frac{1}{n}[w^{-1}]F(w)(G^{-n}(w))' = \{\text{by (3.10)}\}\\ &= \frac{1}{n}[w^{-1}]F'(w)G^{-n}(w) = \{\text{by (3.12)}\}\\ &= \frac{1}{n}[w^{-1}]F'(w)\phi^n(w)w^{-n}\\ &= \frac{1}{n}[w^{n-1}]F'(w)\phi^n(w).\end{aligned}$$

When $n = 0$ (3.13) can be rewritten as

$$\begin{aligned}[x^0]F(w) &= [w^{-1}]G^{-1}(w)F(w)G'(w)\\ &= [w^{-1}]G^{-1}(w)F(w)(w\phi^{-1}(w))'\\ &\qquad\qquad \{\text{by (3.12) and the product rule}\}\\ &= [w^{-1}]\phi(w)w^{-1}F(w)(\phi^{-1}(w) - w\phi^{-2}(w)\phi'(w))\\ &= [w^{-1}](w^{-1}F(w) - F(w)\phi^{-1}(w)\phi'(w))\\ &= [w^{-1}](w^{-1}F(w)) - [w^{-1}]F(w)\phi^{-1}(w)\phi'(w)\\ &= [w^0]F(w) - [w^{-1}]F(w)(\ln\phi(w))' = \{\text{by (3.10)}\}\\ &= [w^0]F(w) - [w^{-1}]F'(w)\ln(\phi(w)\phi^{-1}(0)).\end{aligned}$$

This ends the proof of Lagrange's theorem.

REMARKS

1. Please note that by selecting $F(y) = y$ we obtain the explicit expression for the coefficients of the series w that satisfies the functional equation $w = x\phi(w)$.
2. One may use Lagrange's theorem to expand a series $F(x)$ in terms of another series $G(x)$ with valuation 1. This is done by writing $G(x) = x\phi^{-1}(x)$, with ϕ a series in $R[[x]]_1$, and by making the substitution

$z = G(x)$. Then x in the resulting functional equation $x = z\phi(x)$ may be viewed as an unknown series in z. Lagrange's theorem expresses $F(x)$ as a power series in z.

3.12

In this section we illustrate the inversion theorem of Lagrange by way of several examples.

Example 1. Assume that w, a series in x, satisfies the functional equation $w = xe^w$. Our task is to find w.

We can apply Lagrange's theorem with $\phi(y) = e^y$ and seek the series expansion of $F(w) = w$. The theorem gives

$$[x^n]w = \frac{1}{n}[y^{n-1}](1 \cdot e^{ny}) = \frac{1}{n}\frac{n^{n-1}}{(n-1)!} = \frac{n^{n-1}}{n!}, \qquad \text{for } n \geq 1.$$

Since the valuation of w is positive the above computation comprises all the coefficients of w. Therefore,

$$w(x) = \sum_{n \geq 1} \frac{n^{n-1}}{n!} x^n, \tag{3.14}$$

which is the explicit answer we seek.

By choosing $F(w) = w^c$ we may compute any power of w (with c a nonzero integer). Indeed,

$$[x^n]w^c = \frac{1}{n}[y^{n-1}](cy^{c-1}e^{ny}) = \frac{1}{n}c\frac{n^{n-c}}{(n-c)!} = c\frac{n^{n-c-1}}{(n-c)!},$$

for $n \neq 0$ and $n \geq c$. And for $n = 0$ we obtain

$$[x^0]w^c = [y^0]y^c + [y^{-1}](cy^{c-1}\ln(e^y \cdot 1)) = 0 + [y^{-1}]cy^c$$
$$= \begin{cases} -1 & \text{if } c = -1 \\ 0 & \text{otherwise.} \end{cases}$$

In conclusion,

$$(w(x))^c = \sum_{\substack{n \geq c \\ n \neq 0}} c\frac{n^{n-c-1}}{(n-c)!} x^n, \qquad \text{for } c \neq -1 \tag{3.15}$$

and

$$w^{-1}(x) = x^{-1} - 1 - \sum_{n \geq 1} \frac{n^n}{(n+1)!} x^n.$$

Note that (for positive c) we may compute w^c the usual way, by convolution, using the explicit form of w as written in (3.14). By doing so we obtain

$$(w(x))^c = \left(\sum_{n \geq 1} \frac{n^{n-1}}{n!} x^n \right)^c = \sum_{n \geq c} \left(\sum_{\substack{n_i \geq 1 \\ \sum n_i = n}} \prod_{i=1}^{c} \frac{n_i^{n_i-1}}{n_i!} \right) x^n.$$

Comparing with (3.15) we obtain the identity

$$\sum \prod_{i=1}^{c} \frac{n_i^{n_i-1}}{n_i!} = c\frac{n^{n-c-1}}{(n-c)!}, \qquad c \geq 1.$$

On the right-hand side the sum is taken over all vectors $(n_1, n_2, \ldots, n_c)$ that satisfy $n_i \geq 1$, and $\sum_{i=1}^{c} n_i = n$.

All this was derived from the initial knowledge that w satisfies $w = xe^w$. One is thus left with the impression that within a functional equation is locked a great deal of information. At least part of it can be made available through the theorem of Lagrange, as we just saw.

Example 2. To show how a series may be expanded in terms of another, we consider e^{xz} (regarded as a power series in z, with x an indeterminate) and expand it in powers of ze^{-z}. (Note that ze^{-z} is a series in $R[[z]]_0$, so that expansions in terms of its powers are indeed well defined.) Let $t = ze^{-z}$, and regard this substitution as a functional equation for z (interpreted as a series in t). Thus $z = te^z$, and Lagrange's theorem [with $\phi(z) = e^z$ and $F(z) = e^{xz}$] yields

$$[t^n]e^{xz} = \frac{1}{n}[u^{n-1}](xe^{xu}e^{nu}) = \frac{1}{n}[u^{n-1}]xe^{(x+n)u}$$

$$= \frac{x}{n}\frac{(x+n)^{n-1}}{(n-1)!}, \qquad \text{for } n \geq 1.$$

The constant term in e^{xz} (as a series in t) is clearly 1. Hence

$$e^{xz} = 1 + \sum_{n\geq 1} \frac{1}{n!}x(x+n)^{n-1}z^n e^{-nz},$$

expresses e^{xz} as a series in powers of ze^{-z} $(= t)$.

Expanding e^{xz}, e^{yz}, and $e^{(x+y)z}$ in terms of ze^{-z} as above, and equating the coefficients of $z^n e^{-nz}$ on both sides of the equation $e^{xz}e^{yz} = e^{(x+y)z}$ yields the identity

$$xy\sum_{k=0}^{n}\binom{n}{k}(x+k)^{k-1}(y+n-k)^{n-k-1}$$

$$= (x+y)(x+y+n)^{n-1}; \qquad n \geq 0.$$

This is (a form of) Abel's identity. [The reader may wish to compare it to that which appears in the proof of (vi) in Section 3.3. Indeed, adding the expressions for $x^{-1}(x+y+n)^n$ and $y^{-1}(x+y+n)^n$ on that page leads to the form that appears above.]

In general, Lagrange's theorem is a powerful method for inventing as well as explaining many identities.

4 MULTIVARIATE LAURENT SERIES

3.13

We now extend to several dimensions the theory presented in Section 3. For the most part the generalization comes in a natural way, yet certain important differences exist (such as the technique of proof of the multivariate change of variable lemma).

Let us begin by introducing the notation. By $\mathbf{x}$ we understand the vector $(x_1, x_2, \ldots, x_m)$, with x_i's indeterminates. The symbol $\mathbf{x}^{\mathbf{n}}$ stands for the product $x_1^{n_1}x_2^{n_2} \cdots x_m^{n_m}$. We call $\mathbf{n} = (n_1, \ldots, n_m)$ the *exponent* of $\mathbf{x}$. Exponents are added componentwise and we write $\mathbf{n} \geq \mathbf{0}$ if $n_i \geq 0$ for all i. (Naturally $\mathbf{n} \ngeq \mathbf{0}$ informs us that at least one component of $\mathbf{n}$ is negative.) The n_i's are integers.

By $R[[\mathbf{x}]]$ we denote the algebra of series $\sum_{\mathbf{n} \geq \mathbf{0}} a_{\mathbf{n}} \mathbf{x}^{\mathbf{n}}$ with real numbers $a_{\mathbf{n}}$ as coefficients. As in the univariate case, $R[[\mathbf{x}]]_1$ consists of series with nonzero constant term, and $R[[\mathbf{x}]]_0$ of series with zero constant term. Most importantly, $R((\mathbf{x}))$ *denotes the algebra of (multivariate Laurent) series* $\sum_{\mathbf{n}} a_{\mathbf{n}} \mathbf{x}^{\mathbf{n}}$ *in which* $\mathbf{n} \ngeq \mathbf{0}$ *for at most finitely many exponents* $\mathbf{n}$. [Clearly $R((\mathbf{x}))$ contains $R[[\mathbf{x}]]$, and $R[[\mathbf{x}]]_1$ and $R[[\mathbf{x}]]_0$ partition $R[[\mathbf{x}]]$ just like in the univariate case.] Composition of two series can be done only if the coefficient of each power of $\mathbf{x}$ in the composed series can be written as a finite sum of monomials. Addition, multiplication by scalars, and multiplication by convolution parallel the corresponding operations in the univariate case, as does the existence of the inverse and of the logarithm. The notion of *valuation* of a series F (if it exists) is defined in the following way: it is the vector $\mathbf{k}$ (if any) that allows us to write $F(\mathbf{x}) = \mathbf{x}^{\mathbf{k}} G(\mathbf{x})$, with G a series in $R[[\mathbf{x}]]_1$.

Examples of multivariate Laurent series are: $\sum_{\mathbf{n} \geq \mathbf{0}} (\mathbf{x}^{\mathbf{n}}/\mathbf{n}!)$, where $\mathbf{n}! = \prod_{i=1}^{m} n_i!$. (When $m = 2$ this expansion becomes

$$\sum_{n_1, n_2 \geq 0} \frac{1}{n_1! n_2!} x_1^{n_1} x_2^{n_2};$$

it has valuation $\mathbf{0}$.) Or, $\mathbf{x}^{-\mathbf{1}} \sum_{\mathbf{n} \geq \mathbf{0}} \mathbf{x}^{\mathbf{n}} = \sum_{\mathbf{n} \geq -\mathbf{1}} \mathbf{x}^{\mathbf{n}}$. (In case $\mathbf{x} = (x_1, x_2, x_3)$ we obtain

$$\sum_{\substack{n_i \geq -1 \\ i=1,2,3}} x_1^{n_1} x_2^{n_2} x_3^{n_3}.)$$

The notation $-\mathbf{1}$ stands for $(-1, -1, \ldots, -1)$. This last series has valuation $-\mathbf{1}$. A typical situation in which the valuation does not exist is illustrated by the (finite) series $x_1^{-1}x_2^{-2} + x_1^{-2}x_2^{-1}$. Most multivariate series do not have valuations, in sharp contrast to the univariate case.

3.14

In proving the multivariate version of Lagrange's theorem we need a change of variable lemma, in itself a most important result. The operator $[\mathbf{y}^{\mathbf{n}}]$ extracts the coefficient of $\mathbf{y}^{\mathbf{n}}$ when applied to a series. Of particular importance remains $[\mathbf{y}^{-1}]$, which extracts the coefficient of $\mathbf{y}^{-1}$ $(= y_1^{-1}y_2^{-1} \cdots y_m^{-1})$ and which we call the *multivariate residue operator*.

The multivariate change of variable lemma follows.

∗ *Let* $F(\mathbf{x})$ *and* $H_1(\mathbf{x}), \ldots, H_m(\mathbf{x})$ *be series in* $R((\mathbf{x}))$, *and let* H_i *have valuation* $(v_{i1}, \ldots, v_{im}) \geq \mathbf{0}$ *with* $\sum_j v_{ij} > 0$, *for* $1 \leq i \leq m$. *Then*

$$|V|[\mathbf{x}^{-1}]F(\mathbf{x}) = [\mathbf{z}^{-1}]\{F(\mathbf{H}(\mathbf{z})) \cdot J(\mathbf{H}(\mathbf{z}))\}. \tag{3.16}$$

[*Here* $V = (v_{ij})$, $\mathbf{H} = (H_1, \ldots, H_m)$, *and* $J(\mathbf{H}(\mathbf{z})) = |\partial H_i/\partial z_j|$, *with* $\partial/\partial z_j$ *denoting the formal (partial) derivative with respect to* z_j. *By* $|M|$ *we denote the determinant of the square matrix* M. *The matrix* $J(\mathbf{H}(\mathbf{z}))$ *is called the Jacobian of* $\mathbf{H}$ *with respect to* $\mathbf{z}$. *Both* $J(\mathbf{H}(\mathbf{z}))$ *and* V *are* $m \times m$ *matrices*.]

We devote the remainder of Section 3.14 to proving this result. The key property used in the proof is the multilinearity of determinants. For the reader's convenience we first review this property: it is a formula for the determinant of a sum of matrices.

For simplicity let us look at the sum $A + B$ of 2×2 matrices. If

$$A = \begin{bmatrix} a_{11} & a_{12} \\ a_{21} & a_{22} \end{bmatrix} \quad \text{and} \quad B = \begin{bmatrix} b_{11} & b_{12} \\ b_{21} & b_{22} \end{bmatrix},$$

then, by expanding successively on rows,

$$\begin{vmatrix} a_{11}+b_{11} & a_{12}+b_{12} \\ a_{21}+b_{21} & a_{22}+b_{22} \end{vmatrix} = \begin{vmatrix} a_{11} & a_{12} \\ a_{21}+b_{21} & a_{22}+b_{22} \end{vmatrix} + \begin{vmatrix} b_{11} & b_{12} \\ a_{21}+b_{21} & a_{22}+b_{22} \end{vmatrix}$$

$$= \begin{vmatrix} a_{11} & a_{12} \\ a_{21} & a_{22} \end{vmatrix} + \begin{vmatrix} a_{11} & a_{12} \\ b_{21} & b_{22} \end{vmatrix} + \begin{vmatrix} b_{11} & b_{12} \\ a_{21} & a_{22} \end{vmatrix} + \begin{vmatrix} b_{11} & b_{12} \\ b_{21} & b_{22} \end{vmatrix}.$$

We write $\sum_\alpha |C(\alpha)|$ for this last sum, with α a subset of $\{1, 2\}$ and $C(\alpha)$ the matrix formed by the rows of A with index in α and the remaining rows from B. [Thus

$$C(\{1\}) = \begin{bmatrix} a_{11} & a_{12} \\ b_{21} & b_{22} \end{bmatrix} \quad \text{and} \quad C(\{\phi\}) = \begin{bmatrix} b_{11} & b_{12} \\ b_{21} & b_{22} \end{bmatrix} = B,$$

for example.] If A and B are $m \times m$ matrices then, in an analogous way,

$$|A + B| = \sum_\alpha |C(\alpha)|, \tag{3.17}$$

with α running over the 2^m subsets of $\{1, 2, \ldots, m\}$, and $C(\alpha)$ being the matrix formed by the rows of A with index in α and the remaining rows from B. We refer to (3.17) as the property of multilinearity for determinants, or just *multilinearity* for short.

There is a natural generalization from a sum of two matrices to a sum of any finite number. Thus, if $A_1, A_2, \ldots, A_n$ are n matrices of size m each we obtain

$$|A_1 + A_2 + \cdots + A_n| = \sum_\alpha |C(\alpha)|. \tag{3.18}$$

To explain what the summation is over we may conveniently think of the m^2 elements of A_i being colored with color "i." Then α is a vector of m positions with colors (chosen with possible repetitions out of the n available) as entries. By $C(\alpha)$ we understand the matrix with its rows colored as the entries of the vector α indicate. There will be n^m choices for $C(\alpha)$'s, and thus the right-hand side of (3.18) contains that many terms.

To take this one step further, consider countably many matrices of size m (say $A_1, A_2, \ldots$) with entries of A_i colored "i." Then we may still write

$$|A_1 + A_2 + A_3 + \cdots| = \sum_{\alpha} |C(\alpha)|, \tag{3.19}$$

with α running over (the infinity of) vectors of length m formed with m choices out the infinitude of available colors. Expansion (3.19) is proved in the same way that (3.18) is. This slightly more exotic expansion we still call multilinearity.

To prove the multivariate change of variable lemma we first observe that the composition $F(\mathbf{H}(\mathbf{x}))$ exists in $R((\mathbf{x}))$, since the valuation $(v_{i1}, \ldots, v_{im})$ of H_i is a nonnegative vector, and since $\Sigma_j v_{ij} > 0$. Write $F(\mathbf{x}) = \Sigma_{\mathbf{k}} c_{\mathbf{k}} \mathbf{x}^{\mathbf{k}}$ and denote $\partial/\partial z_j$ by D_j. Then

$$\begin{aligned} [\mathbf{z}^{-1}] F(\mathbf{H}(\mathbf{z})) \cdot J(\mathbf{H}) &= [\mathbf{z}^{-1}] \sum_{\mathbf{k}} c_{\mathbf{k}} \mathbf{H}^{\mathbf{k}}(\mathbf{z}) |D_j H_i(\mathbf{z})| \\ &= [\mathbf{z}^{-1}] \sum_{\mathbf{k}} c_{\mathbf{k}} \prod_{i=1}^{m} H_i^{k_i}(\mathbf{z}) |D_j H_i(\mathbf{z})| \\ &= [\mathbf{z}^{-1}] \sum_{\mathbf{k}} c_{\mathbf{k}} |H_i^{k_i}(\mathbf{z}) D_j H_i(\mathbf{z})|. \end{aligned} \tag{3.20}$$

The last sign of equality is obtained by multiplying row i of Matrix $(D_j H_i(\mathbf{z}))$ by $H_i^{k_i}(\mathbf{z})$.

The $m \times m$ matrix $(H_i^{k_i}(\mathbf{z}) D_j H_i(\mathbf{z}))$ depends on $\mathbf{k}$ and we denote it by $A_{\mathbf{k}}(\mathbf{z})$. Naturally $H_i^{k_i}(\mathbf{z}) D_j H_i(\mathbf{z})$, the (i, j) th entry of $A_{\mathbf{k}}(\mathbf{z})$, equals

$$\frac{1}{k_i + 1} D_j H_i^{k_i+1}(\mathbf{z}) \qquad \text{if } k_i \neq -1.$$

And if $k_i = -1$, by writing $H_i(\mathbf{z}) = \beta_i z_1^{v_{i1}} z_2^{v_{i2}} \cdots z_m^{v_{im}} G_i(\mathbf{z})$ with $\beta_i \neq 0$ and $G_i(\mathbf{z})$ in $R[[\mathbf{z}]]_1$, $H_i^{-1}(\mathbf{z}) D_j H_i(\mathbf{z})$ becomes (upon a straightforward use of the product rule for differentiation)

$$\frac{v_{ij}}{z_j} + D_j \ln G_i(\mathbf{z}).$$

In conclusion,

$$A_{\mathbf{k}}(\mathbf{z}) = \left[\frac{v_{ij}}{z_j} \delta_{k_i, -1} + \sum_{\mathbf{l}_i} \frac{l_{ij}}{z_j} a_i(\mathbf{l}_i, k_i) \mathbf{z}^{\mathbf{l}_i} \right],$$

where

$$\sum_{\mathbf{l}_i} a_i(\mathbf{l}_i, k_i)\mathbf{z}^{\mathbf{l}_i} = \begin{cases} \dfrac{1}{k_i + 1} H_i^{k_i+1}(\mathbf{z}) & \text{if } k_i \neq -1 \\ \ln G_i(\mathbf{z}) & \text{if } k_i = -1, \end{cases}$$

and $\mathbf{l}_i = (l_{i1}, l_{i2}, \ldots, l_{im})$. By δ_{st} we mean 1 if $s = t$ and 0 if $s \neq t$. (The coefficient of $\mathbf{z}^{\mathbf{l}_i}$ depends on $\mathbf{l}_i$ primarily, but also on k_i. This is why we write it as $a_i(\mathbf{l}_i, k_i)$.)

Think of $A_{\mathbf{k}}(\mathbf{z})$ as the sum of two matrices

$$\left[\sum_{\mathbf{l}_i} \frac{l_{ij}}{z_j} a_i(\mathbf{l}_i, k_i)\mathbf{z}^{\mathbf{l}_i}\right] + \left[\frac{v_{ij}}{z_j}\delta_{k_i, -1}\right]$$

and use multilinearity, as it appears in (3.17), to write

$$|A_{\mathbf{k}}(\mathbf{z})| = \sum_{\alpha} |C(\alpha)|, \tag{3.21}$$

with the (i, j)th entry of $C(\alpha)$ equal to $\Sigma_{\mathbf{l}_i}(l_{ij}/z_j)a_i(\mathbf{l}_i, k_i)\mathbf{z}^{\mathbf{l}_i}$ if the row index $i \in \alpha$, and equal to $(v_{ij}/z_j)\delta_{k_i,-1}$ if $i \notin \alpha$. The matrix $C(\alpha)$ may be viewed as a sum of countably many matrices, since each of its entries is a series (if an entry is a finite series we can always add zeros at the end to make it an infinite series). Using the multilinear expansion (3.19) for each $C(\alpha)$ that appears in (3.21) we obtain

$$A_{\mathbf{k}}(\mathbf{z}) = \sum_{t=0}^{m} \sum_{\alpha = \{\alpha_1, \ldots, \alpha_t\}} \sum_{\mathbf{l}_{\alpha_1}} \cdots \sum_{\mathbf{l}_{\alpha_t}} \mathbf{z}^{\mathbf{l}_{\alpha_1} + \cdots + \mathbf{l}_{\alpha_t} - \mathbf{1}} |B| \prod_{s=1}^{t} a_{\alpha_s}(\mathbf{l}_{\alpha_s}, k_{\alpha_s}),$$

where the $m \times m$ matrix $B = (b_{ij})$ has entries

$$b_{ij} = \begin{cases} l_{ij} & \text{if } i \in \alpha \\ v_{ij}\delta_{k_i, -1} & \text{if } i \notin \alpha \end{cases}.$$

For a nonzero contribution to the coefficient of $\mathbf{z}^{-1}$ we must therefore have $\mathbf{l}_{\alpha_1} + \cdots + \mathbf{l}_{\alpha_t} = \mathbf{0}$. But this describes a linear dependence among t rows of B and thus $|B| = 0$, unless $t = 0$. In this latter case $b_{ij} = v_{ij}\delta_{k_i,-1}$, and B will have a row of zeros (thus $|B| = 0$), unless all the k_i's are -1. This conclusion we express by writing

$$|B| = |v_{ij}| \prod_{s=1}^{m} \delta_{k_s, -1} = |V| \prod_{s=1}^{m} \delta_{k_s, -1} = [\mathbf{z}^{-1}]A_{\mathbf{k}}(\mathbf{z}).$$

Therefore $[\mathbf{z}^{-1}]A_{\mathbf{k}}(\mathbf{z}) = 0$, unless $\mathbf{k} = -\mathbf{1}$ in which case $[\mathbf{z}^{-1}]A_{-\mathbf{1}}(\mathbf{z}) = |V|$. We showed that

$$\begin{aligned} [\mathbf{z}^{-1}]\{F(\mathbf{H}(\mathbf{z})) \cdot J(\mathbf{H})\} &= \sum_{\mathbf{k}} c_{\mathbf{k}}[\mathbf{z}^{-1}]|A_{\mathbf{k}}(\mathbf{z})| \qquad \text{[see (3.20)]} \\ &= c_{-\mathbf{1}}|V| = |V|[\mathbf{x}^{-1}]F(\mathbf{x}), \end{aligned}$$

as was to be shown.

3.15

Recall that we set δ_{ij} to be 1 if $i = j$ and 0 if $i \neq j$, and that $|M|$ denotes the determinant of the square matrix M. From the change of variable lemma it now takes but a small step to establish the following theorem.

Theorem (Multivariate Lagrange). *Suppose that* $w_i = x_i\phi_i(\mathbf{w})$ *for* $i = 1, \ldots, m$, *where* $\mathbf{w} = (w_1, \ldots, w_m)$ *and* $\boldsymbol{\phi} = (\phi_1, \ldots, \phi_m)$, *with* $\phi_i(\mathbf{y}) \in R[[\mathbf{y}]]_1$. *Then for any* $F(\mathbf{y}) \in R((\mathbf{y}))$ *we may write*

$$[\mathbf{x}^{\mathbf{n}}]F(\mathbf{w}(\mathbf{x})) = [\mathbf{y}^{\mathbf{n}}]F(\mathbf{y})\boldsymbol{\phi}^{\mathbf{n}}(\mathbf{y})\left|\delta_{ij} - y_i\phi_i^{-1}(\mathbf{y})\frac{\partial\phi_i(\mathbf{y})}{\partial y_j}\right|.$$

Proof. As in the univariate case, we show that there exists a unique series $w_i(\mathbf{x}) \in R[[\mathbf{x}]]_0$ such that $w_i = x_i\phi_i(\mathbf{w})$, for $i = 1, \ldots, m$. To compute the coefficients of $F(\mathbf{w})$ we observe that $[\mathbf{x}^{\mathbf{n}}]F(\mathbf{w}) = [\mathbf{x}^{-1}]\mathbf{x}^{-(\mathbf{n}+1)}F(\mathbf{w})$. Substituting $x_i = w_i\phi_i^{-1}(\mathbf{w})$ and using the multivariate change of variable lemma (3.16) we obtain

$$\begin{aligned}[\mathbf{x}^{\mathbf{n}}]F(\mathbf{w}) &= [\mathbf{x}^{-1}]\mathbf{x}^{-(\mathbf{n}+1)}F(\mathbf{w}) \\ &= [\mathbf{w}^{-1}]\mathbf{w}^{-(\mathbf{n}+1)}(\boldsymbol{\phi}(\mathbf{w}))^{\mathbf{n}+1}F(\mathbf{w})J(\mathbf{x}),\end{aligned} \tag{3.22}$$

where $J(\mathbf{x}) = |\partial x_i/\partial w_j|$. [Observe that $|V| = |I_m| = 1$, since V is the identity matrix. Indeed, the valuation of $x_i = w_i\phi_i^{-1}(\mathbf{w})$ as a function of $\mathbf{w}$ is $(0, \ldots, 0, 1, 0, \ldots, 0)$ with 1 in the ith position only. For this reason $|V|$ does not appear in (3.22).]

We may simplify the expression for $J(\mathbf{x})$ by writing

$$\begin{aligned}J(\mathbf{x}) &= \left|\frac{\partial x_i}{\partial w_j}\right| = \left|\frac{\partial\left(w_i\phi_i^{-1}(\mathbf{w})\right)}{\partial w_j}\right| = \left|\delta_{ij}\phi_i^{-1}(\mathbf{w}) - w_i\phi_i^{-2}(\mathbf{w})\frac{\partial\phi_i(\mathbf{w})}{\partial w_j}\right| \\ &= \left|\phi_i^{-1}(\mathbf{w})\left(\delta_{ij} - w_i\phi_i^{-1}(\mathbf{w})\frac{\partial\phi_i(\mathbf{w})}{\partial w_j}\right)\right| \\ &= (\boldsymbol{\phi}(\mathbf{w}))^{-1}\left|\delta_{ij} - w_i\phi_i^{-1}(\mathbf{w})\frac{\partial\phi_i(\mathbf{w})}{\partial w_j}\right|.\end{aligned}$$

[The third sign of equality follows by using the product and chain rules. And the last equality sign is explained by factoring $\phi_i^{-1}(\mathbf{w})$ out of row i and taking it outside the determinant.] This proves the theorem.

REMARK. Sometimes it is convenient to replace $|\delta_{ij} - w_i\phi_i^{-1}(\mathbf{w})(\partial\phi_i(\mathbf{w})/\partial w_j)|$ by $|\delta_{ij} - w_j\phi_i^{-1}(\mathbf{w})(\partial\phi_i(\mathbf{w})/\partial w_j)|$. The two determinants are the same, for the latter is obtained from the former by dividing row i by w_i and multiplying column j by w_j. Since we do this for all i and all j the value of the determinant remains unchanged.

3.16 MacMahon's Master Theorem

A special choice of ϕ_i's and F in Lagrange's theorem leads to a famous result of MacMahon. Select

$$\phi_i(\mathbf{y}) = 1 + \sum_{j=1}^{m} a_{ij} y_j \quad \text{and} \quad F(\mathbf{y}) = |\delta_{ij} - y_i\phi_i^{-1}(\mathbf{y})(\partial\phi_i(\mathbf{y})/\partial y_j)|^{-1}.$$

It is clear that $\phi_i(\mathbf{y}) \in R[[\mathbf{y}]]_1$ and $F(\mathbf{y}) \in R((\mathbf{y}))$. Making the substitution $x_i = w_i\phi_i^{-1}(\mathbf{w})$, for $i = 1, \ldots, m$, we obtain by Lagrange's theorem

$$\begin{aligned}[\mathbf{x}^{\mathbf{n}}]F(\mathbf{w}) &= [\mathbf{y}^{\mathbf{n}}]\{F(\mathbf{y})\phi^{\mathbf{n}}(\mathbf{y})D\} \\ &= [\mathbf{y}^{\mathbf{n}}]\{D^{-1}\phi^{\mathbf{n}}(\mathbf{y})D\} = [\mathbf{y}^{\mathbf{n}}]\phi^{\mathbf{n}}(\mathbf{y}) \\ &= [\mathbf{y}^{\mathbf{n}}]\prod_{i=1}^{m}\left(1 + \sum_{j=1}^{m} a_{ij}y_j\right)^{n_i} = [\mathbf{y}^{\mathbf{n}}]\prod_{i=1}^{m}\left(\sum_{j=1}^{m} a_{ij}y_j\right)^{n_i}.\end{aligned}$$

By D we denoted the determinant $|\delta_{ij} - y_i\phi_i^{-1}(\mathbf{y})(\partial\phi_i(\mathbf{y})/\partial y_j)|$ and our assumption is that $F(\mathbf{y}) = D^{-1}$.

Direct computation gives, however,

$$\begin{aligned}[\mathbf{x}^{\mathbf{n}}]F(\mathbf{w}) &= [\mathbf{x}^{\mathbf{n}}]\left|\delta_{ij} - w_i\phi_i^{-1}(\mathbf{w})\frac{\partial\phi_i(\mathbf{w})}{\partial w_j}\right|^{-1} \\ &= \left\{\text{since } w_i\phi_i^{-1}(\mathbf{w}) = x_i, \text{ and } \frac{\partial\phi_i(\mathbf{w})}{\partial w_j} = a_{ij}\right\} \\ &= [\mathbf{x}^{\mathbf{n}}]|\delta_{ij} - x_i a_{ij}|^{-1} = [\mathbf{x}^{\mathbf{n}}]|I - XA|^{-1},\end{aligned}$$

where X is the diagonal matrix with diagonal entries $x_1, x_2, \ldots, x_m$, and $A = (a_{ij})$.

We have thus demonstrated the following:

MacMahon's Master Theorem. *Let* $A = (a_{ij})$ *and* $X = \operatorname{diag}(x_1, \ldots, x_m)$ *be* $m \times m$ *matrices. Then*

$$[\mathbf{x}^{\mathbf{n}}]\prod_{i=1}^{m}\left(\sum_{j=1}^{m} a_{ij}x_j\right)^{n_i} = [\mathbf{x}^{\mathbf{n}}]|I - XA|^{-1},$$

where $\mathbf{n} = (n_1, \ldots, n_m)$.

Due to the special (linear) choice of $\phi_i(\mathbf{y}) = 1 + \sum_{j=1}^{m} a_{ij}y_j$, we may view MacMahon's master theorem as the linear case of the multivariate inversion theorem of Lagrange.

3.17

Many are the consequences of these multivariate results. MacMahon has explored the potential of his master theorem principally with regard to permutations with restricted positions, yet (considering the more general results available) this must be but the tip of the iceberg. Research of a more recondite character is likely to follow.

As is not uncommon with abstract theories, it is the skill of unleashing their true powers in particular instances that is often lacking and should be cultivated to a higher degree. At the moment we confine attention only to examples that illustrate the master theorem, classical examples that MacMahon himself cites.

Example 1. Perhaps the simplest illustration is obtained by examining the coefficient of $x_1^p x_2^q$ in the product $(x_1 + x_2)^p (x_1 + x_2)^q$. Direct computation finds this coefficient to be

$$\begin{aligned}&[x_1^p x_2^q](x_1 + x_2)^p (x_1 + x_2)^q \\ &\quad = [x_1^p x_2^q]\left(\sum_i \binom{p}{i} x_1^{p-i} x_2^i\right)\left(\sum_j \binom{q}{j} x_1^j x_2^{q-j}\right) \\ &\quad = [x_1^p x_2^q]\sum_{i,j} \binom{p}{i}\binom{q}{j} x_1^{j+p-i} x_2^{i+q-j} = \sum_i \binom{p}{i}\binom{q}{i}.\end{aligned}$$

Applying the master theorem we obtain

$$[x_1^p x_2^q](x_1 + x_2)^p (x_1 + x_2)^q = [x_1^p x_2^q]|I - XA|^{-1},$$

where

$$X = \operatorname{diag}(x_1, x_2) \qquad \text{and} \qquad A = \begin{bmatrix} 1 & 1 \\ 1 & 1 \end{bmatrix}.$$

Thus,

$$|I - XA|^{-1} = \begin{vmatrix} 1 - x_1 & -x_1 \\ -x_2 & 1 - x_2 \end{vmatrix}^{-1} = \frac{1}{1 - x_1 - x_2} = \sum_{k \geq 0} (x_1 + x_2)^k.$$

Consequently,

$$\begin{aligned}[x_1^p x_2^q] \sum_{k \geq 0} (x_1 + x_2)^k &= [x_1^p x_2^q]\left(1 + (x_1 + x_2) + (x_1 + x_2)^2 + \cdots\right) \\ &= [x_1^p x_2^q](x_1 + x_2)^{p+q} = \binom{p+q}{p}.\end{aligned}$$

We thus proved that $\sum_i \binom{p}{i}\binom{q}{i} = \binom{p+q}{p}$, an easy Vandermond-like identity. [Of course, by writing $(x_1 + x_2)^p (x_1 + x_2)^q = (x_1 + x_2)^{p+q}$, and comparing coefficients of $x_1^p x_2^q$ on both sides, the identity surfaces right away without any recourse to the master theorem.]

The next example convincingly displays the power of this result.

Example 2. Symbols $x_1, x_2, \ldots, x_m$ are said to be in *standard position* if they form a sequence of length mn in which x_1 appears in the first n positions, x_2 in the following n, and so on. Thus $x_1 x_1 x_2 x_2 x_3 x_3$ describes the standard position for x_1, x_2, x_3; $m = 3$, $n = 2$. *We wish to count permutations that are* *such that in the places occupied in the standard position by x_s there are found x_s or x_{s+1}* (when $s = m$, x_1 takes the place of x_m). In case $m = 3$ and $n = 2$ there are ten such permutations; an example is $x_2 x_1 x_3 x_2 x_1 x_3$.

The problem suggests considering a product of m linear forms, each raised to power n:

$$(a_{11}x_1 + a_{12}x_2 + \cdots + a_{1m}x_m)^n (a_{21}x_1 + a_{22}x_2 + \cdots + a_{2m}x_m)^n \cdots (a_{m1}x_1 + \cdots + a_{mn}x_m)^n.$$

In the ith bracket set the coefficients of x_i and x_{i+1} to 1 and let the rest be zero. From bracket i it is now only possible to select x_i or x_{i+1}. In fact the above product becomes

$$(x_1 + x_2)^n (x_2 + x_3)^n \cdots (x_m + x_1)^n.$$

The number of permutations that we wish to count is simply the coefficient of $x_1^n x_2^n \cdots x_m^n$ in the expansion of this product (as is clear, upon a moment's reflection).

MacMahon's master theorem can be well adapted to this situation by forming the $m \times m$ matrix

$$A = \begin{bmatrix} 1 & 1 & & & & \\ & 1 & 1 & & 0 & \\ & & 1 & 1 & & \\ & & & 1 & 1 & \\ & 0 & & & \ddots & \\ 1 & & & & & 1 \end{bmatrix}$$

and setting $X = \operatorname{diag}(x_1, \ldots, x_m)$. It informs us that

$$\begin{aligned} &[x_1^n x_2^n \cdots x_m^n](x_1 + x_2)^n (x_2 + x_3)^n \cdots (x_m + x_1)^n \\ &\quad = [x_1^n x_2^n \cdots x_m^n] |I - XA|^{-1}. \end{aligned}$$

In this case the determinant $|I - XA|$ can be easily computed. Expanding according to the $m!$ permutations we find only two terms that do not vanish. Specifically,

$$|I - XA| = \begin{vmatrix} (1 - x_1) & -x_1 & & 0 & \\ & (1 - x_2) & -x_2 & & \\ & & (1 - x_3) & -x_3 & \\ & 0 & & \ddots & \\ -x_m & & & & (1 - x_m) \end{vmatrix}$$

$$= \prod_{i=1}^{m} (1 - x_i) - \prod_{i=1}^{m} x_i.$$

Thus the coefficient of $x_1^n x_2^n \cdots x_m^n$ in the product $(x_1 + x_2)^n (x_2 + x_3)^n$

$\cdots (x_m + x_1)^n$ is equal to that of $x_1^n x_2^n \cdots x_m^n$ in

$$\frac{1}{\prod_{i=1}^{m}(1 - x_i) - x_1 x_2 \cdots x_m}. \tag{3.23}$$

Denote $\prod_{i=1}^{m}(1 - x_i)$ by v and $x_1 x_2 \cdots x_m$ by u. Then fraction (3.23) may be expanded as follows:

$$\begin{aligned}\left(\prod_{i=1}^{m}(1 - x_i) - x_1 x_2 \cdots x_m\right)^{-1} &= (v - u)^{-1} \\ &= v^{-1}\left(1 - \frac{u}{v}\right)^{-1} = v^{-1} \sum_{k \geq 0} \left(\frac{u}{v}\right)^k \\ &= \sum_{k \geq 0} \frac{u^k}{v^{k+1}}.\end{aligned}$$

Interest lies in the coefficient of $x_1^n x_2^n \cdots x_m^n = u^n$. Out of this series we therefore attempt to isolate the terms that are powers of u. We proceed in doing so:

$$\begin{aligned}\sum_{k \geq 0} \frac{u^k}{v^{k+1}} &= \sum_{k \geq 0} u^k v^{-k-1} = \sum_{k \geq 0} u^k \prod_{i=1}^{m} (1 - x_i)^{-k-1} \\ &= \sum_{k \geq 0} u^k \left[\prod_{i=1}^{m} \left(\sum_{j \geq 0} \binom{k+j}{j} x_i^j\right)\right].\end{aligned}$$

Part of the series in powers of u only is

$$\begin{aligned}&u^0\{1 + u + u^2 + u^3 + \cdots\} \\ &\quad + u\left\{1 + \binom{2}{1}^m u + \binom{3}{2}^m u^2 + \binom{4}{3}^m u^3 + \cdots\right\} \\ &\quad + u^2\left\{1 + \binom{3}{1}^m u + \binom{4}{2}^m u^2 + \binom{5}{3}^m u^3 + \cdots\right\} \\ &\quad + u^3\left\{1 + \binom{4}{1}^m u + \binom{5}{2}^m u^2 + \binom{6}{3}^m u^3 + \cdots\right\} \\ &\quad + \cdots\end{aligned}$$

wherein the coefficient of u^n is

$$1 + \binom{n}{1}^m + \binom{n}{2}^m + \binom{n}{3}^m + \cdots + \binom{n}{n}^m.$$

This provides the answer to our original question. (By returning to the case $m = 3$, $n = 2$, which the reader may have initially examined, one verifies that $1 + \binom{2}{1}^3 + \binom{2}{2}^3 = 10$, as expected.)

Example 3. An identity of Dixon is much quoted in connection with the master theorem. It concerns finding a closed form for the sum $S =$

$\sum_{k=0}^{n}(-1)^k\binom{n}{k}^3$. One may recognize that S occurs as the coefficient of $x^n y^n z^n$ in the product $(x-y)^n(y-z)^n(z-x)^n$. [Indeed,

$$(x-y)^n(y-z)^n(z-x)^n$$
$$= \left(\sum_i(-1)^i\binom{n}{i}x^i y^{n-i}\right)\left(\sum_j(-1)^j\binom{n}{j}y^j z^{n-j}\right)\left(\sum_k(-1)^k\binom{n}{k}z^k x^{n-k}\right)$$
$$= \sum_{i,j,k}(-1)^{i+j+k}\binom{n}{i}\binom{n}{j}\binom{n}{k}x^{n-k+i}y^{n-i+j}z^{n-j+k}.$$

When $i=j=k$ the sum S appears as the coefficient of $x^n y^n z^n$.] This valuable observation allows immediate recourse to the master theorem:

$$S = [x^n y^n z^n](x-y)^n(y-z)^n(z-x)^n = [x^n y^n z^n]|I - XA|^{-1},$$

where

$$X = \operatorname{diag}(x, y, z), \qquad \text{and} \qquad A = \begin{bmatrix} 1 & -1 & 0 \\ 0 & 1 & -1 \\ -1 & 0 & 1 \end{bmatrix}.$$

A simple calculation yields

$$|I - XA|^{-1} = \frac{1}{1 + xy + yz + zx}.$$

Therefore,

$$S = [x^n y^n z^n]\frac{1}{1 + xy + zy + zx} = [x^n y^n z^n]\sum_{i \geq 0}(-1)^i(xy + yz + zx)^i$$
$$= (-1)^{r+s+t}\frac{(r+s+t)!}{r!s!t!},$$

where r, s, and t are integers satisfying $r+s=s+t=t+r=n$. This last expression clearly reveals that the coefficient in question must be zero, unless $r=s=t=n/2$ is an integer (m, say) in which case it equals $(-1)^m(3m)!(m!)^{-3}$. In conclusion,

$$\sum_{k=0}^{n}(-1)^k\binom{n}{k}^3 = \begin{cases}(-1)^m(3m)!(m!)^{-3} & \text{if } n = 2m \\ 0 & \text{otherwise}\end{cases}.$$

EXERCISES

1. Which of the following are well-defined Laurent series?

(a) $\exp(\sin x)$. (b) $\sin(\exp x)$.

(c) $(x + y(1-x))^n$. (d) $(1 + xy^{-1})^{-1}$.

(e) $\exp(1 - \cos(xy) + x - z)$. (f) $\exp(xy + x^{-1})$.

2. What are the valuations of the following series?

(a) $\sum_{n \geq -3} n^2 x^n$. (b) $\dfrac{1+x}{x^2(1-x)^3}$.

(c) $\sum_{n_1, n_2 \geq 2} \dfrac{n_2}{n_1!} x_1^{n_1 - 1} x_2^{n_2}$. (d) $x^{-1}y(x + y^{-1} + 1)^n$.

(e) $x^{-1}(1 + xy)^{-1} + y^{-1}(1 + x)$. (f) $2x^{-1}z(xy - xz + 3)^{-1}$.

3. Series $F(x)$ satisfies the functional equation

$$F = x(e^F - F).$$

Find the explicit form of F.

4. Using Lagrange's inversion theorem, or otherwise, show that

$$\sum_{n \geq k} \binom{2n}{n-k} x^n = (1-4x)^{-1/2} \left\{ \frac{1 - 2x - (1-4x)^{1/2}}{2x} \right\}^k.$$

5. Find the explicit form of the series C that satisfies the functional equation $C^2 - C + x = 0$. By way of Lagrange's theorem write out the series e^C, $\ln(1 + C)$, and C^k.

6. Relying on the multivariate form of Lagrange's theorem find the series $G(x, y)$ that satisfies the functional equation

$$G = x\{(1-G)^{-1} + (y-1)G\}.$$

7. Compute the coefficient of $x^n y^n z^n$ in the product

$$(y+z)^n (x+z)^n (x+y)^n.$$

8. Expand $(1 - xy)^{-1}$ in powers of xe^x.

5 THE ORDINARY GENERATING FUNCTION

The skill of transferring combinatorial information into algebraic operations with generating functions lies at the very center of successful counting and enumeration. Establishing recurrence relations and then deriving the associated generating functions is a traditional method commonly used.

But a most elegant way to attack problems of this kind is by setting forth precise and direct ways of converting usual set theoretic operations with combinatorial structures into algebraic symbolism. Such approaches often yield immediately functional equations for the enumerative generating function of those combinatorial structures, and the technical aspect of extracting explicit solutions rests with the several inversion theorems that we have studied (of which the most powerful seems to be Lagrange's). It is this latter, and seemingly quite novel, point of view that we now attempt to share with the reader.

3.18

In the following discussions the word *configuration* is a general descriptive term for any combinatorial object in which we may take interest. A configuration can be a permutation, a graph, a partition of an integer, or a color pattern (which the fallen autumn leaves display). We are concerned with a (usually infinite) set S of configurations. To each configuration c in S we attach a nonnegative integer $w(c)$, which we call its *weight*. The function w thus defined is called a *weight function*. One may arbitrarily assign a weight to a configuration but in specific settings choices of weights present themselves in natural ways. The weight of a configuration often reflects a numerical attribute intrinsically attached to that configuration. (Thus the weight of a graph may be the number of its edges; the weight of a permutation may be the number of its fixed points; the weight of a tree may be the number of its vertices of degree 1, etc.) Selecting the weight of a configuration in such a way is sufficiently important to motivate general nomenclature: in a configuration we thus identify the features of interest, which we call *opals*. The weight of the configuration then equals the number of opals that it contains. (For example, the opals of a graph may be its edges; the opals of a permutation may be its fixed points, and so on.)

Given a set of configurations interest lies in identifying (be that counting or enumerating) the subset of configurations of a given weight.

3.19

Let S be a set of configurations and w a weight function defined on S. The *ordinary generating function* of S is defined as follows:

$$S(x) = \sum_{c \in S} x^{w(c)}, \tag{3.24}$$

where x is an indeterminate. [For convenience we always denote by $S(x)$ the generating function of set S.]

By collecting like powers of x in (3.24) we observe immediately that

$$S(x) = \sum_{c \in S} x^{w(c)} = \sum_{n \geq 0} \left(\sum_{\substack{c \in S \\ w(c)=n}} 1 \right) x^n = \sum_n a_n x^n$$

with a_n counting the number of configurations of weight n in S.

[For instance, let the set S of configurations consist of sequences of any finite length made with symbols 0, 1, 2. We choose as the weight of a sequence its length. By $w(c)$, therefore, we understand the length of sequence c. The generating function of S may now be written as follows:

$$S(x) = \sum_{c \in S} x^{w(c)} = \sum_n a_n x^n,$$

where a_n denotes the number of such sequences of length n, i.e., $a_n = 3^n$.]

3.20

Perhaps the most important combinatorial operation is that of establishing weight preserving bijections between two sets of configurations. It is through such bijections that functional equations for the generating functions in question are put into evidence.

To define this concept we start with two sets of configurations S and T with associated weight functions w and v. A bijection f from S to T is called *weight preserving* if $v(f(s)) = w(s)$, for all configurations s in S. We write

$$S \sim T$$

to indicate the presence of a weight preserving bijection between S and T.

[With S the set of all sequences of finite length made with symbols $0, 1, 2$ and weight $w(c)$, the length of sequence c, we define

$$T = \{e\} \cup \{0,1,2\} \times S \qquad \text{(disjoint union)}.$$

Here e is simply the empty sequence (i.e., the sequence consisting of no symbols whatsoever). We select the weight v on T to be $v((i, c)) = w(i) + w(c) = 1 + w(c)$, with $i \in \{0,1,2\}$ and $c \in S$; also $v(e) = 0$. There exists a weight preserving bijection f from S to T that can be described as follows: for a sequence c different from e write $c = is$ (where $i \in \{0,1,2\}$ is the first symbol in c, and s is simply the rest of c); then set $f(c) = (i, s)$, and $f(e) = e$. It is clear that f is a weight preserving bijection. Thus

$$S \sim \{e\} \cup \{0,1,2\} \times S. \tag{3.25}$$

We soon see the combinatorial implication of (3.25).]

The reader will observe that the set T is really obtained by cleverly manipulating the initial set S. Such manipulating skills turn out to be of the essence in combinatorial enumeration. We encourage the serious student to develop and enrich them.

The point we wish to stress in this section is the following:

> *If there exists a weight preserving bijection between two sets of configurations, then the generating functions of the two sets are the same.* [Or, in symbols, $S \sim T$ implies $S(x) = T(x)$.] (3.26)

To see this, merely write

$$T(x) = \sum_{t \in T} x^{v(t)} = \sum_{f(s) \in T} x^{v(f(s))} = \sum_{s \in S} x^{w(s)} = S(x),$$

where f is the weight preserving bijection in question.

3.21

Now we are off to establish a close interplay between combinatorial operations with sets of configurations on the one hand and algebraic operations with generating functions on the other.

The Disjoint Union

Let S be a set of configurations with weight function w and A and B be disjoint subsets of S. We assert as follows:

Lemma (Sum). *The generating function of the disjoint union of two subsets equals the sum of the generating functions of the two subsets.*

Denoting (as agreed) by $A(x)$, $B(x)$, and $(A \cup B)(x)$ the generating functions of A, B, and $A \cup B$, respectively, we obtain

$$\begin{aligned}(A \cup B)(x) &= \sum_{c \in A \cup B} x^{w(c)} = \sum_{c \in A} x^{w(c)} + \sum_{c \in B} x^{w(c)} \\ &= A(x) + B(x),\end{aligned}$$

as stated.

The Cartesian Product

Let A and B be sets of configurations with weight functions w and v, respectively. Form the Cartesian product $A \times B$ and endow it with the weight function p, defined by $p(a, b) = w(a) + v(b)$. With background thus prepared we may state the following.

Lemma (Product). *The generating function of a Cartesian product of two sets is equal to the product of the generating functions of the two sets.*

To wit,

$$\begin{aligned}(A \times B)(x) &= \sum_{c \in A \times B} x^{p(c)} = \sum_{(a, b) \in A \times B} x^{p(a, b)} \\ &= \sum_{a \in A,\, b \in B} x^{w(a)+v(b)} = \left(\sum_{a \in A} x^{w(a)}\right)\left(\sum_{b \in B} x^{v(b)}\right) \\ &= A(x)B(x).\end{aligned}$$

Example. We examined on several occasions the set S of sequences of all finite lengths made out of symbols 0, 1, and 2. The weight $w(c)$ of sequence c equals its length. In particular, it was shown in (3.25) that

$$S \sim \{e\} \cup \{0,1,2\} \times S,$$

which informs us that there exists a weight preserving bijection between the two sets. (We remind the reader that in the disjoint union above e is the empty sequence.) Taking the generating functions of both sides, using the sum and product lemmas, and recalling the contents of (3.26), we obtain

$$S(x) = x^0 + (x + x + x)S(x). \tag{3.27}$$

[In detail, $S(x) = (\{e\} \cup \{0,1,2\} \times S)(x) = \{e\}(x) + \{0,1,2\}(x)S(x) = x^0 + (x + x + x)S(x)$.] This is a functional equation for $S(x)$, which is easy to solve. We immediately obtain

$$S(x) = (1 - 3x)^{-1} = \sum_n 3^n x^n,$$

which allows us to conclude that there are 3^n sequences of weight n. This is hardly surprising, for one has three choices in each of the n available positions (a total of 3^n choices). The remarkable fact is that no direct counting was done to arrive at this conclusion. The crucial point was in establishing the weight preserving bijection (3.25) and then effectively translating it into functional equation (3.27) by making use of the sum and product lemmas. This simple example is typical of the point of view we take.

Composition

The richest combinatorial operation is that of composition. It essentially consists of replacing the opals of one configuration by configurations from a new set. We describe it next.

Let A and B be sets of configurations. Identify the opals in the configurations of A. In each configuration of A replace each opal by an element of B. If each of the resulting configurations arises from exactly one configuration of A through this process of replacement we call the resulting set of configurations a *composition* of A with B and denote it by $A \circ B$.

[As an illustration, suppose A is the set of sequences of any finite length made with symbols $0, 1, 2$ in which no string of two or more consecutive 0's occurs. Let B be the set of sequences of any positive finite length made with 0 as the sole symbol. The opals of a sequence in A are the 0's that it contains. One obtains a composition $A \circ B$ by replacing each 0 in a sequence of A by a (not necessarily same) sequence of B. Observe that $A \circ B$ consists of all finite sequences made with symbols $0, 1, 2$. And any sequence in $A \circ B$ comes from exactly one sequence of A through this process of replacement. For example, 10021000220 can only come from 10210220 by replacing the first zero by 00, the second by 000, and the third by 0.]

Lemma (Composition). *Let A be a set of configurations with opals clearly identified and let B be an arbitrary set of configurations. Form a composition $A \circ B$, if possible. Endow the sets A, B, and $A \circ B$ with weight functions as follows: The weight $w(a)$ of a configuration a in A equals the number of opals that a contains; B has an arbitrary weight function u; and for a configuration c of $A \circ B$ obtained by replacing n opals of a configuration of A by configurations $b_1, b_2, \ldots, b_n$ of B we set the weight $v(c) = u(b_1) + u(b_2) + \cdots + u(b_n)$. Then the generating function of $A \circ B$ is*

$$(A \circ B)(x) = A(B(x)).$$

Let us check that this is indeed the case. We write

$$(A \circ B)(x) = \sum_{c \in A \circ B} x^{v(c)} = \sum_{n \geq 0} \sum_{\substack{c=(a;\, b_1,\ldots,\, b_n) \in A \times B^n \\ w(a)=n}} x^{u(b_1)+\cdots+u(b_n)}$$

$$= \sum_{n \geq 0} \sum_{\substack{a \in A \\ w(a)=n}} \left(\sum_{b \in B} x^{u(b)} \right)^n$$

$$= \sum_{a \in A} \left(\sum_{b \in B} x^{u(b)} \right)^{w(a)} = A(B(x)),$$

proving the composition lemma.

Differentiation

We now discuss the concept of differentiation from a combinatorial standpoint.

Let S be a set of configurations, with opals identified in each configuration. Take such a configuration; it has n opals, say. Produce n copies of this configuration and distinguish opal i in the ith copy, $1 \leq i \leq n$. Do this to each configuration in S, then assemble all resulting copies in a new set S'. We call S' the *derivative* of S.

[For instance, let S consist of sequences of length 2 made with symbols 1, 2, 3. We allow any one of the symbols 1, 2, 3 to qualify as an opal. Then $S = \{12, 13, 23\}$ and $S' = \{\bar{1}2, 1\bar{2}, \bar{1}3, 1\bar{3}, \bar{2}3, 2\bar{3}\}$, where we graphically distinguished an opal by placing a bar over it.]

The notion of derivative is thus relatively easy to grasp. Often weight preserving bijections may be difficult to find for the original set but they readily suggest themselves for the derivative; in this important respect the derivative proves to be quite helpful.

Lemma (Differentiation). *Let S be a set of configurations with weight function $w(s)$ equal to the number of opals in configuration s. Consider the derivative S' endowed with the same opals as S and the same weight function w. Then*

$$S'(x) = x\,DS(x),$$

where D is the formal derivative with respect to x.

One can see this as follows. From the definition of S' a configuration of weight n in S generates n copies (i.e., configurations of S') each of weight n. Thus, if a_n configurations of weight n exist in S there will result na_n configurations of weight n in S'. In terms of generating functions,

$$S'(x) = \sum_{n \geq 0} na_n x^n = x \sum_{n \geq 1} na_n x^{n-1} = x\,D\left(\sum_{n \geq 0} a_n x^n \right) = x\,DS(x),$$

as was enunciated.

3.22

We illustrate the general theoretical remarks made so far by a few examples.

Example 1. ***Enumerating Planted Plane Trees.*** Attention is first given to the enumeration of trees of a certain kind. By a *tree* on n vertices we understand a set of $n - 1$ edges with no cycles present (and no loops or multiple edges). It necessarily follows that there exists one and only one path (of edges) between any two vertices of a tree. A *rooted tree* is a tree in which a single vertex is distinguished (by placing a circle around it); the distinguished vertex is called the *root* of the tree.

In this section all trees are drawn in the plane and we call two trees *isomorphic* if one can be obtained from the other by continuous motions in the plane. The set of *plane trees* consists of equivalence classes of nonisomorphic trees.

A Enumerating Trivalent Planted Plane Trees

By the *degree* of a vertex we understand the number of edges emanating from it. A rooted tree is said to be *planted* if its root has degree 1. We call a tree *trivalent* if each of its vertices has degree either 3 or 1. *We propose to find the generating function of trivalent planted plane trees.*

Denote by S the (infinite) set of all trivalent planted plane trees. We define the weight $w(t)$ of such a tree t to be the number of nonroot vertices of degree 1. The following weight preserving bijection seems an evident choice:

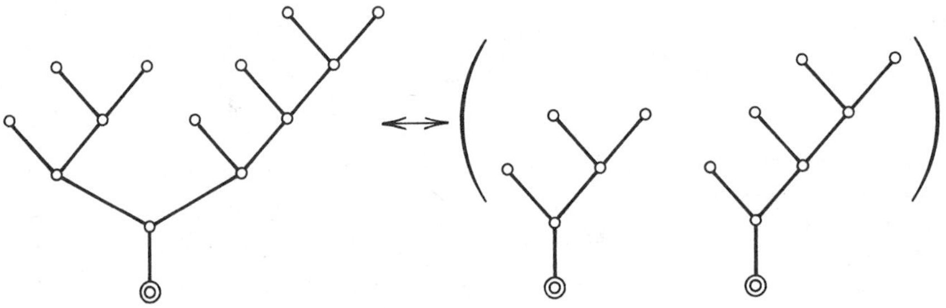

That is, to a trivalent planted plane tree t we attach its two "branches" (t_1, t_2) that immediately suggest themselves: The bijection is thus $t \to (t_1, t_2)$. Conversely, from two branches (t_1, t_2) we may uniquely assemble a tree t by identifying the roots of t_1 and t_2 in one (nonroot) vertex and then append it with a root. This is a bijection, indeed, if we presume the trees under discussion to have at least one nontrivial branch, that is, if we preclude from discussion the trivial tree: ⚲, which we denote by e (this is an important observation!).

We thus established a weight preserving bijection between $S - \{e\}$ and $S \times S$ and may, therefore, write

$$S - \{e\} \sim S \times S.$$

Applying the sum and product lemmas we deduce that the generating function $S(x)$ of the set S of trivalent planted plane trees satisfies the functional equation

$$S(x) - x = (S(x))^2.$$

[Observe that $w(e) = 1$, since e has one nonroot vertex of degree 1; thus $\{e\}(x) = x^{w(e)} = x^1 = x$.]

Simple rewriting allows the above equation to take the form

$$S(x) = x(1 - S(x))^{-1}. \tag{3.28}$$

A simple application of the inversion theorem of Lagrange (see Section 3.11), with $w(x) = S(x)$, $\phi(y) = (1-y)^{-1}$, and $F(y) = y$, yields easily the coefficients of $S(x)$:

$$\begin{aligned} [x^n]S(x) &= \frac{1}{n}[y^{n-1}]\{1 \cdot (1-y)^{-n}\} \\ &= \frac{1}{n}[y^{n-1}](1-y)^{-n} = \frac{1}{n}\binom{2n-2}{n-1}, \qquad n \geq 1. \end{aligned}$$

If not apparent as written, the last sign of equality may be explained by shuffling back to Chapter 2 and looking up equation (2.2).

Consequently,

$$S(x) = \sum_{n \geq 0} \frac{1}{n+1}\binom{2n}{n} x^{n+1},$$

which informs us that there are $(n+1)^{-1}\binom{2n}{n}$ trivalent planted plane trees with $n+1$ nonroot vertices of degree 1 (i.e., of weight $n+1$).

B Enumerating Planted Trees

Let P be the (infinite) set of all planted plane trees. Define the weight $w(t)$ of a planted tree t to be the number of its nonroot vertices. *We take interest in finding the number of planted plane trees of weight n.* Figuring out a weight preserving bijection seems like the way to go (if we have learned anything from the previous example). There is one bijection that attracts attention: map a nontrivial planted tree t into its branches $(t_1, t_2, \ldots, t_k)$. At least in principle this is the right idea, but it requires some adjustments to put it into good working order. To have the weights preserved we choose the bijection f as follows:

$$t \xrightarrow{f} (t_1, t_2, \ldots, t_k; e).$$

Graphically,

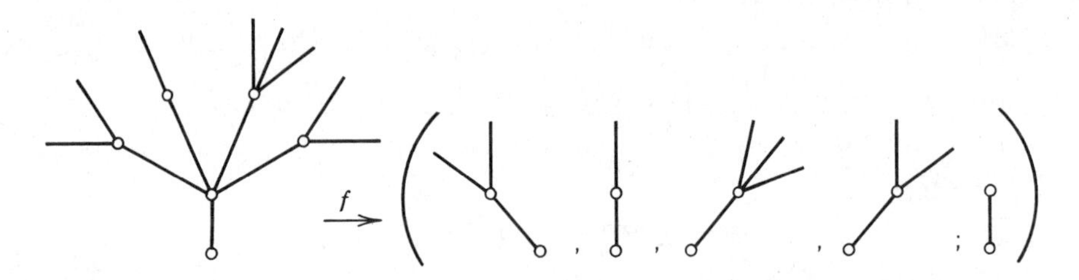

In the definition of f we append the trivial tree e (of weight 1) as the last component so as to make the weights add up right. Decomposing and reassembling a planted plane tree according to f describes now a weight preserving bijection.

This argument shows that

$$P - \{e\} \sim \bigcup_{k \geq 1} \left(P^k \times \{e\} \right) \qquad \text{(disjoint union).} \tag{3.29}$$

Passing to generating functions we obtain

$$P(x) - \{e\}(x) = \sum_{k \geq 1} (P(x))^k x$$

or,

$$P(x) - x = xP(x)(1 - P(x))^{-1}.$$

Rewriting yields the functional equation

$$P(x) = x(1 - P(x))^{-1},$$

which coincides with that in (3.28). Thus

$$P(x) = \sum_{n \geq 0} \frac{1}{n+1} \binom{2n}{n} x^{n+1},$$

allowing the conclusion that there are $(n+1)^{-1}\binom{2n}{n}$ planted plane trees on $n + 1$ nonroot vertices.

REMARK. Observe that the enumeration problems in A and B admit the same numerical solution: the Catalan number $(n+1)^{-1}\binom{2n}{n}$. The weight functions were quite different in the two problems, however. It is interesting to puzzle over the common answer though and try to see directly why there are as many planted plane trees on $n + 1$ nonroot vertices as there are trivalent planted plane trees with $n + 1$ nonroot vertices of degree 1. We will not digress in attempting to explain this, but invite the interested reader to construct an explicit bijection between the two sets of configurations.

C *Planted Plane Trees with Bivalent Vertices*

Bijection (3.29) is useful for other purposes as well. We rely on it now to *count the number of planted plane trees with n nonroot vertices and m vertices of degree* 2 (called *bivalent* vertices). This is an example of a generating function in two indeterminates.

To count such trees we endow the set P of planted trees with a weight function v as follows: the weight $v(t)$ of a tree t will be the ordered pair of integers (n, m), with n being the number of nonroot vertices and m the number of bivalent vertices in t. The generating function of P with this choice of weight function is by definition

$$B(\mathbf{x}) = B(x_1, x_2) = \sum_{t \in P} \mathbf{x}^{v(t)} = \sum_{n, m \geq 0} \left(\sum_{\substack{t \in P \\ v(t) = (n, m)}} 1 \right) x_1^n x_2^m.$$

We seek an explicit form for $B(\mathbf{x})$.

Such an explicit form can be obtained by first observing that bijection (3.29) preserves the weight v for all planted plane trees except those that have one branch only. For a tree in this latter class we are precisely one bivalent vertex short on the right-hand side of (3.29). We simply have to remember this and appropriately modify the resulting functional equation. Indeed (3.29) yields

$$B(\mathbf{x}) - x_1 x_2^0 = x_1 x_2 B(\mathbf{x}) + x_1 B^2(\mathbf{x}) + x_1 B^3(\mathbf{x}) + \cdots. \qquad (3.30)$$

(The needed adjustment is seen in the first term on the right-hand side.) We may rewrite as follows [with B abbreviating $B(\mathbf{x}) = B(x_1, x_2)$]:

$$B - x_1 x_0 = x_1 \left(x_2 B + B^2 + B^3 + \cdots \right).$$

Or,

$$B = x_1 \left[(1 - B)^{-1} + (x_2 - 1) B \right].$$

Lagrange's inversion theorem can be applied now to solve for B. We obtain

$$[x_1^n] B(x_1, x_2) = \frac{1}{n} [y^{n-1}] \left\{ (1 - y)^{-1} + (x_2 - 1) y \right\}^n,$$

which after routine manipulation yields the following coefficient for $x_1^n x_2^m$:

$$\frac{1}{n} \sum_{k \geq 1} \binom{n}{k} \binom{2k - 2}{k - 1} \binom{n - k}{m} (-1)^{n-k-m}. \qquad (3.31)$$

We thus found the explicit form of $B(x_1, x_2)$ and conclude that the number of planted plane trees with n nonroot vertices and m bivalent vertices is given by formula (3.31).

REMARK. Equation (3.30) was derived by using the sum and product lemmas in each indeterminate separately. A moment's reflection convinces us that this is permissible. We chose to present the general theory in one variable only for the sake of simplicity and notational ease.

Example 2. Counting Sequences. The problem we consider can be described as follows:

$*$ *Let A be the set of sequences of finite lengths made with symbols* $0, 1, 2$ *in which no successive* 0*'s occur. We wish to count the number of sequences in A of length n.*

Denote by B the set of sequences $\{0, 00, 000, \dots\}$ formed with the symbol 0 only. Then the composition $A \circ B$ obtained by replacing the 0's in the elements of A by elements of B describes the set of all sequences of finite length made out of symbols $0, 1, 2$. This observation suggests that we may be able to gain information on $A(x)$ by examining the generating functions of the "simpler" sets B and $A \circ B$ (indeed, $A \circ B$ is a less intricate set than A for there are no substantive restrictions placed on the structure of its elements).

For a sequence s (be it in A, B, or $A \circ B$) define its weight $w(s)$ to be the ordered pair (i, j), with i the number of 0's in s and j the number of 1's and 2's in s. Then $i + j$ equals the length of s. The bivariate generating functions for B and $A \circ B$ can easily be written down (observe that B does not contain the empty sequence). They are

$$B(x_1, x_2) = x_1 + x_1^2 + x_1^3 + \cdots = x_1(1 - x_1)^{-1},$$

and

$$(\hat{A} \circ B)(x_1, x_2) = (A \circ B)(\mathbf{x}) = \sum_{s \in A \circ B} \mathbf{x}^{w(s)}$$

$$= \sum_{i, j \geq 0} \binom{i+j}{i} x_1^i x_2^j = \sum_{n \geq 0} \sum_{\substack{i, j \\ i+j=n}} \binom{i+j}{i} x_1^i x_2^j$$

$$= \sum_{n \geq 0} (x_1 + x_2)^n = (1 - x_1 - x_2)^{-1}.$$

The generating function of A is

$$A(x_1, x_2) = A(\mathbf{x}) = \sum_{a \in A} \mathbf{x}^{w(a)} = \sum_{i, j \geq 0} \left(\sum_{\substack{a \in A \\ w(a) = (i, j)}} 1 \right) x_1^i x_2^j.$$

The number of sequences of length n in A is now easily seen to be the coefficient of x^n in $A(x, x)$. To find an explicit form for $A(x_1, x_2)$ first observe that the composition lemma allows us to write

$$(1 - x_1 - x_2)^{-1} = (A \circ B)(x_1, x_2) = A\big(B(x_1, x_2), x_2\big)$$

$$= A\big(x_1(1 - x_1)^{-1}, x_2\big).$$

A simple substitution yields

$$A(x_1, x_2) = \big(1 - x_1(1 + x_1)^{-1} - x_2\big)^{-1},$$

and thus

$$A(x, x) = (1 + x)(1 - x - x^2)^{-1}.$$

To find an explicit form for the coefficient of x^n in $A(x, x)$ we can use a partial fraction decomposition or other methods of expansion. These easy details are omitted.

We used a multivariate version (bivariate, actually) of the composition lemma to find the generating function $A(x_1, x_2)$. The reader may wish to write down and prove the lemma in this more general form. Several of the exercises at the end of this section require the multivariate analogues of the lemmas as well.

Concluding Remarks

Parallel theories can be developed for combinatorial situations that naturally lead to consideration of other types of generating functions: such as the exponential, Dirichlet, or Eulerian. The analogues of the sum product and composition lemmas can be, in turn, developed for each of these.

With S a set of configurations and w a weight function on S we can describe these three types of generating functions as follows:

The *exponential generating function* of S is defined by

$$G_S(x) = \sum_{c \in S} \frac{x^{w(c)}}{w(c)!}.$$

It is helpful when counting configurations in which the opals are tagged or labeled.

The *Dirichlet generating function* of S is

$$D_S(x) = \sum_{c \in S} w(c)^{-x}.$$

One of its many uses in the theory of numbers is the enumeration of (ordered) factorizations of an integer.

By the *Eulerian generating function* of S we understand the series

$$E_S(x, q) = \sum_{c \in S} \frac{x^{w(c)}}{(1 - q)(1 - q^2) \cdots (1 - q^{w(c)})}.$$

This is a natural series to consider when working with the subspaces of a vector space of finite dimension over a finite field. The denominator is related to the Gaussian polynomials that we discuss in some detail in the next section.

EXERCISES

1. Let S be the set of all sequences of finite lengths formed with elements of the set $N_n = \{1, 2, \ldots, n\}$. An opal is any element of a sequence in S. The weight of a sequence is its length. Establish the weight preserving bijection

$$(S - \{e\})' \sim S \times N_n \times S,$$

where e is the empty sequence. Find $S(x)$, the generating function of S.

2. Show that the number of (ordered) ways of obtaining a sum of n when k usual dice are rolled is

$$\sum_{i \geq 0} (-1)^i \binom{k}{i} \binom{n - 6i - 1}{k - 1}.$$

3. The number of integral solutions to $x_1 + x_2 + \cdots + x_k = n$, with $a_i \leq x_i < b_i$ (for $i = 1, 2, \ldots, k$) is the coefficient of x^n in

$$x^{a_1 + \cdots + a_k}(1 - x)^{-k} \prod_{i=1}^{k} (1 - x^{b_i - a_i}).$$

4. Show that the number of ways of covering a $2 \times n$ chessboard with m dominoes (1×2 rectangles) and k 1×1 squares is the coefficient of $x^m y^k$ in

$$(1 - x)[(1 - x - x^2)(1 - x) - (1 + x)y^2]^{-1}$$

where $2m + k = 2n$.

5. Let A be the set of all sequences of finite lengths made with symbols $1, 2, \ldots, n$ with no two successive symbols the same. By establishing a weight preserving bijection, or otherwise, show that the multivariate generating function of A is

$$A(\mathbf{x}) = A(x_1, \ldots, x_n) = \left[1 - \sum_{i=1}^{n} x_i (1 + x_i)^{-1}\right]^{-1}.$$

The number of sequences of A with k_i occurrences of symbol i, $1 \leq i \leq n$, is the coefficient of $\mathbf{x}^{\mathbf{k}}$ in $A(\mathbf{x})$, where $\mathbf{k} = (k_1, \ldots, k_n)$.

6. Show that the number of sequences of length n, made with symbols 0 and 1, in which no adjacent 0's occur and with precisely r pairs of adjacent 1's, is the coefficient of $x^n y^r$ in

$$(1 + x)(1 + (1 - y)x)(1 - yx - x^2)^{-1}.$$

7. Take two copies of each of the symbols $1, 2, \ldots, n$. Prove that there are

$$\sum_{k=0}^{n} (-1)^k 2^{k-n} \binom{n}{k} (2n - k)!$$

permutations on $2n$ positions with no identical symbols adjacent. (By a permutation we mean a sequence of length $2n$ with each of the n available symbols occurring twice.)

8. Let the vector $\mathbf{d} = (d_1, d_2, \ldots)$ have as ith entry the number of nonroot vertices of degree i in a planted plane tree t. We call $\mathbf{d}$ the degree sequence of t. Denote by $c(\mathbf{d})$ the number of planted plane trees with degree sequence $\mathbf{d}$. By suitably selecting a multidimensional weight function on the set of planted plane trees show that

$$c(\mathbf{d}) = \frac{1}{n} \frac{n!}{d_1! \cdots d_n!},$$

if $d_1 + d_2 + \cdots + d_n = n$, and $d_1 + 2d_2 + \cdots + nd_n = 2n - 1$, and $c(\mathbf{d}) = 0$ otherwise.

9. A *2-chromatic tree* is a tree whose vertices are colored with one of two colors such that adjacent vertices have different colors. Let P_i be the set of planted plane 2-chromatic trees with root of color i $(i = 1, 2)$. With appropriate choices of weight functions show that

(a) $P_1 - \{e_1\} \sim \bigcup_{k \geq 1} \{e_1\} \times P_2^k$ (disjoint union)

and

$P_2 - \{e_2\} \sim \bigcup_{k \geq 1} \{e_2\} \times P_1^k$ (disjoint union)

where e_1 and e_2 are the trivial trees in P_1 and P_2, respectively.

(b) Show that there exist

$$\frac{n+m-1}{mn}\binom{n+m-2}{m-1}^2; \qquad m, n \geq 1$$

planted plane 2-chromatic trees on m nonroot vertices of color 1, n nonroot vertices of color 2, and root of color 1.

10. Show that the number of planted plane trees with $2n$ nonroot vertices, $2m + 1$ of which have odd degrees, is

$$\frac{1}{2n+m}\binom{2n}{2m+1}\binom{2n+m}{m}.$$

11. Show that the number of planted plane trees with n nonroot vertices, i of which are of degree $s + 1$, is

$$\frac{1}{n}\sum_{i=k}^{n-1}(-1)^{i-k}\binom{i}{k}\binom{n}{i}\binom{2n-2-(s+1)i}{n-i-1}.$$

12. (a) Prove that there are

$$\frac{m-1}{n-1}\binom{2n-m-2}{n-2}$$

planted plane trees on n nonroot vertices in which the vertex adjacent to the root has degree m.

(b) There are

$$\frac{2}{n-1}\sum_{i=0}^{n-3}(-1)^i(i+1)\binom{2n-3-i}{n}, \qquad n>1$$

planted plane trees on n nonroot vertices with the vertex adjacent to the root being of odd degree.

13. By a *block* of length l of a sequence we understand l successive appearances of the same symbol. The block is called *maximal* if it is not properly contained in another block.

(a) Show that the number of sequences of length m formed with symbols $1, 2, \ldots, n$ in which all maximal blocks have length at least k is the coefficient of x^m in

$$(1-x+x^k)(1-x-(n-1)x^k)^{-1}.$$

(b) Show that the number of such sequences with all maximal blocks of length less than k is the coefficient of x^m in

$$(1-x^k)(1-nx+(n-1)x^k)^{-1}.$$

6 GAUSSIAN POLYNOMIALS

This section is independent of the rest of the chapter. Several occasions arise when we need to work with finite dimensional vector spaces over finite fields. The polynomials that we introduce here are intimately related to the basic arithmetical properties of these spaces. Readers not familiar with finite fields and vectors spaces are referred to Appendix 2.

3.23

For n and k nonnegative integers define polynomials $\left[{n \atop k}\right](x)$ as follows:

$$\left[{n \atop k}\right](x) = \frac{(x^n-1)(x^{n-1}-1)\cdots(x^{n-k+1}-1)}{(x^k-1)(x^{k-1}-1)\cdots(x-1)},$$

with $\left[{n \atop n}\right](x) = 1$ for all n, and $\left[{n \atop k}\right](x) = 0$ for $k > n$ or $k < 0$. The $\left[{n \atop k}\right]$'s so defined are indeed polynomials. We call them *Gaussian polynomials*.

Two important properties of these polynomials may easily be remembered by analogy to those of the binomial numbers.

∗ *The Gaussian polynomials have the following properties*:

(i) $\left[{n \atop k}\right](x) = \left[{n-1 \atop k}\right](x) + x^{n-k}\left[{n-1 \atop k-1}\right](x)$ (the Pascal recurrence).

(ii) $\left[{n \atop k}\right] = \left[{n \atop n-k}\right]$.

Property (i) is reminiscent of Pascal's triangle for binomial numbers. From this property it inductively follows that the $\begin{bmatrix} n \\ k \end{bmatrix}$'s are polynomials. Let us quickly verify (i):

$$\begin{aligned}\begin{bmatrix} n-1 \\ k \end{bmatrix}(x) &+ x^{n-k}\begin{bmatrix} n-1 \\ k-1 \end{bmatrix}(x) \\ &= \frac{(x^{n-1}-1)\cdots(x^{n-1-k+1}-1)}{(x^k-1)\cdots(x-1)} \\ &\quad + \frac{x^{n-k}(x^{n-1}-1)\cdots(x^{n-1-(k-1)+1}-1)}{(x^{k-1}-1)\cdots(x-1)} \\ &= \frac{(x^{n-1}-1)\cdots(x^{n-k+1}-1)}{(x^k-1)\cdots(x-1)}((x^{n-k}-1)+x^{n-k}(x^k-1)) \\ &= \begin{bmatrix} n \\ k \end{bmatrix}(x),\end{aligned}$$

as stated. Property (ii) is even easier to check.

3.24 Combinatorial Interpretations

In direct analogy to the binomial numbers $\binom{n}{k}$, which count the number of subsets of size k of a set with n elements, we have the following interpretation for the Gaussian polynomials $\begin{bmatrix} n \\ k \end{bmatrix}$.

∗ *The number of subspaces of dimension k in a vector space of dimension n over the finite field with q elements is* $\begin{bmatrix} n \\ k \end{bmatrix}(q)$.

Proof. Denote by V the vector space in question. The vector space V contains q^n vectors. We assert that there are $(q^n-1)(q^n-q)\cdots(q^n-q^{k-1})$ ordered selections of k linearly independent vectors in V. [Indeed, we have q^n-1 available choices for the first nonzero vector, q^n-q for the second, ..., q^n-q^{k-1} for the kth. At step r we subtract from q^n the q^{r-1} vectors in the subspace (of dimension $r-1$) generated by our first $r-1$ selections of linearly independent vectors.] A subspace of dimension k is spanned by precisely $(q^k-1)(q^k-q)\cdots(q^k-q^{k-1})$ such ordered selections, the counting being done the same way as above. We therefore conclude that the number of subspaces of dimension k in V is

$$\begin{aligned}&\frac{(q^n-1)(q^n-q)\cdots(q^n-q^{k-1})}{(q^k-1)(q^k-q)\cdots(q^k-q^{k-1})} \\ &\quad = \frac{(q^n-1)(q^{n-1}-1)\cdots(q^{n-k+1}-1)}{(q^k-1)(q^{k-1}-1)\cdots(q-1)} \\ &\quad = \begin{bmatrix} n \\ k \end{bmatrix}(q).\end{aligned}$$

This ends the proof.

Several combinatorial interpretations are possible for the coefficient of a power of x in $\left[{n \atop k}\right](x)$. We mention three:

* *The number of nondecreasing planar lattice paths between* $(0,0)$ *and* $(k, n-k)$ *for which the area enclosed by the path, the x-axis, and the vertical line* $x = k$ *is* m, *equals the coefficient of* x^m *in* $\left[{n \atop k}\right](x)$.

* *The number of partitions of m into at most k parts each less than or equal to* $n - k$ *is the coefficient of* x^m *in* $\left[{n \atop k}\right](x)$.

* *The number of sequences of* 0's *and* 1's, *with* k 1's, $n-k$ 0's, *and* m *inversions, equals the coefficient of* x^m *in* $\left[{n \atop k}\right](x)$. [Recall that by an *inversion* in the sequence we mean a pair of positions $i < j$ with 1 in position i and 0 in position j (contrary to the usual increasing order $0 < 1$).]

There exist visual bijections between the three statements made above. For example, the area underneath a nondecreasing path may be partitioned into vertical strips of width 1 and of positive (integral) heights. These strips then correspond to the parts of the partition. This shows the equivalence of the first two statements.

On the other hand, envision a nondecreasing path as a sequence of 0's and 1's, with 0 indicating a horizontal move and 1 a vertical move. The number of inversions that occur for a 1 in position i equals the number of 0's in the sequence at the right of position i. The number of such 0's may be interpreted as the area of the horizontal strip (of height 1) at the right of that vertical move. The total number of inversions equals the area of all such (disjoint) horizontal strips, which is the whole area underneath the path.

The three statements made above are therefore seen to admit the same numerical solution. Why is this solution the coefficient of x^m in $\left[{n \atop k}\right](x)$? To see this write $b_{m,n,k}$ for the coefficient of x^m in $\left[{n \atop k}\right](x)$. Let $a_{m,n,k}$ denote the number of nondecreasing paths between $(0,0)$ and $(k, n-k)$ with area m underneath. We are asked to prove that $a_{m,n,k} = b_{m,n,k}$. The Pascal recurrence (i) informs us that the b's satisfy the recurrence

$$b_{m,n,k} = b_{m,n-1,k} + b_{m-(n-k),n-1,k-1}.$$

But observe that the a's satisfy the same recurrence:

$$a_{m,n,k} = a_{m,n-1,k} + a_{m-(n-k),n-1,k-1}.$$

[Indeed, there are precisely two positions from which we may reach the point $(k, n-k)$ in one move; these are $(k, n-k-1)$ and $(k-1, n-k)$. The move from $(k, n-k-1)$ adds nothing to the area, since it is a vertical move. The move from $(k-1, n-k)$, being horizontal, adds $n-k$ to the existing area. This explains the recurrence for the a's; geometrically one cannot miss it.] Since the initial values for the a's and b's coincide as well, we inductively conclude that $a_{m,n,k} = b_{m,n,k}$. This ends our discussion of the three combinatorial interpretations.

3.25

We conclude with a brief list of the coefficients of $\begin{bmatrix} n \\ k \end{bmatrix}$. Clearly, $\begin{bmatrix} n \\ 1 \end{bmatrix}(x) = \sum_{i=0}^{n-1} x^i$, for all n. Recall also that $\begin{bmatrix} n \\ k \end{bmatrix} = \begin{bmatrix} n \\ n-k \end{bmatrix}$.

For small values of n and k, the coefficients (when read from left to right) correspond to increasing powers of x:

$\begin{bmatrix} 4 \\ 2 \end{bmatrix}$ 11211

$\begin{bmatrix} 5 \\ 2 \end{bmatrix}$ 1122211

$\begin{bmatrix} 6 \\ 2 \end{bmatrix}$ 112232211

$\begin{bmatrix} 6 \\ 3 \end{bmatrix}$ 1123333211

$\begin{bmatrix} 7 \\ 2 \end{bmatrix}$ 11223332211

$\begin{bmatrix} 7 \\ 3 \end{bmatrix}$ 1123445443211

$\begin{bmatrix} 8 \\ 2 \end{bmatrix}$ 1122334332211

$\begin{bmatrix} 8 \\ 3 \end{bmatrix}$ 112345666543211

$\begin{bmatrix} 8 \\ 4 \end{bmatrix}$ 11235577877553211

$\begin{bmatrix} 9 \\ 2 \end{bmatrix}$ 112233444332211

$\begin{bmatrix} 9 \\ 3 \end{bmatrix}$ 1123457788877543211

$\begin{bmatrix} 9 \\ 4 \end{bmatrix}$ 1123568911 11 12 11 1198653211.

EXERCISES

1. Write the explicit form of $\begin{bmatrix} 3 \\ 2 \end{bmatrix}(x)$, verify that $\begin{bmatrix} 3 \\ 2 \end{bmatrix}(2) = 7$, and interpret the result on an ordinary cube (which plays the role of a vector space of dimension 3 over the field $GF(2)$ with two elements).

2. Draw a 3×3 grid, interpret it as a vector space of dimension 2 over the field $GF(3)$ with three elements, and verify that there are four one-dimensional subspaces. List them.

3. By sorting out the k-dimensional subspace of a vector space of dimension $m + n$ over $GF(q)$ according to the dimension of the intersection with a fixed m-dimensional subspace ($k \le m$), prove the Vandermonde-like identity

$$\begin{bmatrix} m+n \\ k \end{bmatrix}(q) = \sum_{i=0}^{k} q^{(m-i)(k-i)} \begin{bmatrix} m \\ i \end{bmatrix}(q) \begin{bmatrix} n \\ k-i \end{bmatrix}(q).$$

4. Let V and W be vector spaces of dimensions k and n over $GF(q)$, respectively. Find the number of injective (vector space) homomorphisms from V to W. Find the number of surjective homomorphisms.

5. Consider the Diophantine system of equations

$$\left.\begin{aligned} x_0 + x_1 + x_2 + \cdots + x_{n-k} &= k \\ x_1 + 2x_2 + \cdots + (n-k)x_{n-k} &= m \end{aligned}\right\}.$$

Show that the number of its nonnegative integral solutions is the coefficient of x^m in $\begin{bmatrix} n \\ k \end{bmatrix}(x)$.

6. What is the degree of the polynomial $\begin{bmatrix} n \\ k \end{bmatrix}$? If d is the degree of $\begin{bmatrix} n \\ k \end{bmatrix}(x)$ show that x^m and x^{d-m} have the same coefficient. [*Hint*: Think in terms of areas. The degree of $\begin{bmatrix} n \\ k \end{bmatrix}$ is then clearly seen to be $k(n-k)$, etc.]

NOTES

The first two sections contain classical work on inversion. We generally followed the treatment given in [1] and [2]. While the analytic version of the theorem of Bürmann and Lagrange is well known, its formal (algebraic) version appears to be of a recent origin. We refer the reader to the book of Goulden and Jackson [3, p. 26] for detailed information. The point of view toward enumeration taken in Section 5 follows [3] as well; so do the exercises in that section. Combinatorial interpretations for the Gaussian polynomials were known to MacMahon [5]; they appear also in Andrews' book [4].

REFERENCES

[1] C. Berge, *Principles of Combinatorics*, Academic Press, New York, 1971.

[2] J. Riordan, *Combinatorial Identities*, Wiley, New York, 1968.

[3] I. P. Goulden and D. M. Jackson, *Combinatorial Enumeration*, Wiley, New York, 1983.

[4] G. E. Andrews, *The Theory of Partitions*, Addison-Wesley, Reading, MA, 1976.

[5] P. A. MacMahon, *Combinatory Analysis*, 2 vols., Cambridge University Press, Cambridge, 1915 (reprinted by Chelsea, New York, 1960).

CHAPTER 4

Graphs

She must be seen to be appreciated.

Old Saint Pauls' Book I, W. H. AINSWORTH

Many situations may be graphically summarized by marking down a finite set of points and drawing lines between those pairs of points that are related in some way, or share some common attribute. The systematic study of such structures is known as graph theory. To a graph we attach one or more matrices. It is the spectra of these matrices that interest us most. In principle, we seek to understand the relationship that exists between the graph and the eigenstructure of these matrices. Such an approach is often called algebraic graph theory.

Our first section is not algebraic, however. In it we present well-known, classical results due to Euler, Ramsey, and Turán. Gradually we find ourselves leaning toward the algebraic side by discussing strongly regular graphs and then graphs with extreme spectral behavior. This last aspect is of particular interest because of its connections to the construction of efficient and optimal statistical designs. Eigenvalues of the matrix can be directly interpreted as variances of certain estimates. Extreme spectral behavior represents then minimization of variance (or maximization of precision). In this sense the chapter is closely related to, and strongly motivated by, the chapter on optimal statistical design (Chapter 8). The interested reader may wish to study these two chapters simultaneously. Important connections are also found with the material on block designs that appears in Chapter 7.

Let us begin, as is usual and necessary, with a description of the fundamental concepts.

1 CYCLES, TRAILS, AND COMPLETE SUBGRAPHS

4.1

Among n people relationships of friendship may exist. If these relationships are known to an observer, he or she may choose to conveniently display them by drawing n points on a piece of paper (representing the n people) and connecting two points by a line if and only if the two people in question are friends. What results is a figure that displays the relationships of friendship and that we call a *graph*. The n points are called the *vertices* of the graph and the joining lines are called *edges*.

A graph thus consists of vertices and edges. We allow more than one edge to be drawn between two vertices; and a vertex may be joined to itself (such an edge being called a *loop*). Generally a graph is understood to contain multiple edges and multiple loops. Often we wish to forbid these, however; we call a graph *simple* if at most one edge exists between any two vertices and if no loops are present.

4.2 More Terminology

By the *degree* of a vertex we understand the number of edges that emanate from that vertex (a loop contributing 1 to the degree). A *path* is a sequence $v_0e_1v_1e_2v_2 \cdots e_kv_k$ of distinct vertices and edges in which edge e_i has vertices v_{i-1} and v_i for its endpoints. The initial vertex v_0 is called the beginning point of the path and v_k is called the endpoint. Two (distinct) vertices that occur in a path are said to be connected by that path. A graph is called *connected* if any two distinct vertices are connected by a path.

A *cycle* is a sequence $v_0e_1v_1e_2v_2 \cdots e_kv_0$ of vertices and edges with all edges e_i being distinct. (One may say that a cycle is a path with its beginning and end points being the same.) For graphical displays we refer the reader to Section 6 of Chapter 2.

A *tree* is a set of edges that contains no cycles. By a *spanning tree* we mean a tree with $n - 1$ edges, where n is the number of vertices in the graph.

4.3 A Result of Euler

Let G be a graph. An *Eulerian trail* in G is a sequence

$$T = v_0e_1v_1e_2v_2 \cdots e_{m-1}v_{m-1}e_mv_0,$$

in which each edge e_i of G appears exactly once (while vertices v_i may be repeated). Intuitively an Eulerian trail describes a way of "walking" along the edges of a graph, starting at some vertex v_0, traversing each edge exactly once, and ending the walk at the starting place v_0. Not every graph has an Eulerian trail. A characterization of the graphs that do is perhaps the oldest known result in graph theory, and is due to Euler. It is stated as follows.

$*$ *A connected graph admits an Eulerian trail if and only if all its vertices are of even degree.*

Proof. Assume that the graph G has an Eulerian trail T. As we traverse the trail each occurrence of a vertex contributes 2 to the degree of that vertex.

Since each edge occurs exactly once in T, the degree of a vertex will be a sum of 2's, and is therefore even.

We now prove the converse, by contradiction. Assume that G is a connected graph with each vertex of even degree and a *minimum* number of edges, for which an Eulerian trail does not exist. Since the vertices are of even degree a cycle C exists in G. (It may be located by walking along the edges without retracing.) If there are no edges outside C, then C is an Eulerian trail of G, a contradiction. If edges outside C exist we can still construct an Eulerian trail for G as follows: Start at a vertex v_0 of degree 4 or more of C (which exists by connectivity), walk along the edges of C back to v_0, and continue on along an Eulerian trail (which exists by the minimality of G) in the graph the same as G but with edges of C removed. What results is an Eulerian trail of G, contradicting one of our assumptions. This ends the proof.

The characterization given above is a convenient one. To decide whether or not an Eulerian trail exists it suffices to examine the parities of the degrees of vertices. If one vertex has odd degree, there will be no such trail, and thus all attempts to find one will fail.

Another well-known problem, and much of the same spirit, is deciding whether a cycle containing all the vertices of a graph exists. Such a cycle is called *Hamiltonian*. We ask for a convenient characterization of those graphs that admit Hamiltonian cycles. No satisfactory characterization is known, although many sufficient conditions for existence are available in the literature. We refer the reader to [1] and [2] for more information.

4.4 Turán's Theorem

A *subgraph* of a graph G is simply a subset of vertices and edges of G. By an *induced subgraph* we understand a subset S of vertices of G having as edges the edges of G that have both ends in S.

A graph on n vertices is called *complete* if any two distinct vertices are joined by exactly one edge. We denote such a graph by K_n, and observe that K_n has $\binom{n}{2}$ edges. Another graph of interest to us may be constructed as follows. Partition a set of $n = n_1 + n_2 + \cdots + n_r$ vertices into r classes, the ith class containing n_i vertices. Produce a simple graph $K(n_1, n_2, \ldots, n_r)$ by joining each vertex in class i to all vertices outside class i and to none in class i; do this for all i. The resulting graph is called the *complete r-partite* graph with r classes, the ith of which is of size n_i. When the class sizes n_i differ by at most 1 (i.e., they are as nearly equal as possible) we use the simpler notation $K(n, r)$ for this graph. [A display of $K(5, 2)$ will be found in Exercise 4 at the end of Section 2.16.]

The number of edges in $K(n, r)$ is denoted by $e(n, r)$. We can, of course, write out an explicit expression for $e(n, r)$ in terms of n and r, but that would not be necessary in what is to follow.

One class of problems concerning graphs that has received much attention asks for the maximal number of edges that a simple graph on n vertices can have without containing a subgraph of a prescribed kind. When the subgraph in question is the complete graph K_r we have the answer in the following theorem due to Turán.

* *A simple graph on n vertices not containing the complete graph K_r as a subgraph has at most $e(n, r-1)$ edges, and $K(n, r-1)$ is the unique graph on n vertices and $e(n, r-1)$ edges not containing a K_r.*

The proof is by induction on n. Fix $r \geq 3$ and assume that the theorem is true for all graphs on less than n vertices. We show that it is also true for graphs on n vertices, $n \geq r$.

Suppose that the simple graph G has n vertices, $e(n, r-1)$ edges, and contains no K_r. Since $K(n, r-1)$ is a maximal graph without a K_r (in that no edge can be added to it without creating a K_r) the inductive step follows if we show that G is in fact $K(n, r-1)$.

To show this, let x be a vertex of minimal degree [$d(x)$, say] in G. Since the degrees in $K(n, r-1)$ are as equal as possible, $d(x)$ is less than or equal to the smallest degree in $K(n, r-1)$. Take x (and all edges connected to it) away from G. The number of edges in the graph $G - \{x\}$ so obtained is

$$e(G - \{x\}) = e(G) - d(x) \geq e(K(n-1, r-1)) = e(n-1, r-1),$$

where $e(H)$ denotes the number of edges in the graph H. The inductive hypothesis allows us now to conclude that $G - \{x\}$ is a $K(n-1, r-1)$. This implies, in particular, that the degree $d(x)$ equals the minimum degree of $K(n, r-1)$. So what can G itself be? The only way of putting x back to $G - \{x\}$ [which is a $K(n-1, r-1)$] and not produce a K_r is by forcing G to be a $K(n, r-1)$. This ends our proof.

* * *

In Sections 4.5 through 4.9 we discuss a result of Ramsey. It is a theorem of interest to logicians as well as graph theorists. We initially motivate it as a theorem on coloring the edges of a complete graph (each edge being colored either red or blue), but a more general statement and proof are in fact given.

4.5 A Theorem of Ramsey

It is self-evident that when $n + 1$ pigeons fly into n pigeon holes, one of the holes contains two or more pigeons. This fact (often called the "pigeon-hole" principle) can take many much less obvious forms.

Consider, for instance, coloring (in red and blue) the 15 edges of a complete graph on 6 vertices. No matter how this is done we always have either a red triangle or a blue triangle. Though on the surface a much less obvious statement, it is (upon careful examination) equivalent to the "pigeon-hole" principle.

4.6

Ramsey's theorem is a vastly generalized version of this principle. Fix positive integers p and q and ask yourself: "*Can I find a* (large) *complete graph that, no matter how I color its edges in red and blue, will contain either a red complete subgraph of size p or a blue one of size q*?" The answer is always *yes*, and the actual size of such a large graph has to do with the so-called Ramsey numbers.

By the *size* of a complete graph we mean the number of its vertices. By a *blue complete subgraph* we understand a complete subgraph with all its edges colored blue (similarly for red). If $p = q = 3$, our introductory remarks allow us to conclude that the (large) graph in question can be the complete graph on six (or more) vertices. (Can it be the complete graph on five vertices? Why not?)

4.7

Instead of working with 2-subsets of a set (i.e., simple graphs) we formulate Ramsey's theorem for the r-subsets of a set. For a set S with n elements color each of its $\binom{n}{r}$ subsets of size r either red or blue. [That is, line up the r-subsets into $\binom{n}{r}$ columns and color each column either red or blue.] Call a subset of S *red* if its cardinality is at least r and all its r-subsets are colored red (parallel definition goes for blue).

Throughout this section the symbol $(p, q; r)$ indicates that p, q, and r are positive integers with both p and q larger than or equal to r.

Let a triple $(p, q; r)$ be given. The *Ramsey number* $R(p, q; r)$ is the smallest positive integer so that if the r-subsets of a set S with at least $R(p, q; r)$ elements are colored in red and blue (in any way whatsoever), then there exists either a red subset of size p of S or a blue one of size q. [In the introductory lines of this section we remarked that $R(3, 3; 2) \leq 6$. Show that we actually have equality.] In general it is not clear that the Ramsey numbers (as defined above) even exist. It is Ramsey's theorem that assures their existence.

A few values for $R(p, q; 2)$ are listed below:

p \ q	2	3	4	5	6	7
2	2	3	4	5	6	7
3	3	6	9	14	18	23
4	4	9	18			
5	5	14				
6	6	18				
7	7	23				

4.8

We now prove the following theorem due to Ramsey.

* *Given any integers p, q, and r, with $p, q \geq r \geq 1$, there exists a number $R(p, q; r)$ such that for any set S with at least $R(p, q; r)$ elements and any coloring in red and blue of the r-subsets of S, either there is a red p-subset of S or there is a blue q-subset of S.*

The proof is in steps.

Step 1. $R(p, q; 1) = p + q - 1$.

Since $r = 1$, we color in red and blue the 1-subsets (or elements) of S. If $|S| \geq p + q - 1$ and we have p or more red elements, these red elements form the required red subset. If we have less than p red elements, we necessarily have at least q blue ones, and they form the blue subset. On the other hand, if $|S| \leq p + q - 2$ we can color $p - 1$ elements red and $q - 1$ blue and no red nor blue subsets of required sizes exist.

Step 2. $R(p, r; r) = p$ and $R(r, q; r) = q$.

We show that $R(p, r; r) = p$, the other assertion having a parallel proof. Let $|S| \geq p$. If we have one blue r-subset of S we are done. Else all r-subsets of S are red and thus S itself is a red set; since $|S| \geq p$ we are done again. If $|S| \leq p - 1$, we can color all r-subsets of S red yet (naturally) we will not find a red subset of size p, since $|S| \leq p - 1$.

Step 3 (The Main Step). *If Ramsey's theorem is true for every triple $(p^*, q^*; r - 1)$ and for the triples $(p - 1, q; r)$ and $(p, q - 1; r)$, then it is true for $(p, q; r)$. In fact, if we let*

$$p_1 = R(p - 1, q; r) \qquad \textit{and} \qquad q_1 = R(p, q - 1; r),$$

then

$$R(p, q; r) \leq R(p_1, q_1; r - 1) + 1.$$

Let S be a set with $R(p_1, q_1; r - 1) + 1$ elements and let its r-subsets be colored red and blue. We have to show that either there exists a red subset A of S with p elements or a blue subset B of S with q elements. Fix an element x of S. There is a "natural" coloring of the $(r - 1)$-subsets of $S - \{x\}$ induced by the coloring of the r-subsets of S: To decide the color of an $(r - 1)$-subset σ of $S - \{x\}$ form $\sigma \cup \{x\}$; the color of σ is the color of $\sigma \cup \{x\}$.

Since $|S - \{x\}| = R(p_1, q_1; r - 1)$ either there exists $\bar{A}$ ($|\bar{A}| = p_1$), a subset of $S - \{x\}$ with all its $(r - 1)$-subsets colored red, or there exists a subset $\bar{B}$ ($|\bar{B}| = q_1$) with all its $(r - 1)$-subsets colored blue. Suppose the first possibility occurs. Since $|\bar{A}| = p_1 = R(p - 1, q; r)$ either there exists a subset B ($|B| = p$) of $\bar{A}$ with all the r-subsets of B being blue (in which case B is the required subset), or there exists a subset $\tilde{A}$ ($|\tilde{A}| = p - 1$) of $\bar{A}$ with all the r-subsets of $\tilde{A}$ colored red (in which case we can take $A = \tilde{A} \cup \{x\}$ as the required subset). We deal with the second possibility in the same way.

Step 4. If Ramsey's theorem is true for every triple $(p^*, q^*; r-1)$ *and for every triple* $(p_0, q_0; r)$ *with* $p_0 + q_0 \leq m - 1$, *then it is true for any triple* $(p, q; r)$ *for which* $p + q = m$.

Indeed, if $p + q = m$, then by hypothesis Ramsey's theorem is true for $(p-1, q; r)$ and $(p, q-1; r)$. The conclusion follows now from Step 3.

Step 5. If Ramsey's theorem is true for every triple $(p^*, q^*; r-1)$, *then it is true for any triple* $(p, q; r)$.

By Step 2, $R(r, r; r) = r$. Hence Ramsey's theorem is true for the triple $(r, r; r)$ with $p + q = r + r = 2r$ (recall that we assume at all times $p, q \geq r$). The statement follows from Step 4 by induction on $p + q$.

Step 6. From Step 1 Ramsey's theorem is true for triples $(p, q; 1)$, with $r = 1$. The general validity of Ramsey's theorem follows now from Step 5 by induction on r. This ends our proof.

(*Aside*: This is the most "nonconstructive" proof I know, with the exception of the next.)

4.9

So far we have been using two colors only. A multicolored version exists and it is stated as follows.

The Multicolored Version of Ramsey's Theorem. *Given integers* $p_1, p_2, \ldots, p_k$ *and* r, *with* $p_i \geq r \geq 1$ (*for all* i), *there exists a number* $R(p_1, p_2, \ldots, p_k; r)$ *such that for any set* S *with at least* $R(p_1, p_2, \ldots, p_k; r)$ *elements and any coloring of the r-subsets of S with colors "1," "2," ..., "k," a subset of S of size* p_i *and color "i" exists for some i.*

(By a subset of color "i" we understand, of course, a subset all of whose r-subsets are colored "i.")

The proof is by induction. We proved the theorem for $k = 2$ in Section 4.8. Suppose that it has been shown true for k less than m. To prove it for $k = m$, let S be a set of size $R(p_1, R(p_2, p_3, \ldots, p_k; r); r)$ (this expression involves two Ramsey numbers, both of which exist by the inductive assumption). Ramsey's theorem with $k = 2$ allows us to conclude that either there exists a subset of S of size p_1 with all its r-subsets colored "1" (in which case we are done), or there exists a subset of size $R(p_2, p_3, \ldots, p_k; r)$ with none of its r-subsets colored "1." In this latter case we are done by induction (since we have one color less).

EXERCISES

1. Can a graph with an even number of vertices and an odd number of edges contain an Eulerian trail? Explain.

2. Show that a simple graph on n vertices and more than $n^2/4$ edges must contain a triangle.

3. We call a graph *regular* if each vertex has the same degree. A complete graph K_3 will be called a *triangle*. Prove the following: A regular simple graph on mn vertices with $\binom{m}{2}n^2$ edges has at least $\binom{m}{3}n^3$ triangles. If it has precisely $\binom{m}{3}n^3$ triangles, then it necessarily is the m-partite graph $K(mn, m)$, which has m parts of size n each.

4. A simple graph on n vertices and m edges contains at least

$$m(4m - n^2)/3n$$

triangles. Prove this.

5. Show that a simple graph on n vertices and m edges, with $n^2/4 \leq m \leq n^2/3$, contains at least $n(4m - n^2)/9$ triangles.

6. Show that $R(4,4;2) \leq 24$, and $R(3,3;2) \geq 6$.

7. Show that $R(3,3,3,3;2) \leq 66$ and that $R(4,3;2) \geq 8$.

8. Prove that $R(2,\ldots,2;1) = k + 1$, if we have k 2's.

9. Show that $R(p,q;2) \leq R(p-1,q;2) + R(p,q-1;2)$.

10. The Ramsey numbers $R(p,q;2)$ satisfy $R(p,q;2) \leq \binom{p+q-2}{p-1}$. Prove this.

11. Every sequence of $n^2 + 1$ distinct integers contains either an increasing subsequence of length $n + 1$ or a decreasing sequence of length $n + 1$. Show this.

12. (Erdös and Szekeres.) For any given integer $k \geq 3$ there exists an integer $n = n(k)$, such that any n points in a plane (no three on a line) contain k points that form a convex k-gon. Prove this, and also show that $n(k) \leq R(k,5;4)$. Can you prove that, in fact, $n(k) \leq \binom{2k-4}{k-2} + 1$?

13. Denote by R_k the Ramsey number $R(p_1, p_2, \ldots, p_k; 2)$ with $p_i = 3$, for all i. Show that $R_k \leq k(R_{k-1} - 1) + 2$.

14. (Schur.) Let $S_1, S_2, \ldots, S_k$ be any partition of the set of integers $\{1, 2, \ldots, R_k\}$ (see Exercise 13 for the definition of R_k). Then, for some i, S_i contains three integers x, y, and z (not necessarily distinct) satisfying the equation $x + y = z$.

2 STRONGLY REGULAR GRAPHS

4.10

Important connections exist between the contents of this section and Section 7 of Chapter 7 on association schemes.

A graph is called *regular* if all its vertices have the same degree. By a *strongly regular* graph we understand a regular simple graph with the property that the number of vertices adjacent to v_1 and v_2 ($v_1 \neq v_2$) depends only on whether or not vertices v_1 and v_2 are adjacent. (Two vertices are *adjacent* if an edge exists between them.) The *parameters* of a strongly regular graph are (n, d, a, c), where n is the total number of vertices, d is the degree of a vertex, and a or c denote the number of vertices adjacent to v_1 and v_2, according to whether v_1 and v_2 are adjacent or not.

4.11 Examples and Constructions of Strongly Regular Graphs

The graphs we just defined are beautiful indeed. Somewhat trivial examples are cycles, complete graphs, and complete multipartite graphs with parts (or classes) of the same size. Due to large cardinalities we draw only the strongly regular graph with parameters $(10, 3, 0, 1)$, known as Petersen's graph:

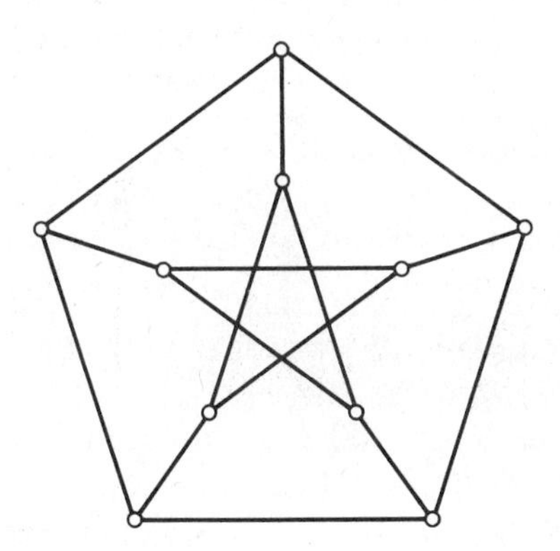

A nontrivial strongly regular graph on nine vertices exists; it has parameters $(9, 4, 1, 2)$ and it is displayed in Exercise 5 at the end of Section 2.16.

Let v_1 and v_2 be two adjacent vertices in a strongly regular graph with parameters (n, d, a, c). It is not difficult to see that there are precisely $d - c - 1$ vertices adjacent to one but not the other of the two vertices v_1 and v_2; and that there are $(n - d - 1) - (d - c - 1)$ vertices adjacent to neither. Similar statements hold true for a pair of nonadjacent vertices.

With these remarks in mind we now define the *complement* of a simple graph G to be the graph $\overline{G}$ with the same vertices as G, two vertices being adjacent in $\overline{G}$ if and only if they are not adjacent in G. By what we just said,

G is strongly regular if and only if $\overline{G}$ is.

Several classes of such graphs are now described.

(a) The *triangular graph* $T(m)$ has as vertices the subsets of cardinality 2 of a set with m elements, $m \geq 4$. Two vertices are adjacent if and only if the corresponding 2-subsets are not disjoint. [The Petersen graph is the complement of $T(5)$.] The graph $T(m)$ has parameters $\left(\binom{m}{2}, 2(m-2), m-2, 4\right)$.

(b) The *lattice graph* $L_2(m)$ is the graph with vertex set $S \times S$ where S is a set of m elements, $m \geq 2$. Two vertices are joined if and only if they have a common coordinate. Its parameters are $(m^2, 2(m-1), m-2, 2)$. The graph in Exercise 5 at the end of Section 2.16 is $L_2(3)$.

(c) Consider the finite field $GF(q)$, with q elements, where q is a prime power equal to 1 modulo 4. Define a graph whose vertices are the elements of the field, two elements being adjacent if and only if their difference is a nonzero square. (The element y is a square if $y = x^2$ for some element x of the field.) We denote the resulting strongly regular graph by $P(q)$ and call it the *Paley graph*. It has parameters $(q, \frac{1}{2}(q-1), \frac{1}{4}(q-5), \frac{1}{4}(q-1))$.

(d) A *disjoint union of m complete graphs* each with r vertices yields a strongly regular graph. Its complement is a multipartite graph (m-partite to be exact) with parts of equal size (each with r vertices).

(e) A quite general way of constructing strongly regular graphs is through special actions of groups on sets. We recommend Appendix 1 and the references given there to readers not familiar with such notions.

Let group G act on set S. Assume that G induces just one orbit on S (such action being called *transitive*). We may naturally extend the action of G to $S \times S$ componentwise by writing $g(x, y) = (g(x), g(y))$, for g in G and (x, y) in $S \times S$. The set $S \times S$ is thus partitioned into orbits, one of which is (by transitivity) the diagonal $D = \{(x, x) : x \in S\}$. We call G a *rank 3 group* if there are precisely two other orbits, O_1 and O_2, on $S \times S$ in addition to the diagonal D. An orbit on $S \times S$ is called symmetric if whenever (x, y) belongs to it so does (y, x). *From a rank 3 group, with symmetric* (nondiagonal) *orbit O_1, we obtain a strongly regular graph as follows: The vertices are elements of S, two vertices x and y being adjacent if and only if (x, y)* [and hence also (y, x)] *belongs to O_1.* [For a rank 3 group it is clear that if O_1 is symmetric so is O_2, and vice versa. If the pairs in O_1 form a strongly regular graph, then the pairs in O_2 produce the complementary graph. One last remark: an easy way to ensure that O_1 is symmetric is to work with a rank 3 group G of even order. An element of order two exists then in G and this group element sends necessarily x to y and y to x, for some x and y in S. The orbit of (x, y) is then obviously symmetric.] All regular graphs mentioned so far may be obtained from suitable rank 3 groups in the manner just described. The Petersen graph arises by acting with the group S_5 of permutations on five symbols on the set S of subsets of cardinality 2 of $\{1,2,3,4,5\}$; $|S_5| = 120$, $|S| = \binom{5}{2} = 10$. Write out the details in this case, to firm up the abstractions.

4.12 The Eigenstructure

The *adjacency matrix* of a simple graph G is a vertex versus vertex matrix with 0's on the diagonal, a 1 in position (i,j) if there is an edge between i and j, and a 0 in position (i,j) otherwise. We write $A = (a_{ij})$ for the adjacency matrix.

As we just mentioned, the entries a_{ij} of A are 0 or 1, and A is a symmetric matrix, that is, $a_{ij} = a_{ji}$ for all i, j. Furthermore, if we denote by J the square matrix with all entries 1, a simple graph is regular of degree d if and only if its adjacency matrix A satisfies

$$AJ = JA = dJ. \tag{4.1}$$

In addition, if the graph is strongly regular with parameters (n, d, a, c), then the (i,j)th entry of A^2 is equal to the number of vertices adjacent to i and j; this number is d, a, or c according as i and j are the same, adjacent, or nonadjacent. This may be conveniently expressed as follows:

$$A^2 = dI + aA + c(J - I - A), \tag{4.2}$$

where I is the identity matrix.

Think of I, J, A as linear maps (of an n-dimensional vector space over the real numbers to itself). The vector $\mathbf{1}$, with all entries 1, is an eigenvector for all three maps [see (4.1)]. Select now an orthogonal basis $\mathbf{1}, x_2, \dots, x_{n-1}$ consisting of eigenvectors of A (since A is symmetric such a basis exists). In this basis the mappings A, J, and I are all represented by diagonal matrices. Equation (4.2) allows us now to compute the eigenvalues of A. As (4.1) indicates $A\mathbf{1} = d\mathbf{1}$, so d is the eigenvalue corresponding to the eigenvector $\mathbf{1}$. Any other eigenvector x with eigenvalue λ is perpendicular on $\mathbf{1}$ (thus $Jx = \mathbf{0}$), and thus applying (4.2) to x yields

$$\lambda^2 x = dx + a\lambda x + c(\mathbf{0} - x - \lambda x).$$

Comparing the coefficients of the vector x on both sides we obtain a quadratic equation for λ:

$$\lambda^2 = (d - c) + (a - c)\lambda.$$

There are therefore at most two distinct eigenvalues attached to the eigenvectors of A perpendicular on $\mathbf{1}$:

$$\lambda_1, \lambda_2 = \tfrac{1}{2}\left\{(a - c) \pm \left[(a - c)^2 + 4(d - c)\right]^{1/2}\right\}.$$

If f_1 and f_2 are the respective multiplicities of λ_1 and λ_2, we may determine them from the two equations that they must satisfy:

$$1 + f_1 + f_2 = n \qquad \text{(the dimension)}$$

$$1 \cdot d + f_1\lambda_1 + f_2\lambda_2 = 0 \qquad \text{(the trace of } A\text{)}.$$

We summarize as follows:

∗ *The adjacency matrix of a strongly regular graph with parameters* (n, d, a, c) *has eigenvalues* d, λ_1, λ_2 *with multiplicities* $1, f_1, f_2$ *where*

$$\lambda_1, \lambda_2 = \tfrac{1}{2}\left\{(a - c) \pm \left[(a - c)^2 + 4(d - c)\right]^{1/2}\right\}$$

and (4.3)

$$f_1, f_2 = \tfrac{1}{2}\left\{n - 1 \pm [(n - 1)(c - a) - 2d]\left[(a - c)^2 + 4(d - c)\right]^{-1/2}\right\}.$$

The multiplicities must, of course, be integral and this offers useful arithmetical conditions that the parameters of a strongly regular graph must satisfy.

4.13 Characterization by Parameters

The parameters of a strongly regular graph determine the eigenvalues of the adjacency matrix and their multiplicities, as we have just seen.

Knowledge of the dimension and the eigenvalues alone determines in turn the parameters. Indeed, if d, λ_1, λ_2 are the eigenvalues, then n and d are already determined and a simple computation based on (4.3) gives $a = d + \lambda_1 + \lambda_2 + \lambda_1\lambda_2$ and $c = d + \lambda_1\lambda_2$.

However, a more interesting question is whether the parameters themselves determine uniquely the strongly regular graph. Generally this is not the case. Yet for several well-known families of parameters a unique strongly regular graph with given parameters exists. These questions were addressed by Bose and some of his students. We illustrate the method of proof by characterizing the triangular graphs:

∗ *A strongly regular graph with the same parameters as* $T(m)$ *must be* $T(m)$, $m \neq 8$.

Proof. Assume $m > 8$. The parameters in question are

$$\left(\binom{m}{2}, 2(m - 2), m - 2, 4\right).$$

Let G be a strongly regular graph with these parameters. For a vertex x in G denote by $G(x)$ the *induced subgraph* whose vertex set consists of the $2(m - 2)$ vertices adjacent to x. The graph $G(x)$ is regular of degree $m - 2$, since any vertex y in $G(x)$ is adjacent to x and there are precisely $m - 2$ vertices adjacent to both y and x [all of which are in $G(x)$]. Let y, z be a pair of nonadjacent vertices in $G(x)$ and let p be the number of vertices in $G(x)$ adjacent to both y and z. Since $c = 4$, and since x is adjacent to y and z, we must have $p \leq 3$. There are $c = m - 2$ vertices in $G(x)$ adjacent to x and y; of these p are adjacent also to z and $m - z - p$ are not. We conclude that there are $m - z - p$ vertices in $G(x)$ adjacent to y but not to z, $m - z - p$ vertices adjacent to z but not to y, and $(2(m-2) - 2) - 2(m - 2 - p) - p = p - 2$ vertices adjacent to neither. Necessarily $p - 2 \geq 0$ and hence $p \geq 2$. If $p = 3$, let w be the (sole) vertex adjacent to neither y nor z. Then every vertex of $G(x)$ adjacent to w is adjacent to either y or z, and hence $m - 2 =$ degree of $w \leq 3 + 3 =$ (number of vertices in $G(x)$ adjacent to y and w) + (same for z and w). This implies $m \leq 8$, a contradiction to our assumption that $m > 8$. Hence $m = 2$, that is, there exist precisely two vertices in $G(x)$ adjacent to both vertices of any given pair of nonadjacent vertices of $G(x)$.

We now determine uniquely the graph $\overline{G(x)}$, the complement of $G(x)$. The graph $\overline{G(x)}$ is regular of degree $2(m-2) - 1 - (m-2) = m - 3$ and it contains no triangles [since $p = 2$ and there are $p - 2 = 0$ vertices nonadjacent to a pair of nonadjacent vertices in $G(x)$]. Suppose cycles of odd lengths exist in $\overline{G(x)}$. Let q be the *minimal* length of an odd cycle. By its minimality this cycle is in fact an induced subgraph. We know that $q > 3$, and in fact $q = 5$. [Otherwise there will be more than two vertices adjacent to a pair of nonadjacent vertices in $G(x)$.] Denote this cycle by $C = (x_0, x_1, x_2, x_3, x_4)$. There are $m - 5$ vertices adjacent to x_0 in $\overline{G(x)}$, other than x_1 and x_4; all of these must be nonadjacent to both x_1 and x_4 (else a triangle occurs). For the same reasons there are $m - 5$ vertices nonadjacent to x_1 and x_3. Since x_1 is nonadjacent to $m - 4$ vertices outside C, there are at least $m - 6$ vertices outside C nonadjacent to both x_3 and x_4. (At this point the reader should draw a picture of what is happening.) Since x_3 and x_4 are adjacent, there are $p = 2$ vertices nonadjacent to both, and hence we must have $m - 6 \leq 2$, or $m \leq 8$ (a contradiction to the assumption $m > 8$). We conclude that $\overline{G(x)}$ contains no odd cycles. The fact that $\overline{G(x)}$ is regular of degree $m - 3$ with no cycles of odd lengths forces $\overline{G(x)}$ to be the bipartite graph with $n - 2$ vertices in each part, and each vertex adjacent to all vertices but one in the opposite part.

The graph $G(x)$ contains therefore two disjoint complete graphs on $m - 2$ vertices each. It thus follows that every vertex in G lies in two complete subgraphs, each of size $m - 1$, and that any edge of G is in a unique complete subgraph of size $m - 1$. The complete subgraphs of size $m - 1$ are called *grand cliques*. In all there are $2\binom{m}{2}(m-1)^{-1} = m$ grand cliques, any two having exactly one vertex in common. We can now identify each vertex by the two grand cliques to which it belongs, two vertices being adjacent if there is a common grand clique that occurs in both. That is, if we label the grand cliques $1, 2, \ldots, m$ we obtain the graph G by taking as vertices the $\binom{m}{2}$ subsets of size two with an edge between two vertices if and only if the 2-subsets in question are not disjoint. This shows that G is $T(m)$.

For $m < 8$ we can analyze each case individually to obtain the same results. In case $m = 8$, however, three exceptional graphs arise; they were discovered by Chang. This ends the proof.

Srikhande proved the uniqueness of $L_2(m)$, for all $m \neq 4$. In case $m = 4$ a single exceptional graph exists. It consists of 16 vertices produced by 3 parallel classes of lines on the torus. We end Section 4.13 by stating Srikhande's result:

* *A strongly regular graph with parameters the same as* $L_2(m)$ *must be* $L_2(m)$, $m \neq 4$. *Apart from* $L_2(4)$ *an exceptional strongly regular graph with the same parameters as* $L_2(4)$ *exists and is drawn below*:

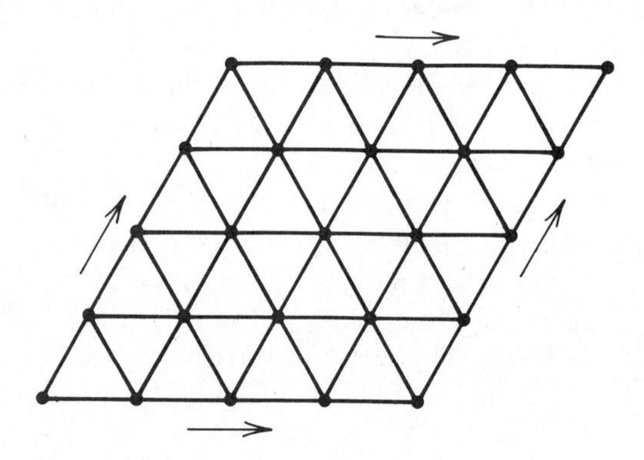

(The graph is, as we mentioned, drawn on a torus so opposite sides should be identified as shown.) The proof of this result may be obtained by arguments similar to those used in the triangular case. In fact Bose [3] gives a unifying result on such characterizations; a result that comprises, among other things, the triangular and L_2 types of strongly regular graphs.

An example of a family of strongly regular graphs not characterized by its parameters consists of the Paley graphs $P(q)$ mentioned in part (c) of Section 4.11.

EXERCISES

1. Find the parameters of the graph complementary to a strongly regular graph with parameters (n, d, a, c).
2. Display the Paley graph $P(13)$.
3. Let G be the group of all 5! permutations on the symbols $\{1, 2, 3, 4, 5\}$ and let S be the $\binom{5}{2} = 10$ 2-subsets of $\{1, 2, 3, 4, 5\}$. The group G acts on S by

sending $\{i, j\}$ to $\{g(i), g(j)\}$, where g is an arbitrary group element. Show that G acts transitively on S and that it is in fact a rank 3 group. Its orbits on $S \times S$ are symmetric. What is the strongly regular graph that results from one of them?

4. Let G be a strongly regular graph on $2m$ vertices whose eigenvalues of the adjacency matrix have multiplicities $1, m - 1, m$. Then:

(a) G or $\overline{G}$ consists simply of a set of parallel edges, or

(b) The parameters of G or those of $\overline{G}$, are $(4s^2 + 4s + 2, s(2s + 1), s^2 - 1, s^2)$, for some positive integer s.

[*Hint*: Look at the traces of A, A^2, and A^3.]

5. Construct a strongly regular graph as follows. Take a vector space V of dimension 2 over $GF(q)$. Partition the $q + 1$ subspaces of dimension 1 into two subsets P and Q of size $\frac{1}{2}(q + 1)$ each. As vertices take the q^2 vectors of V, two vectors x and y are adjacent if and only if $\langle x - y \rangle \in P$, that is, if the subspace generated by $x - y$ is in P. Show that the resulting graph has the same parameters as the Paley graph on q^2 vertices. Is it the same as the Paley graph?

6. The *Clebsch* graph has as vertices all subsets of a set with five elements. Two subsets A and B are adjacent whenever their symmetric difference $(A - B) \cup (B - A)$ has cardinality 4. Show that the Clebsch graph is strongly regular. Compute its parameters. Show that it is the unique strongly regular graph with these parameters.

7. (Delsarte, Goethals, Seidel.) Let G be a strongly regular graph on n vertices having the property that G and $\overline{G}$ are both connected. Show that $n \leq \frac{1}{2}f(f + 3)$, where f $(f > 1)$ is the multiplicity of an eigenvalue of the adjacency matrix of G.

3 SPECTRA, WALKS, AND ORIENTED CYCLES

In this section we discuss some basic connections that exist between the spectrum of a graph and certain graph theoretic features such as the number of cycles, spanning trees, and walks.

To keep the vocabulary simple, and at no real loss of generality, we assume that the graphs in question contain no loops. The *adjacency matrix* of a graph G on n vertices is a $n \times n$ vertex versus vertex matrix $A(G) = (a_{ij})$ with $a_{ii} = 0$ and a_{ij} equal to the number of edges between vertices i and j, $i \neq j$. (In case G is a simple graph $a_{ij} = 0$ or 1, according to whether vertices i and j are adjacent or not.)

Another important matrix attached to the graph G is the $n \times n$ vertex versus vertex matrix $C(G) = (c_{ij})$, with c_{ii} being the degree of vertex i, and c_{ij}

being the negative of the number of edges between vertices i and j, $i \neq j$. We call $C(G)$ the *Kirchhoff matrix* of the graph.

To a graph G we thus attach two symmetric matrices: the adjacency matrix $A(G)$ and the Kirchhoff matrix $C(G)$.

4.14 Recalling Some Matrix Theory

We suggest at the very outset that the reader visualize a square matrix of dimension n as both an array of numbers and a linear map from a n-dimensional vector space (over the real numbers) to itself.

Let A be a symmetric matrix of dimension n. A scalar λ and a nonzero vector v are called, respectively, an *eigenvalue* and *eigenvector* of A if $Av = \lambda v$. Composition of linear maps corresponds to multiplication of matrices; thus A^k signifies the kth power of A or the k-fold composition of A with itself. We observe at once that *if* λ *and* v *are an eigenvalue and an eigenvector of* A, *then* λ^k *and* v *are an eigenvalue and an eigenvector of* A^k. [Indeed, $Av = \lambda v$ implies $A^2v = A(Av) = A(\lambda v) = \lambda Av = \lambda\lambda v = \lambda^2 v$, and iteratively $A^k v = \lambda^k v$, for any nonnegative integer k.] The eigenvalues of a symmetric matrix are real numbers.

Write a_{ij} for the entries of A. The *determinant* of A is then defined as the scalar $\Sigma_\sigma(\pm 1)\Pi_{i=1}^n a_{i\sigma(i)}$, where σ ranges over all permutations on the index set $1, 2, \ldots, n$ and the sign ± 1 is positive for even permutations and negative for odd ones. We write $|A|$ for the determinant of A. (It turns out that the determinant is an attribute of the linear map, for it remains the same upon a change of basis.) One remarkable feature of the determinant is its multiplicative behavior, that is, $|AB| = |A|\,|B|$.

For indeterminates $s_1, s_2, \ldots, s_n$ we define their kth *elementary symmetric sum* as a sum of $\binom{n}{k}$ monomials, each monomial being the product of k distinct indeterminates out of the n available, $0 \leq k \leq n$. [Thus the first elementary symmetric sum is simply the sum $s_1 + s_2 + \cdots + s_n$, while the nth is their product $s_1 s_2 \cdots s_n$. As a further example, with four indeterminates the second elementary symmetric sum is $s_1s_2 + s_1s_3 + s_1s_4 + s_2s_3 + s_2s_4 + s_3s_4$, a sum of $\binom{4}{2}$ monomials.] We wish to study the elementary symmetric sums of eigenvalues of a symmetric matrix. The 0th elementary symmetric sum is by definition taken to be 1.

To do so denote by I the identity matrix of dimension n and write $ch_A(x)$ for the determinant $|A - xI|$. By the definition of the determinant $ch_A(x)$ is a polynomial of degree n in x, which we call the *characteristic polynomial* of (the symmetric matrix) A. Note that if $B = PAP^{-1}$, then $ch_B(x) = ch_A(x)$. [Indeed, $ch_B(x) = |B - xI| = |PAP^{-1} - xI| = |P(A - xI)P^{-1}| = |A - xI|\,|P|\,|P^{-1}| = |A - xI| = ch_A(x)$.] In particular, let P be the matrix of eigenvectors of A, so that B will be the diagonal matrix of eigenvalues of A. Then $ch_B(x) = |B - xI| = \Pi_{i=1}^n(\mu_i - x)$, with the μ_i's the eigenvalues of A. The coefficient of x^k in $ch_B(x)$ is precisely the $(n-k)$th elementary symmetric sum of the μ_i's. A direct computation of $ch_A(x)$ reveals, however, that the coefficient of x^k in $ch_A(x)$ equals the sum of the determinants of the $\binom{n}{n-k}$ principal minors of dimension $n - k$ in A. [This can be seen by observing that the coefficient of x^k in $|A - xI|$ surfaces upon picking x out of precisely k diagonal elements of $A - xI$ and completing this selection in $(n - k)!$ ways according to the expression of the determinant of the principal minor of A complementary to the rows and columns of the k initial diagonal choices for x.]

Since $ch_A(x) = ch_B(x)$ the coefficient of x^{n-k} is the same on both sides and we conclude that

The kth elementary symmetric sum of eigenvalues equals the sum of determinants of the $\binom{n}{k}$ principal minors of size k. (4.4)

Particular cases of this result are of note. For $k = 1$ we are informed that the sum of eigenvalues of A is equal to the sum of the diagonal entries of A; this common value is called the *trace* of A. When $k = n$ we conclude that the product of the eigenvalues of A is equal to the determinant of A.

Another particular case of interest is obtained by applying the result in (4.4), with $k = 1$, to A^r. As we emphasized in the opening passage of this section, the eigenvalues of A^r are μ_i^r, where the μ_i's are the eigenvalues of A. We thus conclude as follows:

The sum of the rth powers of eigenvalues equals the sum of the diagonal entries of the rth power of the matrix. (4.5)

[In other words $\Sigma_i \mu_i^r = \text{trace}(A^r)$, where the μ_i's are the eigenvalues of A.]

4.15 The Adjacency Matrix

The adjacency matrix $A(G)$, being symmetric, has real eigenvalues that we denote by $\lambda_0(G) \le \lambda_1(G) \le \cdots \le \lambda_{n-1}(G)$. A remarkable feature of $A(G)$ is that its entries are nonnegative integers. Since the remarks we are about to make are true for any graph G, we abbreviate by writing A for $A(G)$ and λ_i for the ith eigenvalue of $A(G)$.

We wish to draw attention to several graph theoretic interpretations of certain symmetric functions of eigenvalues. One of these concerns the elementary symmetric sums, the other the sums of powers. The notions of oriented cycles and walks must be introduced in order to accomplish this.

By a *walk* in a graph we understand a sequence $v_0 e_1 v_1 e_2 v_2 \cdots v_{k-1} e_k v_k$ of (possibly repeated) vertices and (possibly repeated) edges in which vertices v_i and v_{i+1} are the endpoints of edge e_i. We call v_0 the *starting point* of the walk and v_k the *endpoint*. The number of not necessarily distinct edges that occur (or are traversed) in a walk is known as its *length*. A *closed walk* is a walk in which the beginning point and endpoint are the same. (A path is a special kind of a walk, one in which there are no repeated vertices or edges. And a cycle is a closed walk with no repeated vertices or edges, except for the beginning point and endpoint.) Two walks are *the same* if they are described by the same sequence of vertices and edges.

We now state and prove the following:

The number of walks of length r from vertex i to vertex j equals the (i, j)th *entry of* A^r. (4.6)

Proof. Denote by a_{ij}^s the (i, j)th entry of A^s. The proof is by induction on r. For $r = 1$ the statement is easily verified.

Suppose that the result is true for $r - 1$. A walk of length r from i to j consists of a walk of length $r - 1$ from i to some vertex k (say) plus the edge joining k and j. By the inductive assumption there are a_{ik}^{r-1} walks of length $r - 1$ from i to k. Thus the number of walks of length r from i to j is

$$\sum_{\substack{k \\ \{k,j\} \text{ is} \\ \text{an edge}}} a_{ik}^{r-1} = \sum_{\substack{k \\ \text{all vertices } k}} a_{ik}^{r-1} a_{kj} = a_{ik}^{r},$$

as stated. This ends our proof.

Particular cases of this result soon command our attention. For instance, the ith diagonal entry of A^r (written a_{ii}^r) counts the number of closed walks beginning and ending at i. Thus the sum $\Sigma_i a_{ii}^r$ is an expression for all closed walks of length r in the graph. One recognizes in this last sum the trace of A^r, however, and since this trace equals also the sum $\Sigma_{i=0}^{n-1}\lambda_i^r$ of the rth powers of eigenvalues of A [cf. (4.5)] we conclude as follows:

The sum $\Sigma_{i=0}^{n-1}\lambda_i^r$ *of* rth *powers of eigenvalues of the adjacency matrix is equal to the number of closed walks of length r in the graph.* (4.7)

4.16

The other observation we wish to make regards the elementary symmetric sums of eigenvalues of the adjacency matrix. As we know [see (4.4)], the kth elementary symmetric sum of eigenvalues equals the sum of the $\binom{n}{k}$ determinants of the principal minors of size k in A. Since the adjacency matrix A is a vertex versus vertex matrix it immediately follows that:

* *A principal minor of A is itself the adjacency matrix of the induced subgraph whose vertices correspond to the rows and columns of that principal minor.*

We thus understand the graph theoretical meaning of the kth elementary symmetric sum of eigenvalues if we can interpret the determinant of an arbitrary adjacency matrix of a graph on k vertices.

To interpret such a determinant first replace each edge of the graph by two arcs oriented in opposite directions; thus edge $\{i, j\}$ is replaced by two arcs:

one pointing from i to j, the other from j to i. (We place an arrow on an arc to indicate its orientation.) An *oriented cycle* is a cycle with all its arcs having the same orientation. Think now of the determinant as an expansion in accordance with the $k!$ permutations on the k vertices of the graph. An arbitrary monomial in this expansion is

$$\pm a_{1\sigma(1)} \cdots a_{k\sigma(k)},$$

where σ is a permutation. Visualize σ as a product of disjoint cycles (cf. Section 1.8). As is visible, the cycles of σ can be directly interpreted as oriented cycles in the graph. The above monomial is zero, unless $a_{j\sigma(j)}$ is nonzero for all j. (To simplify matters, think initially of the graph as being simple.) Summing over all permutations σ and all principal minors of size k in A we obtain the expression

$$\sum_{\substack{S \\ S \in C_k}} (-1)^{e(S)},$$

where C_k is the set of all induced subgraphs on k vertices, each of which is simply a disjoint union of nontrivial oriented cycles, and $e(S)$ counts the number of cycles of even length in S. The number of terms in the above sum exceeds $k!$ in general, especially if the graph contains multiple edges. (All we are saying here, really, is that each disjoint union of oriented cycles on k vertices contributes $+1$ or -1 to the value of the determinant, according as the cycle decomposition in question describes an even or odd permutation on the k vertices.) Observe, in addition, that $e(S)$ and the parity of the permutation that S induces are the same modulo 2.

In conclusion:

* *The* kth *elementary symmetric sum of eigenvalues of the adjacency matrix equals*

$$\sum_{\substack{S \\ S \in C_k}} (-1)^{e(S)},$$

where C_k is the set of all induced subgraphs S on k vertices, each of which is a disjoint union of nontrivial oriented cycles, and $e(S)$ counts the number of cycles of even length in S.

4.17 The Kirchhoff Matrix

The $n \times n$ matrix $C(G) = (c_{ij})$, with c_{ii} the degree of vertex i and c_{ij} the negative of the number of edges between vertices i and j $(i \neq j)$ has been studied in detail in section 6 of Chapter 2. The material in that section should be familiar to the reader; it is freely used in the remainder of this section.

We remind the reader that the matrix $C(G)$ is symmetric nonnegative definite, with zero row sums. The vector $\mathbf{1}$, with all entries 1, is therefore always in the kernel of $C(G)$. These and other related facts were proved in Section 2.15.

Denote by $0 = \mu_0(G) \leq \mu_1(G) \leq \cdots \leq \mu_{n-1}(G)$ the eigenvalues of $C(G)$. [The eigenvalue $\mu_0(G) = 0$ reflects the permanent presence of the vector **1** in the kernel.] Perhaps the most significant feature of the matrix $C(G)$ is its nonnegative definiteness. Since the structure of the graph G is not of importance in this section we omit the suffix G and simply write C for $C(G)$ and μ_i for the ith eigenvalue of $C(G)$.

In several areas of applicability (such as experimental design, electrical engineering, chemistry, and information science) the spectrum of the Kirchhoff matrix C is of importance. Extremization of certain kinds of functions on the spectrum (especially Schur-convex functions) is often of significant interest.

Graph theoretic meaning can be given to the kth elementary symmetric sum of eigenvalues of the $n \times n$ Kirchhoff matrix C. The interpretation relies on statement (4.4), which equates the symmetric sum in question to a sum of determinants of principal minors.

It is easy to see that the nth elementary symmetric sum is just the product of the n eigenvalues, of which one (namely μ_0) is zero; hence the nth elementary symmetric sum of eigenvalues of C is zero.

Of central importance is the interpretation of the $(n-1)$st symmetric sum. (On it rests, as we shall see, the interpretation of the general kth symmetric sum.) Since $\mu_1 = 0$, the $(n-1)$st symmetric sum of eigenvalues is simply the product $\prod_{i=1}^{n-1}\mu_i$. Its graph theoretic interpretation relies on the general observation found in (4.4), which states that this symmetric sum is equal to the sum of the determinants of the n principal minors of order $n-1$ in C. But we devoted Section 2.16 (Chapter 2) to proving that all of these principal minors have the same determinant, and the common value equals the number of spanning trees in the graph. We thus conclude that

The number of spanning trees in a graph on n vertices equals $n^{-1}\prod_{i=1}^{n-1}\mu_i$. (4.8)

Or, equivalently, that the $(n-1)$st elementary symmetric sum of eigenvalues of C is n times the number of spanning trees in the graph.

Moving on to the interpretation of the kth elementary symmetric sum we quote (4.4) again and focus on interpreting the determinants of the $\binom{n}{k}$ principal minors of dimension k in C. It turns out that each such determinant is equal to the number of spanning trees in a graph easily derivable from our initial graph. To be specific, if the rows and columns of a principal minor correspond to vertices $v_1, v_2, \ldots, v_k$ of the initial graph G, then produce a new graph $G(v_1, v_2, \ldots, v_k)$ on $k+1$ vertices by identifying (or *amalgamating*) all of the remaining $n-k$ vertices of G into one single vertex (and subsequently removing all the loops that may be generated). The graph $G(v_1, v_2, \ldots, v_k)$ has $k+1$ vertices and the same number of edges as G, if we count in the loops; upon removing the loops $G(v_1, v_2, \ldots, v_k)$ in general has fewer edges than G. Vertices of $G(v_1, v_2, \ldots, v_k)$ are $v_1, v_2, \ldots, v_k$ and a vertex v in which all the

$n - k$ remaining vertices of G were amalgamated. Denote by $C(v_1, v_2, \ldots, v_k)$ the $(k+1) \times (k+1)$ Kirchhoff matrix of $G(v_1, v_2, \ldots, v_k)$. It is clear by construction that upon erasing the row and column corresponding to the amalgamated vertex v in $C(v_1, v_2, \ldots, v_k)$ we obtain the $k \times k$ principal minor of C that corresponds to vertices $v_1, v_2, \ldots, v_k$. The determinant of this principal minor, when viewed as a minor in $C(v_1, v_2, \ldots, v_k)$, equals the number of spanning trees in $G(v_1, v_2, \ldots, v_k)$ (as the basic result of Section 2.16 in Chapter 2 states).

We illustrate the process of amalgamation below:

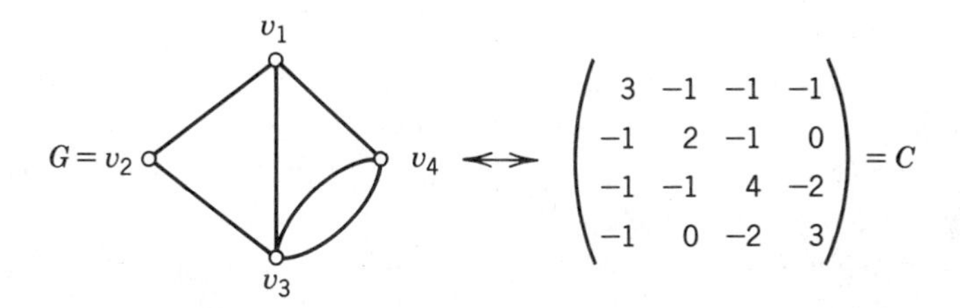

$v_1 \equiv v_2 \equiv v$

$$G(v_3, v_4) = \quad v_4 \quad \longleftrightarrow \quad \begin{pmatrix} 3 & -2 & -1 \\ -2 & 4 & -2 \\ -1 & -2 & 3 \end{pmatrix} = C(v_3, v_4).$$

v_3

The determinant of the minor $\begin{pmatrix} 4 & -2 \\ -2 & 3 \end{pmatrix}$ of C, which (as we observed) occurs also as a minor of $C(v_3, v_4)$, is equal to the number of spanning trees in the amalgamated graph $G(v_3, v_4)$.

Two conclusions surface out of our discussion. One regards the determinant of an arbitrary principal minor of C:

∗ *The determinant of the $k \times k$ principal minor of the Kirchhoff matrix that corresponds to vertices $v_1, v_2, \ldots, v_k$ of the graph G is equal to the number of spanning trees in the amalgamated graph $G(v_1, v_2, \ldots, v_k)$ on $k+1$ vertices.*

The other conclusion concerns the graph theoretical interpretation we sought for the kth elementary symmetric sum.

∗ *The kth elementary symmetric sum of eigenvalues of the Kirchhoff matrix is equal to the total number of spanning trees to be found in the $\binom{n}{k}$ amalgamated subgraphs on $k+1$ vertices of the original graph.*

4.18 In Search of Nice Graphs

Considering the diverse subjective elements that may enter our undertakings, the objects we search for could easily prove to be illusive for, as the saying goes, beauty is in the eye of the beholder. By a uniform process of selection, however, we attempt to single out a "nicest" graph among all graphs with a given numer of vertices and edges. When the setting allows for the existence of strongly regular graphs our method often (and possibly always) uncovers them.

The fundamental motivation rests with the larger problem of designing optimal statistical experiments. A specific model of response (that with varieties and blocks, and with no interactions) leads to the question we now consider.

Let $S(n, m)$ denote the set of all graphs on n vertices having at most m edges. For a graph G in $S(n, m)$ we denote by $C(G)$ its Kirchhoff matrix, and by $\mu(G) = (\mu_1(G), \mu_2(G), \dots, \mu_{n-1}(G))$ the eigenvalues (i.e., spectrum) of $C(G)$ in increasing order. [Note that since $\mu_0(G) = 0$, for all G, we dispense with this eigenvalue altogether when listing the spectrum.] We take interest in minimizing over $S(n, m)$ certain kinds of functions on the spectrum. [When the graph G is arbitrary we write μ_i for $\mu_i(G)$.]

Two important examples of functions to be minimized over $S(n, m)$ are $\Sigma_{i=1}^{n-1}\mu_i^{-1}$ and $\Sigma_{i=1}^{n-1} \ln \mu_i^{-1}$ (with the understanding that μ_i^{-1} is infinity when $\mu_i = 0$). The two choices correspond to the A- and D-optimality criteria in statistical design, respectively. Thus a graph in $S(n, m)$ will be called A- (or *trace*) *optimal* if it minimizes the first function, and D- (or *determinant*) *optimal* if it minimizes the second. [It is clear that $\Sigma_{i=1}^{n-1} \ln \mu_i^{-1}$ is minimal if and only if $\Pi_{i=1}^{n-1}\mu_i$ is maximal. But [cf. (4.8)] $n^{-1}\Pi_{i=1}^{n-1}\mu_i$ equals the number of spanning trees in the graph, and therefore *a graph is D-optimal if and only if it has a maximal number of spanning trees among all graphs in* $S(n, m)$.]

To discuss the problem a bit more generally we consider minimizing over $S(n, m)$ functions of the form

$$\sum_{i=1}^{n-1} f(\mu_i), \tag{4.9}$$

where f satisfies the following two conditions:

(a) Either $f(\mu + x) = c(x)f(\mu x^{-1} + 1)$, or $f(\mu + x) = c(x) + f(\mu x^{-1} + 1)$, for c is a function of x only.

(b) f is differentiable any number of times in a neighborhood of 1, with $(-1)^j f^{(j)}(1)$ being nonnegative (and $f^{(j)}$ denoting the jth derivative of f).

In the case of the A-criterion we have $f(y) = y^{-1}$, the choice $f(y) = -\ln y$ yielding the D-criterion. The selections $f(y) = y^{-p}$, for $p > 0$, are acceptable as well, and statisticians will recognize in them the ϕ_p-criteria introduced by

Kiefer. [In all of these special cases the function c (which appears in (a)) is in fact f itself.] From now on whenever we consider expression (4.9) we assume that f possesses properties (a) and (b) listed above.

The general problem in which we take interest may now be stated as follows:

* *Find the graph(s) on n vertices and at most m edges that minimize*

$$\sum_{i=1}^{n-1} f(\mu_i), \tag{4.10}$$

for some (or all) *functions f of the type described above.*

The particular case obtained by taking just one function $f(y) = -\ln y$ reduces to identifying the graph with a maximal number of spanning trees among all graphs on n vertices and (at most) m edges. This problem has also been discussed at the end of Section 2.16.

REMARK. Questions may be raised as to the appropriateness of our choice of functions to be minimized on the spectrum. We feel that any interesting class should at least include $-\ln y$ and y^{-p}, for $p > 0$. Often the functions of interest are in fact convex (or Schur-convex). Our choice was somewhat arbitrary, motivated by the asymptotic expansions to follow.

4.19 Asymptotics

To gain insight into problem (4.10) we first restrict the minimization over a subclass of graphs of $S(n, m)$.

Start out with the set $\overline{S}(n, e)$ of all simple graphs on n vertices having at most e edges. To a simple graph $\overline{G}$ in $\overline{S}(n, e)$ associate its Kirchhoff matrix $C(\overline{G})$ with spectrum $0 = \bar{\mu}_0 \le \bar{\mu}_1 \le \cdots \le \bar{\mu}_{n-1}$. Superimpose x copies of the complete graph K_n on $\overline{G}$. What results is a graph G with n vertices and at most $x\binom{n}{2} + e$ edges. The Kirchhoff matrix of G is $C(G) = C(\overline{G}) + x(nI - J)$. A set of orthogonal eigenvectors of $C(\overline{G})$ diagonalizes $C(G)$. We can thus relate the spectrum of $C(G)$ to that of $C(\overline{G})$ in a simple manner: $\mu_i = \bar{\mu}_i + x$, for $1 \le i \le n - 1$, and $\mu_0 = 0$.

The operation of superimposing x copies of complete graphs yields the association $\overline{G} \to G$. We denote by $S^*(n, x\binom{n}{2} + e)$ the set of graphs G obtained by superimposing x copies of K_n upon each of the graphs $\overline{G}$ of $\overline{S}(n, e)$. It is clear that $S^*(n, x\binom{n}{2} + e)$ consists of all graphs on n vertices having m edges, where $x\binom{n}{2} \le m \le x\binom{n}{2} + e$. [Unless $x = 0$, the set $S^*(n, x\binom{n}{2} + e)$ is a proper subset of $S(n, x\binom{n}{2} + e)$, the set of all graphs with n vertices and at most $x\binom{n}{2} + e$ edges.]

For a function f on the spectrum of $C(G)$, where G is a graph in $S^*(n, x\binom{n}{2} + e)$, we can write (recall conditions (a) and (b) in Section 4.18):

$$\begin{aligned}\sum_i f(\mu_i) &= \sum_i f(\bar{\mu}_i + x) = \sum_i c(x) f(\bar{\mu}_i x^{-1} + 1) \\ &= c(x) \sum_i \sum_{j \geq 0} \frac{f^{(j)}(1)}{j!} (\bar{\mu}_i x^{-1})^j \\ &= c(x) \sum_{j \geq 0} \left(\sum_i \bar{\mu}_i^j \right) \frac{f^{(j)}(1)}{j!} x^{-j}.\end{aligned} \tag{4.11}$$

[A similar expression exists for functions f that satisfy $f(\bar{\mu}_i + x) = c(x) + f(\bar{\mu}_i x^{-1} + 1)$.]

Expression (4.11) reveals that for large x we can find the graph(s) that minimize $\Sigma_i f(\mu_i)$ over $S^*(n, x\binom{n}{2} + e)$ by minimizing one after the other the coefficients of the powers of x: first of x^0, then of x^{-1}, of x^{-2}, and so on. And since $f^{(j)}(1)$ has (by assumption) the same sign as $(-1)^j$, this is equivalent to minimizing first $(-1)^0 \Sigma_i \mu_i^0$, then $(-1)^1 \Sigma_i \mu_i$, then $(-1)^2 \Sigma_i \mu_i^2$, and so on. We are thus led to define as follows.

Let $\bar{S}_0(n, e)$ stand for $\bar{S}(n, e)$. For $j \geq 1$, we let $\bar{S}_j(n, e)$ consist of those simple graphs in $\bar{S}_{j-1}(n, e)$ that minimize $(-1)^j \Sigma_{i=1}^{n-1} \bar{\mu}_i^j$ [i.e., $(-1)^j \text{trace}(C(\bar{G})^j)$]. Clearly $\bar{S}_0 \supseteq \bar{S}_1 \supseteq \bar{S}_2 \supseteq \cdots$.

In a spectral sense $S_n(n, e)$ consists of one graph only; that is, any two graphs in $\bar{S}_n(n, e)$ have the same spectrum of the Kirchhoff matrix. (This is true because the two graphs have equal sums of rth powers of eigenvalues for all r, $0 \leq r \leq n$. They thus have equal elementary symmetric sums of eigenvalues. Hence they have identical characteristic polynomials of the Kirchhoff matrix and therefore equal eigenvalues.) Two graphs with the same spectrum (of the Kirchhoff matrix, in this case) are called *cospectral*. In general cospectral graphs are not isomorphic. The simple graphs in $\bar{S}_n(n, e)$, however, not only have equal sums of rth powers of eigenvalues but for each r these sums have minimal value. This latter (strong and) restrictive property was not used at all when deriving cospectrality; it is for this reason (in addition to computational evidence) that we suspect the following stronger result to be true:

Up to isomorphism $\bar{S}_n(n, e)$ consists of one graph only.

Let us summarize what we achieved so far:

> *For any x exceeding a sufficiently large constant, by superimposing x complete graphs on a simple graph $\bar{G}$ in $\bar{S}_n(n, e)$ we obtain a graph G that minimizes over $S^*(n, x\binom{n}{2} + e)$ any finite number of functions of the form $\Sigma_{i=1}^{n-1} f(\mu_i)$* [with f as in (4.9)]. (4.12)

[A natural question to raise, but seemingly difficult to answer, is whether such a graph G minimizes the same functions over the larger set of graphs

$S(n, x\binom{n}{2} + e)$. How large should x be? Surprisingly, in certain settings x can in fact be 0.]

Lastly, computational evidence suggests that for j much smaller than n the subset $\bar{S}_j(n, e)$ consists of one graph only (up to isomorphism).

4.20

We now examine settings in which $\bar{S}(n, e)$ contains regular graphs with e edges. Graph theoretic interpretation can in this case be given to the subsets $\bar{S}_j(n, e)$.

For a regular graph the Kirchhoff matrix C and the adjacency matrix A are related in a simple way: $C = dI - A$, where d is the degree of any vertex of the graph. Thus $A = dI - C$ and hence

$$A^j = (dI - C)^j = \sum_{k=0}^{j} d^{n-k}(-1)^k C^k.$$

Taking traces on both sides we obtain

$$\begin{aligned} \operatorname{trace} A^j &= \sum_{k=0}^{j} d^{n-k} \operatorname{trace}(-1)^k C^k \\ &= (-1)^j \operatorname{trace} C^j + \sum_{k=0}^{j-1} d^{n-k} \operatorname{trace}(-1)^k C^k. \end{aligned} \tag{4.13}$$

Recall that $\bar{S}_0(n, e) = \bar{S}(n, e)$ and that $\bar{S}_j(n, e)$ consists of those graphs in $\bar{S}_{j-1}(n, e)$ that minimize $(-1)^j \operatorname{trace} C^j$. In particular, $\bar{S}_1(n, e)$ is the set of graphs on n vertices with exactly e edges and $\bar{S}_2(n, e)$ is the set of regular graphs on v vertices having e edges. With this in mind identity (4.13) reveals at once that, for $j \geq 2$, $(-1)^j \operatorname{trace} C^j$ is minimized if and only if $\operatorname{trace} A^j$ is minimized. We can thus describe $\bar{S}_j(n, e)$ as consisting of those graphs in $\bar{S}_{j-1}(n, e)$ that minimize $\operatorname{trace} A^j$, $j \geq 2$. But, as was mentioned in (4.6), $\operatorname{trace} A^j$ equals the number of closed walks of length j in the graph (with adjacency matrix A).

In conclusion,

If regular simple graphs on n vertices and with e edges exist, then

$\bar{S}_0(n, e)$ *is the set of all simple graphs on v vertices having at most e edges,*

$\bar{S}_1(n, e)$ *is the set of all simple graphs on v vertices having exactly e edges,*

and, for $j \geq 2$, (4.14)

$\bar{S}_j(n, e)$ *is the set of regular graphs in* $\bar{S}_{j-1}(n, e)$ *having a minimal number of closed walks of length j.*

The above is a graph theoretical description of the sets $\bar{S}_j(n, e)$, and in particular of (the spectrally, if not isomorphically, unique) $\bar{S}_n(n, e)$, which possesses the distinguished extremal spectral property mentioned in (4.12).

The "sieving" process described in (4.14) leads to nice graphs, nice from a spectral as well as graph theoretical point of view.

4.21

Motivated by (4.14) we now describe a method for selecting nice graphs out of all graphs on n vertices having at most m edges. The selection is carried out in steps. At step j we (iteratively) select the subset $S_j(n, m)$ described below:

$S_0(n, m)$ *is the set of all graphs on n vertices having at most m edges.*

$S_1(n, m)$ *is the subset of graphs on n vertices having exactly m edges.*

$S_2(n, m)$ *is the subset of graphs in $S_1(n, m)$ whose degrees differ by at most* 1 *and with the number of edges between any two vertices differing by at most* 1.

and, for $j \geq 3$,

$S_j(n, m)$ *is the subset of graphs in $S_{j-1}(n, m)$ having a minimal number of closed walks of length j.*

[In particular, $S_2(n, m)$ describes the nearly regular graphs with nearly the same number of edges between any two vertices. And $S_3(n, m)$ consists of the nearly regular graphs in $S_2(n, m)$ having a minimal number of triangles.] In all known cases, for some $j \leq n$ the set $S_j(n, m)$ consists of one graph only (up to isomorphism). The resulting graph is nice; usually strongly regular whenever the setting permits.

Example 1. When $m = k\binom{n}{2}$ the set $S_2\left(n, k\binom{n}{2}\right)$ consists of one graph only: k superimposed copies of the complete graph K_n.

Example 2. If $n = pq$ and $m = \binom{p}{q}q^2$ the set $S_3(n, m)$ contains a single graph: the complete multipartite graph with p parts of size q each.

Example 3. By taking $n = 10$ and $m = 15$ we find that the sole graph in $S_5(10, 15)$ is the familiar Petersen's graph displayed in Section 4.11.

Example 4. The sole graph in $S_n(n, n-1)$ is a path of length $n - 1$ (which may be thought of as the nicest spanning tree).

Example 5. The sole graph in $S_n(n, n)$ is the cycle with n edges.

Verifying the statements made in the above examples, while not very difficult, does require some time and counting abilities on the reader's part.

EXERCISES

1. Find an explicit formula for the number of spanning trees of a strongly regular graph with parameters (n, d, a, c).
[*Hint*: Compute the eigenvalues of the Kirchhoff matrix $C = dI - A$; the eigenvalues of A were computed in Section 4.12.]

2. Show that the eigenvalues of a symmetric matrix are real. Show also that a nonnegative definite matrix has nonnegative eigenvalues.

3. Let $P(x)$ be a polynomial in x. Prove that if μ is an eigenvalue of A, then $P(\mu)$ is an eigenvalue of $P(A)$. Furthermore, show that the eigenvectors of A are also eigenvectors of $P(A)$.

4. A *forest* with k trees (or k components) of a graph G is a partition of the set of vertices of G into k classes, each class being a tree in G. Show that the kth elementary symmetric sum of eigenvalues of the $n \times n$ Kirchhoff matrix $C(G)$ is equal to

$$\sum p(F),$$

the sum being over all forests F of G with $n - k$ components, and $p(F)$ signifying the product of the number of vertices in the $n - k$ components of F.

5. Prove the assertions made in the five examples given at the end of Section 4.21.

6. Are all the graphs in $\bar{S}_n(n, e)$ isomorphic?

7. Find lower bounds on x in statement (4.12).

8. If strongly regular graphs on n vertices and m edges exist is (are) the graph(s) in $S_n(n, m)$ necessarily strongly regular?

4 GRAPHS WITH EXTREME SPECTRA

The class of problems that we bring forth in this section, like those in the last, arise from considerations of statistical efficiency.

4.22

We restrict attention to simple graphs only. The general question is as follows:

* *Fix a class S of simple graphs on v vertices and a convex or concave function f defined on the spectra of adjacency matrices of the graphs in S. Find the graph(s) in S that minimize* (or maximize) *f over S.*

We cannot allow too much freedom of choice in the selection of f. Actually the most interesting choices for f are the minimum eigenvalue, the maximum eigenvalue, the determinant, and the harmonic mean, that is λ_1, λ_v, $\Pi_i \lambda_i$, and $\Sigma_i \lambda_i^{-1}$.

Two interesting choices for S suggest themselves:

(a) All simple graphs on v vertices.

(b) Simple graphs on v vertices having e edges.

We address the question in (a) for minimizing λ_1, the minimal eigenvalue. Complete answers are obtained for this case. (Let us inform in passing that the question of maximizing the maximal eigenvalue λ_v over all simple graphs on v vertices is not difficult to answer: the maximum is $v - 1$ and is attained if and only if the graph is the complete graph K_v.)

Extremization over simple graphs with v vertices and e edges appears to be a more refined problem and more difficult to settle. The answer for a general setting is likely to be elusive, but special settings (which admit multipartite graphs with equal parts, for example) may be manageable. We invite the (intelligent and) enthusiastic reader to give it a try.

4.23

In Sections 4.23 through 4.26 we rely on three well-known results in matrix theory. These we list below; the reader is referred to [9] for details.

1. *The Courant–Fischer Theorem. For a symmetric matrix A of dimension n we have $\lambda_1 = \min_{\|x\|=1} x^tAx$ and $\lambda_n = \max_{\|x\|=1} x^tAx$, where λ_1 and λ_n are the minimal and maximal eigenvalues of A, respectively.*

(A similar minmax expression for the ith eigenvalue exists. By $\|x\|$ we understand the square root of the sum of squares of the entries of x.)

2. *The Perron–Frobenius Theorem. A matrix with nonnegative entries has a nonnegative eigenvalue that is the largest in absolute value among all other eigenvalues.*

3. *The Interlacing Lemma. Let A be a symmetric matrix of dimension n and let M be a principal minor of dimension $n - 1$. Denote by $\lambda_1 \leq \cdots \leq \lambda_n$ the eigenvalues of A and by $\mu_1 \leq \cdots \leq \mu_{n-1}$ the eigenvalues of M. Then*

$$\lambda_1 \leq \mu_1 \leq \lambda_2 \leq \mu_2 \leq \cdots \leq \mu_{n-1} \leq \lambda_n.$$

(Iteratively we may proceed to minors of A of smaller sizes.)

We now begin studying the minimization of λ_1 (the minimal eigenvalue) over the class of all simple graphs.

4.24

For a simple graph G we denote by $A(G)$ its *adjacency matrix*. If G has v vertices, then $A(G) = (a_{ij})$ is a $v \times v$ symmetric matrix with rows and columns labeled by the vertices of G, with $a_{ii} = 0$ for all i and $a_{ij} = 1$ if

$\{i, j\}$ is an edge of G and 0 otherwise. We denote by $\lambda_1(G) \leq \lambda_2(G) \leq \cdots \leq \lambda_v(G)$ the eigenvalues of $A(G)$. For an arbitrary symmetric $v \times v$ matrix A we denote by $\lambda_1(A)$ the minimal eigenvalue of A. As we mentioned, $\lambda_1(A) = \min_{\|x\|=1} x^tAx$ and $\lambda_v(A) = \max_{\|x\|=1} x^tAx$.

As G ranges over the collection of all simple graphs on v vertices one often needs to know how small does $\lambda_1(G)$ ever get. We are searching, therefore, for graphs with extreme spectral behavior. Such questions arise when searching for optimal statistical designs (E-optimal designs in particular). They also arise in statistical mechanics, in the so-called Hückel theory, as well as in other fields.

The problems that arise most often in the applied fields mentioned above involve knowledge of a spectral structure *as well as* the fact that that spectral structure minimizes (or maximizes) a known function of eigenvalues over a given class of graphs, say. With this information one then proceeds in discovering the combinatorial structure (if any) that has that spectral property.

We need the following notation: K_v denotes the complete graph on v vertices, K_{v_1, v_2} is the complete bipartite graph on $v_1 + v_2$ vertices, J is the matrix with all entries 1, $\mathbf{1}$ is the column vector with all entries 1, and I denotes the identity matrix.

4.25

We begin by proving the following result:

Proposition 4.1. *A simple graph on v vertices has adjacency matrix of rank 2 if and only if it is isomorphic to K_{v_1, v_2} plus an additional $v - (v_1 + v_2)$ isolated vertices.*

Proof. It is easy to see that all graphs of the form K_{v_1, v_2} plus an additional $v - (v_1 + v_2)$ isolated vertices have adjacency matrix of rank 2. Conversely, suppose G is a simple graph with rank $A(G) = 2$. By a suitable relabeling of vertices line up the rows (and columns) of $A(G)$ in decreasing order of row sums. Let $x = (x_1, \ldots, x_v)$ be the first row of $A(G)$ (note that $x \neq \mathbf{0}$ and that it is the row with the largest number of 1's). Select another row $y = (y_1, \ldots, y_v)$ $(\neq \mathbf{0})$ of $A(G)$ such that x and y are linearly independent. We show that any row of $A(G)$ is either $\mathbf{0}$, x, or y.

Let z be a row of $A(G)$ different from x and y. By the assumption on the rank of $A(G)$, $z = \alpha x + \beta y$. Since $x \neq y$, there exists an index i such that $x_i = 1$ and $y_i = 0$. This implies $\alpha = 1$. We now separate the proof into two (exhaustive) cases: *Case 1.* x and y do not share a 1 in the same position. *Case 2.* There exists i with $x_i = y_i = 1$.

Under Case 1 there exists j such that $y_j = 1$ and $x_j = 0$. This implies $\beta = 1$ and hence $z = x + y$. But then $x + y$ has *strictly* more 1's than x, which contradicts the arrangement of rows of $A(G)$ in decreasing order of row sums. Hence $x + y$ is not a row of $A(G)$. In Case 2 we must have $\beta = -1$, giving $z = x - y$. Any row of $A(G)$ is in this case either x, y, $x - y$, or $\mathbf{0}$.

Assume that $x - y$ is a row of $A(G)$. Then $x_i = 0$ implies $y_i = 0$. And since $x_1 = 0$, we have $y_1 = 0$ as well as $x_1 - y_1 = 0$. This means that the first *column* of $A(G)$ consists of 0's only! By the symmetry of $A(G)$ we conclude $x = 0$, which is contradictory. Hence $x - y$ cannot be a row and we conclude that the only rows of $A(G)$ are x, y, and possibly $\mathbf{0}$.

Suppose x occurs v_1 times, y occurs v_2 times, and $\mathbf{0}$ occurs $v - (v_1 + v_2)$ times as rows of $A(G)$. Since $A(G)$ has 0's on the diagonal this forces $A(G)$ to be of the form

$$\begin{array}{c|c|c} 0 & J & 0 \\ \hline J & 0 & 0 \\ \hline 0 & 0 & 0 \end{array}$$

[upon a possible reshuffling of rows and (same) columns]. This shows that G is isomorphic to K_{v_1, v_2} plus a number of isolated vertices. This ends the proof.

[A shorter graph theoretic proof of Proposition 4.1 is as follows: Let the rank of $A(G)$ be 2 and assume, without loss, that G is connected. If G is not bipartite it has an odd cycle and hence an induced (or cordless) odd cycle. Then $A(G)$ has a principal submatrix of odd order at least 3 that is invertible. This implies $A(G)$ has rank 3 or more, a contradiction. If G is not complete bipartite one easily finds three independent rows in $A(G)$.]

For any simple graph G on v vertices and m edges it is useful to be constantly aware of the following simple facts:

(a) $\operatorname{Trace} A(G) = 0 = \sum_{i=1}^{v} \lambda_i(G)$.
(b) $\operatorname{Trace} A(G)^2 = 2m = \sum_{i=1}^{v} \lambda_i(G)^2$.

We now prove the following:

Proposition 4.2. (i) *The minimal eigenvalue of the adjacency matrix of a simple graph on* $2n$ *vertices is never less than* $-n$. *It equals* $-n$ *if and only if the graph is isomorphic to* $K_{n,n}$.

(ii) *The minimal eigenvalue of the adjacency matrix of a simple graph on* $2n + 1$ *vertices is never less than* $-\sqrt{n(n+1)}$. *It equals* $-\sqrt{n(n+1)}$ *if and only if the graph is isomorphic to* $K_{n,n+1}$.

Proof. (i) Let G be a simple graph with $2n$ vertices and m edges. It is well known that the largest eigenvalue of a symmetric matrix equals $\max_{\|x\|=1} x^t A x$. Taking the special value $x = (1/\sqrt{2n})\mathbf{1}$ we thus obtain the inequality $\lambda_{2n}(G) \geq (1/2n)\mathbf{1}^t A(G)\mathbf{1} = (1/2n)2m = m/n$. From (b) above we conclude that $\lambda_1(G)^2 + \lambda_{2n}(G)^2 \leq 2m$. Combining these two inequalities leads us to

$$\lambda_1(G)^2 + \frac{m^2}{n^2} \leq 2m.$$

If $\lambda_1(G) < -n$ this last inequality becomes $n^2 + m^2/n^2 < 2m$, or $(n - m/n)^2 < 0$, which is a contradiction. Hence $\lambda_1(G) \geq -n$. If $\lambda_1(G) = -n$, then by above $(n - m/n)^2 = 0$, or $m = n^2$. This implies that $\lambda_1(G) = -n$, $\lambda_{2n}(G) = n$, and the other eigenvalues are 0. We thus conclude that the rank of $A(G)$ is 2 and now Proposition 4.1 informs us that G must be complete bipartite.

(ii) Let us turn our attention to simple graphs on an odd number of vertices. We let G be a simple graph on $2n + 1$ vertices with m edges. Our aim is to prove that $\lambda_1(G) \geq -\sqrt{n(n+1)}$ (irrespective of m). We separate the proof into two cases: $m \leq n(n+1)$ and $m \geq n(n+1) + 1$.

The first case is not so difficult to handle. If $\lambda_1(G) < -\sqrt{n(n+1)}$, then necessarily $\lambda_{2n+1} \leq \sqrt{n(n+1)}$, else the assumption on m and (b) above lead to $2n(n+1) \geq 2m \geq \lambda_1(G)^2 + \lambda_{2n+1}(G)^2 > n(n+1) + n(n+1) = 2n(n+1)$, a contradiction. But if $\lambda_{2n+1}(G) \leq \sqrt{n(n+1)}$, then Perron–Frobenius' theorem forces $|\lambda_1(G)| \leq \lambda_{2n+1}(G) \leq \sqrt{n(n+1)}$. This contradicts the assumption $\lambda_1(G) < -\sqrt{n(n+1)}$. We must therefore have $\lambda_1(G) \geq -\sqrt{n(n+1)}$.

In the second case we assume that $m \geq n(n+1) + 1$. It easily follows by induction on n (or as a form of Turán's theorem) that a simple graph on $2n + 1$ vertices and at least $n(n+1) + 1$ edges must contain a triangle (i.e., a K_3).

Then $m \geq n(n+1) + 1$ implies the existence of a triangle in G. Since $A(K_3)$ has eigenvalues -1, -1, and 2 it follows that $\lambda_2(G) \leq -1$ (by a well-known result on the interlacing of eigenvalues of principal minors in symmetric matrices, or by ad hoc arguments). Assume $\lambda_1(G) < -\sqrt{n(n+1)}$ (we contradict this). We have $\lambda_{2n+1}(G) \geq (1/(2n+1))\mathbf{1}^t A(G)\mathbf{1} = 2m/(2n+1)$ and, as we just saw, $\lambda_2(G) \leq -1$. Now fact (b) leads to $2m = \sum_{i=1}^{2n+1}\lambda_i^2(G) > n(n+1) + 1 + (2m/(2n+1))^2$. Writing $m = n(n+1) + w$, for $w \geq 1$, this last inequality becomes

$$2n^2 + 2n + 2w > n^2 + n + 1 + (2n^2 + 2n + 2w)^2(4n^2 + 4n + 1)^{-1}.$$

Upon simplification this becomes equivalent to

$$-3n^2 - 3n - (4w^2 - 2w + 1) > 0, \tag{4.15}$$

which is a contradiction. Hence $\lambda_1(G) \geq -\sqrt{n(n+1)}$, as desired. [We should mention that the argument used above cannot be successfully carried out without observing that $\lambda_2(G) \leq -1$.]

The lower bound $-\sqrt{n(n+1)}$ is the best possible, since $\lambda_1(K_{n,n+1}) = -\sqrt{n(n+1)}$. Assume that for a simple graph G on $2n + 1$ vertices (and m edges) we have $\lambda_1(G) = \sqrt{n(n+1)}$. We show that G is $K_{n,n+1}$. If $m \geq n(n+1) + 1$, then, knowing that G must contain a triangle, we arrive at the contradiction (4.15) just as we did before. Hence we must have $m \leq n(n+1)$. By the Perron–Frobenius theorem we must also have $\lambda_{2n+1}(G)$

$\geq \sqrt{n(n+1)}$. But now fact (b) gives a contradiction: unless $m = n(n+1)$, $\lambda_{2n+1}(G) = \sqrt{n(n+1)}$ and the remaining $2n-1$ eigenvalues of $A(G)$ are all 0, that is, $A(G)$ has rank 2. Proposition 4.1 and $m = n(n+1)$ now force G to be isomorphic to $K_{n,n+1}$. This ends our proof.

4.26

Proposition 4.2 has the following matrix-theoretic consequence:

Corollary. (i) *Let* $A = (a_{ij})$ *be a* $2n \times 2n$ *symmetric matrix with* $a_{ii} \geq n$ *and* $0 \leq a_{ij} \leq 1$, *for* $i \neq j$; *then A is positive semidefinite. A is singular if and only if* $A = A(G) + nI$, *where* $A(G)$ *is the adjacency matrix of a graph G isomorphic to* $K_{n,n}$.

(ii) *Let* $A = (a_{ij})$ *be a* $(2n+1) \times (2n+1)$ *symmetric matrix with* $a_{ii} \geq \sqrt{n(n+1)}$ *and* $0 \leq a_{ij} \leq 1$, *for* $i \neq j$; *then A is positive semidefinite. A is singular if and only if* $A = A(G) + \sqrt{n(n+1)}\, I$, *where* $A(G)$ *is the adjacency matrix of a graph G isomorphic to* $K_{n,n+1}$.

Proof of the Corollary. Standard convexity arguments show that any $v \times v$ symmetric matrix $B = (b_{ij})$ with $b_{ii} = 0$ for all i and $0 \leq b_{ij} \leq 1$, for $i \neq j$ is a convex combination of adjacency matrices of simple graphs on v vertices. Therefore, if we denote by D the diagonal matrix with ith diagonal entry a_{ii}, we can write $A - D$ as a convex combination of adjacency matrices. Let this convex combination be written as $A - D = \Sigma_i \alpha_i A_i$.

Assume A is $2n \times 2n$. We have $\lambda_1(A - D) = \min_{\|x\|=1} x^t(A - D)x = \min_{\|x\|=1} x^t(\Sigma_i \alpha_i A_i)x \geq \Sigma_i \alpha_i \min_{\|x\|=1} x^t A_i x \geq \Sigma_i \alpha_i(-n) = -n$. Proposition 4.2 (i) explains the last inequality and also informs us that this last inequality becomes equality if and only if $A - D = A(K_{n,n})$. We conclude that $A - D + nI \geq 0$ or $A \geq D - nI \geq 0$, which establishes part of the corollary. In addition, we know that $A - D + nI$ is singular if and only if $A - D = A(K_{n,n})$. If $A - D + nI$ is indeed singular, then its kernel is of dimension one and spanned by the vector $(1/\sqrt{2n})(-1,\ldots,-1,1,\ldots,1)^t$ with first n entries -1 and the rest 1; this is also the eigenvector of $A(K_{n,n})$ with $-n$ as the corresponding eigenvalue.

Suppose $x^tAx = 0$ for some nonzero vector $x = (x_1,\ldots,x_{2n})^t$ with $\|x\| = 1$. We can rewrite this as $0 = x^tAx = x^t(A - D + nI)x + x^t(D - nI)x$. Since both forms on the right are positive semidefinite this implies that $x^t(A - D + nI)x = 0$ and $x^t(D - nI)x = 0$. However $x^t(A - D + nI)x = 0$ for $x \neq 0$ implies that $A - D + nI$ is singular. Then, as we saw,

$$x = \left(1/\sqrt{2n}\right)(-1,\ldots,-1,1,\ldots,1)^t.$$

The condition $(1/2n)(-1,\ldots,-1,1,\ldots,1)(D - nI)(-1,\ldots,-1,1,\ldots,1)^t = 0$ forces now $D - nI$ to be the zero matrix. This proves part (a) of the corollary. Part (b) is proved analogously. [We mention in passing that the eigenvector of $A(K_{n,n+1})$ associated with the eigenvalue $-\sqrt{n(n+1)}$ has its first n entries equal to $-1/\sqrt{2n}$ and the remaining $n+1$ entries equal to $1/\sqrt{2(n+1)}$.] This ends the proof.

We now give an alternate version of the corollary that puts the emphasis on the nonnegativity of the entries of the matrix. Apart from trivial adjustments its proof parallels that of the corollary, so we omit it.

For a square matrix $A = (a_{ij})$ let $M = \max_{i,j,i \neq j}|a_{ij}|$ and $m = \min_i |a_{ii}|$. We can assert the following:

Proposition 4.3. (i) *If A is a $2n \times 2n$ symmetric matrix with nonnegative entries, then $\lambda_1(A) \geq m - Mn$. Equality occurs if and only if $A = mI + MA(G)$, where G is isomorphic to $K_{n,n}$.*

(ii) *If A is a $(2n+1) \times (2n+1)$ symmetric matrix with nonnegative entries, then $\lambda_1(A) \geq m - M\sqrt{n(n+1)}$. Equality occurs if and only if $A = mI + MA(G)$, where G is isomorphic to $K_{n,n+1}$.*

REMARK. Simpler proofs can be given for the results in Sections 4.25 and 4.26. The idea is to reveal the structure of matrices for which the greatest lower bound for the minimal eigenvalue is attained. This lower bound is obviously attained by some matrix A in the indicated class and some unit vector x. Since $x^tAx = \Sigma_{i,j} a_{ij}x_ix_j$ it is clear that we must have $a_{ij} = 0$ whenever $x_ix_j > 0$ and $a_{ij} = 1$ whenever $x_ix_j < 0$, else we could decrease the objective function below the asserted minimum. Without loss of generality we may suppose $a_{ij} = 0$ if $x_i = 0$ or $x_j = 0$. Let P be a permutation matrix such that

$$x^tP^t = (+ + \cdots + - - \cdots - 0 \cdots 0)$$

(i.e., with first m_1 entries positive, the next m_2 entries negative, and the rest of the entries 0). Then PAP^t becomes the adjacency matrix of the complete bipartite graph K_{m_1,m_2} plus $n - (m_1 + m_2)$ isolated vertices. An easy calculation reveals that the minimum eigenvalue of PAP^t (and hence also of A itself) is $-\sqrt{m_1m_2}$. In order, then, that this minimum eigenvalue be as small as possible, we must have $m_1 + m_2 = n$ (no zero entries in x) and m_1 and m_2 be as nearly equal as possible ($m_1 = m_2 = n/2$ if n is even, or $m_1 = (n-1)/2$ and $m_2 = (n+1)/2$ if n is odd). This verifies the asserted bounds and cases of equality.

NOTES

In this chapter we have concerned ourselves, for the most part, with identifying graphs with extreme spectral properties. These problems are a natural outgrowth of designing efficient (if not optimal) experimental strategies.

The material in the first section is available in most books on graph theory: we recommend [1] and [2]. Strongly regular graphs are also known as association schemes with two classes. The last section of Chapter 7 offers close connections. A good text to read on this subject is [3] (the chapter on strongly regular graphs). The same text emphasizes much more the existing interconnections between block designs and strongly regular graphs. In this sense it complements our more separatist approach rather well. On graph spectra we mention the recent book [4]. Extremization of Schur-convex functions over classes of graphs leads to many open questions: some were stated in the text, others are mentioned in exercises. We wish to bring this interesting type of optimization to the attention of graph theorists, especially those working on the algebraic side. The approach originated with the works of J. C. Kiefer and it could be more vigorously pursued. Parts of Sections 3 and 4 are due to Cheng and co-workers [5], Patterson [8], and the author [6, 7].

REFERENCES

[1] J. A. Bondy and U. S. R. Murty, *Graph Theory with Applications*, The MacMillan Press Ltd., London, 1976.

[2] R. C. Bose and B. Manvel, *Introduction to Combinatorial Theory*, Wiley, New York, 1984.

[3] P. J. Cameron and J. H. van Lint, *Graphs, Codes, and Designs*, Cambridge University Press, Cambridge, 1980.

[4] D. M. Cvetković, M. Doob, and H. Sachs, *Spectra of Graphs*, Academic Press, New York, 1979.

[5] C.-S. Cheng, D. Masaro, and C.-S. Wong, Do nearly balanced multigraphs have more spanning trees?, *J. Graph Th.*, **8**, 342–345, (1985).

[6] G. Constantine, On the optimality of block designs, *Ann. Inst. Statist. Math.*, **38**, 161–174 (1986).

[7] G. Constantine, Lower bounds on the spectra of symmetric matrices with nonnegative entries, *Lin. Alg. and Applic.*, **65**, 171–178 (1985).

[8] L. J. Patterson, Circuits and efficiency in incomplete block designs, *Biometrika*, **70**, 215–225 (1983).

[9] M. Marcus and H. Minc, *A Survey of Matrix Theory and Matrix Inequalities*, Allyn and Bacon, Boston, 1964.

CHAPTER 5

Flows in Networks

So they roughed him in a bag, tied it to a heavy millstone, and decided to throw him in the deepest waters of the Danube. "Where is the most water?" Some said it must be uphill 'cause it always flows from there, others said it ought be down in the valley 'cause it gathers there.

Păcală, ROMANIAN FOLK TALE

The combinatorial problems and algorithms herein described can be viewed as part of the larger problem of linear programing: that of maximizing a linear function over a set of points in the n-dimensional space that satisfy some (finite but usually large number of) linear constraints. An algorithm commonly used to perform this task is the so-called simplex algorithm. Our interest is in a subclass of such problems that can be solved by combinatorial and graph theoretical means. The algorithms that accompany the problems in this subclass are known to be very efficient in terms of computing time. We do not dwell much on questions of efficiency, however, but focus instead on the underlying combinatorial and graph theoretical techniques.

The well-known result of Birkhoff and von Neumann that identifies the extreme points of the set of doubly stochastic matrices is our starting point. We present two proofs: one based on graph theoretical arguments, the other within the context of linear programming involving unimodular matrices. Matching and marriage problems are the subject of the second section, with results of König and P. Hall. Minty's arc coloring lemmas are in Section 3. The next two sections contain the max-flow min-cut theorems along with some of their corollaries: Menger's theorem on connectivity, a result on lattices by Dilworth, and yet another by P. Hall. The "out of kilter" algorithm is the subject of Section 6. We end the chapter with a discussion of matroids and the greedy algorithm.

1 EXTREMAL POINTS OF CONVEX POLYHEDRA

5.1

A set in the usual Euclidean n-dimensional space is called *convex* if together with two distinct points of the set the whole line segment joining the two points lies within that set. (With x and y two given points, a point on the line segment joining them can be written as $\alpha x + (1 - \alpha)y$, with $0 \leq \alpha \leq 1$.) By a *convex polyhedron* we understand the set of all points x that satisfy a finite number of linear inequalities, that is, x satisfies a system of inequalities of the form

$$Ax \leq b,$$

where A is a (rectangular) matrix of scalars and b is a vector of scalars. [We remark, in passing, that any system of linear inequalities (and equalities) can be written in the form $Ax \leq b$, by multiplying inequalities of the form $\geq$ by -1, and replacing $=$ by both $\geq$ and $\leq$.]

A convex polyhedron is said to be *bounded* if it can be placed within a (possibly large) sphere.

The unit cube is in many ways a typical example of a bounded convex polyhedron. It is defined by the system of inequalities given below:

$$\begin{bmatrix} 1 & 0 & 0 \\ -1 & 0 & 0 \\ 0 & 1 & 0 \\ 0 & -1 & 0 \\ 0 & 0 & 1 \\ 0 & 0 & -1 \end{bmatrix} \begin{pmatrix} x_1 \\ x_2 \\ x_3 \end{pmatrix} \leq \begin{bmatrix} 1 \\ 0 \\ 1 \\ 0 \\ 1 \\ 0 \end{bmatrix}.$$

A point of a convex polyhedron is called *extreme* if any segment of positive length centered at that point contains points not belonging to the polyhedron. (Geometrically the extreme points are simply the corners of the polyhedron.)

When one wants to maximize a *linear* function over a set of points that form a bounded convex polyhedron, it is easy (and helpful) to observe that the maximum is reached at an extreme point. This follows immediately upon observing two things: that a linear function is convex and that a local maximum is necessarily a global one. Within this context (of linear programming) it becomes of importance to identify the extreme points of bounded convex polyhedra.

In small dimensions it is not hard to "see" the extreme points of a bounded convex polyhedron. Not so in higher dimensions, when the polyhedron might be described by thousands of linear inequalities (possibly with redundant repetitions), each involving hundreds of variables, maybe. The case of doubly stochastic matrices is a classical example; let us examine it next.

5.2

A $n \times n$ matrix (x_{ij}) is called *doubly stochastic* if $x_{ij} \geq 0$ and $\Sigma_i x_{ij} = \Sigma_j x_{ij} = 1$. The set of all doubly stochastic matrices, K, is a bounded convex polyhedron in R^{n^2}; R denotes the set of real numbers. [Its dimension is in fact $(n-1)^2$.] The Birkhoff–von Neumann theorem asserts that the extreme points of K are permutation matrices. It follows as an obvious corollary of the following result:

The Birkhoff–von Neumann Theorem. *Let $p_1, \dots, p_m;\ q_1, \dots, q_n$ be natural numbers and let C be the convex set of all $m \times n$ matrices (x_{ij}) such that*

$$x_{ij} \geq 0, \qquad \sum_i x_{ij} = q_j, \qquad \sum_j x_{ij} = p_i.$$

Then the extreme points of C are matrices in C with integer entries.

Proof. We show that any matrix in C that contains noninteger entries is not an extreme point. Let M be such a matrix. Form a (bipartite) graph whose vertices are the rows of M and the columns of M; an edge joins a "row" and a "column" if the corresponding entry in M is not an integer. Evidently each vertex has degree at least two (a row with one noninteger entry must have at least one other noninteger entry). Decompose the graph into its connected components. No component can be a tree since every finite tree has at least two "leaves" (i.e., vertices of degree 1). (To see this, start at any vertex of a finite tree and "walk along it" placing an arrow on each edge traversed and never reversing direction. One cannot return to the starting vertex since a tree has no loops. Eventually one can go no further, i.e., reaches a vertex of degree 1. To get a second leaf start at the first leaf, retrace all steps in reverse direction, and continue to another leaf.) Hence we get a closed path in our original graph. This corresponds to a "closed path" in the matrix M as shown:

$$M = \begin{bmatrix} 1 & * & 0 & * \\ * & * & 3 & 4 \\ 1 & 2 & 5 & 0 \\ * & 7 & 1 & * \\ 3 & 1 & 2 & 4 \end{bmatrix} \longleftrightarrow D = \begin{bmatrix} 0 & 1 & 0 & -1 \\ 1 & -1 & 0 & 0 \\ 0 & 0 & 0 & 0 \\ -1 & 0 & 0 & 1 \\ 0 & 0 & 0 & 0 \end{bmatrix}.$$

Let D be the $m \times n$ matrix as shown, that is, let D have entries $0, 1, -1$. The nonzero entries occur only at the vertices of the "closed path" in M and alternative values 1, -1 are assigned as we traverse the path. So D has zero row and column sums. Let $M_1 = M + \varepsilon D$, $M_2 = M - \varepsilon D$ for $\varepsilon > 0$. The row and column sums of M_1 and M_2 satisfy the conditions for matrices in C. For ε small enough (less than the least distance from noninteger entries in M to integers), M_1 and M_2 have nonnegative entries (hence they are in C). Moreover $M = \frac{1}{2}(M_1 + M_2)$ with $M_1, M_2 \in C$, $M_1 \neq M_2$. Therefore M is not an extreme point. This ends our proof.

In the above theorem not all integer-entry matrices are extreme points. Take $m = n = 2$, $p_i = q_j = 2$, and note that

$$\begin{bmatrix} 1 & 1 \\ 1 & 1 \end{bmatrix} = \frac{1}{2}\begin{bmatrix} 2 & 0 \\ 0 & 2 \end{bmatrix} + \frac{1}{2}\begin{bmatrix} 0 & 2 \\ 2 & 0 \end{bmatrix}.$$

5.3

Out of the general class of problems dealing with the maximization of a linear function $c^t x$ subject to constraints $Ax \leq b$, the subclass of combinatorial interest is that in which the matrix A (or its transpose) has the following properties: all its entries are -1, 1, or 0, with at most two nonzero entries in a column, and if two nonzero entries occur in a column then they must add up to 0.

We aim to give a proof of the Birkhoff–von Neumann theorem via linear algebra. A general lemma, stated below, plays a central role:

Lemma 5.1 (On Extreme Points of Convex Polyhedra). *Let K be a convex polyhedron in R^n determined by a system $Ax \leq b$ of more than n inequalities. Then any extreme point of K is the unique solution of a subsystem of n equations of the larger* (and possibly inconsistent) *system $Ax = b$.*

Proof. Let e be (the vector in R^n corresponding to) an extreme point of K. Translating by e we may assume, without loss of generality, that $e = 0$. (Note that such a translation, say $y = x - e$, leaves the coefficient matrix A unchanged, i.e., the system $Ax \leq b$ becomes simply $Ay \leq b + Ae$.) With the understanding that $e = 0$, the vector b has nonnegative components (since the extreme point 0 is in K, and thus satisfies $0 = A0 = Ae \leq b$). Suppose that $b_i = 0$ for the first m equations and that $b_i > 0$ for the remaining equations (of $Ax = b$). We claim that the first m equations produce a system of rank n, with 0 ($= e$) as its unique solution. For if this is not true, then there exists a nonzero solution v to the $m \times n$ system. Since all other $b_i > 0$ we can select $\varepsilon > 0$ such that $tv \in K$ for all $-\varepsilon \leq t \leq \varepsilon$ and this contradicts the assertion that 0 is an extreme point. This proves the lemma.

[Geometrically all that Lemma 5.1 says is that an extreme point (i.e., a corner) of K is the intersection of precisely n "faces" of K (if $K \subseteq R^n$).]

The next lemma gives insight into the algebraic structure of the matrix attached to a system of "combinatorial" constraints:

Lemma 5.2. *Suppose that a $n \times n$ matrix M satisfies the following conditions:*

(i) *The entries are* $-1, 1,$ *or* 0.
(ii) *Any column has at most two nonzero entries.*
(iii) *If a column has two nonzero entries, then they add up to zero.*

Then the determinant of M is $-1, 1,$ *or* 0.

Proof. If M has a column with all entries 0, we are done (the determinant is 0). If M has a column with precisely one nonzero entry in it, we expand the determinant by that column and reduce the problem to a $(n-1) \times (n-1)$ matrix with the same properties as M, so we can apply induction. The only other possibility is that all column sums are zero and then the determinant of M is zero (since M^t has $\mathbf{1}$, the vector with all entries 1, in the kernel). This ends our proof.

We conclude Section 5.3 with another proof of the theorem by Birkhoff and von Neumann, stated in Section 5.2. The proof goes as follows.

Note that the convex set C can be described by $-x_{ij} \leq 0$, $\Sigma_i x_{ij} \leq q_j$, $\Sigma_i - x_{ij} \leq -q_j$, $\Sigma_j x_{ij} \leq p_i$, $\Sigma_j - x_{ij} \leq -p_i$. These inequalities describe a convex polyhedron in R^{mn} (as in Lemma 5.1) and each extreme point is obtained by solving a *nonsingular* system $Mx = b$, which is a *subsystem* of the larger (inconsistent) system:

$$\left.\begin{aligned} \sum_i x_{ij} &= q_j \\ \sum_j x_{ij} &= p_i \\ x_{ij} &= 0. \end{aligned}\right\} \tag{5.1}$$

We solve $Mx = b$ by Cramer's rule. All the determinants in question are clearly integers. So it will be sufficient to show that the determinants in the denominators are ± 1, that is, the determinant of M is $+1$ or -1.

Denote by A the $(m + n + mn) \times mn$ matrix of the (inconsistent) system (5.1). Note that

$$A = \begin{bmatrix} B \\ \cdots \\ I \end{bmatrix}$$

with I the $mn \times mn$ identity matrix [corresponding to the mn equations $x_{ij} = 0$ in (5.1)], and B a matrix that satisfies the conditions of Lemma 5.2. The $mn \times mn$ coefficient matrix M equals

$$M = \begin{bmatrix} B^* \\ \cdots \\ I^* \end{bmatrix},$$

with B^* consisting of some rows of B and I^* consisting of some rows of the identity matrix I. Expand the determinant of M by rows of I^* until left with a minor of B^*. The determinant of M (which is not 0 since M is nonsingular) equals (up to sign) the determinant of this minor of B^*. The determinant of the minor is therefore nonzero, and from Lemma 5.2 we conclude that it equals in fact $+1$ or -1. An extreme point of C has therefore *integral* entries.

REMARK. The coefficient matrix A attached to system (5.1) is in general of the form

$$A = \left[\begin{array}{cccccc} -1 & -1 & -1 & & & \\ & & & -1 & -1 & -1 \\ \hline 1 & & & 1 & & \\ & 1 & & & 1 & \\ & & 1 & & & 1 \\ \hline 1 & & & & & \\ & 1 & & & & \\ & & 1 & & & \\ & & & 1 & & \\ & & & & 1 & \\ & & & & & 1 \end{array}\right] \begin{array}{l} m \leftrightarrow \sum_j - x_{ij} = -p_i \\ \\ n \leftrightarrow \sum_i x_{ij} = q_j \\ \\ \\ mn \leftrightarrow x_{ij} = 0. \end{array}$$

In this small example the variables are ($m = 2$, $n = 3$):

$$\begin{matrix} x_{11} & x_{12} & x_{13} \\ x_{21} & x_{22} & x_{23} \end{matrix}.$$

5.4 The Shortest Route Algorithm

Consider a graph as below:

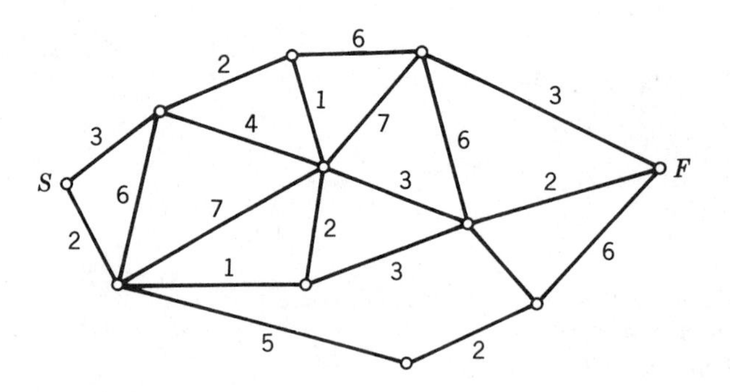

We wish to find the shortest route from S (start) *to F* (finish), *where the numbers on the edges represent distances.* In particular, if all the distances are 1 we get the path with the minimal number of edges. (It is hopelessly inefficient to measure the length of every path from S to F; if we have n vertices the worst possible case requires much more than $(n-2)!$ comparisons.)

The dual problem to this is that of maximizing the distance between S and F subject to the linear constraints that the distances between points do not exceed specified constants along the edges. These constraints can be written as $-d_{ij} \leq x_i - x_j \leq d_{ij}$, where x_i is the coordinate attached to vertex i (upon projecting the graph on a straight line) and d_{ij} is the specified distance between vertices i and j. When the constraints are written in the form $Ax \leq b$, where b is the vector of d_{ij}'s, the matrix A has -1, 1, or 0 as entries, at most two nonzero entries per column, and if two nonzero entries occur in a column they sum up to zero.

The algorithm that solves the shortest route problem is as follows:

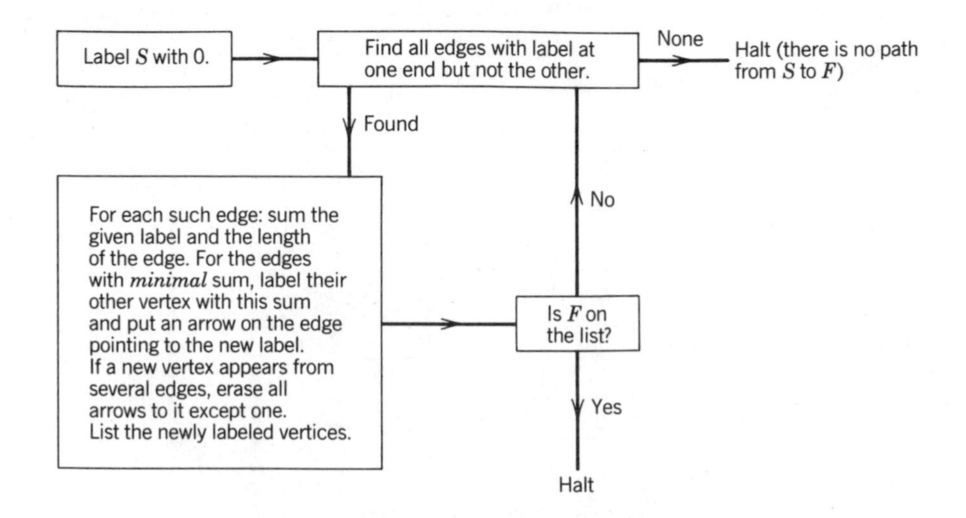

The edges with arrows form a tree. To identify the shortest route, start at F and follow unambiguously back to S against the direction of the arrows.

Looking at this algorithm one can see that we run through the loop less than $n\binom{n}{2}$ times, where n is the number of vertices [and $\binom{n}{2}$ is an upper bound on the number of edges to be checked at each step]. This algorithm is therefore of the order of n^3 (much more efficient than $(n-2)!$, which exceeds 2^n).

[*Aside*: *An impractical practical solution*. Make a string model with each edge the appropriate length, hold S with left hand and F with right hand, pull apart until string is tight—the shortest route is now visible! ("Oh what a tangled web we weave... .") One should appreciate the dynamic interplay between the initial (shortest route) problem and the dual (longest distance) problem which this string model vividly brings forth. It is easy to visualize the shortest route algorithm in terms of the string model: lay the string model flat on the table and slowly lift vertex S straight up. The shortest route becomes visible!]

EXERCISES

1. Write the doubly stochastic matrix $n^{-1}J$ (where J is the $n \times n$ matrix with all entries 1) as a convex combination of a *minimal* number of permutation matrices. Do the same for

$$\frac{1}{12}\begin{pmatrix} 5 & 1 & 6 \\ 0 & 8 & 4 \\ 7 & 3 & 2 \end{pmatrix}.$$

2. Prove that every $n \times n$ doubly stochastic matrix is a convex combination of at most $n^2 - 2n + 2$ permutation matrices. Is this best possible for every n?

3. Work through the shortest path algorithm for the graph

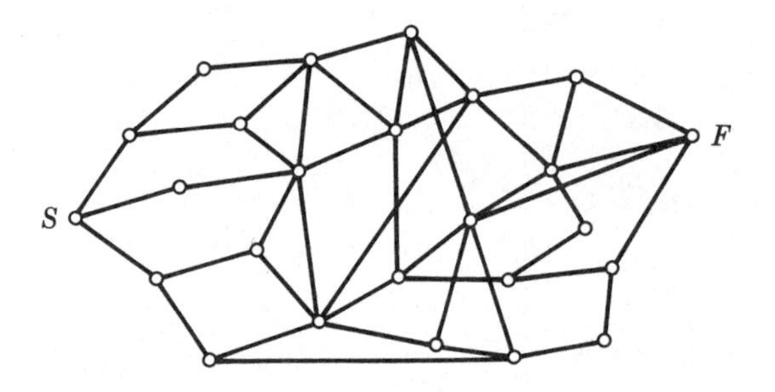

(Distances on all edges are 1.)

2 MATCHING AND MARRIAGE PROBLEMS

The reader should be informed at the very outset that the contents of this section are more pleasant than the title may suggest. Consequences of the theory herein presented are numerous and its applications reach far beyond the narrow scope suggested by the title. On this latter point still we vigorously inform that these results, although magnanimously designed for the enhancement of happiness at large, remain of limited value on an individual basis.

Having said this, and consequently being left with the mathematically inclined only, we describe two qualitatively different types of applications. The first rests on the algorithm for extracting a *maximal* number of matchings between objects of one type (girls, say) and objects of another (boys) given a matrix of "likings" between the two groups. Matching people with jobs, given a matrix of their qualifications regarding these jobs, is another typical example. Applications of another kind involve the so-called marriage lemma. The proof of the existence of a Haar measure on a compact group is one such consequence, and the proof of the existence of finite approximations to measure preserving transformations is another. Those of the former kind are applied assignment problems while the latter are theoretical.

5.5

To start with, consider the problem of assigning roommates in a dormitory (a sorority). Form a graph in which the vertices are the girls and edges are drawn when there is a liking between the girls. A matching is a coloring in red (indicated by r on the drawings below) of some of the edges so that for any vertex there is at most one colored edge adjacent (i.e., no bigamy is allowed).

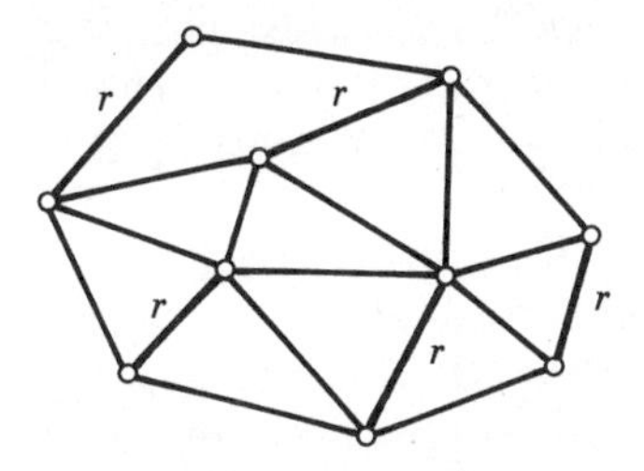

In the diagram above, the red matching is perfect (the assignment problem is completely solved). A perfect matching may not always exist, but we may seek *maximal* matchings, that is, matchings with a largest possible number of edges. Given any matching we then label the vertices U or M to denote unmatched or matched.

Berge observed the following:

∗ *If a matching is not maximal, then the labeled graph contains a path of odd length of the form*

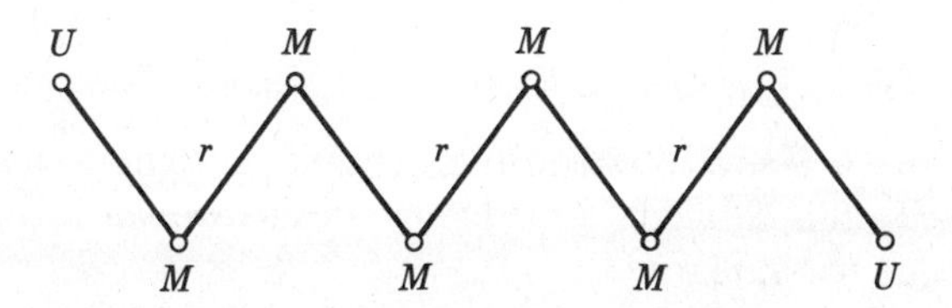

(i.e., a zigzag). (Note that upon locating such a zigzag we can increase by one

the number of matchings by relabeling

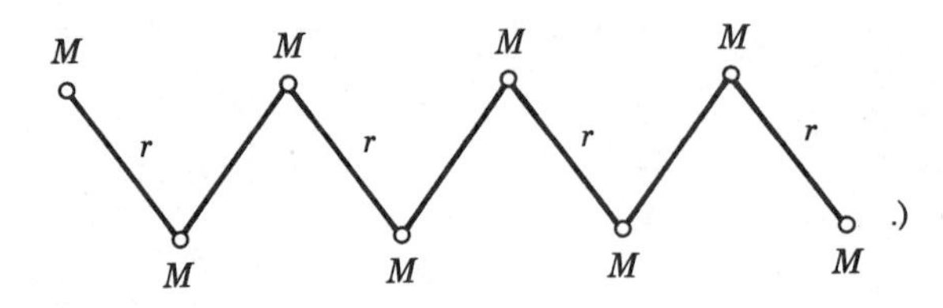

Proof. Suppose that the given red matching is not maximal. Then there is a green matching (say) with a greater number of edges. From the given graph delete all uncolored edges and all doubly colored edges. Decompose the remainder into its connected components and focus attention on the components that contain at least one edge. No vertex can have degree three or more —else that vertex would be a "bigamist" for red or green. Therefore every vertex has degree one or two. There exists, and we choose, a component with more green than red vertices. It must be one of four possible types:

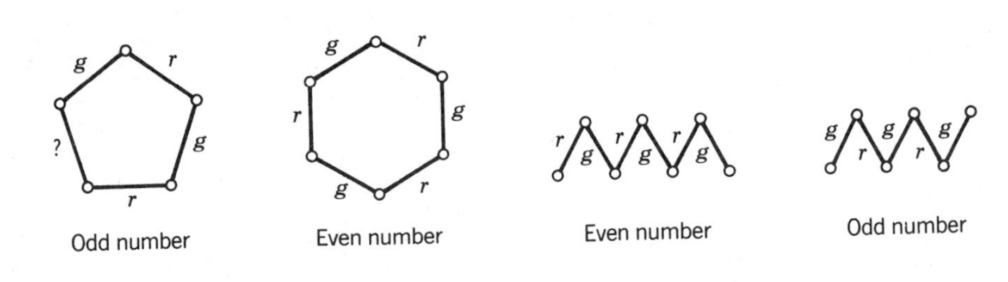

The first case leads to bigamy while the next two cases have an equal number of red and green edges. Hence we must have an odd zigzag with one more green edge than red. It is now enough to show that the ends of the odd zigzag are unmatched in the red matching. But if an end was matched by a red edge, that edge was deleted (or it would still be there!) and so that edge was both red and green—but then we have green bigamy. This ends our proof.

5.6 The Algorithm for Extracting a Maximal Matching

Consider now the marriage problem (i.e., finding a maximal matching) in a *bipartite* graph. In this situation a matching is called a *marriage*. The following method is due to Munkres [4].

Line up the girls in the left column and the boys in the right column (as below) and place an edge between a girl and a boy if there is a liking between them (else place no edge). Marriages are allowed only between liking couples, of course. Marry off as many boys and girls as possible by some means (marry off one couple, say).

Adjoin extra vertices S (start) and F (finish).

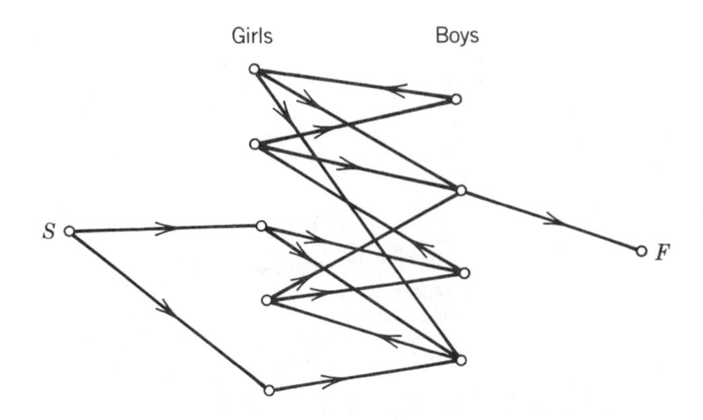

Connect S to the unmarried girls (and direct these edges left to right). Connect the unmarried boys to F (and direct left to right). Direct liking but not married couples left to right. Direct existing marriages *right to left*. (Every edge is now directed.) Use the shortest route algorithm described in Section 5.4 (and assign distance 1 on all edges) to find *any* (in particular the shortest) path from S to F. If there is one, it is a zigzag,

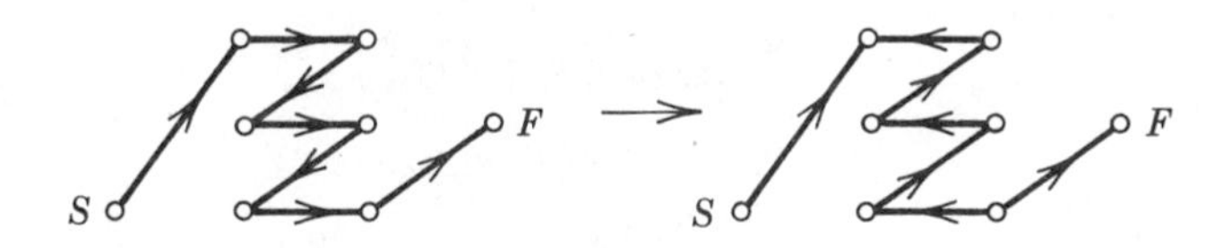

and we can increase the number of marriages by one, as shown. Continue in this fashion until a new path becomes unavailable. The process has now ceased and we ask: Is the resulting matching maximal? Yes. This can be seen by reconsidering Berge's observation (see Section 5.5) in the present context. (Indeed, the existence of an odd zigzag with appropriate directions on edges and S and F adjoined provides a directed path from S to F, contradicting the fact that the process cannot be continued.)

5.7

We now examine more carefully the maximal situation to obtain necessary and sufficient conditions to marry all the girls. At this point it is convenient to introduce matrix notation.

Given m girls and n boys form the $(0, 1)$ matrix M in which a 0 in the (i, j)th entry means girl i likes boy j and 1 denotes dislike. [Note, conversely, that any $m \times n$ $(0, 1)$ matrix M gives a like–dislike graph for m girls and n boys.] We denote marriages by starring some of the 0's. To avoid bigamy we must not have two starred zeros in any row or in any column.

A *set of stars* on the 0 entries is called *admissible* if no two starred zeros occur in the same row or the same column of M. A *set of lines* (that is, rows and/or columns of M) is called *admissible* if their deletion removes all the 0's in M.

Since in particular we have to cross out all starred 0's by an admissible set of lines we conclude that

$$|\textit{admissible stars}| \le |\textit{admissible lines}|,$$
for arbitrary admissible sets.

The main result is due to König:

König's Theorem. *There exists an admissible set of stars and an admissible set of lines of the same cardinality; this admissible set of stars is in fact of maximal cardinality and this admissible set of lines is in fact of minimal cardinality.*

This result surfaces upon a careful examination of what goes on when there is *no path* from S to F. Assume therefore that a number of marriages were made (by the maximal matching algorithm described in Section 5.6 or otherwise) and that we are now at the final stage with *no path* left from S to F.

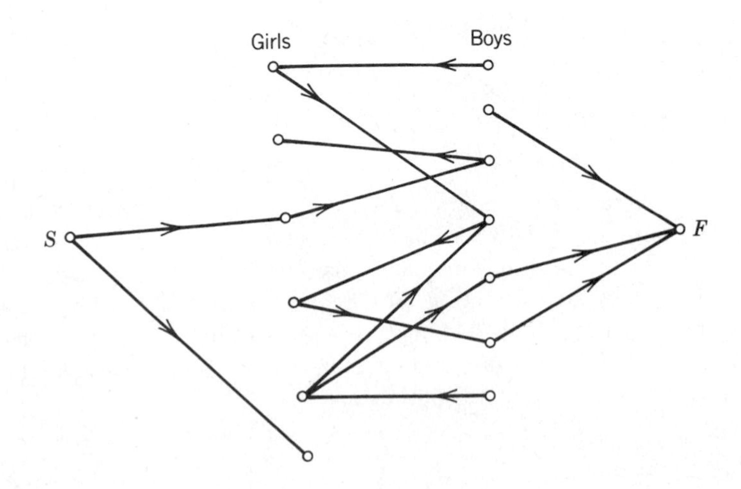

Partition the girls into four classes by married (M) or unmarried (U); accessible by a directed path from S (A), or inaccessible by such a path (I).

Do *exactly* the same for boys. By the construction of the graph all unmarried girls are accessible and since there is no path from S to F no unmarried boy is accessible. The partition is thus as follows:

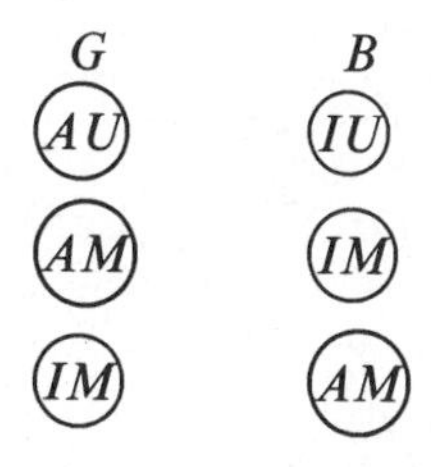

We now want to examine the possible interconnections between these classes by likes (left to right arrows) and marriages (right to left arrows). It is convenient to interpret this in terms of the zeros and the starred zeros in the matrix M. There are nine cases that (upon suitable permutations of rows and columns) are summarized below:

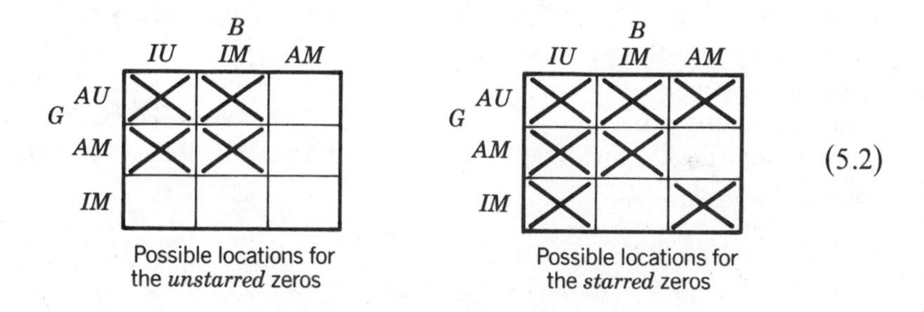

(5.2)

The squares that are crossed out contain no zeros at all (starred or unstarred).

Let us explain in detail the situations summarized in (5.2): For the *unstarred* case there can be no direct link (i.e., no edge) from an accessible girl to an unaccessible boy (else the boy would be accessible). The *starred* zeros represent marriages and so unmarried boys are excluded (i.e., column 1) and, likewise, unmarried girls are excluded (i.e., row 1). There can be no marriage between an *IM* girl and an *AM* boy (else the girl would be accessible)—see below:

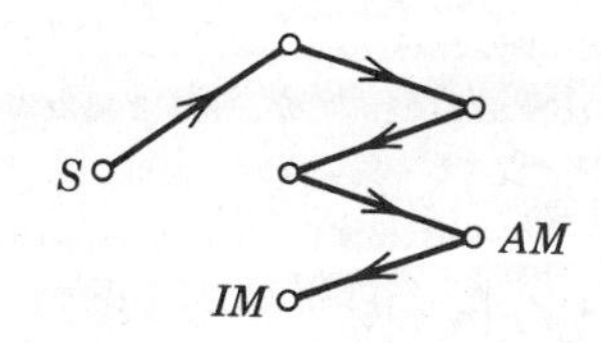

Also, there can be no marriage between an AM girl and an IM boy—for the last link to the AM girl must be the marriage and thus the boy to whom she is married is accessible (see below):

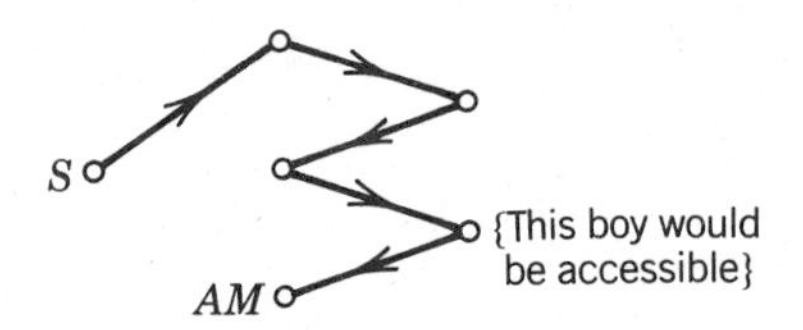

We therefore conclude that the marriages occur only between two inaccessibles or two accessibles [as the summary (5.2) of the possible positions for the starred zeros indicates].

In the matrix M we therefore have exactly one starred 0 in each column under AM (and it occurs in the AM row region), and we have exactly one starred 0 in each row across from IM (and it occurs in the IM column region). Delete these rows and columns (i.e., these lines) and we delete all the zeros in M (starred and unstarred). The number of lines equals precisely the number of starred zeros. (And this set of lines is admissible, as we just saw.) This proves König's theorem.

We actually proved that when a maximal matching is reached we can extract from it an admissible set of stars and an admissible set of lines of the same cardinality.

5.8

By the list of "eligibles" for a girl we mean the subset of boys that she likes. Each girl g has her list of "eligibles" $L(g)$.

The Marriage Lemma (P. Hall). *All the girls can be married if and only if* $|\cup_{g \in T} L(g)| \geq |T|$, *for all subsets* T *of the set of girls.*

Proof. The proof follows easily from König's theorem. Necessity is obvious (for if the joint lists of t girls contain less than t boys, then there is no way to marry all of these t girls). Sufficiency is proved as follows: Suppose that we cannot marry all the girls (which are m in total, say). Set up the matrix M of girls versus boys with a 0 in (i, j)th entry if boy j is on the list of eligibles of girl i. By our supposition the number of starred 0's in M is strictly less than m. By König's theorem there exists an admissible set of less than m lines. Without loss assume that the admissible lines are given by the first r rows and the first c columns of M. So $r + c < m$. Let T be the set of girls in the remaining $m - r$ rows; $|T| = m - r$. The total number of eligible boys for these girls is $\leq c < m - r = |T|$. This ends the proof.

There exist fairly short inductive proofs of the marriage lemma. We selected this approach because of its close interplay with the algorithm for actually extracting a maximal matching.

5.9

By the list of eligibles for a boy we understand the set of girls that he likes. A simple sufficient condition to marry all the girls is described below:

* *If each girl has at least k eligible boys, and if each boy has at most k eligible girls, then all the girls can be married.*

Indeed, let T be a set of girls. Then

$$\begin{aligned} k|T| &\le |\text{the set of edges emanating from } T| \\ &= |\text{the set of edges terminating in } \bigcup_{g \in T} L(g)| \\ &\le |\bigcup_{g \in T} L(g)|k. \end{aligned}$$

Hence $|\cup_{g \in T} L(g)| \ge |T|$ and the marriage lemma now guarantees that all the girls can be married.

EXERCISES

1. With the likings indicated below, can all the girls be married?

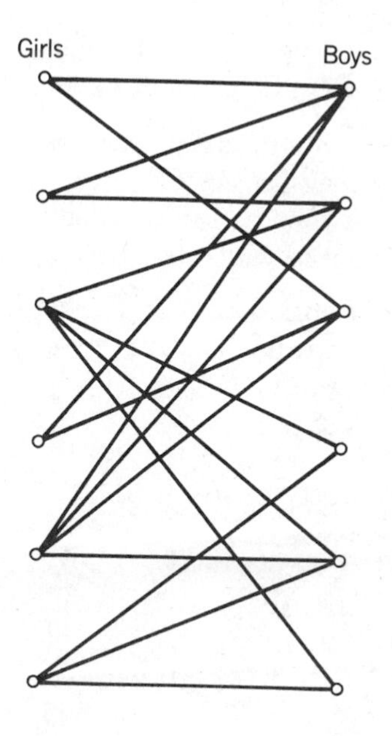

2. With the likings drawn below, can we match up the ten girls in pairs of roommates?

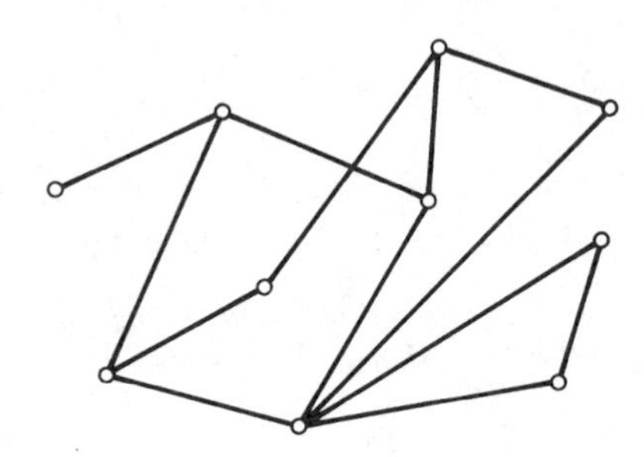

3. Workers A, B, C, D, E are qualified for jobs 1, 2, 3, 4, 5, 6, 7 as shown in the diagram below:

	1	2	3	4	5	6	7		1	2	3	4	5	6	7
A	—	√	√	√	√	√	—	A	√	—	—	√	—	—	√
B	√	—	—	—	—	√	—	B	—	—	√	√	—	√	—
C	√	—	—	—	—	√	—	C	√	—	√	√	—	—	—
D	√	—	√	—	—	—	—	D	√	—	—	√	—	—	—
E	√	—	√	—	—	√	—	E	—	√	—	—	√	—	√

In only one of the two situations all workers can be employed. Which one?

4. An employment agency has nine job openings and many candidates for employment, each qualified for three of the jobs. The agency selects a group of candidates that includes three people qualified for each job. Can all nine jobs be filled with this group of candidates?

5. Twelve students are asked (in some fixed order) to choose their three favorite classes from a long list. As they express preferences, each class is removed from the list as soon as three students have chosen it. Show that each student can be matched with one of his or her favorite classes.

6. Prove that any square matrix with 0 or 1 as entries, and row and column sums equal to k can be written as a sum of k permutation matrices. Try this for

$$\begin{matrix} 0 & 1 & 1 & 1 & 0 & 1 & 0 \\ 0 & 0 & 1 & 1 & 1 & 0 & 1 \\ 1 & 0 & 0 & 1 & 1 & 1 & 0 \\ 0 & 1 & 0 & 0 & 1 & 1 & 1 \\ 1 & 0 & 1 & 0 & 0 & 1 & 1 \\ 1 & 1 & 0 & 1 & 0 & 0 & 1 \\ 1 & 1 & 1 & 0 & 1 & 0 & 0 \end{matrix}$$

3 THE ARC COLORING LEMMAS

5.10

Consider a finite graph G in which each edge is colored green or red. We distinguish two vertices and ask whether there exists a path with green edges between them and, if not, we want to know what *the alternative required feature* of the graph is. It is more convenient to adjoin an extra yellow edge joining the two distinguished vertices. We then ask whether there exists a green closed path in G that includes the yellow edge. It is easy to see that there is a closed path if and only if there is a simple closed path (i.e., a path in which no vertex is passed more than once). We call a simple closed path a *cycle*.

The appropriate alternative object arises naturally in the setting of *planar graphs* (i.e., graphs that can be drawn in the plane, or on a sphere, with no edges intersecting). For such a graph G we may form a *dual graph* G^*—each "country" in G becomes a vertex in G^* and we have an edge between two "countries" if they have a common boundary. (Note that the unbounded complement of G is regarded as a country—one can, equivalently, consider graphs on the surface of a sphere where all countries are bounded.)

Example. G = graph with edges of the form o———o
G^* = graph with edges of the form o- - - - - -o

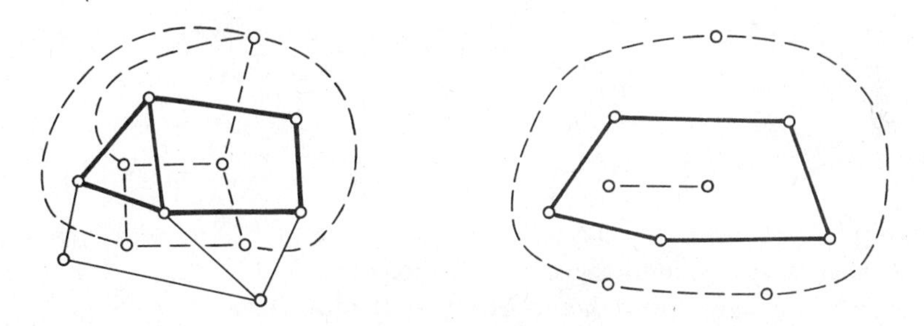

The red cycle in G may be regarded as a cutting out of some edges in G^* that *separates* G^* into two connected components (as shown in the figure at right). Any cycle in G has such effect upon G^*. The cut set of edges in G^* is called a cocycle. From this example one can see that for planar graphs there is a bijective correspondence between the cycles of G and the cocycles of the dual graph G^*.

We may define the concept of a cocycle in *any* graph G (planar or otherwise) as follows: A *separating set* for G is a collection of edges whose deletion increases the number of connected components of G; it is minimal (by inclusion) if no smaller collection of edges serves to increase the number of connected components. A *cocycle* is a minimal separating set.

Minty's Arc Coloring Lemma. *Let G be a colored graph with one yellow edge and the other edges red or green. Then one, and only one, of the following always happens: Either there exists a cycle with all edges green except one yellow edge, or there exists a cocycle with all edges red except one yellow edge.*

Proof. Let S be one of the distinguished vertices (i.e., one endpoint of the yellow edge). Mark all the vertices of G that are accessible from S by green paths. This partitions the vertices into accessible A and inaccessible I. Evidently there is no green edge between A and I. If there is no cycle of the required form, then the yellow edge is between A and I. Delete (temporarily) the yellow edge and all red edges between A and I, and decompose the remaining graph into its connected components A and $I_1, I_2, \ldots, I_k$ (say). All vertices in A remain connected to S and so (upon putting back all the edges we deleted) the picture looks like this:

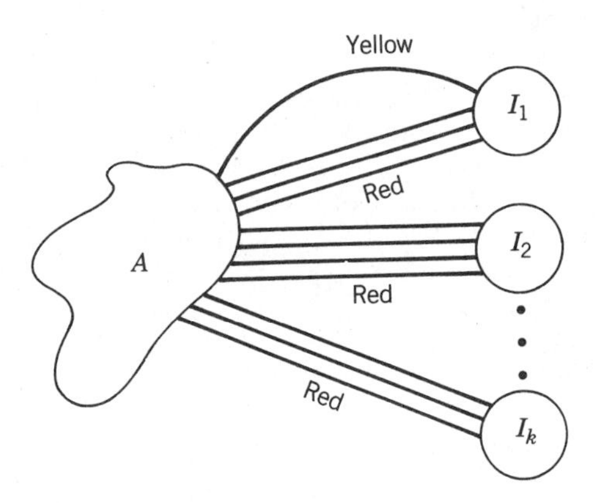

Without loss of generality the yellow edge joins A to I_1. The edges joining A to I_1 provide the required cocycle. This ends our proof.

A useful extension, also due to Minty, is the following:

Lemma 5.3. *Color the edges of a graph green, red, and yellow* (one color per edge). *Place an arrow on each yellow edge and identify one of the yellow edges as the "distinguished" yellow edge. Then precisely one of the two possibilities occurs:*

(a) *There exists a cycle containing the "distinguished" yellow edge with all its edges green or yellow oriented in the direction of the "distinguished" yellow edge.*

(b) *There exists a cocycle containing the "distinguished" yellow edge with all its edges red or yellow oriented in the direction of the "distinguished" yellow edge.*

Proof. Label by S and F the endpoints of the "distinguished" yellow edge so that the arrow points from F to S. Partition the vertices into those accessible from S by a path of green edges and yellow edges with arrows pointing away from S, and those inaccessible by such a path. Denote these two disjoint subsets of vertices by A (accessibles) and I (inaccessibles).

If the situation described in (a) does not occur the edges running between A and I are necessarily red and yellow (pointing from I to A). Since F is in I, the edges between A and the connected component of F in I (call it I_1) form the required cocycle.

4 FLOWS AND CUTS

5.11

By a *directed graph* we mean a connected graph with no loops and with one arrow placed on each edge. A *network* is a directed graph in which there are two distinguished vertices, a source S and a sink F, with each edge being assigned a nonnegative real number called its *capacity*. The vertices are labeled $1, 2, \ldots, m$ and the capacity of edge $\{i, j\}$ is denoted by $c(i, j)$. [If there is more than one edge between i and j we may use $c_k(i, j)$ to denote the capacity of the kth edge between them. This is seldom necessary, however.]

One can think of the capacity of an edge as the maximal amount of liquid (or goods of some kind) possible to transport through that edge in the direction of the arrow. Examples of networks may be highway systems, railroad networks, electrical devices, and so on.

A *network flow* is a function assigning a nonnegative number $f(i, j)$ to each edge $\{i, j\}$ (the flow through that edge) so that $f(i, j) \leq c(i, j)$ and $\Sigma_i f(i, j) = \Sigma_k f(j, k)$, for each vertex j different from S and F. [This latter condition informs us that all the liquid that enters a vertex (not S or F) must necessarily exit it. In electrical network theory this is known as Kirchhoff's law.] The *value* of a network flow is the sum of the flows on edges emanating from S minus the sum of the flows on edges pointing into S (or, equivalently, the sum of the flows on edges pointing into F minus the flow on edges emanating from F). A *maximal flow* is a network flow of maximal value.

Below we illustrate a network flow with value 3. The capacities are written in parentheses:

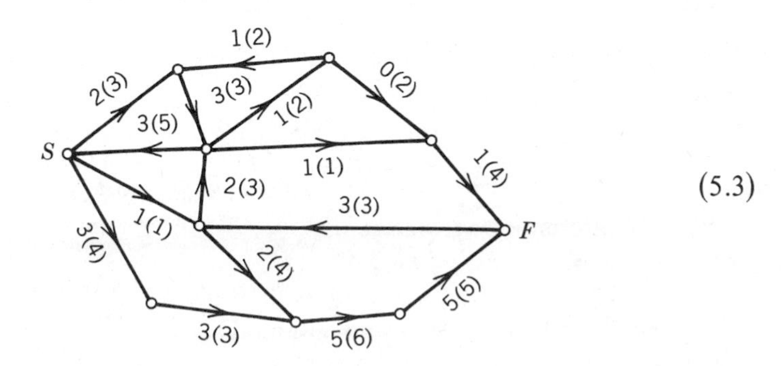

(5.3)

Is this flow maximal? If not, try to find a maximal flow.

The main problem in the theory of network flows is to *determine a maximal flow in a given network*. (The solution is surprising because an obvious necessary condition on capacities turns out to be also sufficient, as we shall see shortly.)

5.12

A *cut* in a network is a subset of edges whose removal disconnects S from F. (By disconnecting S from F we mean that there is no undirected path left between S and F.) When we remove the edges of a cut from a network S and F lie in different connected components of the resulting graph. Let $\bar{S}$ be the connected component of S and $\bar{F}$ that of F ($\bar{S}$ and $\bar{F}$ depend on the choice of the cut, of course). The *capacity of a cut* equals the sum of capacities of those edges of the cut that point *from* $\bar{S}$ *to* $\bar{F}$.

In the figure below we display a cut in the network (5.3) (consisting of six edges in total) with capacity $2 + 1 + 4 + 3 = 10$:

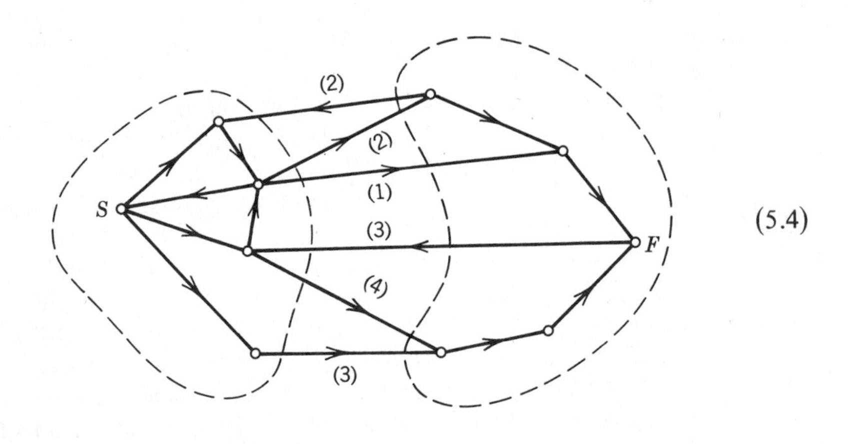

(5.4)

5.13

If P is an undirected path between S and some vertex i of the network, we call an edge of P *forward* if its direction along P is from S to i, and *backward* if its direction along P is from i to S. Given a network flow f we call a path P between S and i *unsaturated* if $f(k, l) < c(k, l)$ for each forward edge $\{k, l\}$ of P, and $f(k, l) > 0$ for each backward edge $\{k, l\}$ of P (the strict inequalities are important in this definition). An unsaturated path between S and F is called *flow augmenting*.

If we can find a flow augmenting path P we can strictly increase the value of the network flow f (thus the terminology). Indeed, look at the list of numbers $c(i, j) - f(i, j)$ for the forward edges $\{i, j\}$ of the path P, and at the list of numbers $f(i, j)$ for the backward edges. All these numbers are strictly positive and finitely many in number. Pick the smallest number from the union of both lists; call it d. Define a new flow g as follows:

$$g(i, j) = \begin{cases} f(i, j) + d, & \text{if } \{i, j\} \text{ is a forward edge of } P \\ f(i, j) - d, & \text{if } \{i, j\} \text{ is a backward edge of } P \\ f(i, j), & \text{if } \{i, j\} \text{ is not an edge of } P. \end{cases}$$

It is easy to check that g is a network flow. Moreover, since the path P starts at S, the value of the flow g exceeds the value of the initial flow f by d (a strictly positive quantity). From this we conclude that if a network flow f admits a flow augmenting path, then f is not a maximal flow. We prefer to summarize this as follows:

If f is a maximal flow in a network, then there are no flow augmenting paths in that network. (5.5)

5.14

Let f be any network flow and C be any cut. The cut C (by its definition) separates S from F. Therefore, the only way that the liquid that emanates from S will reach F is if it passes through the edges of the cut C that point from $\bar{S}$ to $\bar{F}$ (the connected components of S and F that the cut C induces). It thus follows that

The value of any network flow cannot exceed the capacity of any cut. (5.6)

We wish to investigate maximal flows, and the main result in this regard is:

The Max-Flow Min-Cut Theorem (Ford and Fulkerson). *The value of a maximal flow equals the capacity of a minimal cut.*

By a *minimal cut* we understand a cut of minimal capacity.

Proof. From (5.6) we conclude that a maximal flow will have value at most equal to the capacity of a minimal cut. Conversely, assume that f is a maximal flow. Then (5.5) informs us that there is no flow augmenting path in the network. Partition the vertices into two classes: those that can be reached from S by an *unsaturated path*, and those that cannot. Vertices S and F are in different connected components, by assumption; denote their respective components by $\tilde{S}$ and $\tilde{F}$. The edges of the network running between $\tilde{S}$ and $\tilde{F}$ obviously form a cut; call it C. This is the cut we want: Its capacity equals that of f. Indeed, an edge $\{i, j\}$ pointing from $\tilde{S}$ to $\tilde{F}$ has $f(i, j) = c(i, j)$ (or else j would be in $\tilde{S}$). An edge $\{i, j\}$ pointing from $\tilde{F}$ to $\tilde{S}$ has $f(i, j) = 0$ (or else, again, j would be in $\tilde{S}$). The value of f is, therefore, at least equal to $\Sigma c(i, j)$ (the capacity of C), where the sum runs over all edges of C pointing from $\tilde{S}$ to $\tilde{F}$. We thus conclude that f has value *equal* to the capacity of C. (Note that the arc coloring lemmas of Section 3 are implicit in this argument.) This ends the proof.

5.15

There is a so-called vertex version of the max-flow min-cut theorem. To understand it we have to work with a species of networks in which all edges are assigned *infinite* capacity and, in addition, each vertex has a finite capacity. A *flow* in this network is a function f assigning a nonnegative number to each edge (the flow through that edge) such that the sum of the flows through the edges that enter a vertex does not exceed the capacity of that vertex, and the Kirchhoff law (i.e., flow into a vertex = flow out) is satisfied at all vertices other than S and F. The *value* of a flow, and the notion of a *maximal flow*, are defined as before.

A *vertex cut* is a subset of vertices whose removal disconnects S from F. (By removing a vertex we understand the removal of that vertex *and* all edges coming into or going out of that vertex.) The *capacity of a vertex cut* equals the sum of the capacities of the vertices it contains. By a *minimal vertex cut* we understand a vertex cut of minimal capacity.

The vertex version of the max-flow min-cut theorem states the following:

* *In a network with vertex capacities the value of a maximal flow equals the capacity of a minimal vertex cut.*

Proof. The trick in proving this result is to convert the (species of) network with vertex capacities into a network of the more ordinary kind (i.e., a network with edge capacities only).

Do this by "elongating" vertex i into an edge (whose endpoints we label by i_{in} and i_{out}). All edges of the original network that entered vertex i now enter i_{in} and all edges that left vertex i now leave i_{out}. Place an arrow pointing from i_{in} to i_{out} on the (new) edge $\{i_{in}, i_{out}\}$ and define the capacity of this new edge to be the original capacity of vertex i. The process is graphically illustrated below:

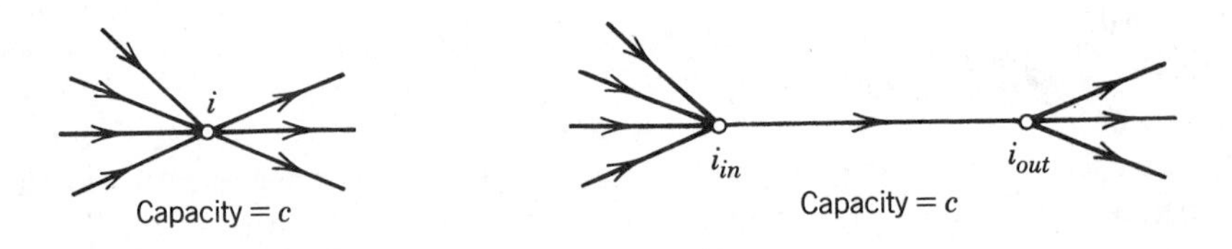

Upon performing this service to each vertex we obtain an ordinary network. In this network the (value of the) maximal flow equals the (capacity of the) minimal cut. Obviously no edges of infinite capacity are in a minimal cut, so the minimal cut corresponds to a vertex cut in the original network. This ends our proof.

5.16

The general problem of finding a maximal flow in a network can be viewed as a problem in linear programming. To see this, attach a variable to each edge: Specifically, write x_{ij} for an edge $\{i, j\}$ with arrow oriented from i to j. Introduce an additional edge between S and F (color it yellow to distinguish it from the rest) and place an arrow on it that points from F to S. Give it infinite capacity and attach to it the variable x_{fs}.

Suppose there are n edges (including the yellow edge) in the network so amended. A flow in the original network corresponds now to a vector with n components with variables x_{kl}'s as entries such that the Kirchhoff law is satisfied at each vertex (*including* S and F).

The task of finding a maximal flow becomes that of maximizing x_{fs} subject to the linear constraints:

$$x_{ij} \leq c(i, j),$$

$$-x_{ij} \leq 0,$$

and

$$\sum_i x_{ik} = \sum_i x_{ki} \qquad \text{(the Kirchhoff law)}$$

for *all* vertices k of the network.

5.17 The Algorithm for Extracting a Maximal Flow

The algorithm described here is based on the observation that

> *If there is no flow augmenting path, then the flow is maximal.* (5.7)

This is the converse of statement (5.5) and is proved as follows: Suppose that no flow augmenting path exists. Partition the vertices into those that can be reached from S by an unsaturated path and those that cannot. Vertices S and F are in different classes ($\tilde{S}$ and $\tilde{F}$, say) of this partition. The flow through any edge pointing from $\tilde{S}$ to $\tilde{F}$ equals the capacity of that edge (by the construction of the partitions). Similarly, flows through edges pointing from $\tilde{S}$ to $\tilde{F}$ are zero. The edges between $\tilde{S}$ and $\tilde{F}$ form a cut whose capacity equals that of the flow. By (5.6) the flow is maximal.

The algorithm we give is for networks with integral capacities (of edges) and in which integral flows only are allowed through the edges. Observation

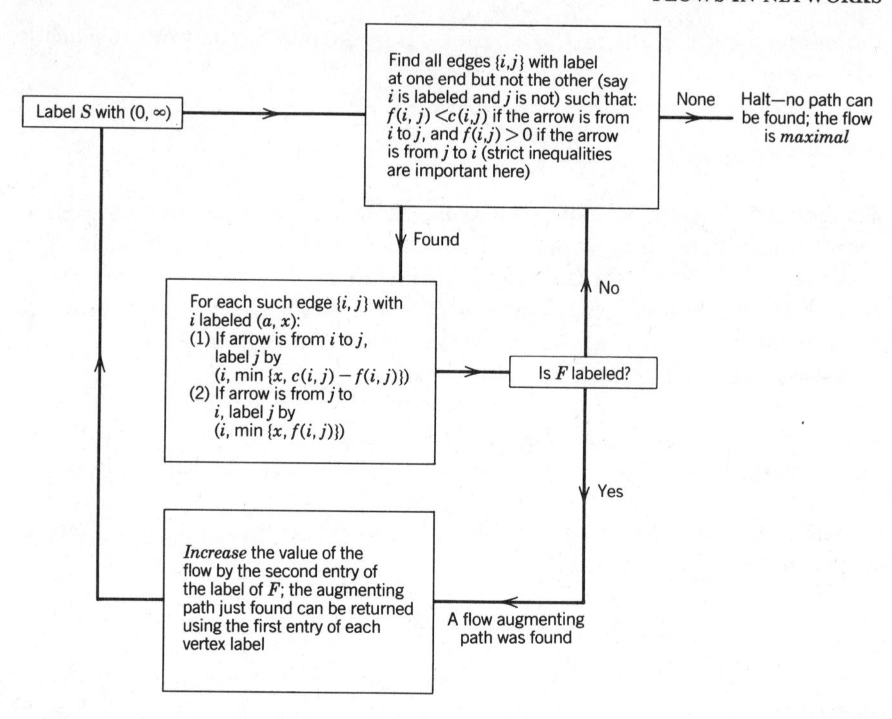

(5.7) will allow us to conclude that when no augmenting path can be found we have reached the maximal flow.

5 RELATED RESULTS

The two max-flow min-cut theorems proved in Section 4 can be used to derive several well-known results on partially ordered sets and on connectivity of graphs. Best known, perhaps, are Menger's theorems on connectivity of graphs obtained in 1927, several decades before the max-flow min-cut results.

5.18

Two paths in a graph are called *edge-disjoint* if they have no common edges (but may have common vertices). By two *vertex-disjoint* paths between two nonadjacent vertices we understand two paths with no vertices in common other than the endpoints. We have the following:

Menger's Results.

(i) *The maximal number of edge-disjoint paths between two nonadjacent vertices equals the minimal number of edges whose removal disconnects the two vertices.*

(ii) *The maximal number of vertex-disjoint paths between two nonadjacent vertices equals the minimal number of vertices whose removal disconnects the two initial vertices.*

(By the removal of a vertex we understand, as before, the removal of that vertex *and* all edges adjacent to it.)

Proof of part (i). Produce a network N out of the given graph G by replacing each edge

by two edges oriented opposite to each other

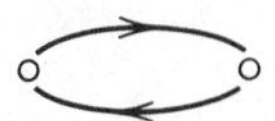

Denote the two nonadjacent vertices by S and F; give each edge of the network capacity 1. Through each edge of N we only allow flows of 0 or 1.

Observe that the value of a maximal flow in N equals the maximal number of edge-disjoint paths in G. [To see this, start at S and walk along the edges with flow 1, never against the direction of an arrow and never retracing. We may find ourselves visiting a vertex several times (and this includes S itself) but if the flow out of S is positive we ultimately reach F. Extract a path out of this walk as indicated below:

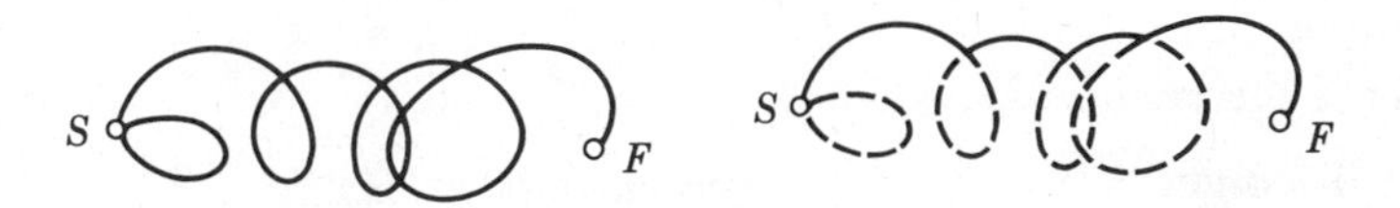

On all edges of this path change the flow from 1 to 0. What results is *another flow* with value precisely 1 less. Repeat the process. The resulting paths will be edge-disjoint because at each stage we only walk along edges with nonzero flow. By this process we extract as many edge-disjoint paths in G as the value of the maximal flow in N.] By the max-flow min-cut theorem there is a minimal cut in N whose capacity equals the maximal number of edge-disjoint paths in G. But its capacity is simply the minimal number of edges of G that disconnects S from F.

Proof of part (ii). To the graph G attach a network N as in part (i) but allow infinite capacity on edges; give each vertex capacity 1. N is a network with vertex capacities. By "elongating" each vertex into an edge with capacity 1 (as in Section 5.17) obtain a new network with only edge capacities of 1 or ∞. The same arguments as in part (i) lead to the stated result. Observe that the resulting cut corresponds to a vertex cut and that the paths we obtain are vertex-disjoint. This ends the proof.

5.19

It is easy to derive König's theorem (proved "from scratch" in Section 5.7) from the vertex-disjoint version of Menger's theorem. Let us explain how this is done.

Line up the m girls in a column at left and the n boys in a column at right. Place an (undirected) edge between girl i and boy j if they like each other; else place no edge. Join all the girls to a new vertex S at left and join all boys to a new vertex F at right. Call the resulting graph (of $m + n + 2$ vertices) G.

Attach also a $m \times n$ matrix M of likings (girls versus boys) with (i, j)th entry 0 if girl i likes boy j, and 1 otherwise.

By a *line* we mean a row or a column of M. In the graph G a line corresponds to a vertex (i.e., a girl or a boy). We indicate a marriage between girl i and boy j by starring the 0 in position (i, j) of M. In the graph G a marriage between girl i and boy j should be viewed as a path (of *three edges*: Si, ij, and jF) between S and F (with the "middle" edge ij indicating the marriage).

We seek a maximal number of marriages. Since no bigamy is allowed we call a set of starred 0's *admissible* if no two are in the same row or column. A maximal number of marriages corresponds to an admissible set of starred 0's of maximal cardinality.

A set of lines is called *admissible* if their deletion removes all the 0's in M. In the graph G an admissible set of lines corresponds to a set of vertices whose removal disconnects S from F (i.e., it leaves no edges between the boys and the girls).

König's theorem is as follows:

König's Theorem. *The cardinality of a maximal admissible set of starred* 0*'s equals the cardinality of a minimal set of admissible lines.*

By Menger's theorem, part (ii) (see Section 5.18), we conclude that the minimal number of admissible lines equals the maximal number of vertex-disjoint paths between S and F. To conclude the proof observe that if the number of vertex-disjoint paths is maximal, then all the paths must necessarily have length 3 (i.e., they are marriages); (this is clear, since any path of the form

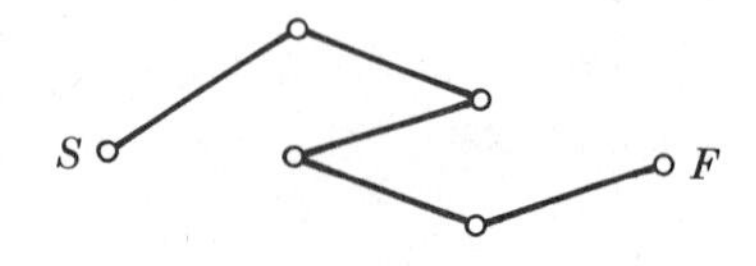

can be replaced by the two paths below:

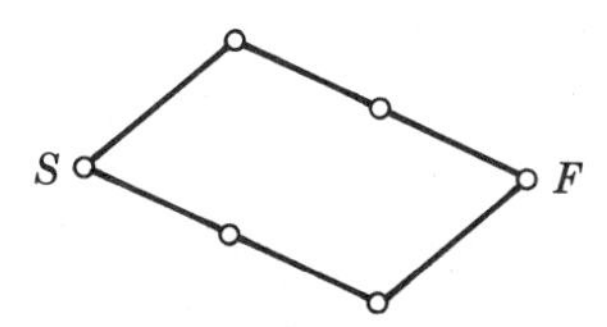

contradicting maximality). We proved that *the maximal number of possible marriages equals the minimal number of vertices whose deletion disconnect S from F*. Or, in matrix language, König's theorem (as stated).

5.20 A Result by Dilworth

König's theorem implies, in turn, a result on the maximal number of disjoint chains in partially ordered sets, first proved by Dilworth. A *partially ordered set* is a pair $(E, \leq)$ consisting of a set of nodes E and a relation $\leq$ of partial order. By definition $\leq$ satisfies: $x \leq x$; $x \leq y$ and $y \leq x$ imply $x = y$; $x \leq y$ and $y \leq z$ imply $x \leq z$. A *chain* is a sequence of nodes $x_1, x_2, \ldots, x_n$ that satisfies $x_1 \leq x_2 \leq \cdots \leq x_n$. Two chains are said to be *disjoint* if they have no common nodes. We say that a set of chains *fills E* if they form a partition for the nodes of E. By a set of *mutually incomparable nodes* we understand a set of nodes no two of which are comparable (in the partial order $\leq$). The result we prove is:

Dilworth's Result. *The minimum number of disjoint chains that fills a partially ordered set equals the maximum number of mutually incomparable nodes.*

It is obvious that we need a separate chain for each of the incomparable nodes (so there are at least as many disjoint chains as there are incomparable nodes).

We establish the other inequality by a "marriage" argument. Represent the partially ordered set E by the usual (upward) Hasse diagram [i.e., place an edge between nodes x and y if $x \leq y$ and there is no z (other than x and y) to satisfy $x \leq z \leq y$]. Now replace each node of E by a pair of vertices $\substack{\bullet \\ \circ}$, the one on top a "girl" and the one on the bottom a "boy." Fill in the "likes" by the partial order $\leq$ from girls to boys (including the transitive links) to obtain a graph G, as shown below:

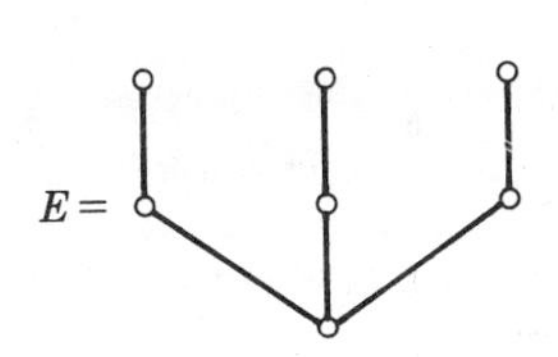

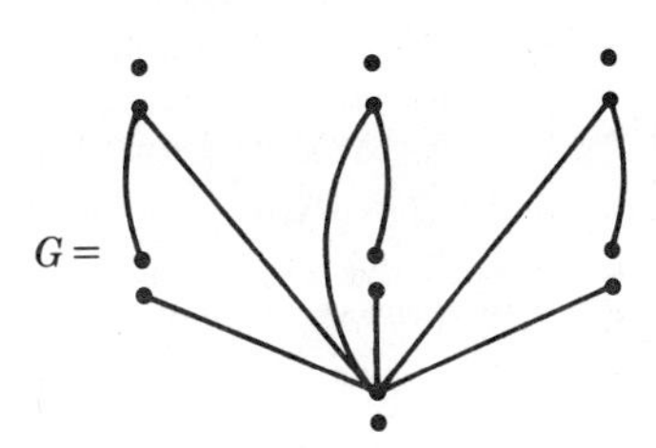

Note that *no edge* is placed between the girl and boy that replace a node of E. [*Aside*: Observe the "princesses" at the very top of G. Apparently none of the boys are quite good enough for them....]

Given a chain $x_1 \le x_2 \le \cdots \le x_n$ (with n distinct nodes) we produce $n - 1$ marriages as follows: marry girl in x_1 to boy in x_2, girl in x_2 to boy in $x_3, \ldots,$ girl in x_{n-1} to boy in x_n. It therefore follows that for a set of chains that *fills* E we have

$$\textit{number of marriages} + \textit{number of chains} = |E|.$$

This last relation now gives:

minimal number of such chains
$= |E|$ − *maximal number of marriages*
$=$ {*by König's theorem*}
$= |E|$ − *minimal number of lines to delete all the* 0*'s in the girls versus boys matrix*
$\le$ *maximal number of incomparable nodes.*

To see the last inequality note that a maximal set of incomparable nodes produces a (square) block of 1's in the girls versus boys matrix. (One should simultaneously view this matrix as a node versus node matrix with 1's on the main diagonal.) Without loss the matrix looks like this:

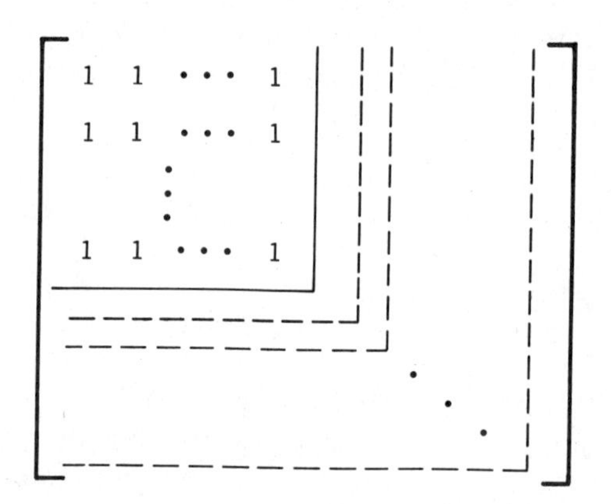

Each of the broken bordered "angles", that is,

must contain a 0 (else we could get a larger set of incomparable nodes). We need a separate line for each such bordered "angle." The proof is now complete.

Example.

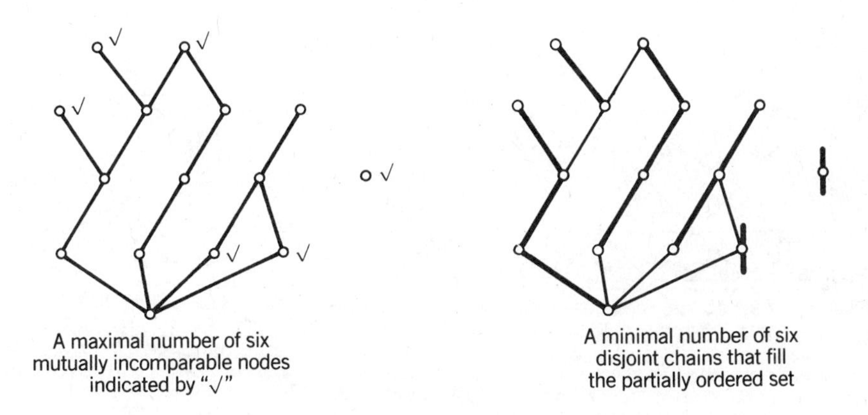

A maximal number of six mutually incomparable nodes indicated by "√"

A minimal number of six disjoint chains that fill the partially ordered set

EXERCISES

1. Consider the following network (with capacities on edges as indicated): Find a maximal flow.

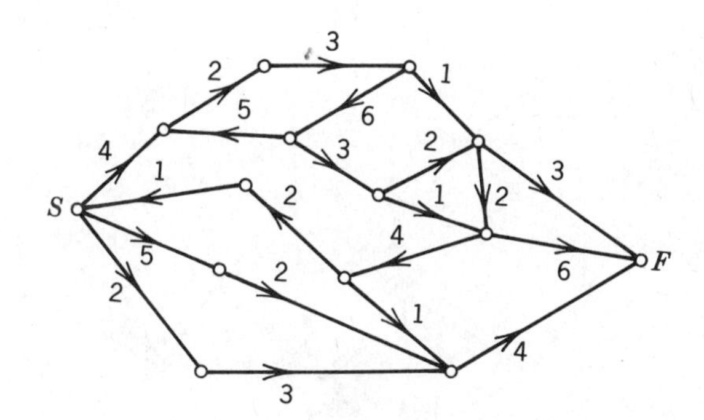

2. For the network with vertex capacities displayed below draw the corre-

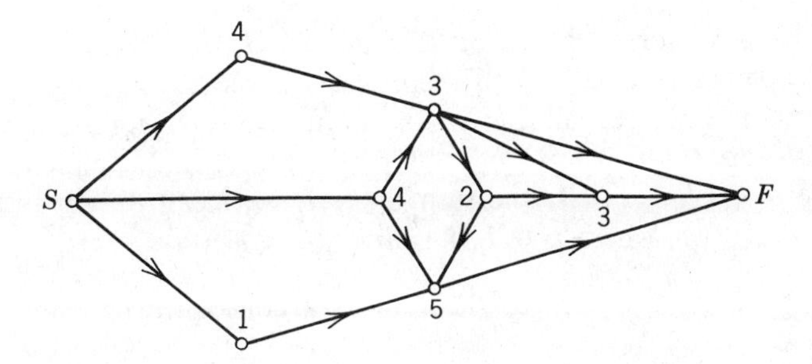

sponding network (with edge capacities only). Find the maximal flow and the corresponding vertex cut.

3. Verify Menger's theorem for the graphs

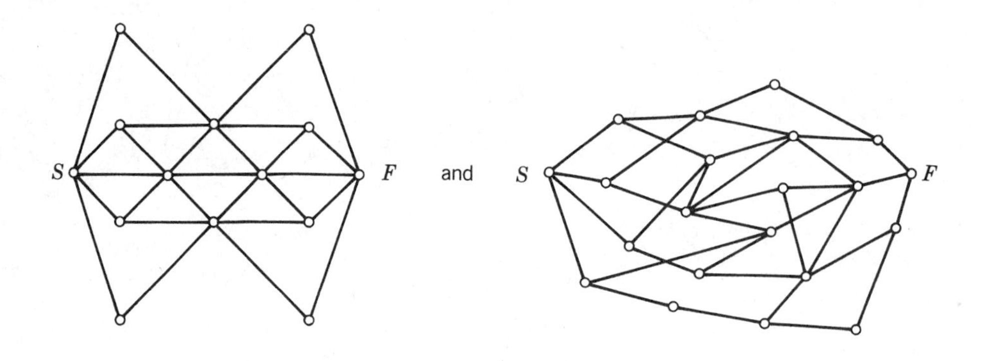

4. What is the minimal number of disjoint chains that fill the following partially ordered set?

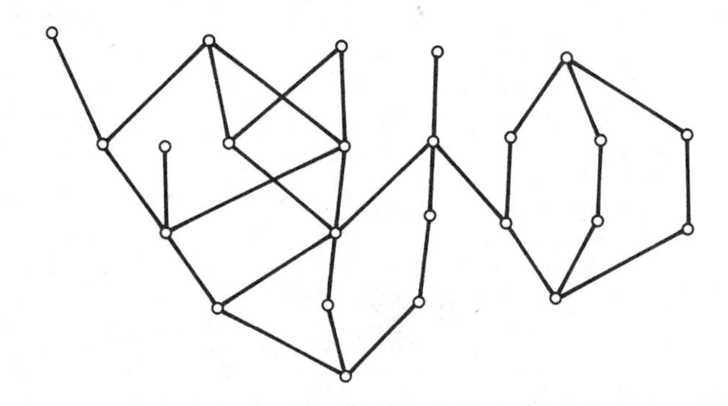

5. Let V be a vector subspace of R^n, and let $I_1, I_2, \ldots, I_n$ be nonempty intervals of the real line. Then there exists a vector $(x_1, x_2, \ldots, x_n) \in V$ such that $x_i \in I_i$, for all $1 \leq i \leq n$, if and only if for each vector of minimal support $(y_1, y_2, \ldots, y_n) \in V^{\perp}$ (the orthogonal complement of V) we have $0 \in \sum_{i=1}^{n} y_i I_i$. (By the sum $\sum_{i=1}^{n} y_i I_i$ we mean the set of all inner products $\sum_{i=1}^{n} y_i z_i$ where $z_i \in I_i$, for all $1 \leq i \leq n$.) Prove this.

6. Derive the max-flow min-cut theorem as a consequence of the result in Exercise 5.

6 THE "OUT OF KILTER" METHOD

To display a certain duality we need to elevate some of the concepts introduced so far to a higher level of algebraic abstraction. Both flows and cuts are viewed as vectors in R^n, where n is the number of edges in the network (including the distinguished "yellow" edge between S and F—see Section 5.15). In such a context we can clearly see the "duality" that exists between flows and cuts.

5.21 The Vector Space Formulation

Let N be a network with edges $E_1, \ldots, E_n$ and vertices $1, 2, \ldots, m$. The edge E_n is the distinguished "extra" edge between S (source) and F (sink). Let $K' \leq R^n$ be the set of possible flows; thus K' is the set of vectors $(x_1, \ldots, x_n)$ such that the directed sum at each vertex of the network is zero. (Note that we take $+x_i$ or $-x_i$ in the sum at j according to whether the arrow on E_i leaves j or enters j.) Evidently K' is a subspace. (In terms of electrical networks the conservation laws at the vertices are the Kirchhoff laws for current.)

Let $K'' \leq R^n$ be the set of vectors $(y_1, \ldots, y_n)$ such that the directed sum around any cycle in the network is 0 (equivalently around any closed path). Evidently K'' is a subspace. We easily construct $y \in K''$ as follows: Assign arbitrary reals $\alpha_1, \ldots, \alpha_m$ to the vertices $1, 2, \ldots, m$. To each edge

assign $\alpha_j - \alpha_i$ (i.e., the directed difference). This produces a vector in K'' since any directed sum around a cycle becomes a "telescoping sum" and is therefore 0. We allow the possibility of changing the direction of an arrow on an edge if we also change the sign of both the flow and the directed difference through it; that is

$$i \bullet \xrightarrow{\quad 2 \quad} \bullet\, j$$

is the same as

$$i \bullet \xleftarrow{\quad -2 \quad} \bullet\, j.$$

Observe that:

∗ *Each $y \in K''$ arises from a labeling of the vertices* (unique up to a constant).

Proof. Select any vertex as base point and label it 0. Label all adjacent vertices to satisfy the directed difference condition for the corresponding y_i.

Continue from these vertices to their adjacent vertices and so on, until all are labeled. There are two possible difficulties. We may move to a vertex already labeled or we may have two (or more) edges leading to a new vertex. In either case we get an associated closed path in N, for example,

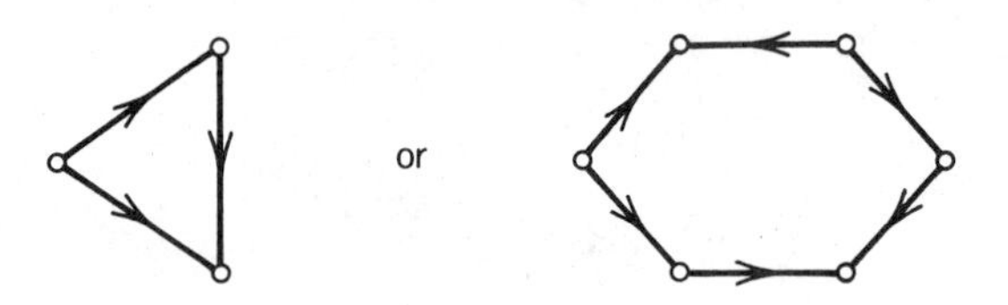

The possible labels *agree*, since the directed sum around the path is zero.

The subspaces K' and K'' are orthogonal complements. Indeed, we have the following proposition:

Proposition 5.1. $R^n = K' \oplus K''$ (orthogonal direct sum).

Proof. Let $x \in K'$, $y \in K''$. Then

$$\begin{aligned}\sum_{k=1}^{n} x_k y_k &= \sum_{k=1}^{n} x_k \left(\alpha_{j(k)} - \alpha_{i(k)}\right) \\ &= \sum_{i=1}^{m} \alpha_i \left(\sum_{\text{at vertex } i} \pm x_{k(i)} \right) \\ &= 0 \text{ (the directed sum at } i \text{ being “correctly” directed).}\end{aligned}$$

(This shows that K' and K'' are orthogonal subspaces.)

We now show that they are orthogonal complements in R^n. Choose a spanning tree T of N. Since N has m vertices, T has $t = m - 1$ edges. We produce t linearly independent vectors in K'' (one for each edge of T) as follows: Pick an edge of T. Remove it to disconnect T. In one of the two resulting connected components label all vertices 0, and in the other label them all 1, and thence construct a vector in K''. All entries in this vector that correspond to edges of T are zero, except for the removed edge (which takes value -1 or 1). We thus obtain t vectors in K''. They are linearly independent; indeed, the vector that takes nonzero value (i.e., value -1 or 1) on an edge of T cannot be a linear combination of the others because they all take value 0 on that edge. Hence $\dim K'' \geq t$.

On the other hand, with each edge not in T we produce a vector in K'. Each such edge produces a cycle with some edges of T. Appropriately assign -1 or 1 to the edges of the cycle and 0 to the other edges, to produce a flow (i.e., a vector in K'). We thus produce $n - t$ linearly independent vectors in K' and conclude that $\dim K' \geq n - t$.

Hence $n = (n - t) + t \leq \dim K' + \dim K''$ and therefore $R^n = K' \oplus K''$. This ends our proof.

5.22

The maximal flow problem now takes the following form: We are given a network N with distinguished edge E_n. For each of the other edges E_i we have a (nonempty) interval of capacities I_i, $1 \leq i \leq n-1$.

We wish to find the maximal value of x_n *subject to* $x \in K'$ *and* $x_i \in I_i$ $(1 \leq i \leq n-1)$. We may change to a more symmetrical formulation by assigning an arbitrary interval I_n to E_n and asking when there is a *solution* to the problem $x \in K'$, $x_i \in I_i$ $(1 \leq i \leq n)$. For example, we may take I_n to consist of a single point and seek the maximal flow through E_n by gradually moving this point to the right (until a solution ceases to exist). [In our initial formulation (see Section 5.10) we took $I_i = [-c_i, c_i]$ with c_i the capacity of edge i, $1 \leq i \leq n-1$.]

A *cocycle* in N is a set of edges whose removal disconnects N (i.e., it separates N into two connected components). As we shall see, some vectors of K'' correspond to cocycles in N. By the support of a vector we understand the set of coordinates with nonzero entries. A vector has minimal support if its support is minimal with respect to inclusion. As is easy to see, the vectors of minimal support *in a subspace* are unique up to scalar multiples (and consequently finitely many in number, up to such multiples).

It turns out that we have a bijective correspondence between cocycles and vectors of minimal support with entries 0, ± 1 in K''. For any cocycle we can produce $y \in K''$ by labeling vertices in one connected component 0, and 1 in the other. Conversely, the (minimal) support of $y \in K''$ corresponds to edges with different end labels. Pick one such edge with labels $0 < 1$. Partition the vertices into those labeled $\leq \frac{1}{2}$ (say A) and those labeled $> \frac{1}{2}$ (say B). The edges running between A and B form the cocycle we want; note that the minimality of the support implies that *all* vertices in A are necessarily labeled 0 and those in B are labeled 1. [A parallel discussion goes for the cycles of N and the vectors with minimal support in K'.] We summarize:

* *There is a bijective correspondence between the cycles (respectively, cocycles) of N and the vectors with entries* 0, ± 1 *of minimal support in* K' *(respectively* K''*).*

To translate the max-flow min-cut theorem in this language we first attach to each edge an interval I_i $(1 \leq i \leq n-1)$ of the form $[-c_i, c_i]$. A cut is a cocycle that omits E_n, the distinguished "yellow" edge. Evidently the maximal flow does not exceed the minimal sum of cut set capacities. We actually achieve equality by "lining up" all the arrows on a cut set and taking $I_n = \{-s\}$, where $s = \Sigma_i c_i$, with c_i the capacities of the cut set that maximizes s.

5.23 Problem Data for the "Out of Kilter" Algorithm

We are given a network with m vertices $1, 2, \ldots, m$ and n edges $E_1, E_2, \ldots, E_n$ (and we *do not* distinguish any of the edges at this stage). For each edge we have an increasing function (defined on some interval of the real line) *with vertical lines allowed*. By an increasing function we actually mean a discretized "straw" graph as follows:

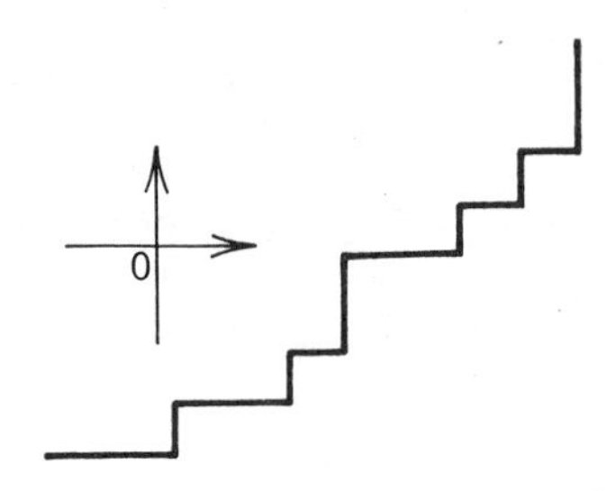

[This generalizes the case described in Section 7.22 in which the graphs look like this:

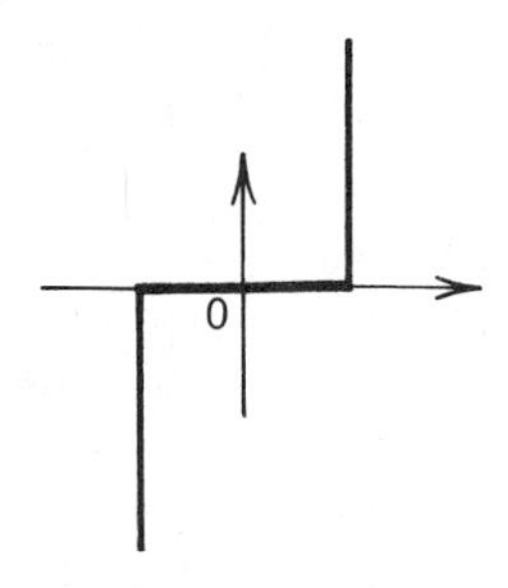

—that is, we have (in essence) n nonempty intervals $I_1, I_2, \ldots, I_n$, one for each edge. It helps to parallelly interpret what we do next in this important special case.]

Let C_i be the "straw" graph attached to edge i. Annex to this background the subspaces K' and K'' of flows and "cocycles."

PROBLEM. *Find $x \in K'$ and $y \in K''$ such that $(x_i, y_i) \in C_i$, $i = 1, 2, \ldots, n$ (if any such x and y exist).*

The problem (as stated) treats K' and K'' on equal footing. This formulation permits us to treat questions on flows and cuts simultaneously and in duality. To find vectors x and y as in the *problem*, we initially start with a pair of vectors (in K' and K'', respectively) that do not quite satisfy $(x_i, y_i) \in C_i$, for all i. We can always move "a step closer," however, by being able to find *either* a cycle in K' or a *cocycle* in K''; this crucial fact is precisely Minty's arc coloring lemma working at its full potential (and it is in this context that that lemma was discovered).

REMARK. Assume that the curves C_i correspond to *intervals*. The task of finding a maximal flow x_n through E_n is implied by our *problem*. For if we have a way of finding solutions to $x \in K'$, $y \in K''$, and $(x_i, y_i) \in C_i$, we can in particular seek a solution in which the coordinate x_n in x is maximal (and discard y completely). Conversely, a solution $(x_1, \ldots, x_n) = x \in K'$ with $x_i \in I_i$ (or C_i, equivalently) and x_n maximal allows us to always take $(x_1, \ldots, x_n) = x \in K'$, $0 \in K''$, and then $(x_i, 0) \in C_i$, for all i.

5.24 A Connection to Young's Inequality

The *problem*, as stated in Section 5.23, is directly linked to two separate optimization problems. To understand the connection we need first recall Young's inequality for increasing functions. This inequality is so "graphical" that we can both state and prove it by looking at the following graphical display:

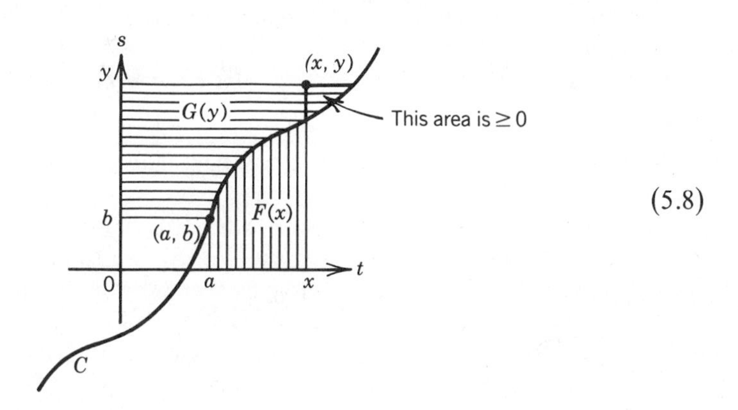

(5.8)

Here the point (a, b) is on the curve C and

$$F(x) = \int_a^x C(t)\,dt \text{ and } G(y) = \int_b^y C^{-1}(s)\,ds$$

are the antiderivatives of the increasing functions C and C^{-1} (the inverse function of C), respectively.

The inequality we mentioned is as follows:

Young's Inequality. $F(x) + G(y) + ab - xy \geq 0$ *with equality if and only if* $(x, y) \in C$.

[The statement is apparent from display (5.8) since $F(x) + G(y) + ab - xy$ is the area marked "this area is ≥ 0."]

Suppose now that *there exist* $x \in K'$ and $y \in K''$ such that $(x_i, y_i) \in C_i$ for *given* increasing curves C_i, $1 \leq i \leq n$. Select points (a_i, b_i) on the curves C_i, define antiderivatives F_i and G_i, and consider the function

$$H(x, y) = \sum_{i=1}^{n} F_i(x_i) + G_i(y_i) + a_i b_i - x_i y_i.$$

By Young's inequality $H(x, y) \geq 0$, with equality if and only if $(x_i, y_i) \in C_i$, for all i. Denote by $\overline{H}$ the restriction of H to $K' \times K''$. Since K' and K'' are orthogonal subspaces for $(x, y) \in K' \times K''$ we have $\sum_{i=1}^{n} x_i y_i = 0$ and hence $\overline{H}(x, y) = \sum_{i=1}^{n} F_i(x_i) + G_i(y_i) + a_i b_i$. Observe now that $\overline{H}(x, y) \geq 0$, and it equals 0 if and only if $(x, y) \in K' \times K''$ is such that $(x_i, y_i) \in C_i$ for all i. Moreover, since the variables are separate in $\overline{H}(x, y)$ we can conclude that $\overline{H}(x, y)$ attains the minimal value of 0 if and only if x in K' minimizes $\sum_{i=1}^{n} F_i(x_i)$ and y in K'' minimizes $\sum_{i=1}^{n} G_i(y_i)$. In conclusion we state the following:

* *Let n increasing curves C_i with antiderivatives F_i and G_i be given. There exist vectors $x \in K'$ and $y \in K''$ such that $(x_i, y_i) \in C_i$, for all i, if and only if there exists x in K' that minimizes $\sum_{i=1}^{n} F_i(x_i)$ and there exists y in K'' that minimizes $\sum_{i=1}^{n} G_i(y_i)$.*

A solution to our *problem* can thus be found by solving two *separate* minimization problems.

To each increasing curve C_i we attach a (nonempty) interval I_i on the horizontal axis, namely its *projection* on that axis, and similarly a (nonempty) interval J_i on the vertical axis. An obvious *necessary* condition for our *problem* to have a solution is that vectors $x \in K'$ and $y \in K''$ exist, such that $x_i \in I_i$ and $y_i \in J_i$, for all i. As it turns out this necessary condition is actually sufficient and this is explained in a constructive manner by the "out of kilter" algorithm.

5.25 The "Out of Kilter" Algorithm

Let a network N be given with an increasing "straw" curve attached to each one of its n edges. We adopt the following fundamental *convention*. We may change the direction of an arrow on an edge provided that we also do three other things along with that: Change the *sign of the flow* through that edge, change the *sign of the directed difference* through that edge, and also *rotate by 180°* the increasing curve attached to that edge. (This change corresponds to a simple change of variable—an actual change of sign, in fact.)

Denote the edges of N by $E_1, \ldots, E_n$ and the associated increasing "straw" curves by $C_1, \ldots, C_n$. Also let $I_1, \ldots, I_n$ and $J_1, \ldots, J_n$ be the intervals obtained by projecting the curves on the horizontal and vertical axes, respectively. *Assume* there exist "starting points" $x \in K'$ and $y \in K''$ such that $x_i \in I_i$ and $y_i \in J_i$, for all $1 \leq i \leq n$. If in fact $(x_i, y_i) \in C_i$ for all i, *we are done*. A solution to our *problem* (see Section 5.23) was found. If not, there exists at least one edge (edge i, say) for which (x_i, y_i) is not on the curve C_i (we say that edge i is out of kilter, hence the terminology). Focus attention on edge i: Color it yellow and think of it as being the distinguished yellow edge. The essence of what we do is two-fold: We bring edge i in kilter *and* keep in kilter all edges that were already in kilter. [The rigorous definition of an edge (say, edge j) being in *kilter* is the existence of x in K' and y in K'' such that $(x_j, y_j) \in C_j$.]

The general step is graphically illustrated below:

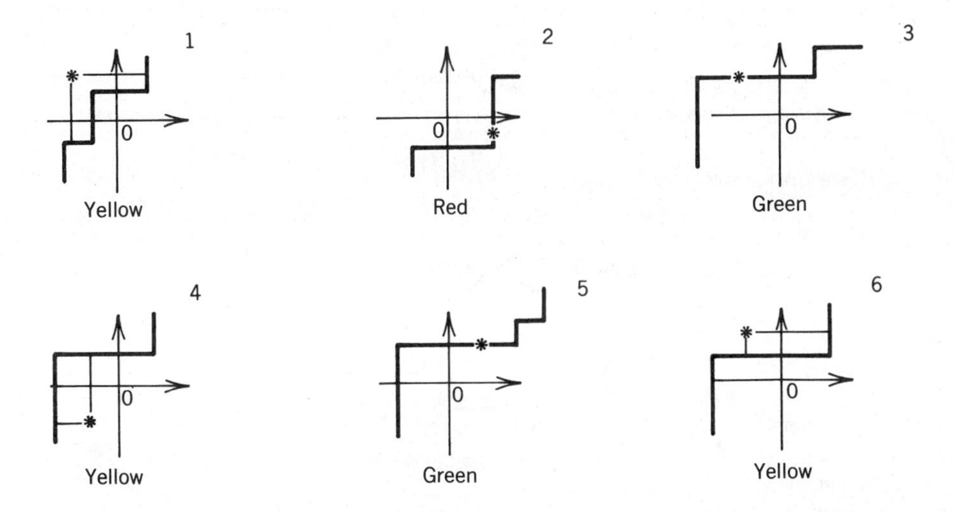

We have $n = 6$ edges, and the six "straw" curves C_i (trails of honey) are drawn in somewhat thicker lines. The asterisk (a *bear* in Minty's colorful orations) on edge i is the point (x_i, y_i), with $x = (x_1, \ldots, x_6) \in K'$ and $y = (y_1, \ldots, y_6) \in K''$. Edges 2, 3, and 5 are in kilter (and they will continue to remain so). *Color* the edges of the network as follows: If the local picture (indicated in the lighter lines for edges not in kilter) is:

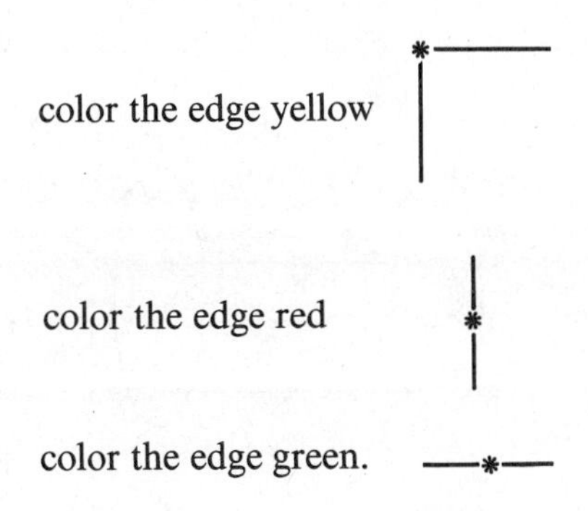

(If we have ⊥ then change the arrow on that edge; accordingly change the sign of the flow and the directed difference through it and *rotate* the graph attached to that edge by 180°. Now the local picture becomes ⊤ and we color the edge *yellow*. Edge 4 above offers this opportunity.) Select an edge that is out of kilter (it will be a yellow edge) and think of it as the "distinguished" yellow edge; edge 1 will be our choice.

Each edge of the network is now colored yellow, red, or green (with edge 1 being distinguished yellow). Minty's arc coloring lemma, see Lemma 5.3, *assures* the existence of *precisely one* of the following two possibilities:

(a) A *cycle* containing the distinguished yellow edge with all its edges green or yellow oriented in the direction of the distinguished yellow edge.
(b) A *cocycle* containing the distinguished yellow edge with all its edges red or yellow oriented in the direction of the distinguished yellow edge.

If we find a cycle as in (a), say

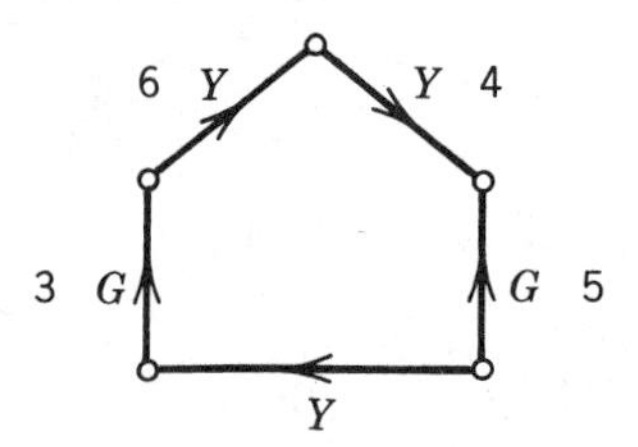

we move the asterisks to the *right* on the graphs attached to edges 1, 3, 6, and 4, but to the *left* on the graph attached to edge 5 (because its arrow is against the direction of the distinguished yellow edge—equivalently we could first change the direction of the arrow on edge 5, thus *rotating* the graph of edge 5 by 180°, and then move *all* the asterisks to the right). Note that upon performing these changes to the graphs of the edges that occur in the cycle all edges that were in kilter *are still in kilter*, and that the asterisk on the graph of the distinguished yellow edge (edge 1 in our case) is definitely *one step closer* to its thicker curve (curve C_1 in our case).

If we find a cocycle as in (b), say the one drawn below (for the sake of conversation),

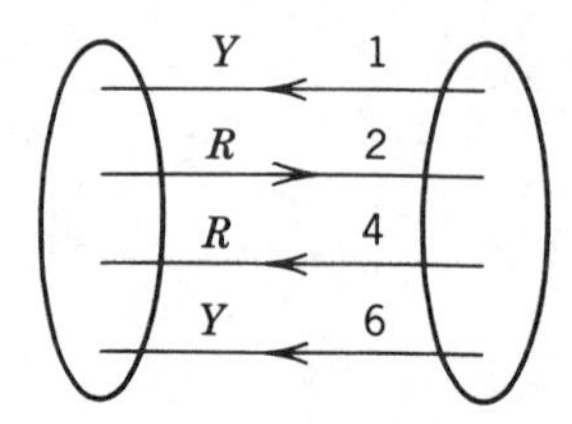

we move the asterisks *down* on the graphs attached to edges 1, 4, and 6, but up on the graph attached to edge 2 (because its arrow is against the direction of our distinguished yellow edge—equivalently we could first change the direction of the arrow on edge 2, thus *rotating* the graph of edge 2 by 180°, and then move *all* the asterisks down). Note that upon performing these changes to the graphs of the edges that occur in the cocycle all edges that were in kilter *are still in kilter*, and that the asterisk on the graph of the distinguished yellow edge (edge 1 in our case) is *definitely one step closer* to its thicker curve (curve C_1 in our case).

The motions of the asterisks thus induced generate new vectors $(x_1, \ldots, x_6) = x \in K'$ and $(y_1, \ldots, y_6) = y \in K''$ such that for the distinguished yellow edge (edge 1 in our case) we have the asterisk (x_1, y_1) *one step closer* to C_1 than before (and such that all edges that were in kilter before remain in kilter).

Repeat this process until edge 1 (the distinguished yellow edge) is in kilter; then, if there are other edges out of kilter (edge 2, say), call edge 2 the distinguished yellow edge and bring *it* in kilter, until all edges are in kilter. We thus found $x \in K'$ and $y \in K''$ such that $(x_i, y_i) \in C_i$, $1 \leq i \leq n$ $(= 6)$, in our example. This ends the description of the algorithm.

A Concluding Remark

All that we went through was done under the assumption that a solution to the "projected problem" exists, that is, that there exist $x \in K'$ and y in K'' such that $x_i \in I_i$ and $y_i \in J_i$, for all i.

On occasion such a solution may be apparent from the general context of the problem and we can use it as our starting point for the "out of kilter" algorithm. If not, start with $0 \in K'$ and $0 \in K''$ and apply the "out of kilter" algorithm to the same network but with curves C_i replaced by their projections I_i; stop when a cocycle appears (then there is no solution), else obtain a point x in K' with $x_i \in I_i$, $1 \leq i \leq n$. Do the same for the J_i's and stop when a cycle appears. If no cycle occurs obtain a point y in K'' with $y_i \in J_i$, for all i. The vectors x and y are starting points for the "out of kilter" algorithm.

7 MATROIDS AND THE GREEDY ALGORITHM

Notions such as linear independence, trees, or matchings in a graph can be treated in a unified axiomatic way through what has become known as matroid theory. There exists also an algorithmic characterization of matroids: It states (roughly speaking) that a combinatorial optimization problem should be solved using a "greedy" algorithm if and only if a matroid is lurking somewhere in the background.

5.26 The Definition of a Matroid

A *matroid* (P, S) consists of a set P of points with a set S of distinguished subsets of P called *independent sets* satisfying the two axioms below:

If I is in S, then all subsets of I are in S. The empty set is in S. (5.9)

If I and J are subsets of S with n and $n + 1$ points, respectively, then there exists a point e in J but not in I such that $I \cup \{e\}$ is in S. (5.10)

These axioms are motivated by three examples that we now bring to attention.

Example 1 (linear Independence). Let V be a finite-dimensional vector space over an arbitrary field, finite or infinite. Take the vectors in V as points. The set S of distinguished subsets consists of all linearly independent sets of vectors.

We take as our premise that the empty set belongs to S. Using freely the definitions and results of linear algebra, axiom (5.9) is immediately verified (it simply states that any subset of a linearly independent set of vectors is linearly independent). Axiom (5.10) is a little more difficult to verify directly. But it is a well-known result called the exchange principle. [Using familiar concepts in linear algebra, if I and J are as in the axiom, and if each vector in J is a linear combination of vectors in I, then J is part of the subspace generated by I. This is not possible, since I generates a subspace of dimension n, and J consists of $n + 1$ independent vectors. Axiom (5.10) is therefore satisfied.]

Example 2 (Cycle-Free Subsets of a Graph). P is the set of edges of a graph. A subset of edges is distinguished (and thus belongs to S) if it contains no cycles. One may verify that the pair (P, S) with P and S so defined is a matroid. Verifying axiom (5.10) can be a little tricky but not exceedingly so.

Example 3 (The Matching Matroid). Let the set of points consist of the vertices of a graph. The distinguished subsets are those subsets of points for which a matching covering all the vertices of the subset exists. We refer to Section 5.5 for the definition of a matching. This defines a matroid. Details of verification are omitted.

5.27

We now explore some consequences of our definition. Let (P, S) be a matroid. Fix a subset A of P (independent or not). Set A contains independent subsets, since the empty set is independent. We assert that *all independent subsets of A maximal with respect to inclusion have the same cardinality.* [Indeed, let I and J be two maximal subsets of A. Assume that $|I| < |J|$. Select a subset of J with $|I| + 1$ points; call it K. Axiom (5.9) informs us that K, being a subset of J, is independent. Axiom (5.10) assures the existence of a point e in K but not in I such that $I \cup \{e\}$ is independent. The set $I \cup \{e\}$ contains I strictly, thus contradicting the assumed maximality of I. We thus conclude that $|I| < |J|$ cannot occur, and neither can $|J| < |I|$ for completely analogous reasons. Thus $|I| = |J|$, as stated.] This observation allows us to define the *rank* of A as the cardinality of any of its maximal independent subsets. We write $r(A)$ for the rank of A.

Some properties of the notion of rank are easy to perceive. It is clear, for instance, that $r(\varnothing) = 0$. Furthermore, *for a subset A and a point e not in A we have either $r(A \cup \{e\}) = r(A)$ or $r(A \cup \{e\}) = r(A) + 1$.* [To see this, let T be a maximal independent set of $A \cup \{e\}$. If $T \subseteq A$, then clearly $r(A \cup \{e\}) = |T| = r(A)$. If not, then necessarily $T = N \cup \{e\}$, with N a maximal independent set of A. In this latter case $r(A \cup \{e\}) = |T| = |N| + 1 = r(A) + 1$.]

Observe yet another property of the rank. *If e_1 and e_2 are points not in A and if $r(A \cup \{e_1\}) = r(A \cup \{e_2\}) = r(A)$, then $r(A \cup \{e_1\} \cup \{e_2\}) = r(A)$.* [This is easy to establish by observing that neither e_1 nor e_2 can be part of a maximal independent set of $A \cup \{e_1\} \cup \{e_2\}$ (otherwise the rank of either $A \cup \{e_1\}$ or $A \cup \{e_2\}$ will exceed the rank of A, contrary to our assumptions). The maximal independent sets of A and $A \cup \{e_1\} \cup \{e_2\}$ are therefore the same, and so then are their ranks.]

It turns out that a function r defined on the subsets of a finite set and possessing the properties of the rank mentioned above is in fact the rank function of a matroid with independent sets described as follows: $S = \{I : r(I) = |I|\}$. Apart from this, many other characterizations of a matroid are known.

To readers interested in more work with axioms we suggest proving the following result due to Nash and Williams:

* *If (P_1, S_1) is a matroid and f is a function from P_1 to P_2, then $(P_2, f(S_1))$ is also a matroid.* (*By $f(S_1)$ we understand* $\{f(T) : T \in S_1\}$.)

5.28 The Result of Rado and Edmonds

Let (P, S) be a matroid. To each point e of P we assign a nonnegative weight $w(e)$. The *weight* of a finite subset of P equals (by definition) the sum of weights of its points. *Our objective is to find an independent set of maximum weight.*

We take interest in a special ordering induced by the weight function w on the independent sets: the *lexicographic* ordering. It is this ordering that usually suggests itself at the outset in many maximization problems, for it accommodates well the impulses of a "greedy" mind. We can describe it as follows:

Let $I = \{a_1, a_2, \ldots, a_n\}$ and $J = \{b_1, b_2, \ldots, b_m\}$ be two independent sets with points listed in the order of weight $w(a_1) \geq w(a_2) \geq \cdots \geq w(a_n)$ and $w(b_1) \geq w(b_2) \geq \cdots \geq w(b_m)$. We say that I is lexicographically greater than or equal to J if there exists an index k such that $w(a_i) = w(b_i)$ for $i = 1, 2, \ldots, k-1$ and $w(a_k) > w(b_k)$; or else $n \geq m$ and $w(a_i) = w(b_i)$ for $i = 1, 2, \ldots, m$. Any two independent sets are comparable (but it may happen that I is lexicographically greater than or equal to J and J is lexicographically greater than or equal to I without I and J being actually equal). An independent set that is not lexicographically less than any other set is said to be lexicographically maximum.

The greedy algorithm is based on the following result:

The Rado–Edmonds Theorem.

(a) *If (P, S) is a matroid, then*:

For any nonnegative weighing of the points in P, a lexicographically maximum set in S has maximum weight. (5.11)

(b) *If (P, S) is a finite structure of points and subsets satisfying* (5.11) *and axiom* (5.9), *then (P, S) is a matroid.*

Proof. We prove part (a) first. Let $I = \{a_1, a_2, \ldots, a_n\}$ be a lexicographically maximum set, where $w(a_1) \geq w(a_2) \geq \cdots \geq w(a_n)$. Let $J = \{b_1, b_2, \ldots, b_m\}$ be any independent set, where $w(b_1) \geq w(b_2) \geq \cdots \geq w(b_m)$. We assert that $w(a_i) \geq w(b_i)$, for all i. [Indeed, if $w(a_k) < w(b_k)$ for some k, then focus attention on the sets $\{a_1, a_2, \ldots, a_{k-1}\}$ and $\{b_1, b_2, \ldots, b_k\}$. By axiom (5.10) there exists a point in the latter set (b_j, say) such that $\{a_1, a_2, \ldots, a_{k-1}, b_j\}$ is an independent set. But this set is then lexicographically greater than I, a contradiction.] Since $w(a_i) \geq w(b_i)$ for all i, we conclude that I is an independent set of maximum weight.

Part (b) is proved next. Assume that (P, S) is a finite structure of points and subsets satisfying (5.11) and axiom (5.9). We also assume that (P, S) is not a matroid and derive a contradiction.

First, we assert that there exists a subset A of P and two subsets I and J in S both maximal in A with respect to inclusion such that $|I| < |J|$. [For if not $|I| = |J|$ for all maximal subsets I and J of any subset A of P. Then let Q and R be in S, of cardinalities n and $n+1$, respectively. Let $A = Q \cup R$. Since $|Q| = n$ and $|R| = n+1$, Q cannot be maximal in A. Hence there must exist a point e in $R - Q$ such that $Q \cup \{e\}$ is in S. But this is what axiom (5.10) states. It thus follows that (P, S) is a matroid, contrary to one of our assumptions.]

With subset A, and subsets I and J thus identified, assign weights as follows: Each point in I has weight $1 + \varepsilon$ (for ε an as yet unspecified positive number), each point in $J - I$ has weight 1, and all of the remaining points in

P have weight 0. If ε is chosen positive and suitably small the set I is contained in a lexicographically maximum set whose weight is less than that of J. This violates assumption (5.11). The finite structure (P, S) must therefore be a matroid. This ends our proof.

In complete analogy to what we just accomplished one may characterize a matroid in terms of lexicographically minimum sets and sets of minimum weight. All definitions and arguments parallel the ones used above.

5.29 The Greedy Algorithm

Based on the result of Rado and Edmonds we can now describe the greedy algorithm.

We have before us a matroid (P, S) and a weight function w defined on the set of points P. Our objective is to find an independent set of maximum weight.

Observe that we may assume without loss of generality that each point is an independent set of cardinality 1. [For if some point is not, then (by axiom (5.9)) this point cannot occur in any of the independent sets. Therefore we might as well omit it altogether from the set P of the initial points.]

The Algorithm

Select a point e_1 of maximal weight. Let $S_1 = \{e_1\} \in S$. Then choose a point e_2 not in S_1 such that $S_1 \cup \{e_2\}$ is an independent set of maximal weight. Let $S_2 = S_1 \cup \{e_2\}$. Choose a point e_3 not in S_2 such that $S_2 \cup \{e_3\}$ is an independent set of maximal weight. Proceed. . . .

After a finite number of steps (r steps, say) a lexicographically maximal independent set S_r will be found. By the result of Rado and Edmonds in Section 5.28, S_r is in fact an independent set of maximal weight. End of algorithm.

The problem of locating an independent set of minimum weight is treated in a completely analogous way.

The algorithm is clearly a "greedy" one for it maximizes weight at each step of the selection process. But if the underlying structure is a matroid this greedy procedure is in fact the desired one.

5.30

Generally a "greedy" form of selection does not produce maximum weight. In the simplest of situations one can see the severe drawbacks of such an attitude:

Being the matchmaker that I am, Larry offers me \$10 for a date with Mary and \$8 for a date with Sherry; Harry gives me \$8 for a date with Mary and \$1 for a date with Sherry. If I am greedy in the sense that I wish to maximize weight (income in this case) at each and every step, I will first take the \$10 from Larry and pair him off with Mary; then I must pair up Harry with Sherry for \$1, for a total income of \$11. (I could have cashed in \$16 though by having Larry date Sherry and Harry date Mary.) Following a "greedy" procedure results in a loss of \$5.

One would not expect the greedy algorithm to be of much use in most optimization problems, and rightly so. The maximum matching algorithm (of Section 5.6) will in all likelihood be easier to adapt to a general optimization problem. But the Rado–Edmonds theorem tells us precisely which optimization problems can be solved by a greedy algorithm: those to which a matroid can be attached in a natural way. Let us examine two such problems.

PROBLEM 1. Given a finite set of vectors in a finite-dimensional vector space find a set of linearly independent vectors whose sum of Euclidean lengths is maximum.

There is a matroid (P, S) that awaits notice: P is the set of vectors available and S the linearly independent subsets of P. The solution to this problem can be obtained by a greedy algorithm: Select the longest vector, then the next longest linearly independent of the first, the third longest linearly independent of the first two, and so on. The resulting linearly independent set will be of maximal total length (for this is what the Rado–Edmonds theorem asserts).

PROBLEM 2. Given a connected graph with weighted edges, find a spanning tree of minimum weight (i.e., a spanning tree with minimum sum of weights on its edges).

A matroid is quietly waiting to be noticed here as well. It is (P, S), with P the set of edges of the graph and S the cycle-free subsets of edges. The greedy algorithm tells us how to find such a spanning tree. First select the lightest edge, then the next lightest that forms no cycle with the first, then the third lightest that forms no cycles with the first two, and so on. This greedy process leads in the end to a spanning tree of minimum weight. Surprised? No, because of the matroid lurking in the background and because the Rado–Edmonds theorem is at work.

EXERCISES

1. Let (P, S) be a matroid. The *span* of a subset A of P is the maximal set containing A and having the same rank as A (by maximal we mean maximal with respect to inclusion). Show that the span of a subset is unique.

2. Take as points the elements of the set $\{1,2,3,4,5,6,7\}$. Let the maximal independent sets consist of all subsets of three points except $\{1,2,4\}$,

$\{2,3,5\}$, $\{3,4,6\}$, $\{4,5,7\}$, $\{5,6,1\}$, $\{6,7,2\}$, and $\{7,1,3\}$. Is this finite structure a matroid?

3. A maximal independent set of a matroid is called a *basis*. Let $\bar{B}$ be the set of bases of a matroid. Then:

(a) $\bar{B} \neq \varnothing$, and no set in $\bar{B}$ contains another properly.

(b) If B_1 and B_2 are in $\bar{B}$ and e_1 is a point in B_1, then there exists a point e_2 in B_2 such that $(B - \{e_1\}) \cup \{e_2\}$ is also in $\bar{B}$.

Show that, conversely, if $(P, \bar{B})$ is a finite structure satisfying (a) and (b) above, then (P, S) is a matroid, where

$$S = \{I : I \subseteq B, \text{ for some } B \text{ in } \bar{B}\}.$$

4. Let (P, S_1) and (P, S_2) be two matroids on the same set of points. By $sp_i(A)$ we denote the span of set A in the matroid (P, S_i), $i = 1, 2$ (for a definition of span read Exercise 1). Let I and J be subsets each belonging to both S_1 and S_2. Prove that there exists K, a subset belonging to both S_1 and S_2, such that $K \subseteq I \cup J$, $sp_1(I) \subseteq sp_1(K)$ and $sp_2(J) \subseteq sp_2(K)$.

5. Let (P_1, S_1) and (P_2, S_2) be matroids. Define P and S as follows: $P = P_1 \cup P_2$, and $S = \{I : I = I_1 \cup I_2,\ I_1 \in S_1, \text{ and } I_2 \in S_2\}$. Show that (P, S) is a matroid.

NOTES

The material presented in this chapter is now available in several books. For more complete information we refer the reader to [1], [2], and [3]. The point of view and much of the details took shape in my mind during the numerous informal discussions with George Minty, a specialist on network flows. Most of the material comes from a course in combinatorial theory that we taught jointly in the spring of 1983. With sadness I wish to mention that Professor Minty has recently passed away; he will be fondly remembered.

Unimodular matrices go back a long way, and Lemma 5.2 can be traced to the work of Poincaré. The maximum matching algorithm in bipartite graphs was apparently first given by Munkres [4]. Section 2 contains the classical results of König and P. Hall on matchings in bipartite graphs. All books on the subject include them. Short inductive proofs can be given but we preferred the narrative, more "constructive" approach. The arc coloring lemmas are Minty's, they being the cornerstone of his "out of kilter" method [5] described in Section 6. In Section 4 the main result is the well-known max-flow min-cut theorem; it was discovered by Ford and Fulkerson (see [6]). Matroids were invented by Whitney and several of the characterizations we mentioned are due to him. The greedy algorithm was so dubbed by Edmonds; it appears in [7].

REFERENCES

[1] E. Lawler, *Combinatorial Optimization: Networks and Matroids*, Holt, Rinehart, and Winston, New York, 1976.

[2] R. T. Rockafeller, *Network Flows and Monotropic Optimization*, Wiley, New York, 1984.

[3] R. C. Bose and B. Manvel, *Introduction to Combinatorial Theory*, Wiley, New York, 1984.

[4] J. Munkres, Algorithms for the assignment and transportation problems, *J. Soc. Indust. Appl. Math.*, **5**, 32–38 (1957).

[5] G. J. Minty, Monotone networks, *Proc. Royal Soc. London, Ser. A*, **257**, 194–212 (1960).

[6] L. R. Ford, Jr. and D. R. Fulkerson, Maximal flow through a network, *Canad. J. Math.*, **8**, 399–404 (1956).

[7] J. Edmonds, Matroids and the greedy algorithm, *Math. Programming*, **1**, 127–136 (1971).

CHAPTER 6

Counting in the Presence of a Group

A mathematician, like a painter or poet, is a maker of patterns. If his patterns are more permanent than theirs, it is because they are made with ideas.

A Mathematician's Apology, G. H. HARDY

We discuss in this chapter a theory that originated in the algebraic works of Frobenius and Burnside [1] and culminated with those of Redfield [2], Pólya [3], and DeBruijn [4]. *The subject matter concerns counting the number of orbits (or orbits of a specific type) that arise by the action of a group on a set.* (When the set in question consists of functions the orbits are usually called patterns.)

6.1

Due to the highly intuitive nature of the subject it is best to introduce the general questions that arise by way of a simple example.

Consider coloring the vertices of a square (labeled $\begin{smallmatrix}1 & 2\\ 3 & 4\end{smallmatrix}$) with colors A and B. There are 16 ways of doing this, for we have 4 vertices with 2 choices of color for each. However, one may wish to say that the colorations

$$\begin{matrix}A & B\\ B & B\end{matrix} \quad \text{and} \quad \begin{matrix}B & A\\ B & B\end{matrix}$$

are in effect the same, for the latter is obtained from the former by a mere (clockwise) rotation of 90 degrees.

To be specific, suppose we allow planar rotations of 0, 90, 180, and 270 degrees, and call two colorations *equivalent* if one is obtained from the other by such a rotation. We now raise the question: How many *nonequivalent* colorations are there? Upon some thought the reader will undoubtedly find the following six representatives:

$$\begin{matrix}A & A\\ A & A\end{matrix},\ \begin{matrix}B & B\\ B & B\end{matrix},\ \begin{matrix}A & B\\ B & B\end{matrix},\ \begin{matrix}B & A\\ A & A\end{matrix},\ \begin{matrix}A & A\\ B & B\end{matrix},\ \begin{matrix}A & B\\ B & A\end{matrix}.$$

By allowing this kind of equivalence (induced by the cyclic group Z_4 of planar rotations) we have only 6 (essentially different) colorations, and not 16.

In addition, we become even more permissive and introduce another equivalence by saying that two colorations are in fact the *same* if one is obtained from the other by interchanging the colors A and B. Thus

$$\begin{matrix} A & B \\ B & B \end{matrix} \quad \text{and} \quad \begin{matrix} B & A \\ A & A \end{matrix}$$

become indistinguishable, and so do

$$\begin{matrix} A & A \\ A & A \end{matrix} \quad \text{and} \quad \begin{matrix} B & B \\ B & B \end{matrix}.$$

Upon which performance we are left with only four nonequivalent representatives:

$$\begin{matrix} A & A \\ A & A \end{matrix}, \quad \begin{matrix} A & B \\ B & B \end{matrix}, \quad \begin{matrix} A & A \\ B & B \end{matrix}, \quad \begin{matrix} A & B \\ B & A \end{matrix}.$$

Generally we visualize a coloration as a function from a domain D (of the four vertices of the square—listed in the order $\begin{smallmatrix} 1 & 2 \\ 3 & 4 \end{smallmatrix}$) to a range R of colors A and B. We allow a group G to act on D (the group Z_4 of planar rotations) and another group H to act on R (the group Z_2 that interchanges A and B). Call two colorations (or functions) f_1 and f_2 *equivalent* if

$$f_2 g = h f_1,$$

for some rotation g in G and some (possible interchange of colors) h in H. The fundamental question that we raise is: *How many nonequivalent functions (or colorations) are there*? The nonequivalent functions are called *patterns*.

With the subject thus introduced, let us close our preliminary discussions by explicitly exhibiting the equivalence of colorations

$$f_1\colon \begin{matrix} A & A \\ A & B \end{matrix} \quad \text{and} \quad f_2\colon \begin{matrix} A & B \\ B & B \end{matrix}.$$

Indeed,

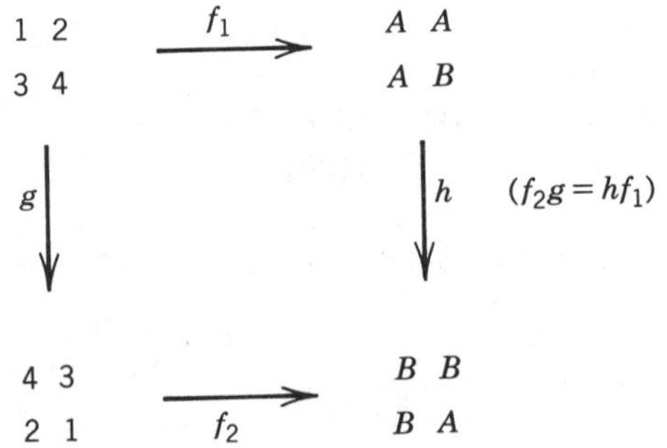

$$\begin{array}{ccc} \begin{matrix} 1 & 2 \\ 3 & 4 \end{matrix} & \xrightarrow{f_1} & \begin{matrix} A & A \\ A & B \end{matrix} \\ g \downarrow & & \downarrow h \\ \begin{matrix} 4 & 3 \\ 2 & 1 \end{matrix} & \xrightarrow[f_2]{} & \begin{matrix} B & B \\ B & A \end{matrix} \end{array} \qquad (f_2 g = h f_1)$$

where the domain is D: $\begin{smallmatrix} 1 & 2 \\ 3 & 4 \end{smallmatrix}$ (the vertices of the square displayed in this fixed order) and the range is R consisting of colors A and B. (A function from D to R is a specification of a color at each vertex, i.e., a coloration.)

1 THE GENERAL THEORY

We start out by reminding the reader of a few preliminaries on permutation groups.

6.2 Permutations on a Set

Let Ω be a set. A bijection $\sigma : \Omega \to \Omega$ is called a *permutation*. By Sym Ω we denote the set of all permutations on Ω. If $\Omega = \{1, 2, \ldots, n\}$ we write S_n for Sym Ω; $|S_n| = n!$. Under composition of functions Sym Ω is a group, called the *symmetric group*. In Sym Ω we multiply (i.e., compose) permutations from right to left, just as we work with composite functions. That is, if α and β are permutations, then $\beta\alpha$ is the permutation that carries x into $\beta(\alpha(x))$.

Any permutation σ can be decomposed into disjoint cycles: pick a symbol (1, say), form the cycle (from right to left)

$$(\cdots \sigma(\sigma(\sigma(1)))\sigma(\sigma(1))\sigma(1)1)$$

and close it when 1 is reached again; repeat the process to the remaining symbols until none are left. For example, the permutation

$$\sigma : \begin{matrix} 1 & 2 & 3 & 4 & 5 & 6 \\ 3 & 2 & 4 & 1 & 6 & 5 \end{matrix}$$

has cycle decomposition (6 5)(2)(4 3 1). Note that we list the elements of a cycle from *right to left*. Cycles of length 1 are usually omitted. Thus σ above is written (6 5)(4 3 1). The decomposition into disjoint cycles is *unique*, up to the order in which the disjoint cycles are listed. Observe that disjoint cycles commute.

A *transposition* is a cycle of the form (ba). We can write $(m \ldots 3\,2\,1) = (m\,1) \cdots (3\,1)(2\,1)$, and thus express a cycle of length m as a product of $m - 1$ transpositions. We now define the *parity* of a permutation. A permutation $\sigma = c_k \cdots c_2 c_1$ (with c_i disjoint cycles) is said to be *even* or *odd* according to the parity of the total number of transpositions obtained upon expanding the cycles. [A cycle of odd length being an even permutation and a cycle of even length being an odd permutation (see above), we conclude that the parity of an arbitrary permutation equals the parity of the number of cycles of even length which appear in its cycle decomposition.]

The *inverse* of $(m \cdots 3\,2\,1)$ is $(1\,2\,3 \cdots m)$. For two permutations α and β, we denote the product $\beta\alpha\beta^{-1}$ by α^β and call it the *conjugate* of α by β. Conjugation preserves cycle structure. In fact α^β can be obtained from α by replacing a position in α by its image under β, for example,

$$\alpha = (8\,7)(9\,6\,5\,2)(4\,3\,1)$$
$$\updownarrow \quad (8\,7\,4)(6\,1) = \beta$$
$$\alpha^\beta = (4\,8)(9\,1\,5\,2)(7\,3\,6).$$

Two permutations are conjugates in Sym Ω if and only if they have the same cycle structure.

The set of even permutations is a subgroup of index 2 in Sym Ω. We call it the *alternating group* and denote it by Alt Ω. If $\Omega = \{1, 2, \ldots, n\}$ we write A_n instead of Alt Ω; $|A_n| = n!/2$.

6.3 Group Actions on Sets

If G is a subgroup of Sym Ω we say that G *acts* on the set Ω. A homomorphism $g \to \hat{g}$ of an (abstract) group G into Sym Ω is called a *permutation representation* of G on Ω. The representation is *faithful* if it has kernel 1, that is, if it is injective.

Let G act on Ω. Each element g of G can be therefore visualized as a permutation $\hat{g}$ on Ω. For notational simplicity we write $g(x)$ instead of $\hat{g}(x)$ for the image of x under the permutation $\hat{g}$.

The subset $\{g(x) : g \in G\}$ is called the *orbit* of x (under G) and is denoted by Gx. The orbit of x is a subset of Ω. Evidently Ω is a disjoint union of orbits. (To see this, pick a point in Ω and form its orbit; if this orbit does not exhaust Ω, pick a new point outside of the first orbit and form its orbit; proceed until every point in Ω is in an orbit. Two orbits are, of course, disjoint.)

The *stabilizer* of a point x in Ω is the subgroup of G defined as follows: $\{g : g(x) = x\}$; we denote it by G_x.

It turns out that the cardinality (or length) of an orbit divides the order of the group. The precise relationship is

$$|Gx|\,|G_x| = |G| \quad \textit{(or, in words, the length of an orbit equals the index of the stabilizer)}. \tag{6.1}$$

Indeed, let y be a point in the orbit Gx of x. If h is an element of G that sends x into y, then a description of *all* elements that send x into y is the coset hG_x. We thus establish a *bijection* between the cosets of G_x in G and the points in orbit Gx of x. There are $|G|/|G_x|$ such cosets and hence $|Gx| = |G|/|G_x|$, as stated.

If x and y are points in the same orbit we naturally have $|Gx| = |Gy| =$ cardinality of the orbit. By (6.1) we conclude that also

$$|G_x| = \frac{|G|}{|Gx|} = \frac{|G|}{|Gy|} = |G_y|.$$

To summarize:

$$\textit{If } x \textit{ and } y \textit{ belong to the same orbit, then } |G_x| = |G_y|. \tag{6.2}$$

6.4 On Two-Way Counting

Throughout this chapter, and especially the next, a certain technique of counting becomes prevalent; we call it two-way counting.

Let S be a subset of the Cartesian product $V \times W$. Define

$$(v, \cdot) = \{w \in W : (v, w) \in S\}$$

and

$$(\cdot, w) = \{v \in V : (v, w) \in S\}.$$

Then

$$\sum_{v \in V} |(v, \cdot)| = |S| = \sum_{w \in W} |(\cdot, w)|.$$

This self-evident fact expresses the cardinality of S in two ways: first by initially fixing the first coordinate and summing over the second, then by initially fixing the second and summing over the first.

An example illustrates this:

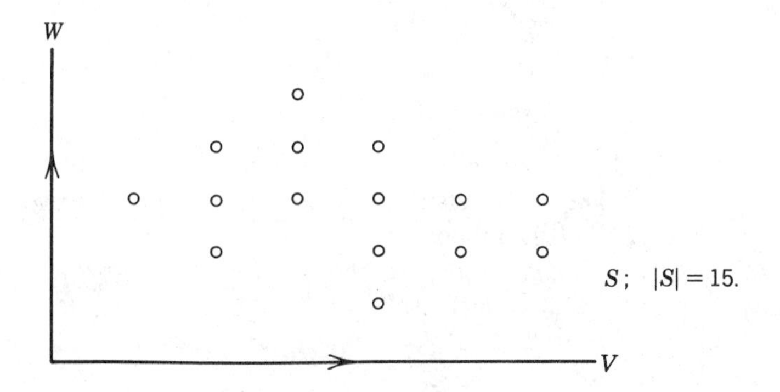

Here $\Sigma_{v \in V}|(v, \cdot)| = 1 + 3 + 3 + 4 + 2 + 2 = |S| = 1 + 4 + 6 + 3 + 1 = \Sigma_{w \in W}|(\cdot, w)|$.

Often the region S is rectangular, in which case we have

$$|(v, \cdot)| = r \text{ (say)}, \qquad \text{for all } v \in V$$

and

$$|(\cdot, w)| = k \text{ (say)}, \qquad \text{for all } w \in W.$$

Then

$$|V|r = |S| = k|W|.$$

6.5 Frobenius' Results

Let the group G act on the set Ω. Denote by t the number of orbits induced by G. In addition, let $F(g) = \{x \in \Omega : g(x) = x\}$; that is, $F(g)$ is the set of points fixed by the group element g.

Burnside's lemma informs us that:

Burnside's Lemma.

$$t = \frac{1}{|G|} \sum_{g \in G} |F(g)|.$$

(*Or, in words, the number of orbits equals the average number of points left fixed by the elements of the group.*)

Proof. We count in two ways the cardinality of the set of pairs $B = \{(g, x): g(x) = x\}$, with g in G and x in Ω.

Observe that in this situation $(g, \cdot)$ is simply $F(g)$ and that $(\cdot, x)$ is G_x. Therefore,

$$\begin{aligned}\sum_{g \in G} |F(g)| = |B| = \sum_{x \in \Omega} |G_x| &= \{\text{by (6.2), upon sorting by orbits}\}\\ &= |G_{x_1}|\,|Gx_1| + |G_{x_2}|\,|Gx_2| + \cdots + |G_{x_t}|\,|Gx_t|\\ &= \{\text{by (6.1)}\} = |G| + |G| + \cdots + |G| = t|G|.\end{aligned}$$

Here the x_i's are representatives from distinct orbits. This ends the proof.

6.6 Groups Acting on Sets of Functions

What was described in the introductory passage (Section 6.1) we now consider in more generality. The reader should refer to Section 6.1 whenever (if ever) the theory seems difficult to follow.

Let D (called the *domain*) and R (called the *range*) be sets. Let also G and H be groups; assume that G acts on D and H acts on R. Denote by $F(D, R)$ the set of functions from D to R. [If $|D| = m$ and $|R| = n$, then $|F(D, R)| = n^m$.] We summarize:

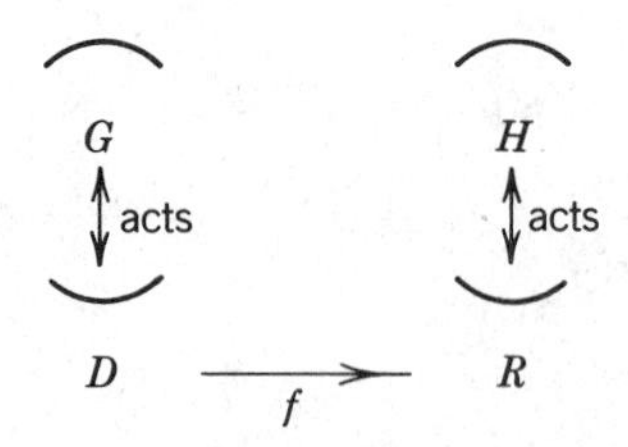

The actions of G on D and of H on R induce a natural action of the direct sum $G \oplus H$ on *the set of functions* $F(D, R)$ as follows:

$$(gh)(f) = hfg^{-1}, \tag{6.3}$$

with $gh \in G \oplus H$ [$g \in G$, $h \in H$, and $f \in F(D, R)$].

We call the functions f_1 and f_2 *equivalent*, and write $f_1 \sim f_2$, if they are in the *same orbit* of the action of $G \oplus H$ on $F(D, R)$. Thus $f_1 \sim f_2$ if $f_2 = (gh)(f_1) = hf_1g^{-1}$, for some $g \in G$ and $h \in H$. This can also be written as follows:

$$f_1 \sim f_2 \qquad \text{if} \quad f_2 g = hf_1,$$

for some g in G and h in H (see figure below and compare to that in Section 6.1).

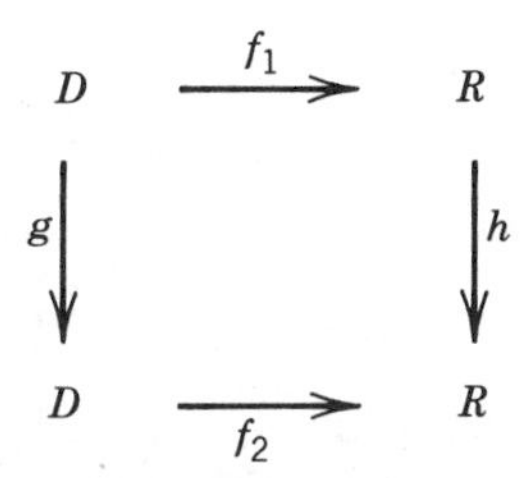

By $f_2g = hf_1$ we mean $f_2(g(d)) = h(f_1(d))$, for all d in the domain D.

It is, of course, important to check that what we have in (6.3) is indeed an *action*. We are required to show that

$$\widehat{g_2h_2g_1h_1} = \widehat{g_2h_2}\widehat{g_1h_1}; \qquad g_1, g_2 \in G \text{ and } h_1, h_2 \in H$$

where, in general, $\hat{g}$ denotes the permutation representation of the group element g. This is indeed true, for $g_2h_2g_1h_1$ equals $g_2g_1h_2h_1$ and it sends f into $h_2h_1f(g_2g_1)^{-1} = h_2h_1fg_1^{-1}g_2^{-1}$. On the other hand g_1h_1 sends f into $h_1fg_1^{-1}$, and g_2h_2 sends $h_1fg_1^{-1}$ into $h_2h_1fg_1^{-1}g_2^{-1}$; hence $\widehat{g_2h_2g_1h_1} = \widehat{g_2h_2}\widehat{g_1h_1}$.] The orbits of $G \oplus H$ on $F(D, R)$ are called *patterns*. Our principal task is to count how many patterns there are.

6.7 DeBruijn's Results

We do actually want something more than just to count the number of patterns. If possible, we would prefer to know the number of patterns of a certain type, for example. Or maybe the number of patterns that involve only a certain number of elements of the range R (i.e., in Section 6.1 the number of colorations with precisely two colors—there are two such). All this indicates that we should perhaps separate the patterns by *weight*.

Let $W: F(D, R) \to A$ be a function that satisfies $W(f_1) = W(f_2)$ whenever $f_1 \sim f_2$. [Observe that W is by its definition *well-defined on the set of patterns*. For a pattern T we can therefore unambiguously write $W(T)$, where $W(T) =$

$W(f)$ for any f in T. The function W is called a *weight function* and $W(T)$ is the *weight* of pattern T.] One would have expected the range of W to be the set of rational numbers, but it is better to make A consist of polynomials in indeterminates $y_1, \ldots, y_s$ (say) with rational or real coefficients. The weight of a pattern could therefore be something like $9y_6^7 - \frac{2}{3}y_1^4 y_5$ (it could, of course, also be its length, or whatever rational number one prefers). The choice of polynomials offers the opportunity to introduce generating functions, a potential tool.

We now prove a result of DeBruijn:

DeBruijn's Result.

$$\sum_{\substack{T \\ (T \text{ pattern})}} W(T) = \frac{1}{|G|\,|H|} \sum_{g \in G} \sum_{h \in H} \sum_{\substack{f \\ (gh)(f) = f}} W(f) \tag{6.4}$$

(The very last sum is over all functions f fixed by gh.)

Proof. Let y be an element in the range of W. Look at all patterns T of weight y; denote this collection by T_y. The group $G \oplus H$ acts on T_y since T_y is a union of orbits (or patterns). By Burnside's lemma we have:

$$|T_y| = \frac{1}{|G \oplus H|} \sum_{gh \in G \oplus H} |F_y(gh)| = \frac{1}{|G|\,|H|} \sum_{g \in G} \sum_{h \in H} |F_y(gh)|,$$

with $F_y(gh)$ the set of functions f in T_y fixed by gh. Sum now over y in the range of W to obtain

$$\begin{aligned}
\sum_{\substack{T \\ (T \text{ pattern})}} W(T) &= \sum_y y|T_y| = \sum_y \frac{y}{|G|\,|H|} \sum_{g \in G} \sum_{h \in H} |F_y(gh)| \\
&= \frac{1}{|G|\,|H|} \sum_{g \in G} \sum_{h \in H} \sum_y y|F_y(gh)| \\
&= \frac{1}{|G|\,|H|} \sum_{g \in G} \sum_{h \in H} \sum_{\substack{f \\ (gh)(f) = f}} W(f).
\end{aligned}$$

The last sign of equality is explained as follows:

$$\begin{aligned}
\sum_y y|F_y(gh)| &= \sum_y y \left(\text{number of functions } f \text{ fixed by } gh \text{ in } T_y\right) \\
&= \sum_{\substack{f \\ (gh)(f) = f}} W(f),
\end{aligned}$$

with summation over all f in $F(D, R)$ fixed by gh. This proves DeBruijn's result.

6.8 Results by Redfield and Pólya

We gradually specialize DeBruijn's result. First, let $H = 1$ (i.e., there is no group acting on the range). The group G (visualized as $G \oplus 1$, if necessary) acts on $F(D, R)$ as follows:

$$g(f) = fg^{-1},$$

for g an element of G [a special case of (6.3)]. Two functions f_1 and f_2 are equivalent (written still $f_1 \sim f_2$) if $f_2 g = f_1$, for some g in G.

Secondly, we restrict attention to special kinds of weight functions W. Let $w: R \to A$ be a function from the range R into A, the set of polynomials in indeterminates $y_1, \ldots, y_s$ with rational coefficients. For r an element of the range R we call $w(r)$ its *weight*. The function w induces a weight function W on $F(D, R)$ by setting

$$W(f) = \prod_{d \in D} w(f(d)). \tag{6.5}$$

[If $f_1 \sim f_2$ then $f_1 = f_2 g$, for some g in G, and then

$$W(f_1) = W(f_2 g) = \prod_d w(f_2(g(d))) = \{\text{since } g \text{ is a bijection}\}$$

$$= \prod_d w(f_2(d)) = W(f_2),$$

which shows that W is indeed well defined on patterns.]

DeBruijn's result (6.4) can now be written

$$\sum_{\substack{T \\ (T \text{ pattern})}} W(T) = \frac{1}{|G|} \sum_{g \in G} \sum_{\substack{f \\ fg = f}} W(f), \tag{6.6}$$

where W is as defined in (6.5).

The last sum $\sum_{f, fg=f} W(f)$ takes, in fact, a more explicit form determined chiefly by the cycle structure of the group element g in its action on D. Write $g = c_k \cdots c_2 c_1$, with c_i's the disjoint cycles of g. The condition $f = fg$ means (for all d in D) that $f(d) = f(g(d)) = f(g^2(d)) = f(g^3(d)) = \cdots$, which implies that f is constant on each cycle c_i of g (but it may take different values on different cycles). [We remind that $g^2(d)$ means $g(g(d))$, etc.] Conversely, every function f constant on the cycles c_i of g obviously satisfies $fg = f$. (In what follows by $|c_i|$ we denote the length, or cardinality, of the cycle c_i.)

With this characterization of the functions f that satisfy $fg = f$ we can write

$$\prod_{i=1}^{k} \sum_{r \in R} (w(r))^{|c_i|}$$

$$= \left(\sum_r (w(r))^{|c_1|}\right)\left(\sum_r (w(r))^{|c_2|}\right) \cdots \left(\sum_r (w(r))^{|c_k|}\right)$$

$$= \left[\begin{array}{l}\text{Expand by picking a term from each of the } k \text{ factors} \\ \text{and sum over all such choices; visualize now the } i\text{th} \\ \text{factor above as the } i\text{th cycle } c_i \text{ of } g\text{; then picking a} \\ \text{term from each factor means defining a function } f \text{ on} \\ D\text{, which is } \textit{constant} \text{ on each of the cycles } c_i \text{ and with} \\ \text{range included in } R.\end{array}\right]$$

$$= \sum_{\substack{(f \text{ is constant on} \\ \text{each cycle} \\ c_i \text{ of } g)}} \prod_{d \in D} w(f(d)) = \sum_{\substack{f \\ fg=f}} W(f)$$

(the next to last step takes some thinking...).

Let $|D| = m$ and suppose that g is of type $1^{\lambda_1}2^{\lambda_2} \cdots m^{\lambda_m}$ (with $\sum_{i=1}^{m} i\lambda_i = m$), that is, g has λ_i cycles of length i. Some of the λ_i's could be 0; we still assume that $g = c_k \cdots c_2 c_1$ with $|c_i| \geq 1$, $1 \leq i \leq k$. Rewriting the previous equation gives

$$\sum_{\substack{f \\ fg=g}} W(f) = \prod_{i=1}^{k} \sum_r (w(r))^{|c_i|}$$
$$= \left(\sum_r w(r)\right)^{\lambda_1}\left(\sum_r (w(r))^2\right)^{\lambda_2} \cdots \left(\sum_r (w(r))^m\right)^{\lambda_m}. \tag{6.7}$$

Substituting (6.7) in (6.6) we obtain Pólya's result:

$$\sum_T W(T) = \frac{1}{|G|} \sum_{g \in G} \left[\prod_{i=1}^{m} \left(\sum_{r \in R} (w(r))^i\right)^{\lambda_i}\right] \tag{6.8}$$

where $1^{\lambda_1}2^{\lambda_2} \cdots m^{\lambda_m}$ *denotes the cycle type of the group element g in its action on the domain.*

Expression (6.8) strongly suggests that the *cycle structure* of the group elements (represented on the domain) is the determining feature of the group that comes into play. We take advantage of this by cleverly choosing our notation.

6.9 Just Notation

Recall that $|D| = m$. To the group G (in its action on the domain D) we associate now a polynomial P_G in m indeterminates $x_1, x_2, \ldots, x_m$ called the *cycle index* of G. Specifically,

$$P_G(x_1, x_2, \ldots, x_m) = \frac{1}{|G|} \sum_{g \in G} x_1^{\lambda_1(g)} x_2^{\lambda_2(g)} \cdots x_m^{\lambda_m(g)},$$

where $\lambda_i(g)$ denotes the number of cycles of length i of g.

By substituting $(w(r))^i$ for x_i in P_G we can rewrite (6.8) as follows:

Pólya's Theorem.

$$\sum_{\substack{T \\ (T \text{ pattern})}} W(T) = P_G\left(\sum_r w(r), \sum_r (w(r))^2, \ldots, \sum_r (w(r))^m\right) \qquad (6.9)$$

where P_G is the cycle index of G in its action on the domain D.

In particular, if $w(r) = 1$ for all r in the range R, we obtain:

$$\textit{The total number of patterns} = P_G(|R|, |R|, \ldots, |R|). \qquad (6.10)$$

There is a lot of information locked in Pólya's theorem (6.9) and even more in DeBruijn's result given in (6.4). In the remaining pages of this chapter we have the opportunity to unleash the power of these results in many interesting special cases. To a large extent this also enhances the understanding of some of the finer points in the proof we just gave.

REMARK. It is instructive to compare Burnside's lemma with Pólya's enumeration theorem. At first it appears that the latter is essentially a special case of the former when the group acts on the set of functions from D to R. And this is indeed true, as the proof we just saw indicates. Due to the large cardinality of the set of functions it is usually difficult to apply Burnside's lemma directly to this set and calculate for each group element the number of functions that it fixes. Pólya's theorem reduces drastically the size of the problem by only studying the group in its original representation on the *domain D* of these functions (a much smaller set). One needs to know the representation on the domain rather well, however: knowing the number of points fixed by each group element on the domain is not enough; rather, a list of the actual cycle structure of each group element (the cycle index) must be made available. This is the price paid by working on the domain only.

Pólya's theorem may be helpful, however, in understanding certain "naturally" induced representations of the group in question on subsets of a fixed cardinality of the domain. Specifically, let G be a group of permutations on the elements of D. Then G acts on the subsets of D of cardinality k in a natural way, that is, $g\{d_1, \ldots, d_k\} = \{g(d_1), \ldots, g(d_k)\}$.

Assume that the cycle index of G on D is known and denote it by $P_G(x_1, x_2, \ldots, x_m)$. We make the following assertion:

$*$ *The number of orbits of G in its action on the subsets of cardinality k of D equals the coefficient of y^k in $P_G(1 + y, 1 + y^2, \ldots, 1 + y^m)$.*

To demonstrate this, look at the set of functions from D to $R = \{0, 1\}$. A *function* from D to R is easily identified with the *subset* of elements of D mapped into 1. Let G act on the set of functions as follows:

$$g(f(d)) = f(g(d)).$$

Select a weight function w on R by setting $w(0) = 1$ and $w(1) = y$. Then f has weight y^k if and only if f maps k elements of D into 1 [see (6.5), if necessary]. It is now clear that the number of orbits that G induces on the subsets of cardinality k of D equals the number of patterns of weight k that thus arise. Pólya's theorem informs us now that the number of such patterns is simply the coefficient of y^k in the polynomial

$$P_G(1 + y, 1 + y^2, \ldots, 1 + y^m).$$

In particular, the number of orbits of G on the elements of D (which is what Burnside's lemma gives) equals the coefficient of y in this polynomial.

2 RECIPE FOR PÓLYA'S THEOREM

6.10

Consider again the problem of coloring the vertices of the square $\begin{smallmatrix}1 & 2\\ 3 & 4\end{smallmatrix}$ with colors A and B (as in Section 6.1). Our domain consists therefore of the four vertices of the square, and the range consists of the two colors A and B. Call two colorations equivalent if they are the same apart from a planar rotation. In Section 6.1 we have found six patterns, a representative from each being listed below:

$$\begin{matrix}A & A\\ A & A\end{matrix}, \quad \begin{matrix}B & B\\ B & B\end{matrix}, \quad \begin{matrix}A & B\\ B & B\end{matrix}, \quad \begin{matrix}B & A\\ A & A\end{matrix}, \quad \begin{matrix}A & A\\ B & B\end{matrix}, \quad \begin{matrix}A & B\\ B & A\end{matrix}. \tag{6.11}$$

In particular observe that precisely two patterns (the last two, as displayed above) are colored with two A's and two B's.

Let us show how we can derive this information from Pólya's theorem.

First, extract the cycle structure of the group G of the four planar rotations (of 0, 90, 180, and 270 degrees) on the domain. The identity (i.e., rotation by 0 degrees) fixes all vertices of the square; thus its cycle structure is $1^4 2^0 3^0 4^0$, which we abbreviate by the monomial $x_1^4 x_2^0 x_3^0 x_4^0 = x_1^4$. The rotation by 90 degrees is (4 3 2 1), which we write $x_1^0 x_2^0 x_3^0 x_4^1 = x_4$, and the rotation by 270 degrees is (1 2 3 4) with same cycle structure $x_1^0 x_2^0 x_3^0 x_4^1 = x_4$. Lastly, the rotation by 180 degrees is (4 2)(3 1), described by the monomial $x_1^0 x_2^2 x_3^0 x_4^0 = x_2^2$. The cycle index of the group G of rotations of the square is therefore

$$P_G(x_1, x_2, x_3, x_4) = \tfrac{1}{4}(x_1^4 + x_2^2 + 2x_4).$$

(Now toss away the group; we do not need it anymore.)

Pólya's theorem informs us that the total number of patterns equals $P_G(|R|, |R|, |R|, |R|) = P_G(2, 2, 2, 2) = \frac{1}{4}(2^4 + 2^2 + 2 \cdot 2) = \frac{1}{4}24 = 6$, as expected.

Further, suppose we become interested in the number of patterns with precisely two A's and two B's. The ability to select weights as general in nature as indeterminates comes in very handy at this time. We wish to clearly distinguish between the A's and the B's that occur in a pattern and thus assign weights $w(A) = y_1$ and $w(B) = y_2$, with y_1 and y_2 indeterminates. Then the weight of a pattern containing a A's and b B's is $y_1^a y_2^b$, with $a + b = 4$ [see (6.5)]. We can therefore read the number of patterns with a A's and b B's in the coefficient of $y_1^a y_2^b$. Pólya's theorem (6.9) explicitly gives

$$\sum_{\substack{T \\ (T \text{ pattern})}} W(T) = P_G\left(\sum_{r \in R} w(r), \sum_{r \in R} (w(r))^2, \ldots, \sum_{r \in R} (w(r))^m \right)$$

$$= P_G\left(y_1 + y_2, y_1^2 + y_2^2, y_1^3 + y_2^3, y_1^4 + y_2^4\right)$$

$$= \tfrac{1}{4}\left((y_1 + y_2)^4 + \left(y_1^2 + y_2^2\right)^2 + 2\left(y_1^4 + y_2^4\right)\right).$$

We seek the number of patterns with two A's and two B's, and upon inspecting the coefficient of $y_1^2 y_2^2$ find the answer $\frac{1}{4}(6 + 2) = 2$, in agreement with what (6.11) displays.

What we have achieved in the case of the square can be carried over to more general circumstances. We describe the procedure in Section 6.11.

6.11 The Recipe

We have before us a domain D with elements $d_1, d_2, \ldots, d_m$ and a range R of colors $r_1, r_2, \ldots, r_n$. Available before us is also a group G, which allows itself to be represented as a group of permutations on the domain D.

Interest is manifested in coloring the elements of the domain D with colors from the range R. Furthermore, two such colorations are said to be *equivalent* if one is carried into the other by a group element g in G. That is, colorations f_1 and f_2 are equivalent if $f_1 = f_2 g$, for some g in G.

The nonequivalent colorations are called *patterns*, and we ask:

(i) How many patterns are there?

(ii) How many patterns are there with

a_1 occurrences of color r_1

a_2 occurrences of color r_2

⋮

a_n occurrences of color r_n? ($\sum_{i=1}^n a_i = m = |D|$.)

Pólya's theorem offers the answers in three steps, the first of which is by far the most difficult and time consuming.

Step 1. Find the cycle index of the group G in its action on the domain D.

[We remind ourselves that the cycle index is a polynomial P_G in indeterminates $x_1, x_2, \ldots, x_m$ defined as follows:

$$P_G(x_1, x_2, \ldots, x_m) = \frac{1}{|G|} \sum_{g \in G} x_1^{\lambda_1(g)} x_2^{\lambda_2(g)} \cdots x_m^{\lambda_m(g)},$$

where $\lambda_i(g)$ is the number of cycles of length i of the group element g in its representation as a permutation on the domain D. (*Aside*: Upon extracting the cycle index throw away the group; we do not need it anymore.)]

Step 2. Obtain the total number of patterns by substituting the number of colors (i.e., $|R|$) *for each of the variables* $x_1, x_2, \ldots, x_m$ *in the cycle index* P_G. That is:

The total number of patterns $= P_G(|R|, |R|, \ldots, |R|)$.

Step 3. The number of patterns with

a_1 *occurrences of color* r_1
a_2 *occurrences of color* r_2
⋮
a_n *occurrences of color* $r_n (\sum_{i=1}^n a_i = m)$

is the coefficient of $y_1^{a_1} y_2^{a_2} \cdots y_n^{a_n}$ *in the polynomial*

$$P_G\left(\sum_{i=1}^n y_i, \sum_{i=1}^n y_i^2, \ldots, \sum_{i=1}^n y_i^m \right).$$

[This is indeed the case, for if we assign weight y_i to color r_i (with y_i indeterminates), then the weight of a pattern in which r_i occurs a_i times, $1 \le i \le n$, is $\prod_{i=1}^n y_i^{a_i}$ (conform (6.5)). Thus the coefficient of this monomial is the answer we seek, and Pólya's theorem (6.9) leads us to it by way of a suitable substitution (i.e., substitute $\sum_{i=1}^n y_i^k$ for x_k) in the cycle index.]

We now illustrate the recipe by several examples.

3 EXAMPLES FOLLOWING THE RECIPE

6.12 The Case of the Trivial Group

Assume that we have a domain D consisting of m elements and a range R of n colors that we denote by $r_1, r_2, \ldots, r_n$. We begin with the simplest possible situation: that in which the group G acting on D consists of only one element (the identity). Two colorations are now equivalent if they are absolutely *identical*.

It is obvious that there are n^m patterns in all (for we have n choices of color for each of the m elements of the domain). The number of patterns with a_1

elements of color r_1, a_2 elements of color $r_2, \ldots, a_n$ elements of color r_n (with $\sum_{i=1}^{n} a_i = m$) is just the multinomial coefficient

$$\frac{m!}{a_1! a_2! \cdots a_n!}.$$

Our recipe indeed gives:

Step 1. The cycle index of G is simply

$$P_G(x_1, \ldots, x_m) = \tfrac{1}{1}(x_1^m + 0 + \cdots + 0) = x_1^m.$$

Step 2. The total number of patterns is

$$P_G(n, n, \ldots, n) = n^m.$$

Step 3. The number of patterns with a_i elements of color r_i $(i = 1, 2, \ldots, n)$ is the coefficient $y_1^{a_1} y_2^{a_2} \cdots y_n^{a_n}$ in

$$P_G\left(\sum y_i, \sum y_i^2, \cdots\right) = \left(\sum y_i\right)^m.$$

This coefficient is indeed the multinomial number written above. [For example, if $n = 2$ we obtain the binomial coefficient $\binom{m}{a_1}$ for an answer, as expected.]

6.13 Coloring the Cube

Who can possibly forget the pleasant childhood pastime of rolling colored cubes? Or the quiet frustration of observing a nicer color pattern displayed on your playmate's cube? To those whose interest in the charming enterprise was at the time less than mathematical we offer the opportunity for a change of emphasis.

Let us dust off our favorite childhood cube (or any ordinary cube, really) and equip ourselves with a selection of attractive colors. Naturally we proceed to try our hand at coloring the object. Guided by principles of good taste we can color either: (a) its 8 vertices, (b) its 12 edges, or (c) its 6 faces. In each case, however, we say that two colorations are essentially the same (i.e., equivalent) if one can be obtained from the other by a *rotation* of the cube.

The (abstract) group G that induces equivalence is therefore that of the $6 \times 4 = 24$ rotations. [There are indeed 24 rotations of the cube, for we have 6 places to send a face to and 4 further (independent) choices for an adjacent face.]

We examine in detail the essentially different colorations in each of the three cases mentioned above.

Let us make a list (by way of a geometrical description) of the 24 rotations of the cube:

(i) The identity.

(ii) Three 180° rotations around lines connecting the centers of opposite faces.

(iii) Six 90° rotations around lines connecting the centers of opposite faces.
(iv) Six 180° rotations around lines connecting the midpoints of opposite edges.
(v) Eight 120° rotations around lines connecting opposite vertices.

This may not appear as the most natural way in the world to classify rotations but it is if the object is computation of cycle indexes.

Coloring the Vertices of the Cube

The group G of rotations acts now on the eight vertices of the cube, which form our domain D.

Step 1. Extract the cycle index of G in its action on the vertices of the cube. The cycle structure of the group elements as outlined in (i) through (v) is summarized below:

(i) $\leftrightarrow x_1^8$
(ii) $\leftrightarrow 3x_2^4$
(iii) $\leftrightarrow 6x_4^2$
(iv) $\leftrightarrow 6x_2^4$
(v) $\leftrightarrow 8x_1^2x_3^2$

And so $P_G = \frac{1}{24}(x_1^8 + 9x_2^4 + 6x_4^2 + 8x_1^2x_3^2)$.

Step 2. If we are equipped with n colors the total number of essentially different colorations is the value of P_G at $x_1 = x_2 = \cdots = x_8 = n$. That is, we have $\frac{1}{24}(n^8 + 9n^4 + 6n^2 + 8n^4)$ distinct patterns in all. [Note that irrespective of n this number is always an integer, an amusing feature in itself. In the event of two available colors, for example, there are precisely $\frac{1}{24}(2^8 + 9 \cdot 2^4 + 6 \cdot 2^2 + 8 \cdot 2^4) = 13$ distinct color patterns.]

Step 3. Let the colors be $r_1, \ldots, r_n$. The number of patterns with a_i vertices colored with color r_i $(i = 1, 2, \ldots, 8)$ is the coefficient of $y_1^{a_1}y_2^{a_2}\cdots y_n^{a_n}$ in

$$P_G\left(\sum y_i, \sum y_i^2, \ldots, \sum y_i^8\right)$$

$$= \frac{1}{24}\left[\left(\sum y_i\right)^8 + 9\left(\sum y_i^2\right)^4 + 6\left(\sum y_i^4\right)^2 + 8\left(\sum y_i\right)^2\left(\sum y_i^3\right)^2\right].$$

[If we color with just colors red and blue, and wish to know how many essentially distinct colorations (patterns) there are with three red vertices and five blue, we look at the coefficient of $y_1^3y_2^5$ in

$$\frac{1}{24}\left[(y_1 + y_2)^8 + 9(y_1^2 + y_2^2)^4 + 6(y_1^4 + y_2^4)^2 + 8(y_1 + y_2)^2(y_1^3 + y_2^3)^2\right].$$

This coefficient is

$$\tfrac{1}{24}\left[\binom{8}{3} + 8 \cdot 2\right] = \frac{1}{8 \cdot 3}(8 \cdot 7 + 8 \cdot 2) = 3.$$

The reader can easily verify that this is true by exhibiting geometrically the three configurations on the cube.]

Coloring the Edges of the Cube

The group of rotations permutes the 12 edges of our cube. Edges are now the elements of our domain.

Step 1. Extract the cycle index of the group of rotations in its action on the 12 edges of the cube. We summarize again, from (i) to (v):

(i) $\leftrightarrow x_1^{12}$
(ii) $\leftrightarrow 3x_2^6$
(iii) $\leftrightarrow 6x_4^3$
(iv) $\leftrightarrow 6x_1^2x_2^5$
(v) $\leftrightarrow 8x_3^4$

The cycle index is

$$P_G = \tfrac{1}{24}\left(x_1^{12} + 3x_2^6 + 6x_4^3 + 6x_1^2x_2^5 + 8x_3^4\right).$$

Step 2. The total number of patterns is

$$P_G(n, n, \dots, n) = \tfrac{1}{24}(n^{12} + 3n^6 + 6n^4 + 6n^7 + 8n^4),$$

if we use n colors. [In case of two colors we obtain $P_G(2, 2, \dots, 2) = 218$ patterns in all.]

Step 3. The number of patterns with a_i edges colored r_i $(i = 1, 2, \dots, 12)$ equals the coefficient of $\prod_{i=1}^{12} y_i^{a_i}$ in the expansion of $P_G(\Sigma y_i, \Sigma y_i^2, \dots, \Sigma y_i^{12})$. For example, there are 27 essentially distinct colorations with four blue and eight red edges.

Coloring the Faces of the Cube

Step 1. The rotation group certainly permutes the six faces and its cycle structure is

(i) $\leftrightarrow x_1^6$
(ii) $\leftrightarrow 3x_1^2x_2^2$
(iii) $\leftrightarrow 6x_1^2x_4^1$
(iv) $\leftrightarrow 6x_2^3$
(v) $\leftrightarrow 8x_3^2$

Hence the cycle index of the rotation group represented on the six faces of the cube is

$$P_G = \tfrac{1}{24}\left(x_1^6 + 3x_1^2x_2^2 + 6x_1^2x_4 + 6x_2^3 + 8x_3^2\right).$$

Step 2. The total number of color patterns on faces possible to generate with n colors is

$$P_G(n, n, \ldots, n) = \tfrac{1}{24}(n^6 + 3n^4 + 6n^3 + 6n^3 + 8n^2).$$

[If two colors are used we obtain $P_G(2, 2, \ldots, 2) = 10$ patterns in all.]

Step 3. Let us become curious and ask how many patterns there are with three distinct colors per pattern and with each of the three colors occurring twice? Assume that we have n colors available; $n \geq 3$.

Choose three out of the n available colors [call them 1, 2, 3 (say)]. Give weight y_l to color l, with y_l an indeterminate. Then the number of patterns colored with colors 1, 2, 3, with each color occurring twice, is the coefficient of $y_1^2y_2^2y_3^2$ in $P_G(\Sigma y_i, \Sigma y_i^2, \ldots, \Sigma y_i^6)$. The intrinsic symmetry of the problem and the symmetric nature of the substitution made in the cyclic index P_G allow us to conclude that there are $\binom{n}{3}$ times the coefficient of $y_1^2y_2^2y_3^2$ such patterns. The coefficient of $y_1^2y_2^2y_3^2$ in $\frac{1}{24}[(\Sigma y_i)^6 + 3(\Sigma y_i)^2(\Sigma y_i^2)^2 + 6(\Sigma y_i)^2(\Sigma y_i^4) + 6(\Sigma y_i^2)^3 + 8(\Sigma y_i^3)^2]$ is

$$\frac{1}{24}\left[\frac{6!}{2!\,2!\,2!} + 3(2 + 2 + 2) + 0 + 6\frac{3!}{1!\,1!\,1!} + 0\right] = 6,$$

and the answer to our question is therefore $6\binom{n}{3}$. Check this result for $n = 3$ by a direct examination of patterns.

6.14 Patterns on a Chessboard

How many essentially distinct configurations can we make by placing 16 checkers on an ordinary chessboard? Two configurations are called essentially distinct if one cannot be obtained from the other by a *rotation* of the chessboard. The alternating black-white coloring of the squares of the chessboard plays no role in what we do; the reader should thus ignore it and assume that the whole chessboard is white. Apart from the initial question we raised, the techniques developed so far answer other natural questions as well, and we shall point these out as we go along.

Step 1. The group in question is that of rotations by 0°, 90°, 180°, and 270°. Our domain consists of the $8 \times 8 = 64$ squares of the chessboard. In its action on the domain our group has cycle structure as indicated below:

$$\begin{aligned} \text{Identity} &\leftrightarrow x_1^{64} \\ \text{Rotation by } 180° &\leftrightarrow x_2^{32} \\ \text{Rotation by } 90° &\leftrightarrow x_4^{16} \\ \text{Rotation by } 270° &\leftrightarrow x_4^{16} \end{aligned}$$

We conclude that the cycle index of the group G (of rotations of the board), in its action on the 64 squares of the chessboard, has cycle index:

$$P_G = \frac{1}{4}\left(x_1^{64} + x_2^{32} + 2x_4^{16}\right).$$

Step 2. The act of placing checkers on the chessboard should be viewed as a coloration of the domain with two colors: *checker* = *red*, and *no checker* = *blue*.

Pólya's theorem informs us now that the *total number of patterns* = $P_G(2,2,2,2) = \frac{1}{4}(2^{64} + 2^{32} + 2 \cdot 2^{16})$.

Step 3. Assign weight y_1 to the red color (i.e., to a checker) and weight y_2 to blue (i.e., to no checker). Our original question asks for the number of patterns with precisely 16 red and 64 − 16 blue colors. The answer can be found in the coefficient of $y_1^{16}y_2^{64-16}$ in

$$\begin{aligned} &P_G\left(y_1 + y_2, y_1^2 + y_2^2, y_1^3 + y_2^3, y_1^4 + y_2^4\right) \\ &\quad = \frac{1}{4}\left((y_1 + y_2)^{64} + \left(y_1^2 + y_2^2\right)^{32} + 2\left(y_1^4 + y_2^4\right)^{16}\right). \end{aligned}$$

Upon collecting terms we find this coefficient to be

$$\frac{1}{4}\left(\binom{64}{16} + \binom{32}{8} + 2\binom{16}{4}\right).$$

More generally, assume that we have checkers of $n - 1$ colors and ask for the number of patterns with a_1 checkers of color r_1, a_2 checkers of color $r_2, \ldots, a_{n-1}$ checkers of color r_{n-1}. Upon assigning weight y_i to checkers of color r_i $(i = 1, 2, \ldots, n-1)$ and weight y_n to *no checker*, the answer is found in the coefficient of

$$y_1^{a_1} y_2^{a_2} \cdots y_{n-1}^{a_{n-1}} y_n^{a_n} \qquad \left(a_n = 64 - \sum_{i=1}^{n-1} a_i\right)$$

upon expanding

$$\begin{aligned} &P_G\left(\sum y_i, \sum y_i^2, \sum y_i^3, \sum y_i^4\right) \\ &\quad = \frac{1}{4}\left(\left(\sum y_i\right)^{64} + \left(\sum y_i^2\right)^{32} + 2\left(\sum y_i^4\right)^{16}\right). \end{aligned}$$

(In the summation signs above the index i runs between 1 and n.)

We leave to the reader the general case of placing a_i checkers of color r_i $(i = 1, 2, \ldots, n-1)$ on a $m \times m$ chessboard and of answering how many patterns arise. (Note that for odd m the cycle index changes its form, because the square in the center remains fixed under all four rotations.)

6.15 Necklaces

Equipped with n colors and m beads we ask for the number of essentially different necklaces of m colored beads that we can make. Two necklaces are said to be equivalent if one can be obtained from the other by a rotation and/or a flip.

The group that induces equivalence is in this case the dihedral group D_{2m} (of order $2m$) consisting of the m planar rotations plus m flips. For example, D_{10} consists of planar rotations with angles $2\pi k/5$ ($k = 0, 1, 2, 3, 4$) and 5 flips across lines joining a vertex to the middle of the edge completely opposite:

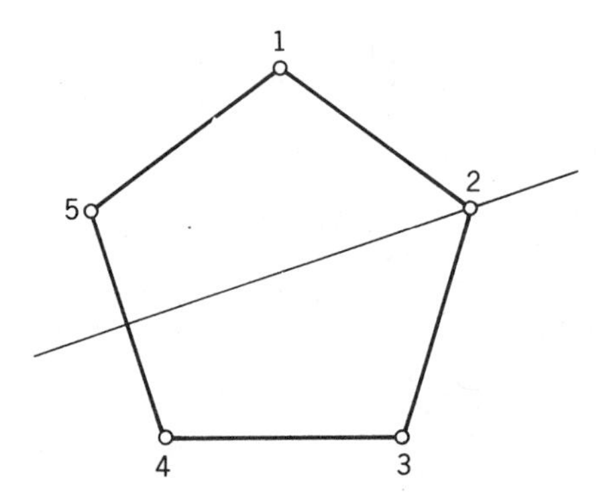

This remains a valid description of the elements of D_{2m} for any odd number m. When m is even we have $m/2$ flips across lines joining the middle points of completely opposite edges, and $m/2$ flips fixing a pair of totally opposite vertices (see below), in addition to the m planar rotations of angles $2\pi k/m$ ($k = 0, 1, \ldots, m - 1$).

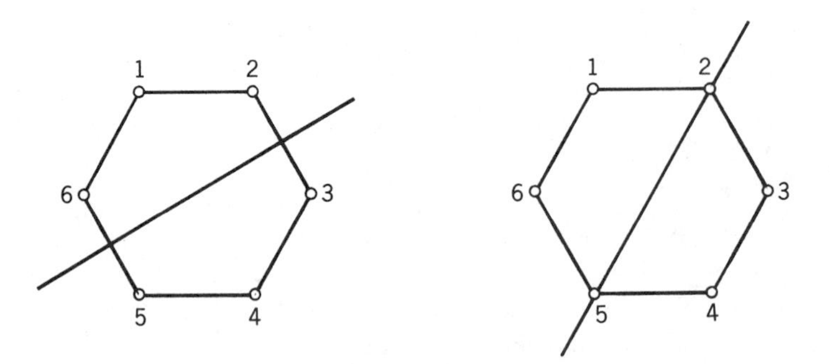

A necklace should be envisioned as a regular m-gon with beads as vertices and the group that induces equivalence is D_{2m} whose elements we just described.

Step 1. The cycle index of D_{2m} in its action on the vertices of the m-gon (i.e., on the m beads of the necklace) depends on the parity of m.

If m is odd, say $m = 2s + 1$, then we have m flips, each of which fixes one vertex and pairs up the remaining $2s$ vertices into s disjoint transpositions. The cyclic subgroup of rotations is generated by a rotation with angle $2\pi/m$. If a rotation has order d, then it consists of m/d disjoint cycles of length d each; and there are $\phi(d)$ such rotations, where $\phi(d)$ denotes the number of positive integers less than d and relatively prime to d. The cycle index of D_{2m} for odd m is therefore

$$P_{D_{2m}} = \frac{1}{2m}\left[m x_1 x_2^{(m-1)/2} + \sum_{\substack{d \\ d|m}} \phi(d) x_d^{m/d} \right].$$

Analogously, for m even we obtain

$$P_{D_{2m}} = \frac{1}{2m}\left[\frac{m}{2} x_2^{m/2} + \frac{m}{2} x_1^2 x_2^{(m-2)/2} + \sum_{\substack{d \\ d|m}} \phi(d) x_d^{m/d} \right].$$

Step 2. The total number of patterns of necklaces with m beads colored with n colors is

$$P_{D_{2m}}(n,\dots,n) = \frac{1}{2m}\left[mnn^{(m-1)/2} + \sum_{\substack{d \\ d|m}} \phi(d) n^{m/d} \right], \qquad \text{for } m \text{ odd}$$

and

$$P_{D_{2m}}(n,\dots,n) = \frac{1}{2m}\left[\frac{m}{2} n^{m/2} + \frac{m}{2} n^2 n^{(m-2)/2} + \sum_{\substack{d \\ d|m}} \phi(d) n^{m/d} \right],$$

for m even.

Step 3. By assigning weight y_i to color r_i $(i = 1,\dots,n)$ we can, as usual, answer how many (patterns of) necklaces with m beads in all there are such that a_1 beads are colored r_1, a_2 are colored $r_2,\dots,$ and a_n are colored r_n. The answer is the coefficient of $y_1^{a_1} y_2^{a_2} \cdots y_n^{a_n}$ in

$$P_{D_{2m}}\left(\sum y_i, \sum y_i^2, \dots, \sum y_i^m\right).$$

If $m = 6$, for example, and if we have two colors (green and red), the number of patterns of necklaces with two green and four red beads is the coefficient of $y_1^2 y_2^4$ in

$$\frac{1}{2\cdot 6}\Big[3(y_1^2+y_2^2)^3 + 3(y_1+y_2)^2(y_1^2+y_2^2)^2 + \phi(1)(y_1+y_2)^6$$
$$+\phi(2)(y_1^2+y_2^2)^3 + \phi(3)(y_1^3+y_2^3)^2 + \phi(6)(y_1^6+y_2^6)\Big].$$

This coefficient is

$$\frac{1}{2\cdot 6}\left[3\binom{3}{2} + 3(1+2) + \phi(1)\binom{6}{2} + \phi(2)\binom{3}{2} + 0 + 0\right]$$
$$= \frac{1}{2\cdot 6}\left[3\cdot 6 + \binom{6}{2} + 1\binom{3}{2}\right] = 3.$$

Indeed, the patterns are:

```
    r r            r r            r g
r       r      r       g  and  r       r .
    g g            g r            g r
```

4 THE CYCLE INDEX

Most of the effectiveness of counting in the presence of a group rests with the ability of understanding the cycle structure of the group as a permutation group on the domain. Having available a list of cycle indexes will thus enhance our computational power. On the other hand, the same abstract group has different cycle indexes when represented on different domains; this latter piece of intelligence renders the idea of a comprehensive list pretty much meaningless.

6.16

We nonetheless mention several cycle indexes that occur fairly often.

The Identity

When represented on a domain with m elements the identity group has cycle index x_1^m.

The Cyclic Group

The cyclic group of the m (planar) rotations of a regular m-gon, represented as a permutation group on the vertices of the m-gon, has cycle index

$$\frac{1}{m}\sum_{\substack{d \\ d|m}} \phi(d) x_d^{m/d},$$

where $\phi(d)$ denotes the positive integers less than d and relatively prime to d.

[The cyclic group discussed above consists of the cycle $(m \cdots 3\,2\,1)$ and all its powers. An examination of the cycle structure of its powers readily reveals the formula for its cycle index written above.]

The Dihedral Group

By the appellation of a dihedral group of order $2m$ we understand the m rotations and m flips of a regular m-gon. These groups appeared in Section

6.15 where we also computed their cycle indexes:

$$\frac{1}{2m}\left[m x_1 x_2^{(m-1)/2} + \sum_{\substack{d \\ d|m}} \phi(d) x_d^{m/d} \right], \qquad \text{for } m \text{ odd}$$

and

$$\frac{1}{2m}\left[\frac{m}{2} x_2^{m/2} + \frac{m}{2} x_1^2 x_2^{(m-2)/2} + \sum_{\substack{d \\ d|m}} \phi(d) x_d^{m/d} \right], \qquad \text{for } m \text{ even.}$$

The Symmetric Group

The cycle index of the $m!$ permutations on m elements (which form the symmetric group on the m elements) is

$$\frac{1}{m!}\left[\sum \frac{m!}{(1!)^{\lambda_1} \cdots (m!)^{\lambda_m} \lambda_1! \cdots \lambda_m!} x_1^{\lambda_1} x_2^{\lambda_2} \cdots x_m^{\lambda_m} \right],$$

where the sum extends over all vectors $(\lambda_1, \dots, \lambda_m)$ that satisfy $1\lambda_1 + 2\lambda_2 + \cdots + m\lambda_m = m$.

[Indeed, we counted on several occasions the number of permutations of cycle type $1^{\lambda_1} 2^{\lambda_2} \cdots m^{\lambda_m}$ (see, e.g., Section 1.8(a)).]

The Alternating Group

As the reader recalls, a cycle of even length is an odd permutation. And hence, a permutation of cycle type $1^{\lambda_1} 2^{\lambda_2} \cdots m^{\lambda_m}$ is even if and only if it has an even number of cycles of even length (or, written in symbols, if and only if $\lambda_2 + \lambda_4 + \lambda_6 + \cdots$ is even).

The alternating group consists of all even permutations on m symbols; its order is $m!/2$. By the remark we just made its cycle index is

$$\frac{m!}{2} \sum \frac{m!}{(1!)^{\lambda_1} \cdots (m!)^{\lambda_m} \lambda_1! \cdots \lambda_m!} x_1^{\lambda_1} \cdots x_m^{\lambda_m},$$

where the sum is taken over all vectors $(\lambda_1, \dots, \lambda_m)$ with $\lambda_2 + \lambda_4 + \lambda_6 + \cdots$ being *even*, and $\sum_{i=1}^{m} i\lambda_i = m$.

The Direct Sum Acting on the Disjoint Union of Sets

We let G and H be (abstract) groups and form their direct sum $G \oplus H$. If G acts on the set X and H acts on the set Y (with X and Y *disjoint*), then we have a natural action of $G \oplus H$ on $X \cup Y$, namely

$$(gh)(z) = \begin{cases} g(z) & \text{if } z \in X \\ h(z) & \text{if } z \in Y. \end{cases}$$

A permutation gh so represented on $X \cup Y$ consists therefore of the permutation of g on X followed by the permutation of h on Y. When translated into cycle indexes this simply says that

$$P_{G \oplus H} = P_G \times P_H.$$

[Or, in words, the cycle index of a direct sum acting on the *union* of two disjoint sets equals the product of the respective cycle indexes.]

The Direct Sum Acting on the Cartesian Product

If G acts on X and H on Y, we can define an action of $G \oplus H$ on $X \times Y$ by letting

$$(gh)(x, y) = (g(x), h(y)).$$

For r and s natural numbers denote by $(r; s)$ their greatest common divisor and by $[r; s]$ their least common multiple. The cycle index of $G \oplus H$ acting in this fashion on $X \times Y$ can now be written as follows:

$$P_{G \oplus H} = \frac{1}{|G|\,|H|} \sum_{g,h} \sum_{r,s} x_{[r;\, s]}^{(r;\, s)\lambda_r(g)\lambda_s(h)}.$$

[To understand this formula, suppose x belongs to a cycle α of G on X, and that y belongs to a cycle β of H on Y. Let r and s denote the lengths of α and β, respectively. Then (x, y) belongs to a cycle of gh of length $[r; s]$. The cycles α and β induce $(r; s)$ cycles of gh on $X \times Y$. Since there are $\lambda_r(g)$ cycles of length r of g and $\lambda_s(h)$ cycles of length s of h, the formula is proved. We suggest that the reader verify this on a small example.]

The Wreath Product

Visualize five cubes as vertices of a regular pentagon. Generate permutations on the $5 \cdot 6 = 30$ positions by rotating each cube individually and rotating the whole pentagon as well. The group of permutations so obtained is called the wreath product of the group of rotations of the pentagon with the group of rotations of the cube. (The five cubes circularly placed form a configuration not unlike a wreath, and this suggests the general nomenclature.) What we do in general the reader is invited to specialize to the case described above.

Let S and T be sets, and let G and H be groups acting on S and T, respectively. (The set S generalizes the pentagon and T is the cube.) We consider the Cartesian product $S \times T$ and construct special permutations on $S \times T$ as follows: Choose g in G, and to each s in S associate an element h_s of H. These elements determine a permutation on $S \times T$ defined by

$$(s, t) \to (g(s), h_s(t)).$$

There are $|G|\,|H|^{|S|}$ such permutations; they form a group we call the *wreath product* of G with H, denoted by $G[H]$. Our objective is to express the cycle index of $G[H]$ in terms of those of G and H. Pólya proved the following result:

$*$ *The cycle index of $G[H]$ is*

$$P_{G[H]}(x_1, x_2, \ldots) = P_G(P_H(x_1, x_2, x_3, \ldots), P_H(x_2, x_4, x_6, \ldots), \ldots),$$

where the right-hand side is obtained by substituting

$$P_H(x_k, x_{2k}, x_{3k}, \ldots) \text{ for } y_k \text{ in } P_G(y_1, y_2, y_3, \ldots).$$

Proof (Following DeBruijn). Denote by m the cardinality of S. Fix an element γ of $G[H]$ by selecting g in G and $h_1, h_2, \ldots, h_m$ in H. We want to determine the cycle type of γ. Let $s_1, \ldots, s_k$ be a cycle of g (in S). We thus have

$$g(s_1) = s_2, g^2(s_1) = s_3, \ldots, g^{k-1}(s_1) = s_k, g^k(s_1) = s_1.$$

It follows that the set of pairs $\{(s_i, t): i = 1, 2, \ldots, k \text{ and } t \in T\}$ is mapped onto itself by γ. We call this set a *block*, and want to find out in what cycles this block splits under the influence of γ. It turns out that this only depends on the product

$$h = h_{s_k} h_{s_{k-1}} \cdots h_{s_2} h_{s_1}$$

in the following way: If h has cycle structure $1^{\lambda_1}2^{\lambda_2}3^{\lambda_3} \cdots$, then the block splits into λ_1 cycles of length k, λ_2 cycles of length $2k$, and so on. This can be seen by writing out what the successive applications of γ do to an element (s_1, t):

$$(s_1, t) \to (s_2, h_{s_1}(t)) \to (s_3, h_{s_2}(h_{s_1}(t))) \to \cdots \to (s_1, h(t)).$$

Now, if l is the order of h, then we close the cycle of (s_1, t) upon kl applications of γ. It follows that (s_1, t) generates a cycle of length kl. Noting that (s_1, t) and (s_1, t') generate the same cycle under γ if and only if t and t' generate the same cycle in T under h, we conclude that our block contains λ_l cycles of length kl, and this holds for each l.

We thus attach the monomial $|H|^{-k}x_k^{\lambda_1}x_{2k}^{\lambda_2}x_{3k}^{\lambda_3} \cdots$ to this block. With g and $s_1, \ldots, s_k$ still fixed, summing over all possible choices for $h_{s_1}, \ldots, h_{s_k}$ we obtain $P_H(x_k, x_{2k}, x_{3k}, \ldots)$. This follows from the fact that if $h_{s_1}, \ldots, h_{s_k}$ all run through H, then their product h runs $|H|^{k-1}$ times through H.

Next we consider all elements γ of $G[H]$ arising from a single g of type $1^{\mu_1}2^{\mu_2}3^{\mu_3} \cdots$. By what we have done it follows that the contribution of these elements to the cycle index $P_{G[H]}$ is

$$[P_H(x_1, x_2, \ldots)]^{\mu_1}[P_H(x_2, x_4, \ldots)]^{\mu_2}[P_H(x_3, x_6, \ldots)]^{\mu_3} \cdots.$$

Summing up over all g in G yields the result.

We conclude this section with an example. Reminiscing once again about the wonderful childhood games with cubes, one cannot fail to observe that rolling one cube may be fun for some, but (as every self-respecting infant knows) it is the multitude of colored cubes that offers true excitement. Let us then take eight cubes, arrange them circularly to form a regular octagon, and use two colors to color their faces. Two colorations are said to form the same pattern if one is obtained from the other upon *rotating the individual cubes and cyclicly rotating all eight cubes* around the shape of the octagon. We ask for the number of patterns.

In this case the domain consists of the $8 \cdot 6 = 48$ faces of the eight cubes, the range consists of two colors (red and blue, say), and the group that induces equivalence is the wreath product $G[H]$, where G is the rotation group of the octagon (of order 8) and H is the familiar rotation group of the cube (of order 24).

Step 1. We just finished investigating the cycle index of $G[H]$. It is

$$P_{G[H]}(x_1, x_2, \ldots) = P_G(P_H(x_1, x_2, \ldots), P_H(x_2, x_4, \ldots), \ldots).$$

In this case

$$\begin{aligned} P_G(y_1, \ldots, y_8) &= \tfrac{1}{8}[\phi(1)y_1^8 + \phi(2)y_2^4 + \phi(4)y_4^2 + \phi(8)y_8] \\ &= \tfrac{1}{8}[y_1^8 + y_2^4 + 2y_4^2 + 4y_8], \end{aligned}$$

where $\phi(d)$ signifies (as usual) the number of positive integers less than and relatively prime to d. From Section 6.13 the cycle index of H is

$$P_H(x_1, \ldots, x_6) = \tfrac{1}{24}[x_1^6 + 3x_1^2x_2^2 + 6x_1^2x_4 + 6x_2^3 + 8x_3^2].$$

Thus

$$\begin{aligned} P_{G[H]} = P_G(&P_H(x_1, x_2, \ldots, x_6), \\ &P_H(x_2, x_4, \ldots, x_{12}), \ldots, P_H(x_8, x_{16}, \ldots, x_{48})). \end{aligned}$$

Step 2. The answer to our problem is $P_{G[H]}(2, 2, \ldots, 2)$. In Section 6.13 we found that $P_H(2, \ldots, 2) = 10$, and hence

$$\begin{aligned} P_{G[H]}(2, 2, \ldots, 2) &= P_G(10, 10, \ldots, 10) \\ &= \tfrac{1}{8}(10^8 + 10^4 + 2 \cdot 10^2 + 4 \cdot 10). \end{aligned}$$

This last (large) figure counts the total number of patterns.

Step 3. We leave it to anyone curious to decide how many patterns with 20 red and 28 blue faces are possible. To do this, an explicit computation of $P_{G[H]}$ is necessary. This is routine but time consuming.

EXERCISES

1. Give an example of two nonisomorphic groups that have the same cycle index.

2. Verify that the cycle index of the symmetric group S_n (in its action on n points) is the coefficient of z^n in the power series

$$\exp\left(zx_1 + \frac{z^2 x_2}{2} + \frac{z^3 x_3}{3} + \cdots \right).$$

3. Equip yourself with n colors and a hollow cube with thin walls.

(a) Represent the rotation group of the cube on the 12 edges of the cube and compute its cycle index.

(b) Represent the rotation group of the cube on the 12 faces (the 6 on the outside plus the 6 on the inside) and compute its cycle index.

(c) Are the representations in (a) and (b) isomorphic?

(d) Do we generate more patterns by coloring the 12 edges or by coloring the 12 (inside and outside) faces with the n available colors?

(e) How many patterns are there with two faces of one color, three faces of another, and seven faces of a third color? Answer the same when the n colors are used to color the edges of the cube.

4. Represent the group of rotations of the octahedron on its vertices, its edges, and then its faces. Derive the cycle index in each case. Then take n colors and find out how many patterns arise in each case. (Compare the results to those in Section 6.13 and explain the "duality.")

5. (a) In the usual setting of Pólya's theorem denote by p the number of patterns that arise under the action of the group G on the domain. Let H be a subgroup of G and denote by q the number of patterns that are induced by H. Show that $q \leq p$.

(b) As an example, examine the case of necklaces with m beads of two colors. Let G be the dihedral group D_{2m} and let H be its cyclic subgroup of M rotations. For what values of m do we have $p = q$?

6. Color the ten vertices of Petersen's graph red, white, and blue. Call two colorations the same if one is obtained from the other upon a permutation of vertices that preserves the edges. How many essentially distinct colorations (i.e., patterns) are possible? How many with two red, three white, and five blue vertices? Petersen's graph is displayed in Section 4.11.

7. The group S_4 of the 24 permutations on 4 vertices induces a natural action on the 6 available edges. Find the cycle index of S_4 in its action on the 6 edges.

8. Let $X = \{1,2,3,4,5,6\}$ and $Y = \{7,8,9\}$ be disjoint sets. Let G be the cyclic group $Z_6 = \langle(1\,2\,3\,4\,5\,6)\rangle$ acting on X, and let H be the symmetric group S_3 acting on Y. Compute the cycle index of $G \oplus H$ in its action on $X \times Y$.

5 MORE THEORY

In the last three sections the focus was on special instances and on many examples connected to Pólya's theorem. We chose that course of exposition mostly for pedagogical reasons: It gave the reader the opportunity to develop a "feel" for what the theorem offers and to practice a little with the cycle index of a group. The main theme of this section is DeBruijn's generalization of Pólya's theorem.

6.17

We look once again at the set $F(D, R)$ of all functions from the domain D to the range R (of colors). A function f in $F(D, R)$ is commonly viewed as a coloration of D, in the sense that it colors element d of D with color $f(d)$. There are two groups present: a group G acting on the elements of D and a group H that permutes the colors of the range R.

The objects of our study are the two induced group actions on the set $F(D, R)$ of functions. Firstly, we let the group G act on $F(D, R)$ by setting $g(f) = fg^{-1}$, with g in G and f in $F(D, R)$. The orbits so induced on $F(D, R)$ are called *G-patterns*. We denote the G-pattern of f by fG.

Secondly, we look at the action of the direct sum $G \oplus H$ on $F(D, R)$ and define $(gh)(f) = hfg^{-1}$, for g in G, h in H, and f in $F(D, R)$. The orbits of this latter action are simply called *patterns* (and we carefully distinguish them from the G-patterns defined above). We denote by HfG the pattern of f.

In addition to these two actions we let H act on the G-patterns by defining $h(fG) = (hf)G$. This is indeed well-defined and indeed an action. Denote the orbit of the G-pattern fG under the action of H by $(Hf)G$. It is important to realize that *the orbits of H on the G-patterns coincide with the patterns that $G \oplus H$ induced on $F(D, R)$ by the action $(gh)(f) = hfg^{-1}$ described above.* [This follows from the law of associativity which composition of functions obeys: $(Hf)G = HfG = H(fG)$.]

6.18

Let w be a *weight function* from R into the set of polynomials with rational coefficients in indeterminates $y_1, y_2, y_3, \ldots$; denote by $w(r)$ the *weight* of color r. We allow w to induce a weight on the functions in $F(D, R)$ by defining

$$W(f) = \prod_{d \in D} w(f(d)). \tag{6.12}$$

[The weight function W is in fact well defined on G-patterns; for if f_1 and f_2 display the same G-pattern, then $f_2 = f_1 g^{-1}$, for some g in G, and $W(f_2) = W(f_1)$ follows from the fact that g is a bijection on D. Define therefore $W(fG) = W(f)$.]

We now fix a permutation h in H and ask: How many G-patterns remain invariant under h? In other words, how many G-patterns fG have the property that $hfG = fG$? (The condition $hfG = fG$ we occasionally express in its equivalent form $hf \in fG$.) We devote Sections 6.19 through 6.21 to answering a slightly more general question. Specifically, *we propose to compute*

$$\sum_S W(S), \tag{6.13}$$

where the sum extends over all G-patterns S that are invariant under h. The weight function W is of the form described in (6.12).

6.19

Let G_f signify the subgroup $\{g \in G : fg^{-1} = f\}$, which we call the *stabilizer* of f in G. We can then write:

$$\begin{aligned}
\sum_{\substack{S\\ hS=S}} W(S) &= \sum_{\substack{f\\ hfG=fG}} \frac{W(f)}{|fG|} = \sum_{\substack{f\\ hfG=fG}} W(f)\frac{|G_f|}{|G|} \\
&= \frac{1}{|G|}\sum_{\substack{f\\ hfG=fG}} W(f)|G_f| = \frac{1}{|G|}\sum_{\substack{(g,f)\\ fg^{-1}=f\\ hfG=fG}} W(f) \\
&= \frac{1}{|G|}\sum_{g\in G}\sum_{\substack{f\\ fg^{-1}=f\\ hfG=fG}} W(f) = \frac{1}{|G|}\sum_{g\in G}\sum_{\substack{f\\ hfg^{-1}=f}} W(f).
\end{aligned}$$

Let us explain in some detail the chain of equalities written above. The first follows by recalling that the weight of the G-pattern fG equals that of any one of the $|fG|$ functions that it contains. Recalling that $|G_f|\,|Gf| = |G|$ [i.e., that the length of an orbit equals the index of the stabilizer—see (6.1)] explains the second. The third is clear. To understand the fourth and fifth equality signs look at the expression in the middle,

$$\frac{1}{|G|}\sum W(f),$$

where the sum is over the set of pairs

$$\{(g,f) : fg^{-1} = f \quad \text{and} \quad hfG = fG\}.$$

First, by fixing f and summing over g one obtains $|G_f|$, and the sum over f explains the fourth sign of equality. By fixing g and summing over f first, one obtains

$$\sum_{\substack{f \\ fg^{-1}=f \\ hfG=fG}} W(f),$$

and then letting g run through G explains the fifth equality sign. Lastly, the last sign of equality is explained by observing that

$$fg^{-1} = f = hfg_0^{-1},$$

and that g runs over G if and only if g_0 does.

For a fixed permutation h of colors we therefore proved that

$$\sum_{\substack{S \\ h(S)=S}} W(S) = \frac{1}{|G|} \sum_{g \in G} \sum_{\substack{f \\ hfg^{-1}=f}} W(f). \tag{6.14}$$

6.20

Formula (6.14) suggests fixing g in G (in addition to h in H, which remains fixed) and then examining the weight $W(f)$ of a function f that satisfies $hfg^{-1} = f$. The weight function W has the form as written in (6.12). This is what we in fact do in this section.

Fix therefore g in G, and let $g = c_k \cdots c_2 c_1$ denote the decomposition of g into disjoint cycles in its action on the domain D. Write $|c_j|$ for the length of the jth cycle. Select an element in the jth cycle c_j of g; call it d_j. The list of the $|c_j|$ elements in the cycle c_j is

$$d_j, gd_j, g^2d_j, \ldots, g^{|c_j|-1}d_j \tag{6.15}$$

[with g^2d_j an abbreviation for $g(g(d_j))$, and so on].

Assume now that f is fixed by gh and thus satisfies $f = (gh)(f) = hfg^{-1}$, or $fg = hf$. This iteratively implies $fg^s = fgg^{s-1} = hfg^{s-1} = hfgg^{s-2} = h^2fg^{s-2} = \cdots = h^sf$, for any power g^s of g.

Applying f to the list of elements in (6.15) we therefore obtain

$$fd_j, hfd_j, h^2fd_j, \ldots, h^{|c_j|-1}fd_j, \tag{6.16}$$

[with h^2fd_j signifying $h(h(f(d_j)))$, for notational simplicity]. [We wish to inform that, while the list in (6.15) consists of $|c_j|$ distinct elements, the list in (6.16) need not consist of $|c_j|$ distinct elements. In general it contains l distinct elements (say), where l is the length of the cycle generated by fd_j under the influence of the group element h. Observe in addition that l divides $|c_j|$, since $h^{|c_j|}fd_j = fg^{|c_j|}d_j = fd_j(= h^lfd_j)$.]

As with regard to the weight $W(f)$ of a function f that satisfies $fg = hf$, taking into account the information obtained so far, we may write

$$W(f) = \prod_{d \in D} w(fd) = \prod_{j=1}^{k} w(fd_j)w(fgd_j) \cdots w\left(fg^{|c_j|-1}d_j\right)$$

$$= \prod_{j=1}^{k} w(fd_j)w(hfd_j) \cdots w\left(h^{|c_j|-1}fd_j\right).$$

Set

$$w(r)w(hr)w(h^2r) \cdots (h^{s-1}r) = p_s(r). \tag{6.17}$$

[The product $p_s(r)$ depends on h, of course, but we omit h as an index for notational simplicity.] The above expression for $W(f)$ takes now the simpler form

$$W(f) = \prod_{j=1}^{k} p_{|c_j|}(r_j), \tag{6.18}$$

where r_j stands for fd_j, $j = 1, 2, \ldots, k$.

6.21

As formula (6.14) demands, our next step is to compute $\Sigma_f W(f)$, where the sum is taken over all functions f that satisfy $(gh)(f) = f$. We still keep g and h fixed.

By selecting an element d_j from cycle c_j of g ($j = 1, 2, \ldots, k$), and denoting $f(d_j)$ by r_j, we completely specify the function f on D. (For let d be an arbitrary element of the domain. Then d is in some cycle c_j of g and thus $d = g^s d_j$, for some power s of g. Consequently $fd = fg^s d_j = h^s fd_j = h^s r_j$.) There is one and only one restriction on the image r_j of d_j under f: If d_j belongs to the cycle c_j, then r_j should satisfy $h^{|c_j|}r_j = r_j$ (in other words r_j should belong to a cycle of h whose length divides that of c_j). This motivates defining

$$R_t = \{r \in R : h^t r = r\}.$$

(The set R_t depends on h, of course, but we omit h as an index for notational simplicity. Observe that the R_t's are *not disjoint* in general. They do cover, however, the whole range R, $t = 0, 1, 2, \ldots$.)

We intend to sum over all f that satisfy $(gh)(f) = f$. It will hence be convenient to fix elements d_1 in c_1, d_2 in $c_2, \ldots, d_k$ in c_k, and write r_j^f for $f(d_j)$, $j = 1, 2, \ldots, k$. The geometrically inclined may wish to think of r_j^f as the "projection" of f on the subset $R_{|c_j|}$ of R. We thus have a bijection

$$f \leftrightarrow \left(r_1^f, r_2^f, \ldots, r_k^f\right),$$

with r_j^f belonging to $R_{|c_j|}$.

Equipped conceptually in such a fashion, and letting f run over all functions that satisfy $(gh)(f) = f$, we now write

$$\sum_f W(f) = \{\text{by (6.18)}\} = \sum_f \prod_{j=1}^{k} p_{|c_j|}(r_j^f)$$

$$= \left[\begin{array}{l}\text{obtain the left side by picking a term in each factor at}\\ \text{right and summing}\end{array}\right]$$

$$= \left(\sum_f p_{|c_1|}(r_1^f)\right)\left(\sum_f p_{|c_2|}(r_2^f)\right)\cdots\left(\sum_f p_{|c_k|}(r_k^f)\right)$$

$$= \prod_{i=1}^{k}\left(\sum_f p_{|c_j|}(r_j^f)\right) = \{\text{by (6.17)}\}$$

$$= \prod_{i=1}^{k} \sum_{r \in R_{|c_j|}} w(r)w(hr)\cdots w(h^{|c_j|-1}r)$$

$$= [\text{upon denoting the above sum over } r \text{ in } R_{|c_j|} \text{ by } p_{|c_j|}(h)]$$

$$= \prod_{i=1}^{k} p_{|c_j|}(h) = \{\text{sorting by the cycle length}\}$$

$$= (p_1(h))^{\lambda_1(g)}(p_2(h))^{\lambda_2(g)}\cdots(p_m(h))^{\lambda_m(g)},$$

where $\lambda_j(g)$ denotes the number of cycles of length j of the group element g, $j = 1, 2, \ldots, m$ (with $m = |D|$).

We proved that

$$\sum_{\substack{f\\ hfg^{-1}=f}} W(f) = (p_1(h))^{\lambda_1(g)}(p_2(h))^{\lambda_2(g)}\cdots(p_m(h))^{\lambda_m(g)}. \quad (6.19)$$

6.22

In view of what (6.19) displays, formula (6.14) becomes

$$\sum_{\substack{S\\ h(S)=S}} W(S) = \frac{1}{|G|}\sum_{g\in G}\prod_{i=1}^{m}(p_i(h))^{\lambda_i(g)}$$

$$= P_G(p_1(h), p_2(h), \ldots, p_m(h)),$$

with P_G the cycle index of G in its action on D. We summarize:

DeBruijn's Result. *The sum of the weights of the G-patterns fixed by the permutation h of colors is*

$$\sum_{\substack{S\\ h(S)=S}} W(S) = P_G(p_1(h), p_2(h), \ldots, p_m(h)) \quad (6.20)$$

with $p_s(h) = \sum_{r,\, h^s r = r} w(r)w(hr)\cdots w(h^{s-1}r)$.

Observe that by taking $h = 1$ (the identity element of H) we have

$$p_s(1) = \sum_{r \in R} w(r)w(r) \cdots w(r) = \sum_{r \in R} w(r)^s.$$

Since the identity fixes all the G-patterns, the result (6.20) yields in this case the well-known result of Pólya:

Pólya's Theorem

$$\sum_{\substack{T \\ (T\ G\text{-pattern})}} W(T) = P_G\left(\sum_r w(r), \sum_r (w(r))^2, \ldots, \sum_r (w(r))^m\right).$$

6.23

Our next objective is to derive a formula for the weight $\Sigma W(T)$ of all the patterns T induced by the action $(gh)(f) = hfg^{-1}$ of $G \oplus H$ on $F(D, R)$. A first requirement is that the weight function W [defined on $F(D, R)$] should be well defined on patterns. Along these lines we remark that weight functions of the form (6.12) *are not well defined on patterns*, in general. [But there are weight functions of this form that are well defined. An important instance is obtained by taking $w(r) = 1$, for all r in R. Then the induced weight function W is well defined on patterns (specifically, $W(T) = 1$ for any pattern T). With this W the sum $\Sigma_T W(T)$ simply equals the total number of patterns.]

Assume in what follows that W is a weight function of the form (6.12) and that *W is well defined on patterns.*

In Section 6.17 we informed the reader that H acts on the G-patterns, and that the orbits induced by H coincide with the patterns of $G \oplus H$ on $F(D, R)$. At this time we take advantage of this point of view and thus *visualize the patterns as orbits of H in its action $h(fG) = (hf)G$ on the G-patterns.* For notational simplicity let us denote by $\bar{f}$ the G-pattern of f (normally written as fG). Then

$$\sum_{\substack{T \\ (T\ \text{pattern})}} W(T) = \sum_{\bar{f}} \frac{W(\bar{f})}{|H\bar{f}|} = \sum_{\bar{f}} W(\bar{f}) \frac{|H_{\bar{f}}|}{|H|}$$

$$= \frac{1}{|H|} \sum_{\bar{f}} W(\bar{f})|H_{\bar{f}}| = \frac{1}{|H|} \sum_{\substack{(h, \bar{f}) \\ h\bar{f} = \bar{f}}} W(\bar{f})$$

$$= \frac{1}{|H|} \sum_{h \in H} \sum_{\substack{\bar{f} \\ h\bar{f} = \bar{f}}} W(\bar{f}) = \{\text{by } 6.20\}$$

$$= \frac{1}{|H|} \sum_{h \in H} P_G(p_1(h), p_2(h), \ldots, p_m(h)).$$

In this chain of equalities $H\bar{f}$ denotes the orbit under H of the G-pattern $\bar{f}$, and $H_{\bar{f}}$ is the stabilizer of $\bar{f}$, that is, $H_{\bar{f}} = \{h \in H : h\bar{f} = \bar{f}\}$, a subgroup of H.

This result is attributed to deBruijn as well. Let us formally state:

DeBruijn's Result. *The sum of weights of the patterns is*

$$\sum_{\substack{T \\ (T\text{-pattern})}} W(T) = \frac{1}{|H|} \sum_{h \in H} P_G(p_1(h), p_2(h), \ldots, p_m(h)) \qquad (6.21)$$

with $p_s(h) = \sum_{r,\, h^s r = r} w(r)w(hr) \cdots w(h^{s-1}r)$ (assuming that W is induced by w and that W is well defined on patterns.)

6 RECIPE FOR DEBRUIJN'S RESULT

6.24

We go back once again to Section 6.1 and examine what information DeBruijn's results give in that simple situation. As the reader recalls, the problem is that of coloring (with colors A and B) the vertices of a square, labeled $\begin{smallmatrix}1 & 2\\ 3 & 4\end{smallmatrix}$.

In this example the group G (acting on the domain) consists of the four planar rotations of the square, and H is the group of order 2, which either fixes each of the two colors or interchanges them. There are six G-patterns that emerge,

$$\begin{matrix}A & A\\ A & A\end{matrix}, \quad \begin{matrix}B & B\\ B & B\end{matrix}, \quad \begin{matrix}A & B\\ B & B\end{matrix}, \quad \begin{matrix}B & A\\ A & A\end{matrix}, \quad \begin{matrix}A & A\\ B & B\end{matrix}, \quad \begin{matrix}A & B\\ B & A\end{matrix}, \qquad (6.22)$$

and four patterns,

$$\begin{matrix}A & A\\ A & A\end{matrix}, \quad \begin{matrix}A & B\\ B & B\end{matrix}, \quad \begin{matrix}A & A\\ B & B\end{matrix}, \quad \begin{matrix}A & B\\ B & A\end{matrix}. \qquad (6.23)$$

We see, in particular, that among the six G-patterns two (the last two) are left invariant when colors A and B are interchanged.

It is possible to reach these conclusions by way of DeBruijn's results. The cycle index of the rotation group of the square is

$$P_G = \tfrac{1}{4}\left(x_1^4 + x_2^2 + 2x_4\right).$$

In this case the group permuting the colors is $H = \{(B)(A), (BA)\}$. For notational simplicity we write 1 (the identity) for $(B)(A)$ and denote (BA) by h, so that $H = \{1, h\}$.

We assign weight y_1 to A and weight y_2 to B, that is, $w(A) = y_1$ and $w(B) = y_2$. Then

$$p_1(1) = \sum_{\substack{r \\ 1r=r}} w(r) = w(A) + w(B) = y_1 + y_2,$$

$$p_2(1) = \sum_{\substack{r \\ 1^2r=r}} w(r)w(1r) = w(A)^2 + w(B)^2 = y_1^2 + y_2^2,$$

$$p_3(1) = \sum_{\substack{r \\ 1^3r=r}} w(r)w(1r)w(1^2r) = w(A)^3 + w(B)^3 = y_1^3 + y_2^3,$$

$$p_4(1) = \sum_{\substack{r \\ 1^4r=r}} w(r)w(1r)w(1^2r)w(1^3r) = w(A)^4 + w(B)^4$$

$$= y_1^4 + y_2^4,$$

and

$$p_1(h) = \sum_{\substack{r \\ hr=r}} w(r) = 0 \qquad \text{(since } h \text{ fixes neither } A \text{ nor } B)$$

$$p_2(h) = \sum_{\substack{r \\ h^2r=r}} w(r)w(hr) = w(A)w(hA) + w(B)w(hB)$$

$$= w(A)w(B) + w(B)w(A) = 2y_1y_2$$

$$\text{(since } h(A) = B \text{ and } h(B) = A)$$

$$p_3(h) = \sum_{\substack{r \\ h^3r=r}} w(r)w(hr)w(h^2r) = 0$$

$$\text{(since } h^3 = h \text{ and it does not fix anything)}$$

$$p_4(h) = \sum_{\substack{r \\ h^4r=r}} w(r)w(hr)w(h^2r)w(h^3r)$$

$$= w(A)w(B)w(A)w(B) + w(B)w(A)w(B)w(A)$$

$$= 2y_1^2y_2^2.$$

A G-pattern consisting of a_1 A's and a_2 B's ($a_1 + a_2 = 4$) has weight $y_1^{a_1}y_2^{a_2}$. Formula (6.20) informs us, therefore, that the number of G-patterns consisting of a_1 A's and a_2 B's that are invariant under h (the permutation that interchanges colors A and B) equals the coefficient of $y_1^{a_1}y_2^{a_2}$ in $P_G(p_1(h), \ldots, p_4(h))$. In particular, the number of G-patterns with two A's and two B's invariant under h equals the coefficient of $y_1^2y_2^2$ in

$$\begin{aligned} P_G(p_1(h),\ldots,p_4(h)) &= P_G(0, 2y_1y_2, 0, 2y_1^2y_2^2) \\ &= \tfrac{1}{4}\left(0^4 + (2y_1y_2)^2 + 2(2y_1^2y_2^2)\right) = 2y_1^2y_2^2. \end{aligned}$$

The answer is 2, in agreement with what (6.22) displays.

Formula (6.20) tells us also that the total number of G-patterns invariant under h can be obtained by setting $w(A) = y_1 = 1$, $w(B) = y_2 = 1$ and then computing $P_G(p_1(h),\ldots,p_4(h))$. We obtain

$$\begin{aligned} P_G(p_1(h),\ldots,p_4(h)) &= P_G(0, 2y_1y_2, 0, 2y_1^2y_2^2) \\ &= 2y_1^2y_2^2 = 2\cdot 1^2\cdot 1^2 = 2, \end{aligned}$$

which compares favorably to display (6.22).

Finally, according to (6.21), the total number of patterns is obtained by taking $w(A) = y_1 = 1$, $w(B) = y_2 = 1$ and then computing

$$\frac{1}{|H|}\sum_{h\in H} P_G(p_1(h),\ldots,p_m(h)).$$

(Observe that the weight function involved is well defined on patterns.)

In this case we obtain

$$\begin{aligned} &\tfrac{1}{2}P_G(p_1(1),\ldots,p_4(1)) + \tfrac{1}{2}P_G(p_1(h),\ldots,p_4(h)) \\ &\quad= \tfrac{1}{2}P_G(2,2,2,2) + \tfrac{1}{2}P_G(0,2,0,2) \\ &\quad= \tfrac{1}{2}\cdot\tfrac{1}{4}(2^4 + 2^2 + 2\cdot 2) + \tfrac{1}{2}\cdot\tfrac{1}{4}(0^4 + 2^2 + 2\cdot 2) = 4, \end{aligned}$$

consistent with (6.23).

6.25 The Recipe

We have before us a domain D of m elements, a range R of colors $r_1, r_2, \ldots, r_n$ and two groups, G and H. The group G acts on D, and H acts on R. A function f from D to R is called a coloration of D [it colors element d of D with color $f(d)$]. The set of all colorations is denoted by $F(D, R)$.

We let G act on $F(D, R)$ by defining $g(f) = fg^{-1}$, with g in G and f in $F(D, R)$; the orbits of this action are called *G-patterns*.

We let $G \oplus H$ act on $F(D, R)$ by defining $(gh)(f) = hfg^{-1}$; with g in G, h in H, and f in $F(D, R)$. The orbits of this action are called *patterns*. (A pattern is in fact a union of G-patterns, as explained in Section 6.17.)

Our interest lies with investigating the patterns and G-patterns. A fair number of questions commonly asked can be answered by following the steps outlined below. [It is not unlikely that answers to many others lie with clever choices of weight functions. Whenever the recipe falls short of the readers' computational objectives we suggest (as a first recourse) a readjustment of the weight function.]

Step 1. Find the cycle index P_G of the group G in its action on the domain D.

[Recall that

$$P_G(x_1,\ldots,x_m) = \frac{1}{|G|}\sum_{g\in G} x_1^{\lambda_1(g)}x_2^{\lambda_2(g)}\cdots x_m^{\lambda_m(g)},$$

where $\lambda_j(g)$ denotes the number of cycles of length j of the group element g in its representation on D.]

Step 2. Assign weight $w(r_i)$ to color r_i $(i = 1,2,\ldots,n)$, and denote (when convenient) $w(r_i)$ by y_i, with y_i an indeterminate.

Step 3. For a permutation of interest h in H, and each s $(s = 1,2,\ldots,m)$, find the set $S(h,s) = \{r \in R : h^s r = r\}$. Compute

$$p_s(h) = \sum_{\substack{r \\ r\in S(h,s)}} w(r)w(hr)w(h^2r)\cdots w(h^{s-1}r).$$

If $S(h,s)$ is the empty set, then define $p_s(h)$ to be 0.

Step 4. The number of G-patterns with

a_1 occurrences of color r_1
a_2 occurrences of color r_2
$\vdots$
a_n occurrences of color r_n $\quad(\sum_{i=1}^{n} a_i = m)$

that are invariant under a permutation h of colors equals the coefficient of $y_1^{a_1}y_2^{a_2}\cdots y_n^{a_n}$ in the polynomial

$$P_G(p_1(h), p_2(h),\ldots, p_m(h)),$$

where the indeterminate y_i stands for $w(r_i)$, $i = 1,2,\ldots,n$.

Step 5. The total number of G-patterns that are invariant under a permutation h of colors is obtained by setting $w(r_i) = y_i = 1$ (for all i) in the polynomial

$$P_G(p_1(h), p_2(h),\ldots, p_m(h)).$$

Step 6. The total number of patterns is obtained by setting $w(r_i) = y_i = 1$ (for all i) in the polynomial

$$\frac{1}{|H|}\sum_{h\in H} P_G(p_1(h), p_2(h),\ldots, p_m(h)).$$

[This step requires computation of $p_s(h)$, for all h in H—often quite a lengthy affair.]

End of recipe.

7 EXAMPLES FOLLOWING THE RECIPE

6.26 Coloring the Faces of a Cube

We have six colors $r_1, r_2, \ldots, r_6$ and take interest in coloring the six faces of a cube. Two colorations are equivalent if they differ by a rotation of the cube. We take special interest in the colorations that remain invariant under the permutation $h = (r_6 r_5)(r_4 r_3)(r_2 r_1)$ of colors.

Step 1. The cycle index of the rotation group G of the cube in its action on the six faces is [cf. Section 6.13]

$$P_G = \tfrac{1}{24}\left(x_1^6 + 3x_1^2 x_2^2 + 6x_1^2 x_4 + 6x_2^3 + 8x_3^2\right).$$

Step 2. Let $w(r_i) = y_i$, $i = 1, 2, \ldots, 6$.

Step 3. We take interest in $h = (r_6 r_5)(r_4 r_3)(r_2 r_1)$. For this permutation of colors we have

$$\begin{aligned}
p_1(h) &= 0\\
p_2(h) &= 2y_1 y_2 + 2y_3 y_4 + 2y_5 y_6\\
p_3(h) &= 0\\
p_4(h) &= 2y_1^2 y_2^2 + 2y_3^2 y_4^2 + 2y_5^2 y_6^2\\
p_5(h) &= 0\\
p_6(h) &= 2y_1^3 y_2^3 + 2y_3^3 y_4^3 + 2y_5^3 y_6^3.
\end{aligned}$$

Step 4. The number of G-patterns with

Two occurrences of color r_1
Two occurrences of color r_2
One occurrence of color r_5
One occurrence of color r_6

that are invariant with respect to h is the coefficient of $y_1^2 y_2^2 y_5 y_6$ in the polynomial

$$\begin{aligned}
P_G(p_1(h), \ldots, p_6(h)) &= \tfrac{1}{24} 6(p_2(h))^3\\
&= \tfrac{1}{4}(2y_1 y_2 + 2y_3 y_4 + 2y_5 y_6)^3.
\end{aligned}$$

This coefficient is $\frac{1}{4}(3!/2!1!)2^2 \cdot 2 = 6$, allowing us to conclude that there are six such colorations.

Step 5. The total number of colorations invariant under h is $P_G(p_1(h), \ldots, p_6(h))$, when $w(r_i) = y_i = 1$, for all i. We obtain

$$\begin{aligned}
P_G(p_1(h), \ldots, p_6(h)) &= \tfrac{1}{24} 6(p_2(h))^3\\
&= \tfrac{1}{4}(2 + 2 + 2)^3 = 2 \cdot 27 = 54
\end{aligned}$$

invariant colorations in all.

6.27 Enumerating Nonisomorphic Graphs—Redfield and Pólya

The domain D consists of all the $\binom{v}{2}$ subsets of size 2 of a set with v elements. For convenience we think of the $\binom{v}{2}$ subsets of size two as the edges of the complete graph K_v on v vertices. We take two colors, red (r) and blue (b), and proceed to color the edges of K_v. The range is thus $R = \{r, b\}$.

The group that acts on the domain is the symmetric group on the v vertices S_v in *its natural permutation representation on the edges of* K_v. (Note that S_v is *not* the symmetric group of D but a much smaller group.)

On the range the group that acts is $H = \{(b)(r), (br)\}$, the group of order 2 that eventually interchanges the two colors.

An S_v-pattern is an isomorphism class of (simple) graphs.

Step 1. We need to find the cycle index of S_v in its representation on the edges of K_v. To accomplish this we take an arbitrary permutation σ of type $x_1^{\lambda_1}x_2^{\lambda_2} \cdots x_v^{\lambda_v}$ on the v vertices and proceed to calculate the cycle type that it induces on the $\binom{v}{2}$ edges of K_v. Select therefore an edge $\{i, j\}$ and investigate what happens to it under the repeated actions of the permutation σ. There are in fact only two cases to consider: (a) the contributions to the cycle index from edges $\{i, j\}$, with i and j in the same cycle of σ, and (b) contributions from edges $\{i, j\}$ with i and j in different cycles.

In case (a) suppose i and j belong to a cycle α of length k. We seek to understand what happens to this whole cycle under the action of σ. If k is odd, say $k = 2n + 1$, then the representation of α on edges of K_v becomes a product of n cycles each of length $2n + 1$. (The reader may wish to quickly check this for $k = 5$, say.) When $k = 2n$ the cycle α decomposes into $n - 1$ cycles of length $2n$ and one cycle of length n on the edges of K_v. [If, e.g., $\alpha = (1\,2\,3\,4\,5\,6)$, then its representation on edges takes the form $(1\,2, 2\,3, 3\,4, 4\,5, 5\,6, 6\,1)(1\,3, 2\,4, 3\,5, 4\,6, 5\,1, 6\,2)(1\,4, 2\,5, 3\,6)$; we abbreviated $\{i, j\}$ by ij.] In summary

$$x_{2n+1} \rightarrow x_{2n+1}^{n},$$

and

$$x_{2n} \rightarrow x_n x_{2n}^{n-1},$$

where the symbols left of the arrow describe the cycle structure on vertices and the symbols at the right describe the cycle structure on edges. Thus if there are λ_k cycles of length k in σ, the pairs of points lying in common cycles contribute as follows:

$$x_{2n+1}^{\lambda_{2n+1}} \rightarrow x_{2n+1}^{n\lambda_{2n+1}},$$

and

$$x_{2n}^{\lambda_{2n}} \rightarrow x_n^{\lambda_{2n}} x_{2n}^{(n-1)\lambda_{2n}},$$

for odd and even cycle lengths, respectively.

Let us address the situation in case (b), for edges $\{i, j\}$ with i and j in different cycles of σ. Consider two cycles of σ, α and β, of lengths r and s, respectively. On the set of edges $\{i, j\}$ with $i \in \alpha$ and $j \in \beta$, the cycles α and β induce exactly $(r; s)$ cycles of length $[r; s]$, where $(r; s)$ denotes the greatest common divisor and $[r; s]$ the least common multiple of r and s. In particular, when $r = s = k$, they contribute k cycles of length k. [We suggest the reader examine the cycles $\alpha\beta = (1\,2\,3\,4\,5\,6)(7\,8\,9\,10)$. They induce two cycles of length 12 on edges ij, with $i \in \alpha$ and $j \in \beta$:

$$(1\,7, 2\,8, 3\,9, 4\,10, 5\,7, 6\,8, 1\,9, 2\,10, 3\,7, 4\,8, 5\,9, 6\,10)$$

$$(1\,8, 2\,9, 3\,10, 4\,7, 5\,8, 6\,9, 1\,10, 2\,7, 3\,8, 4\,9, 5\,10, 6\,7).]$$

Thus the contributions from cycles $\alpha\beta$ with $r \neq s$ is

$$x_r^{\lambda_r} x_s^{\lambda_s} \to x_{[r;\, s]}^{\lambda_r \lambda_s (r;\, s)},$$

and from cycles with $r = s = k$ these contributions amount to

$$x_k^{\lambda_k} \to x_k^{k\binom{\lambda_k}{2}}.$$

The cycle index of S_v on the edges of K_v can now be written out. It is

$$P_{S_v}(x_1, x_2, \dots, x_m) = \frac{1}{v!} \sum_{\lambda} \frac{v!}{\prod_{k=1}^{v} (\lambda_k! k^{\lambda_k})} \prod_{n=0}^{[(v-1)/2]} x_{2n+1}^{n\lambda_{2n+1}}$$

$$\cdot \prod_{n=1}^{[v/2]} x_n^{\lambda_{2n}} x_{2n}^{(n-1)\lambda_{2n}} \cdot \prod_{1 \le r \le s \le p-1} x_{[r;\, s]}^{\lambda_r \lambda_s (r;\, s)} \cdot \prod_{k=1}^{[v/2]} x_k^{k\binom{\lambda_k}{2}}.$$

Here the sum is over all vectors $\lambda = (\lambda_1, \dots, \lambda_v)$ that satisfy $\sum_{i=1}^{v} i\lambda_i = v$, $[t]$ denotes the integral part of the fraction t, and m stands for $\binom{v}{2}$.

For example, $P_{S_4}(x_1, \dots, x_6) = \frac{1}{24}(x_1^6 + 8x_3^2 + 9x_1^2 x_2^2 + 6x_2 x_4)$.

Step 2. Let $w(r) = y_1$ and $w(b) = y_2$.

Step 3. Suppose $h = (b)(r)$, that is, h is the identity. Then $S(h, s) = R$, for all $s = 1, 2, \dots, m$. Consequently,

$$p_s(h) = w(r)^s + w(b)^s = y_1^s + y_2^s.$$

And if $h = (br)$, then $h^3 = h^5 = h^7 = \cdots = h$ and $h^0 = h^2 = h^4 = \cdots = 1$ (the identity in H). Thus

$$S(h, s) = \phi \qquad \text{if } s \text{ is odd}$$

and

$$S(h, s) = R \qquad \text{if } s \text{ is even.}$$

In this case $p_s(h) = 0$, for s odd, and $p_s(h) = w(r)w(hr) \cdots w(h^{s-1}r) + w(b)w(hb) \cdots w(h^{s-1}b) = y_1 y_2 y_1 y_2 \cdots + y_2 y_1 y_2 y_1 \cdots = 2y_1^{(s-1)/2} y_2^{(s-1)/2}$, for s even.

Step 4. The number of isomorphism classes of graphs with a_1 red edges and a_2 blue edges $\left(a_1 + a_2 = \binom{v}{2} = m\right)$ is the coefficient of $y_1^{a_1} y_2^{a_2}$ in

$$P_{S_v}\left(y_1 + y_2, y_1^2 + y_2^2, \cdots, y_1^m + y_2^m\right).$$

The number of isomorphism classes of graphs with a_1 red edges, a_2 blue edges ($a_1 + a_2 = \binom{v}{2} = m$) isomorphic to their complements [which is what invariance with respect to $h = (br)$ means] equals the coefficient of $y_1^{a_1} y_2^{a_2}$ in

$$P_{S_v}(0, 2y_1y_2, 0, 2y_1^2y_2^2, 0, 2y_1^3y_2^3, \dots).$$

Step 5. The total number of (isomorphism classes of) graphs obtained by setting $y_1 = y_2 = 1$ in Step 3 is

$$P_{S_v}(2, 2, \dots, 2).$$

The total number of (isomorphism classes of) graphs isomorphic to their complements equals (again by setting $y_1 = y_2 = 1$)

$$P_{S_v}(0, 2, 0, 2, 0, 2, \dots).$$

Step 6. The total number of patterns equals

$$\tfrac{1}{2}P_{S_v}(2, 2, \dots, 2) + \tfrac{1}{2}P_{S_v}(0, 2, 0, 2, \dots).$$

(Describe a pattern in graph theoretical language.)

EXERCISES

1. How many necklaces with eight beads of (at most) four colors are left invariant by an interchange of two colors? By a cyclic permutation of the four colors?

2. Count the number of patterns of injective functions between a domain D of m elements and a (specified a priori) range R of n colors.

3. A directed graph (digraph, for short) is a set of ordered pairs, called edges, with entries from a finite set of vertices. Two digraphs are isomorphic if one becomes the other upon a permutation of vertices.

(a) Find the number of (isomorphism classes of) digraphs on v vertices.

(b) Find the number of (isomorphism classes of) digraphs with e edges and v vertices.

(c) How many (isomorphism classes of) digraphs on v vertices are isomorphic to their complements?

(d) How many (isomorphism classes of) digraphs with e edges and v vertices are isomorphic to their complements?

(e) Find the number of patterns of digraphs on v vertices.

4. Suppose the edges of the complete graph are colored with n colors. Go through the recipe for DeBruijn's result in this case.

5. Examine what happens if the domain D consists of one element only and the range R has n colors. What are the patterns if the group acting on R is the full symmetric group $\mathrm{Sym}(R)$?

6. How many nonisomorphic classes of spanning trees does the complete graph K_v have? Find at least a recurrence (see Pólya [3]).

7. Solve "le problème des ménages" by making use of Pólya's theorem.

8. Let the group G act transitively on the set Ω (this means that G has one orbit only on Ω). Let x be an element of Ω and G_x the stabilizer of x. Show that the number of orbits of G_x on Ω equals

$$\frac{1}{|G|}\sum_{g\in G}|F(g)|^2,$$

where $F(g)$ is the set of points in Ω fixed by the group element g.

9. Let group G act on set X; $|X| = m$. Define an equivalence relation on ordered pairs of disjoint subsets of X (S_1, S_2) (with S_1 of cardinality r and S_2 of cardinality s) by writing

$$(S_1, S_2) \sim (S_3, S_4)$$

if $S_2 = g(S_1)$ and $S_4 = g(S_3)$ for some group element g of G. Show that the number of equivalence classes that result equals the coefficient of $x^r y^s$ obtained upon substituting $1 + x^k + y^k$ for x_k in the cycle index $P_G(x_1, \dots, x_m)$ of G in its action on X.

10. Denote by $P_{S_v}(x_1, \dots, x_m)$ the cycle index of the symmetric group S_v (of order $v!$) in its action on the $m = \binom{v}{2}$ subsets of size 2 of the set $\{1, 2, \dots, v\}$. Prove that the number of graphs (with multiple edges allowed) on v vertices and r edges is the coefficient of x^r in the polynomial

$$P_{S_v}\big((1-x)^{-1}, (1-x^2)^{-1}, \dots, (1-x^m)^{-1}\big), \qquad m = \binom{v}{2}.$$

NOTES

What we call Burnside's lemma is said to have been known to Frobenius. At any rate, it had been rediscovered by Burnside [1]. Redfield [2] introduced the notion of the group reduction function, which is what we now call the cycle index. His paper anticipates and implicitly contains Pólya's theorem. In his extensive work [3] Pólya explicitly states, demonstrates, and applies his theorem to a host of enumeration problems in chemistry, graph theory, and other fields. The same paper contains asymptotic results in enumeration, such as the asymptotic behavior of the number of isomorphism classes of spanning trees in complete graphs. Noteworthy extensions of Pólya's theorem were obtained by DeBruijn [4]. He introduces equivalence induced by a group on the range of the functions. Harary [5] draws attention to the great enumerative powers of these results (especially to the theory of graphs) by exploiting the notions of power group and exponentiation. Our presentation was influenced significantly by the accounts on this topic given in [4] and [6].

REFERENCES

[1] W. Burnside, *The Theory of Groups of Finite Order*, 2nd ed., Cambridge Univ. Press, 1911 (reprinted by Dover, New York, 1955).

[2] J. H. Redfield, The theory of group-reduced distributions. *Amer. J. Math.*, **49**, 433–455 (1927).

[3] G. Pólya, Kombinatorische Anzahlbestimmungen für Gruppen, Graphen, und Chemische Verbindungen. *Acta Math.*, **68**, 145–254 (1937).

[4] N. G. DeBruijn, Pólya's theory of counting, *Applied Combinatorial Mathematics* (E. F. Beckenbach, Ed.). Wiley, New York, 1964, pp. 144–184.

[5] F. Harary, Applications of Pólya's theorem to permutation groups, *A Seminar on Graph Theory* (F. Harary and L. Beineke, Eds.). Holt, Rinehart and Winston, New York, 1967, pp. 25–33.

[6] C. Berge, *Principles of Combinatorics*, Academic Press, New York, 1971.

CHAPTER 7

Block Designs

One simile that solitary shines
In the dry desert of a thousand lines.

Epilogue to the Satires, ALEXANDER POPE

The collection of all subsets of cardinality k of a set with v elements $(k < v)$ has the property that any subset of t elements, with $0 \leq t \leq k$, is contained in precisely $\binom{v-t}{k-t}$ subsets of size k. The subsets of size k provide therefore a nice covering for the subsets of a lesser cardinality. Observe that the number of subsets of size k that contain a subset of size t depends only on v, k, and t and not on the specific subset of size t in question. This is the essential defining feature of the structures that we wish to study.

The example we just described inspires general interest in producing similar coverings without using all the $\binom{v}{k}$ subsets of size k but rather as small a number of them as possible. The coverings that result are often elegant geometrical configurations, of which the projective and affine planes are examples. These latter configurations form nice coverings only for the subsets of cardinality 2, that is, any two elements are in the same number of these special subsets of size k which we call blocks (or, in certain instances, lines).

A collection of subsets of cardinality k, called blocks, with the property that every subset of size t $(t \leq k)$ is contained in the same number (say λ) of blocks is called a t-design. We supply the reader with constructions for t-designs with t as high as 5. Only recently a nontrivial 6-design has been found and no nontrivial 7-design is known. We then study in more depth the necessary numerical conditions for the existence of a symmetric 2-design contained in a result of Bruck, Ryser, and Chowla. A more recent result, due to Cameron, on extending symmetric 2-designs is included as well.

Special kinds of 2-designs, called Steiner triple systems, were studied in the nineteenth century by Woolhouse, Kirkman, and Steiner. Apart from their implicit connections to multiple transitive groups, 2-designs (known to statisticians as balanced incomplete block designs) arise explicitly from the statistical theories of Sir R. A. Fisher, most notably his analysis of variance. The rich combinatorial content of the theory of experimental design and the analysis of variance, initiated by Fisher and Yates, was further pursued by Bose and many of his students. Much of the material included in this chapter originated in their work. One such instance is the extension of Fisher's inequality from 2-designs to t-designs ($t \geq 2$). We devote a full section to this extension. The latter part of the chapter provides an introduction to association schemes, the Bose–Mesner algebra, and partial designs. Familiarity with finite fields, $GF(q)$, and finite-dimensional vector spaces over such fields is assumed in this chapter. The reader not exposed to these subjects is referred to Appendix 2 and the references given there.

1 THE BASIC STRUCTURE OF t-DESIGNS

7.1

Let $P = \{1, 2, \dots, v\}$ be a set of v elements that we call *points*. A subset of P with k elements is called a *k-subset*. We assume $0 \leq t \leq k < v$.

Definition. A *t-(v, k, λ_t) design* is a pair (P, B), where B is a collection of k-subsets of P (called *blocks*) with the property that each t-subset of P occurs in exactly λ_t blocks. (With less stringent emphasis we call such a pair a *t-design*.)

By $\Sigma_s(P)$ we denote the collection of all s-subsets of P; $|\Sigma_s(P)| = \binom{v}{s}$. For notational ease, however, we simply write σ_s to convey the fact that σ_s is a s-subset.

As we pointed out in the introduction $(P, \Sigma_k(P))$ is a t-$\left(v, k, \binom{v-t}{k-t}\right)$ design, for all $0 \leq t \leq k$; this design is called the *complete design*. Our interest is in studying t-designs that are not complete, that is, t-designs in which not every k-subset is a block. Though by no means abundant, such structures do exist. Several small examples are given at the end of this section. Well-known families of t-designs ($2 \leq t \leq 5$) are described in greater detail in Section 2.

The result we now prove gives insight into the combinatorial structure of a t-design.

Proposition 7.1. *Let (P, B) be a t-(v, k, λ_t) design.*

(a) *(P, B) is also an i-(v, k, λ_i) design, with $\lambda_i = \binom{v-i}{t-i}\binom{k-i}{t-i}^{-1}\lambda_t$, for all $0 \leq i \leq t$.* (Note that $\lambda_0 = |B|$.)

(b) *For σ_i and τ_j such that $0 \leq i + j \leq t$, the number of blocks α such that $\sigma_i \subseteq \alpha$ and $\tau_j \cap \alpha = \varnothing$ is $\binom{v-i-j}{k-i}\binom{v-t}{k-t}^{-1}\lambda_t$.* (We denote this number by λ_i^j and observe that it depends on i and j only and not on the specific choice of the subsets σ_i and τ_j.)

(c) *If* $v \geq k + t$, *then*

$$(P, \{P - \alpha : \alpha \in B\})$$

is a t-$\left(v, v - k, \binom{v-t}{k}\binom{v-t}{k-t}^{-1}\lambda_t\right)$ *design* [called the *complementary design* of (P, B)].

(d) *Let* $a \in P$. *Then*

$$(P - \{a\}, \{\alpha - \{a\} : \alpha \in B \textit{ such that } a \in \alpha\})$$

is a $(t-1)$-$(v-1, k-1, \lambda_t)$ *design* [called the *derived design* of (P, B) at a] *and* $(P - \{a\}, \{\alpha : \alpha \in B \textit{ and } a \notin \alpha\})$ *is a* $(t-1)$-$\left(v-1, k, \binom{v-t}{k-t+1}\binom{v-t}{k-t}^{-1}\lambda_t\right)$ *design* [called the *residual design* of (P, B) at the point a].

Proof. (a) Count in two ways the set of ordered pairs

$$\{(\sigma_t - \sigma_i, \alpha) : \alpha \supseteq \sigma_t \supseteq \sigma_i, \alpha \in B\}.$$

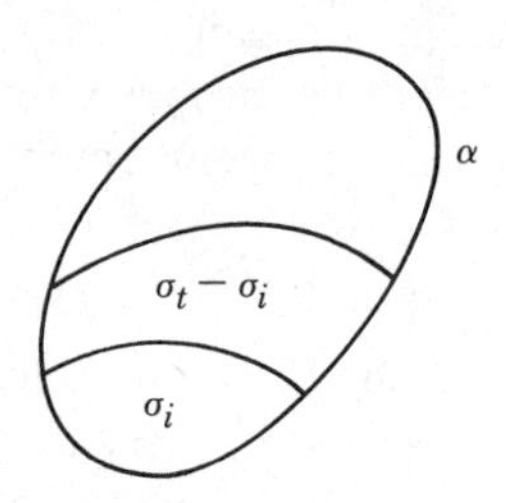

Fix α first and obtain $\binom{k-i}{t-i}\lambda_i$ for an answer (with λ_i denoting the number of blocks containing σ_i). Fix $\sigma_t - \sigma_i$ first and obtain $\binom{v-i}{t-i}\lambda_t$ for an answer. This gives $\lambda_i = \binom{v-i}{t-i}\binom{k-i}{t-i}^{-1}\lambda_t$, an expression that is *independent* of the specific subset σ_i. (The reader should observe that, in fact, $\bigcap_{\sigma_i \subseteq \alpha} \alpha = \sigma_i$ for $0 \leq i < t$—or else $\lambda_i \leq \lambda_{i+1}$, leading to a contradiction.)

(b) Let $X = \{\alpha \in B : \sigma_i \subseteq \alpha\}$; $|X| = \lambda_i$. Let also $\{A_m : m \in \tau_j\}$ be a collection of subsets of X defined as follows:

$$A_m = \{\alpha \in B : \sigma_i \subseteq \alpha \text{ and } \{m\} \subseteq \alpha \cap \tau_j\}.$$

There are j such A_m's. Note that

$$\left|\bigcap_{m \in \sigma_r} A_m\right| = |\{\alpha \in B : \sigma_i \subseteq \alpha \text{ and } \sigma_r \subseteq \alpha \cap \tau_j\}| = \lambda_{i+r} \quad (\sigma_r \subseteq \tau_j).$$

We want the number of blocks in X that are in none of the A_m's. By the principle of inclusion-exclusion (see Section 9.12) this number is

$$\lambda_i^j = \left| \bigcap_{m \in \sigma_r} \bar{A}_m \right| = \sum_{r=0}^{j} (-1)^r \sum_{\substack{\sigma_r \\ (\sigma_r \subseteq \tau_j)}} \left| \bigcap_{m \in \sigma_r} A_m \right|$$

$$= \sum_{r=0}^{j} (-1)^r \sum_{\sigma_r} \lambda_{i+r} = \sum_{r=0}^{j} (-1)^r \binom{j}{r} \lambda_{i+r}.$$

By (a) above $\lambda_s = \binom{v-s}{t-s}\binom{k-s}{t-s}^{-1}\lambda_t$; hence $\lambda_i^j = c\lambda_t$ with

$$c = \sum_{r=0}^{j} (-1)^r \binom{j}{r}\binom{v-i-j}{t-i-r}\binom{k-i-r}{t-i-r}^{-1}.$$

Note that the constant c depends only on the parameters v, k, t and not on the particular design. Taking the design $(P, \Sigma_k(P))$ we find in this case by direct computation $\lambda_t = \binom{v-t}{k-t}$ and $\lambda_i^j = \binom{v-i-j}{k-i}$. We thus obtain the simpler expression $c = \binom{v-i-j}{k-i}\binom{v-t}{k-t}^{-1}$. Thus $\lambda_i^j = \binom{v-i-j}{k-i}\binom{v-t}{k-t}^{-1}\lambda_t$.

(c) The number of blocks in $(P, \{P - \alpha : \alpha \in B\})$ containing a subset of t elements is λ_0^t in (P, B), which is well defined by part (b). In fact

$$\lambda_0^t = \binom{v-t}{k}\binom{v-t}{k-t}^{-1}\lambda_t,$$

which proves (c).

(d) To calculate how many blocks in the derived design contain a subset of $t-1$ points, let $\sigma_{t-1} \in \Sigma_{t-1}(P - \{a\})$. Look at $\sigma_t = \sigma_{t-1} \cup \{a\}$. The subset σ_t is in λ_t blocks of (P, B), all of which contain a, so σ_{t-1} is in λ_t blocks of the derived design, as claimed.

In the case of the residual design, $t-1$ points from $P - \{a\}$ are in precisely λ_{t-1}^1 blocks of (P, B) and

$$\lambda_{t-1}^1 = \binom{v-(t-1)-1}{k-t+1}\binom{v-t}{k-t}^{-1}\lambda_t = \binom{v-t}{k-t+1}\binom{v-t}{k-t}^{-1}\lambda_t;$$

This proves (d) and concludes the proof of our proposition.

7.2

As an example, consider

$$\begin{matrix} 1 & 2 & 3 & 4 & 5 & 6 & 7 & 3 & 4 & 5 & 6 & 7 & 1 & 2 \\ 2 & 3 & 4 & 5 & 6 & 7 & 1 & 5 & 6 & 7 & 1 & 2 & 3 & 4 \\ 4 & 5 & 6 & 7 & 1 & 2 & 3 & 6 & 7 & 1 & 2 & 3 & 4 & 5 \\ 8 & 8 & 8 & 8 & 8 & 8 & 8 & 7 & 1 & 2 & 3 & 4 & 5 & 6 \end{matrix} \tag{7.1}$$

with $P = \{1, 2, 3, 4, 5, 6, 7, 8\}$ and blocks B the columns of (7.1). Then one can directly check that (P, B) is a 3-(8, 4, 1) design. The checking can be facilitated by interpreting the blocks of (7.1) as the six faces, the six diagonal planes, plus the two "skewed" blocks 2 3 5 8 and 4 6 7 1 of the cube below:

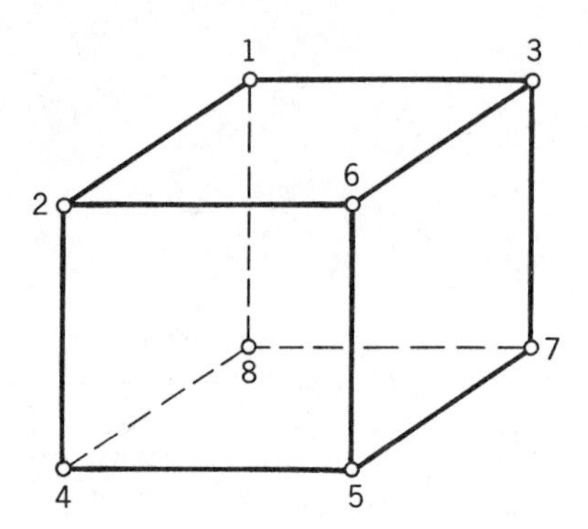

Assertions (a) through (d) made in Proposition 7.1 can also be verified with enough ease by examining the figure above. The design (P, B) turns out to be self-complementary. Its derived design, at $a = 8$, is perhaps better known in its more graphical form

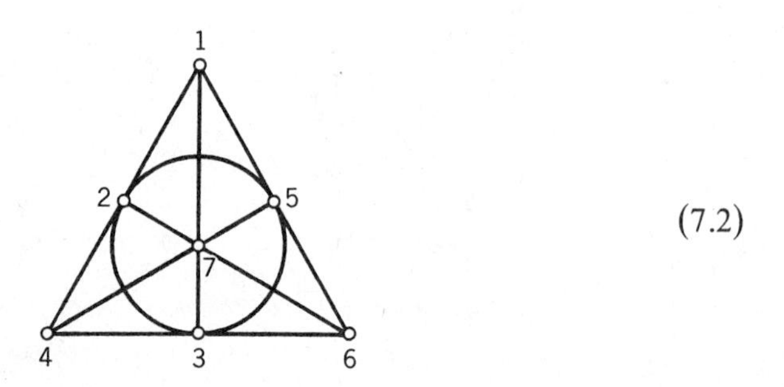

(7.2)

The residual design, at $a = 8$, consists of the last seven blocks of (7.1). This residual design turns out to be, in this case, the complementary design of the derived design of (P, B) at $a = 8$. Hadamard matrices provide us with perhaps the largest supply of 3-designs. The design we just examined is the smallest example of an infinite family of 3-designs that can be obtained from Hadamard matrices. A description of the general method of construction is given in the next section.

Another 3-design, not of the Hadamard kind, is listed below:

$$
\begin{array}{ccccccccccccccc}
1 & 1 & 1 & 1 & 2 & 2 & 2 & 3 & 3 & 4 & 4 & 5 & 1 & 1 & 1 \\
2 & 3 & 5 & 6 & 3 & 6 & 7 & 4 & 7 & 5 & 8 & 6 & 2 & 2 & 2 \\
4 & 8 & 9 & 7 & 5 & 9 & 8 & 6 & 9 & 7 & 9 & 8 & 3 & 5 & 6 \\
T & T & T & T & T & T & T & T & T & T & T & T & 9 & 7 & 8 \\
\\
1 & 1 & 1 & 1 & 1 & 2 & 2 & 2 & 2 & 2 & 3 & 3 & 3 & 4 & 5 \\
3 & 3 & 4 & 4 & 7 & 3 & 3 & 4 & 4 & 5 & 4 & 5 & 6 & 6 & 6 \\
4 & 5 & 5 & 6 & 8 & 4 & 6 & 5 & 7 & 8 & 5 & 7 & 8 & 7 & 7 \\
7 & 6 & 8 & 9 & 9 & 8 & 7 & 6 & 9 & 9 & 9 & 8 & 9 & 8 & 9
\end{array}
\qquad (7.3)
$$

The symbol T abbreviates 10. We invite the reader to examine this design and possibly attach geometrical interpretations to it.

2 CONSTRUCTIONS OF t-DESIGNS

In the next few pages we describe several methods of constructing t-designs with $2 \leq t \leq 5$. Outlines of proofs, along with some references, are given in the latter part of the section.

Large families of 2- and 3-designs can be constructed through Hadamard matrices. A $n \times n$ matrix H with entries 1 and -1 is called a *Hadamard matrix* if $H'H = nI(= HH')$. Here H' denotes the transpose of the matrix H. Observe that multiplying rows and columns of a Hadamard matrix by -1 leads again to a Hadamard matrix.

A Hadamard matrix may exist only if n is a multiple of 4 or if n is 2 or 1. It is not known whether a Hadamard matrix exists for every multiple of 4 but it appears that they do exist in abundance. We quickly mention some known families of Hadamard matrices.

For $A = (a_{ij})$ a $n \times n$ matrix and B a $m \times m$ matrix we denote by $A \otimes B$ the matrix $(a_{ij}B)$ and call it the *tensor product* of A with B; $A \otimes B$ is $nm \times nm$. The matrix $\begin{pmatrix} 1 & 1 \\ -1 & 1 \end{pmatrix}$ is a 2×2 Hadamard matrix. If H and G are Hadamard matrices, then so is $H \otimes G$. By repeatedly tensoring a Hadamard matrix of order 2 with itself we conclude that Hadamard matrices of order 2^m exist for all $m \geq 0$. It is known that a Hadamard matrix of order $q + 1$ exists, where q is a prime power and $q \equiv 3$ (modulo 4). There also exists a Hadamard matrix of order $2(q + 1)$ for q a prime power congruent to 1 modulo 4. In fact it is known that *if there exists a Hadamard matrix of order* $h(h > 1)$, *then there exists a Hadamard matrix of order* $h(q + 1)$, *where* q *is an odd prime power*.

7.3 The Hadamard 2-Designs

Let H be a Hadamard matrix of order n $(n \geq 8)$.

Make 1's in the first row and column of H, by multiplying suitable rows and columns of H by -1.

Delete the first row and column.

In the remaining $(n-1) \times (n-1)$ matrix each row generates a block: it consists of the indices of the columns in which the 1's occur. The resulting structure is a 2-$(n-1, \frac{n}{2}-1, \frac{n}{4}-1)$ design.

This construction is reversible. In other words by starting out with any 2-$(n-1, \frac{n}{2}-1, \frac{n}{4}-1)$ design we can produce a Hadamard matrix of order n by reversing the steps above.

7.4 The Paley Designs

A special case of the above construction are the Paley designs (these are 2-designs). Consider $GF(q)$, the field with q elements, where $q \equiv 3$ (modulo 4) —q is a power of a prime. The points are the elements of $GF(q)$. The blocks are $\{Q + a : a \in GF(q)\}$, where Q is the set of nonzero squares in $GF(q)$. [An element $x \in GF(q)$ is said to be a square in $GF(q)$ if $x = y^2$, for some $y \in GF(q)$.]

As an example, let $q = 11$. Then $Q = \{1, 3, 4, 5, 9\}$ and the resulting 2-design is

$$\begin{array}{ccccccccccc} 1 & 2 & 3 & 4 & 5 & 6 & 7 & 8 & 9 & 10 & 11 \\ 3 & 4 & 5 & 6 & 7 & 8 & 9 & 10 & 11 & 1 & 2 \\ 4 & 5 & 6 & 7 & 8 & 9 & 10 & 11 & 1 & 2 & 3 \\ 5 & 6 & 7 & 8 & 9 & 10 & 11 & 1 & 2 & 3 & 4 \\ 9 & 10 & 11 & 1 & 2 & 3 & 4 & 5 & 6 & 7 & 8. \end{array} \tag{7.4}$$

7.5 The Hadamard 3-Designs

Let H be a Hadamard matrix of order n $(n \geq 8)$.

Make 1's in the first row by multiplying suitable columns by -1.

The points are the (indices of) columns of H.

Blocks are given as follows: each row (other than the first) gives two blocks; the columns in which 1 appears is one, and the (remaining) columns in which -1 appears is the other.

The resulting design is a 3-$(n, \frac{n}{2}, \frac{n}{4}-1)$ design. This construction is reversible as well. The design in (7.1) is the smallest example in this family.

Another way of constructing 2-designs is through projective and affine geometries.

A *projective geometry* over $GF(q)$ is the collection of subspaces of a vector space of finite dimension over $GF(q)$. The *points* of the geometry are the subspaces of dimension 1. A (projective) subspace is identified with the set of points it contains. Points have (projective) dimension 0 and, in general, a (projective) subspace is assigned dimension 1 less than the vector space dimension of the subspace it comes from. Subspaces of projective dimension 1 are called *lines*, those of projective dimension 2 are called *planes*.

An *affine geometry* of dimension n is the collection of cosets of subspaces of a vector space of dimension n over $GF(q)$. The geometric dimension here equals the usual vector space dimension of the underlying subspace. *Points* are just vectors (or cosets of the 0 subspace). An (affine) subspace is identified with the points it contains. The affine subspaces of dimension 1 are called *lines* and those of dimension 2 are called *planes*.

Projective and affine geometries contain 2- and 3-designs. Sections 7.6 and 7.7 give the details of these constructions along with a couple of illustrative examples.

7.6 The Projective Geometries

In a projective geometry of (projective) dimension n over $GF(q)$ the subspaces of a given projective dimension m $(2 \le m \le n-1)$ form a

$$2\text{-}\left(\frac{q^{n+1}-1}{q-1}, \frac{q^{m+1}-1}{q-1}, \begin{bmatrix} n-1 \\ m-1 \end{bmatrix}(q)\right)$$

design, where $\begin{bmatrix} n \\ k \end{bmatrix}(x)$ denotes the Gaussian polynomial.

When $m = n-1$ we obtain a symmetric

$$2\text{-}\left(\frac{q^{n+1}-1}{q-1}, \frac{q^{n}-1}{q-1}, \frac{q^{n-1}-1}{q-1}\right)$$

design. [A 2-design (P, B) is said to be *symmetric* if the number of blocks equals the number of points, i.e., if $\lambda_0 = |B| = |P| = v$.] We denote this symmetric design by $PG(n, q)$. Note that $PG(n, 2)$ is a Hadamard 2-design for all $n \ge 2$.

Example. Let us construct $PG(2, 2)$. We start with a vector space V of dimension 3 over $GF(2)$. The space V contains eight vectors: 000, 100, 010, 001, 110, 101, 011, 111. There are $(2^3 - 1)/(2 - 1) = 7$ subspaces of dimension 2 in V; each is a solution to one of the following equations: $x = 0$, $y = 0$, $z = 0$, $x + y = 0$, $x + z = 0$, $y + z = 0$, and $x + y + z = 0$ (where x, y, and z denote the three coordinate axes). The subspaces are (with 000 omitted):

$$\begin{Bmatrix} 0 & 1 & 0 \\ 0 & 0 & 1 \\ 0 & 1 & 1 \end{Bmatrix}, \begin{Bmatrix} 1 & 0 & 0 \\ 0 & 0 & 1 \\ 1 & 0 & 1 \end{Bmatrix}, \begin{Bmatrix} 1 & 0 & 0 \\ 0 & 1 & 0 \\ 1 & 1 & 0 \end{Bmatrix}, \begin{Bmatrix} 1 & 1 & 0 \\ 0 & 0 & 1 \\ 1 & 1 & 1 \end{Bmatrix},$$

$$\begin{Bmatrix} 1 & 0 & 1 \\ 0 & 1 & 0 \\ 1 & 1 & 1 \end{Bmatrix}, \begin{Bmatrix} 0 & 1 & 1 \\ 1 & 0 & 0 \\ 1 & 1 & 1 \end{Bmatrix}, \text{and} \begin{Bmatrix} 1 & 1 & 0 \\ 0 & 1 & 1 \\ 1 & 0 & 1 \end{Bmatrix}.$$

This gives rise to the following projective picture:

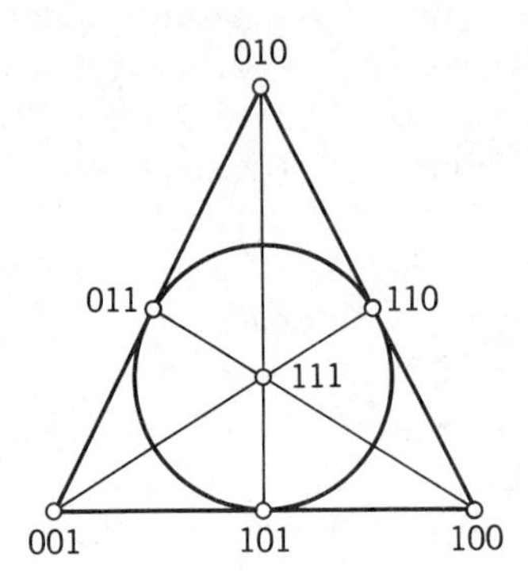

in which the reader will recognize the design displayed in (7.2).

7.7 The Affine Geometries

In an affine geometry of (affine) dimension n over $GF(q)$ the subspaces of a given affine dimension m $(2 \le m \le n-1)$ form a 2-$(q^n, q^m, \left[\begin{smallmatrix} n-1 \\ m-1 \end{smallmatrix}\right](q))$ design, where $\left[\begin{smallmatrix} n \\ k \end{smallmatrix}\right](x)$ denotes the Gaussian polynomial.

When $m = n - 1$ we obtain a 2-$(q^n, q^{n-1}, (q^{n-1} - 1)/(q - 1))$ design. [We denote this design by $AG(n, q)$. An affine plane is $AG(2, q)$.]

In an affine geometry of (affine) dimension n over $GF(2)$ the subspaces of a given affine dimension m $(2 \le m \le n-1)$ form a 3-$(2^n, 2^m, \left[\begin{smallmatrix} n-2 \\ m-2 \end{smallmatrix}\right](2))$ design, with $\left[\begin{smallmatrix} n \\ k \end{smallmatrix}\right](x)$ being the Gaussian polynomial. $AG(n, 2)$ is a Hadamard 3-design. Design (7.1) is such an example, that is, it is $AG(3, 2)$.

Example. Let us construct $AG(2, 3)$, the affine plane over $GF(3)$. We have nine vectors:

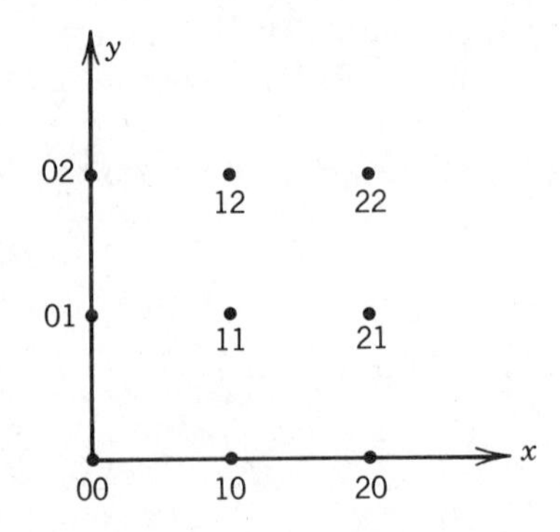

$AG(2,3)$ consists of four subspaces of dimension 1 (each coming with 2 cosets):

$x=0$			$y=0$			$x+2y=0$			$x+y=0$		
00	10	20	00	01	02	00	10	01	00	10	02
01	11	21	10	11	12	11	21	12	12	22	11
02	12	22	20	21	22	22	02	20	21	01	20.

With an immediately obvious relabeling of points (1 through 9) we can write $AG(2,3)$ as follows:

$$\begin{array}{ccc|ccc|ccc|ccc} 1 & 4 & 7 & 1 & 2 & 3 & 1 & 4 & 2 & 1 & 4 & 3 \\ 2 & 5 & 8 & 4 & 5 & 6 & 5 & 8 & 6 & 6 & 9 & 5 \\ 3 & 6 & 9 & 7 & 8 & 9 & 9 & 3 & 7 & 8 & 2 & 7. \end{array} \tag{7.5}$$

$AG(2,3)$ is indeed a 2-$(9,3,1)$ design with the columns of (7.5) as blocks.

7.8 The Multiply Transitive Groups

The t-transitive groups provide yet another way of constructing t-designs, $2 \le t \le 5$. A permutation group G acting on a set P is called *t-transitive* if for any two ordered t-tuples with distinct entries $(x_1, \dots, x_t)$ and $(y_1, \dots, y_t)$ there exists $g \in G$ such that $g(x_1) = y_1, \dots, g(x_t) = y_t$; $x_i, y_i \in P$, $1 \le i \le t$.

Let G act t-transitively on P and let $\alpha \in \Sigma_k(P)$, with $k > t$. Define $B = \{g(\alpha) : g \in G\}$. [We understand by $g(\alpha)$ the k-subset $\{g(x) : x \in \alpha\}$.] Then (P, B) is a t-(v, k, λ_t) design. Since $\binom{v}{t}\binom{k}{t}^{-1}\lambda_t = \lambda_0 = |G|/|G_\alpha|$ we obtain $\lambda_t = \binom{k}{t}\binom{v}{t}^{-1}|G|\,|G_\alpha|^{-1}$; here G_α denotes the (set) stabilizer of α, that is, $G_\alpha = \{g \in G : g(\alpha) = \alpha\}$. Sometimes the designs generated through this process actually turn out to be complete designs.

As nontrivial examples, the Mathieu groups M_{12} and M_{24} act 5-transitively on 12 and 24 points, respectively. They are known to generate the Mathieu designs 5-$(12,6,1)$ and 5-$(24,8,1)$ in the manner described above. We outline a construction and list the Mathieu 5-$(12,6,1)$ design in its entirety in Section 7.9. There exist many 2-transitive groups. Through the process just described they can be used to generate 2-designs.

7.9 Construction through Linear Codes

We illustrate now a method of construction of t-designs through the use of linear codes. Identify a vector space of dimension n over $GF(q)$ with $(GF(q))^n$, that is, n-tuples with entries from $GF(q)$. A subspace C of dimension m in $(GF(q))^n$ is called a (n, m) *linear code*; $|C| = q^m$. Let $x \in C$; the set of nonzero coordinates of x is called its *support*; the number of nonzero coordinates of x is called the *weight* of x and is denoted by $w(x)$.

Let us agree to write the vectors of $(GF(q))^n$ as row vectors. A *generating matrix* $G(C)$ for the (n, m) linear code C is a $m \times n$ matrix whose rows form a basis for C [over $GF(q)$].

Designs can sometimes be obtained by the following process: The distinct supports of the vectors of minimal (nonzero) weight of certain (quite select) linear codes over $GF(q)$ form the blocks of a t-design ($2 \le t \le 5$).

Example. Let C be the linear code of $(GF(2))^8$ with generating matrix

$$G(C) = \begin{bmatrix} 1 & 1 & 1 & 1 & 1 & 1 & 1 & 1 \\ 0 & 1 & 0 & 0 & 1 & 1 & 0 & 1 \\ 0 & 0 & 1 & 0 & 1 & 0 & 1 & 1 \\ 0 & 0 & 0 & 1 & 0 & 1 & 1 & 1 \end{bmatrix}.$$

Apart from the 0 vector and (1 1 1 1 1 1 1 1) all of the 14 remaining vectors in C have minimal (nonzero) weight 4. They are:

$$\begin{array}{cccccccc} 0 & 1 & 0 & 0 & 1 & 1 & 0 & 1 \\ 0 & 0 & 1 & 0 & 1 & 0 & 1 & 1 \\ 0 & 0 & 0 & 1 & 0 & 1 & 1 & 1 \\ 0 & 1 & 1 & 0 & 0 & 1 & 1 & 0 \\ 0 & 1 & 0 & 1 & 1 & 0 & 1 & 0 \\ 0 & 0 & 1 & 1 & 1 & 1 & 0 & 0 \\ 0 & 1 & 1 & 1 & 0 & 0 & 0 & 1 \\ \\ 1 & 0 & 1 & 1 & 0 & 0 & 1 & 0 \\ 1 & 1 & 0 & 1 & 0 & 1 & 0 & 0 \\ 1 & 1 & 1 & 0 & 1 & 0 & 0 & 0 \\ 1 & 0 & 0 & 1 & 1 & 0 & 0 & 1 \\ 1 & 0 & 1 & 0 & 0 & 1 & 0 & 1 \\ 1 & 1 & 0 & 0 & 0 & 0 & 1 & 1 \\ 1 & 0 & 0 & 0 & 1 & 1 & 1 & 0 \end{array} \quad \leftrightarrow \quad \begin{array}{cccc} 2 & 5 & 6 & 8 \\ 3 & 5 & 7 & 8 \\ 4 & 6 & 7 & 8 \\ 2 & 3 & 6 & 7 \\ 2 & 4 & 5 & 7 \\ 3 & 4 & 5 & 6 \\ 2 & 3 & 4 & 8 \\ \\ 1 & 3 & 4 & 7 \\ 1 & 2 & 4 & 6 \\ 1 & 2 & 3 & 5 \\ 1 & 4 & 5 & 8 \\ 1 & 3 & 6 & 8 \\ 1 & 2 & 7 & 8 \\ 1 & 5 & 6 & 7 \end{array}$$

The supports of these vectors give a 3-(8, 4, 1) design; this is the same as design (7.1) in Section 1. [The permutation

$$\begin{pmatrix} 1 & 2 & 3 & 4 & 5 & 6 & 7 & 8 \\ 1 & 5 & 3 & 2 & 4 & 7 & 6 & 8 \end{pmatrix}$$

carries in fact the blocks of this design into those of (7.1).]

Another Example. Consider the (distinguished) (12, 6) linear code C over $GF(3)$, whose elements we write as $-1, 0, 1$, with generating matrix:

$$G(C) = \begin{bmatrix} 1 & 0 & 0 & 0 & 0 & 0 & 1 & 1 & 1 & 1 & 1 & 0 \\ 0 & 1 & 0 & 0 & 0 & 0 & 0 & 1 & -1 & -1 & -1 & -1 \\ 0 & 0 & 1 & 0 & 0 & 0 & -1 & 0 & 1 & -1 & -1 & -1 \\ 0 & 0 & 0 & 1 & 0 & 0 & -1 & -1 & 0 & 1 & -1 & -1 \\ 0 & 0 & 0 & 0 & 1 & 0 & -1 & -1 & -1 & 0 & 1 & -1 \\ 0 & 0 & 0 & 0 & 0 & 1 & 1 & -1 & -1 & -1 & 0 & -1 \end{bmatrix}.$$

The code C is called the *Golay ternary code*. The minimal (nonzero) weight in

C is 6 [it takes a bit of puzzling over $G(C)$ to actually see this]. A computer search finds 264 vectors of weight 6 in C. Since a vector and its negative have the same support, there will be at most 132 distinct supports for the 264 vectors of weight 6. As it turns out there are precisely 132 distinct supports. These 132 supports are the blocks of a 5-(12, 6, 1) design, called the *Mathieu 5-design* on 12 points. The design is listed on page 270, with rows as blocks.

7.10 The Method of Differences

Let G be an Abelian group with addition as group operation. For a subset α of size k of G consider the list of $k(k-1)$ (ordered) differences of its elements. Call a collection of k-subsets $\alpha_1, \alpha_2, \ldots, \alpha_m$ a *set of generating blocks* if among the $mk(k-1)$ differences coming from these k-subsets each nonzero element of G occurs λ times.

If $\alpha_1, \ldots, \alpha_m$ is a set of generating blocks, then the collection of k-subsets $\{\alpha_i + a : a \in G\}, 1 \le i \le m$, is a 2-($v$, k, λ) design, with the elements of G as points. (Here $v = |G|$, $\lambda_1 = mk$, and $\lambda_0 = mv$.)

To illustrate this method of construction we take G to be the Abelian (cyclic) group with 13 elements. The difficulty always rests in selecting a set of generating blocks (especially since sometimes they do not exist). In this case take $\alpha_1 = \{1, 3, 9\}$ and $\alpha_2 = \{2, 6, 5\}$. The list of differences is:

from α_1: $1 - 3 = 11, 1 - 9 = 5, 3 - 9 = 7, 9 - 3 = 6, 9 - 1 = 8, 3 - 1 = 2$

from α_2: $2 - 6 = 9, 2 - 5 = 10, 6 - 5 = 1, 5 - 6 = 12, 5 - 2 = 3, 6 - 2 = 4$.

We conclude that λ equals 1 in this case. By expanding the two blocks we obtain the 2-(13, 3, 1) design listed below:

1	2	3	4	5	6	7	8	9	10	11	12	13
3	4	5	6	7	8	9	10	11	12	13	1	2
9	10	11	12	13	1	2	3	4	5	6	7	8
2	3	4	5	6	7	8	9	10	11	12	13	1
6	7	8	9	10	11	12	13	1	2	3	4	5
5	6	7	8	9	10	11	12	13	1	2	3	4

7.11 Construction through Hypergraphs

A *hypergraph* H is simply a collection of subsets (called *edges*) of a set. We also assume that every element of the set is on at least one edge. An *automorphism group* G of H is a group of permutations on the elements of the set that preserves the edges of H. Nontrivial designs with t as high as 5 (and possibly 6) have been obtained as follows:

Take the edges of H as the points of the design. Then G has a natural action on the subsets of k edges. Unions of certain (carefully selected) orbits under this action form a t-design ($2 \le t < k$).

As a small example, let H be the set of (ordinary) edges of the complete graph on five vertices. Our prospective design will thus have $\binom{5}{2} = 10$ points.

1	2	4	10	11	12
1	3	8	10	11	12
1	5	9	10	11	12
1	6	7	10	11	12
2	3	5	10	11	12
2	6	9	10	11	12
2	7	8	10	11	12
3	4	6	10	11	12
3	7	9	10	11	12
4	5	7	10	11	12
4	8	9	10	11	12
5	6	8	10	11	12
1	2	3	9	11	12
1	2	5	7	11	12
1	2	6	8	11	12
1	3	4	7	11	12
1	3	5	6	11	12
1	4	5	8	11	12
1	4	6	9	11	12
1	7	8	9	11	12
2	3	4	8	11	12
2	3	6	7	11	12
2	4	5	6	11	12
2	4	7	9	11	12
2	5	8	9	11	12
3	4	5	9	11	12
3	5	7	8	11	12
3	6	8	9	11	12
4	6	7	8	11	12
5	6	7	9	11	12
1	2	3	4	6	12
1	2	3	5	8	12
1	2	3	7	10	12
1	2	4	5	9	12
1	2	4	7	8	12
1	2	5	6	10	12
1	2	6	7	9	12
1	2	8	9	10	12
1	3	4	5	10	12
1	3	4	8	9	12
1	3	5	7	9	12
1	3	6	7	8	12
1	3	6	9	10	12
1	4	5	6	7	12

1	4	6	8	10	12
1	4	7	9	10	12
1	5	6	8	9	12
1	5	7	8	10	12
2	3	4	5	7	12
2	3	4	9	10	12
2	3	5	6	9	12
2	3	6	8	10	12
2	3	7	8	9	12
2	4	5	8	10	12
2	4	6	7	10	12
2	4	6	8	9	12
2	5	6	7	8	12
2	5	7	9	10	12
3	4	5	6	8	12
3	4	6	7	9	12
3	4	7	8	10	12
3	5	6	7	10	12
3	5	8	9	10	12
4	5	6	9	10	12
4	5	7	8	9	12
6	7	8	9	10	12
1	2	3	4	5	11
1	2	3	4	7	9
1	2	3	4	8	10
1	2	3	5	6	7
1	2	3	5	9	10
1	2	3	6	8	9
1	2	3	6	10	11
1	2	3	7	8	11
1	2	4	5	6	8
1	2	4	5	7	10
1	2	4	6	7	11
1	2	4	6	9	10
1	2	4	8	9	11
1	2	5	6	9	11
1	2	5	7	8	0
1	2	5	8	10	11
1	2	6	7	8	10
1	2	7	9	10	11
1	3	4	5	6	9
1	3	4	5	7	8
1	3	4	6	7	10
1	3	4	6	8	11

1	3	4	9	10	11
1	3	5	6	8	10
1	3	5	7	10	11
1	3	5	8	9	11
1	3	6	7	9	11
1	3	7	8	9	10
1	4	5	6	10	11
1	4	5	7	9	11
1	4	5	8	9	10
1	4	6	7	8	9
1	4	7	8	10	11
1	5	6	7	8	11
1	5	6	7	9	10
1	6	8	9	10	11
2	3	4	5	6	10
2	3	4	5	8	9
2	3	4	6	7	8
2	3	4	6	9	11
2	3	4	7	10	11
2	3	5	6	8	11
2	3	5	7	8	10
2	3	5	7	9	11
2	3	6	7	9	10
2	3	8	9	10	11
2	4	5	6	7	9
2	4	5	7	8	11
2	4	5	9	10	11
2	4	6	8	10	11
2	4	7	8	9	10
2	5	6	7	10	11
2	5	6	8	9	10
2	6	7	8	9	11
3	4	5	6	7	11
3	4	5	7	9	10
3	4	5	8	10	11
3	4	6	8	9	10
3	4	7	8	9	11
3	5	6	7	8	9
3	5	6	9	10	11
3	6	7	8	10	11
4	5	6	7	8	10
4	5	6	8	9	11
4	6	7	9	10	11
5	7	8	9	10	11

ck size to be 4. The automorphism group in question will be S_5, ric group on the five vertices. The three orbits with orbit representatives $\{12, 14, 23, 34\}$, $\{12, 13, 14, 15\}$, and $\{12, 13, 23, 45\}$ form a 3-(10, 4, 1) design. By drawing out the subgraphs corresponding to these three orbit representatives one can easily check that any three edges of the complete graph on five vertices is contained in exactly one image of one of our initial three subgraphs. The 5-transitivity of S_5 is what simplifies things significantly in this particular example.

The list of construction procedures that we gave is by no means exhaustive. Generating projective planes through complete sets of latin squares is but one example of well-known constructions that we do not include. Of special interest would be constructions that lead to nontrivial t-designs with high values of t. For $t \geq 7$ no such constructions are yet known.

A Justification That the Methods of Construction Described so Far Indeed Generate the t-Designs We Claimed They Do: Hadamard Matrices

In the introductory pages to this section we discussed Hadamard matrices and their existence in particular. A necessary condition for a Hadamard matrix H of order n to exist (with $n \geq 4$) is that n actually be a multiple of 4. Indeed, multiplying suitable rows of H by -1 makes the first column of H consist of all 1's. Look now at the first three columns only. Denote by x the number of rows of the form $(1, 1, 1)$, by y those of the form $(1, 1, -1)$, by z and w those of the form $(1, -1, 1)$ and $(1, -1, -1)$, respectively. (Then $x + y + z + w = n$.) Writing in terms of x, y, z, w the fact that any pair of the three distinct columns are orthogonal, we obtain a system of linear equations with solution $x = y = z = w$ $(= n/4)$. Since x is obviously integral we conclude that 4 divides n, necessarily.

Hadamard matrices are well researched, existence being a central problem. Many constructions are known, of which we mentioned several. We refer the reader to the expository article [17] and the references contained therein for the detailed proofs.

The Hadamard 2-Designs (Section 7.3)

Let the first row and column of H consist of 1's only. Fix the first column and two additional (distinct) columns; the two additional columns correspond to two distinct points in our prospective design. Then the number of blocks containing these two points equals the number of vectors of the form $(1, 1, 1)$ across the three columns, minus one. The "minus one" is the vector $(1, 1, 1)$ from the *first* row of H which has in effect been deleted. From our previous discussion, in which we showed that the dimension of a Hadamard matrix must be a multiple of 4, we know that there are precisely $n/4$ vectors of the form $(1, 1, 1)$. Hence two (arbitrary) points are always in precisely $\frac{n}{4} - 1$ blocks. We thus have a 2-design with $\lambda_2 = \frac{n}{4} - 1$.

The Paley Designs (Section 7.4)

Let Q be the set of nonzero squares in $GF(q)$, with q equal to 3 modulo 4. The first thing to observe is that -1 is not a square in such a field. [If it were we could write $-1 = w^2$ or $1 = w^4$. Since the nonzero elements of $GF(q)$ form a group with respect to multiplication, the order of w must divide the order of the group, that is, 4 must divide $q - 1$. This contradicts the fact that q equals 3 modulo 4.] The second observation is that d ($\neq 0$) is a square if and only if $-d$ is not a square. One can see this by recalling that the multiplication structure of $GF(q)$ is in fact a cyclic group. If ξ generates this group, then the even powers of ξ are the (nonzero) squares in $GF(q)$; there are $\frac{1}{2}(q-1)$ of these. The remaining half are not squares. It is also clear that not both d and $-d$ can be squares because then -1 would also be a square, as their quotient. This tells us that the set of nonsquares of $GF(q)$ is precisely $\{-d : d \text{ a nonzero square in } GF(q)\}$.

The points of our design are the q elements of the field $GF(q)$ and the blocks are the subsets $\{Q + a : a \in GF(q)\}$. To check that we do indeed have a 2-design we have to show that every two (distinct) elements of $GF(q)$ are in the same number of blocks. Let x and y be two such elements. Then

$$\begin{aligned}
&|\{\alpha : \{x, y\} \in \alpha, \alpha \text{ block}\}| \\
&\quad = |\{a \in GF(q) : \{x, y\} \in Q + a\}| \\
&\quad = |\{a \in GF(q) : x - a \in Q \text{ and } y - a \in Q\}| \\
&\quad = |\{a \in GF(q) : x - a = \alpha^2 \text{ and } y - a = \beta^2, \text{with } \alpha, \beta \neq 0\}| \\
&\quad = |\{(\alpha^2, \beta^2) : \alpha^2 - \beta^2 = x - y, \alpha \text{ and } \beta \neq 0\}|.
\end{aligned}$$

Now if $x - y$ is a square, say γ^2, then (dividing out by γ^2) one can see that $\alpha^2 - \beta^2 = \gamma^2$ has the same number of solutions as the equation $s^2 - u^2 = 1$. If $x - y$ is not a square, then we pointed out that $-(x - y)$ must be a square, say δ^2, and $\alpha^2 - \beta^2 = -\delta^2$ has again precisely as many solutions as $u^2 - s^2 = 1$ (or $s^2 - u^2 = 1$) does. This constant number of solutions is therefore independent of x and y and it equals the number of blocks containing two points. We have thus proved that the sets $\{Q + a : a \in GF(q)\}$ form a 2-design. The reader might wish to compute its parameters as functions of q only. [*Hint*: Find $k = |Q|$ first, then λ_1.]

The Hadamard 3-Designs (Section 7.5)

Assume, without loss, that the first row of the Hadamard matrix has all its entries equal to 1. With each row (other than the first) we associate two blocks. Observe that we do not affect these two blocks if we multiply the row by -1.

Select now three columns of H; this corresponds to selecting three distinct points. The number of blocks containing these three points equals the number of vectors of the form $(1, 1, 1)$ or $(-1, -1, -1)$ across the three columns (excepting the first row). Multiply now by -1 all the rows in which

$(-1, -1, -1)$ occurs across the three columns. The number of blocks containing the three points equals now the number of rows having $(1, 1, 1)$ across the three columns, minus one. We subtract one because the first row does not generate blocks. We know (see the proof for Hadamard 2-designs) that there are precisely $n/4$ vectors of the type $(1, 1, 1)$ throughout H in the three columns. Three points are therefore contained in $\frac{n}{4} - 1$ blocks.

The Projective Geometries (Section 7.6)

In a vector space V of dimension $n + 1$ over $GF(q)$ fix two (distinct) subspaces of dimension 1. The two subspaces of dimension 1 necessarily generate a subspace W of dimension 2. The number of subspaces of dimension $m + 1$ (with $2 \leq m \leq n$) containing W equals the number of subspaces of dimension $m - 1$ in the quotient space V/W. The number of such subspaces equals $\begin{bmatrix} n-1 \\ m-1 \end{bmatrix}(q)$, where $\begin{bmatrix} n \\ k \end{bmatrix}(x)$ denotes the Gaussian polynomial (see Section 6 in Chapter 3). Since the number of subspaces of dimension $m + 1$ containing two distinct subspaces of dimension 1 is independent of the choice of the two one-dimensional subspaces [as it always equals $\begin{bmatrix} n-1 \\ m-1 \end{bmatrix}(q)$], we conclude that the structure so defined is a 2-design.

The projective geometries do not generate 3-designs in the same manner. For if we choose three (projective) points they may generate a subspace of (vector space) dimension 2 or 3. In the former case they will be in $\begin{bmatrix} n-1 \\ m-1 \end{bmatrix}(q)$ subspaces of (vector space) dimension $m + 1$, while in the latter case they will be in $\begin{bmatrix} n-2 \\ m-2 \end{bmatrix}(q)$ such subspaces.

We leave to the reader the task of computing the remaining parameters of these 2-designs. After doing so it is easy to check that when $m = n - 1$ we obtain a symmetric design.

The Affine Geometries (Section 7.7)

The affine geometry consists of parallel classes of cosets. When selecting two distinct points this allows us to choose one as the origin, without loss of generality. It is now clear that the number of (affine) subspaces of dimension m containing the origin and the other point (call it x) equals the number of (vector space) subspaces of dimension m containing the one-dimensional subspace generated by x. There are $\begin{bmatrix} n-1 \\ m-1 \end{bmatrix}(q)$ such m-dimensional subspaces, where n is the dimension of the whole vector space [over $GF(q)$]. This number is independent of the original choice of the two points. We thus have a 2-design. The other parameters are even easier to establish and we omit these details.

In a vector space over $GF(2)$ the situation is even nicer (due mostly to the lack of room!). Select three distinct points among which one is the origin (without loss). Then necessarily the other two must be linearly independent (as vectors) and hence must span a subspace of (vector space) dimension 2. The number of affine subspaces of affine dimension m containing the three points equals the number of (vector space) subspaces of dimension m that contain them. There are $\begin{bmatrix} n-2 \\ m-2 \end{bmatrix}(2)$ of these. We thus have a 3-design.

The Multiply Transitive Groups (Section 7.8)

The Mathieu groups M_{12} and M_{24} mentioned in Section 7.8 are 5-transitive groups on 12 and 24 letters, respectively. We refer the reader to [13, pp. 637 and 648] for proofs of 5-transitivity. A general treatment of permutation groups can be found in [14].

Constructions through Linear Codes (Section 7.9)

A result of Assmus and Mattson (see [13, p. 177]) gives sufficient conditions for constructing t-designs from codes. Families of 5-designs have been generated by this method.

The 5-design of Mathieu (on 12 points) that we have listed in its entirety has been generated by the author on a CDC 6600 computer. We list the design as a successive three point extension of the affine plane $AG(2, 3)$, that is, of the 2-(9, 3, 1) design, see (7.5). The reader will observe that the one point extension is the design listed in (7.3), which is a 3-(10, 4, 1) design.

The Method of Differences (Section 7.10)

Let x and y be two distinct elements of G. For a fixed i we refer to $\{\alpha_i + a : a \in G\}$ as the cycle generated by the block α_i. Suppose x and y are in a block β. The block β is in some cycle, say the cycle generated by α_i. Then $\beta = \alpha_i + a$, for some a in G, and in particular $x = c + a$ and $y = d + a$, with c and d elements of α_i. The mapping $\beta \to (c, d)$ is a bijection between the blocks containing x and y and the ordered pairs in the generating blocks whose differences equal $x - y$. There are, by construction, λ such pairs for any nonzero element of G. There are hence λ blocks containing two distinct elements of G. We thus have a 2-design.

Construction through Hypergraphs (Section 7.11)

We mention here the general approach of constructing t-designs through hypergraphs which the reader can find in [16]. Let O_j^k denote the jth orbit in the action of the group G on the k-subsets of edges of H, and let O_i^t denote the ith orbit of G on the t-subsets of edges of H ($2 \leq t < k$). For a t-subset σ_t in O_i^t denote by a_{ij} the number of k-subsets of O_j^k in which σ_t is contained. The number a_{ij} is well defined in the sense that it only depends on O_i^t and O_j^k and not on the specific choice of σ_t in O_i^t. (This is clearly so because of the transitive action of G within an orbit.) Form the matrix $A(t, k) = (a_{ij})$ with the rows indexed by the orbits of the t-subsets of edges. The following statement is now self-evident:

If there exists a vector x with entries 0 and 1 such that $A(t, k)x = \lambda_t x$ then there exists a t-(v, k, λ_t) design. (Hence v is the total number of edges of the hypergraph H.)

It is generally complicated to even determine the dimensions of the matrix $A(t, k)$. This involves the counting of orbits under the action of a group. A substantial theory has been developed in Chapter 6 to handle just this problem. Computing the entries of $A(t, k)$ is no less complicated a task. Much of the work undertaken in this direction involves significant computer interaction.

3 FISHER'S INEQUALITY

7.12

Fisher proved the inequality that bears his name for 2-designs. This inequality informs us that *in a* 2-(v, k, λ_2) *design we must necessarily have at least as many blocks as points*, that is, $\lambda_0 \geq v$.

The original proof, which Fisher gave, goes as follows: Denote by J the matrix with all its entries 1 and by I the identity matrix. Let $N = (n_{ij})$ be the $v \times \lambda_0$ incidence matrix of points versus blocks; that is, $n_{ij} = 1$ if point i belongs to block j and 0 otherwise. Then NN' is a $v \times v$ matrix with all its diagonal entries equal to λ_1 and all its off-diagonal entries equal to λ_2. Write therefore $NN' = (\lambda_1 - \lambda_2)I + \lambda_2 J$. The matrix $NN' - (\lambda_1 - \lambda_2)I = \lambda_2 J$ has rank 1 and its nonzero eigenvalue is $\lambda_2 v$ (the value of the rows sums of $\lambda_2 J$). This tells us that NN' has eigenvalues $(\lambda_1 - \lambda_2) + v\lambda_2 = \lambda_1 + (v - 1)\lambda_2$ of multiplicity 1, and $\lambda_1 - \lambda_2$ of multiplicity $v - 1$. All these eigenvalues are strictly positive [since $\lambda_1 - \lambda_2 = 0$ implies $(v-1)(k-1)^{-1}\lambda_2 = \lambda_2$, or $k = v$, which we do not allow]. Hence the matrix NN' is *nonsingular*. The rank of N must therefore be v and thus N must have at least v columns, that is, $\lambda_0 \geq v$. This ends the proof of Fisher's inequality for 2-designs.

A 2-design in which the number of blocks actually equals the number of points is called *symmetric*. Examples are the Hadamard 2-$(n - 1, \frac{n}{2} - 1, \frac{n}{4} - 1)$ designs and the $PG(n, q)$'s of which the projective planes are special cases. Symmetry has strong geometric implications concerning the intersections of blocks. Our next result shows what these implications are.

Proposition 7.2. *In a* 2-(v, k, λ_2) *design the following conditions are equivalent*:

(i) $\lambda_0 = v$.
(ii) $\lambda_1 = k$.
(iii) *Any two distinct blocks intersect in* λ_2 *points.*

Proof. Observe that $\lambda_0 = v(v-1)k^{-1}(k-1)^{-1}\lambda_2 = vk^{-1}\lambda_1$. From this it directly follows that (i) and (ii) imply each other. Let N denote the incidence matrix between points and blocks. We know that N has rank v (from the proof of Fisher's inequality, which we just gave).

Assume condition (i). This implies that N is a $v \times v$ (nonsingular) matrix. The matrix N satisfies $NJ = JN$ [this simply being condition (ii), which (i) implies]. Since N commutes with J and since $NN' = (\lambda_1 - \lambda_2)I + \lambda_2 J$, we conclude that N commutes also with NN'. Then $N'N = N^{-1}(NN')N = N^{-1}N(NN') = NN' = (\lambda_1 - \lambda_2)I + \lambda_2 J$. The matrix $N'N$ has as (i, j)th entry the cardinality of the intersection of the blocks i and j. The fact that $N'N = (\lambda_1 - \lambda_2)I + \lambda_2 J$ tells us that any two distinct blocks intersect in λ_2 points; this is statement (iii). We thus showed that (i) implies (iii).

To see that (iii) implies (i) think of the blocks as "points" and of the points as "blocks" with a "point" belonging to a "block" if the block contains the respective point. Statement (iii) tells us that any two "points" are in λ_2 "blocks." Our "points" and "blocks" form therefore a 2-$(\lambda_0, \lambda_1, \lambda_2)$ design. Fisher's inequality written for this design gives $v \geq \lambda_0$. In our original 2-(v, k, λ_2) design we have $\lambda_0 \geq v$. Hence $\lambda_0 = v$ and we conclude the proof of our proposition.

The reader should observe the nice *duality* between points and blocks that exists in a symmetric design. Results about points could be dualized into results concerning blocks and conversely.

7.13

The several extensions of Fisher's inequality to t-designs are described in the remaining pages of Section 3. We follow the vector space approach given in [3], although shorter proofs involving matrices exist. To start with we prove Fisher's inequality for t-designs with even t:

$*$ *Let* (P, B) *be a* t-(v, k, λ_t) *design with* $t = 2s$ *(and* $v \geq k + s$*). Then* $\lambda_0 \geq \binom{v}{s}$.

Proof. Let V be a vector space (over the real numbers) with basis indexed by $\Sigma_s(P)$; $\dim V = \binom{v}{s}$. Consider the subspace

$$V(B) = \left\langle \hat{\alpha} = \sum_{\sigma_s \subseteq \alpha} \sigma_s : \alpha \in B \right\rangle$$

spanned by the λ_0 vectors $\hat{\alpha}$, $\alpha \in B$. (In writing $\Sigma_{\sigma_s \subseteq \alpha}\sigma_s$ we identify σ_s in the sum with the basis vector whose index is σ_s.)

We claim that $V(B) = V$. [If we show this, then the number of vectors in the span of $V(B)$ is at least equal to the number of vectors in the basis of V, that is, $|V(B)| = \lambda_0 \geq \binom{v}{s}$, as is to be shown.]

Let $\tau_s \in \Sigma_s(P)$ be fixed. Denote

$$e_i = \sum_{|\sigma_s \cap \tau_s| = s-i} \sigma_s;$$

(note that $e_0 = \tau_s$) and

$$f_i = \sum_{|\alpha \cap \tau_s| = s-i} \hat{\alpha},$$

for $i = 0, 1, \ldots, s$. (Clearly $f_i \in V(B)$, $0 \le i \le s$.) We can express f_r as a linear combination of the e_i's. (The figure below captures the details.) For $\sigma_s \in \Sigma_s(P)$ with $|\sigma_s \cap \tau_s| = s - i$ the coefficient of σ_s in f_r is: the number of blocks α such that $\sigma_s \subseteq \alpha$ and $|\alpha \cap \tau_s| = s - r$.

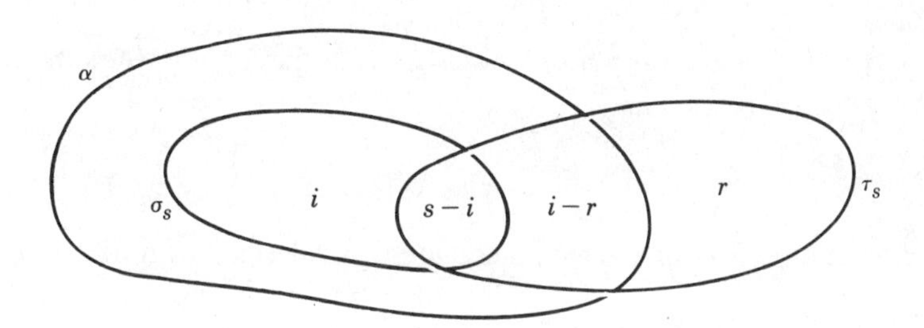

(To see this fix r points in τ_s; there are then $\lambda^r_i + (s - i) + (i - r) = \lambda^r_s - r + i$ blocks α as above that omit these specific r points.) The required coefficient is therefore $\binom{s-(s-i)}{r}\lambda^r_{s-r+i}$. Hence

$$f_r = \sum_{i=r}^{s} \binom{i}{r} \lambda^r_{s-r+i} e_i, \qquad 0 \le r \le s.$$

The associated matrix is $\left(\binom{i}{r}\lambda^r_{s-r+i}\right)_{0 \le r \le s,\, r \le i \le s}$ (the rows being indexed by r). This is an upper triangular matrix; its diagonal elements, λ^i_s, are in fact nonzero since

$$\lambda^i_s = 0 \qquad \text{if and only if} \binom{v-s-i}{k-s}\binom{v-t}{k-t}^{-1}\lambda_t = 0$$
$$\text{if and only if } k + s > v$$

(but our hypothesis assumes $v \ge k + s$). Hence the matrix $\left(\binom{i}{r}\lambda^r_{s-r+i}\right)$ is *nonsingular* and we may solve for $e_0 = \tau_s$ as a linear combination of the f_r's. This shows that $\tau_s \in V(B)$ and hence $V \subseteq V(B)$, which establishes the claim and concludes the proof.

Occasionally Fisher's inequality allows us to conclude the nonexistence of certain t-designs. Consider, for example, whether or not a 6-$(120m, 60m,$

$(20m-1)(15m-1)(12m-1))$ design exists. The numbers

$$\lambda_i = \frac{\binom{120m-i}{6-i}}{\binom{60m-i}{6-i}}(20m-1)(15m-1)(12m-1)$$

are indeed all integers, $0 \le i \le 6$. However, Fisher's inequality requires $\lambda_0 \ge \binom{120m}{3}$. But in this case

$$\lambda_0 = \frac{\binom{120m}{6}}{\binom{60m}{6}}(20m-1)(15m-1)(12m-1)$$
$$= 2(120m-1)(40m-1)(24m-1),$$

which is strictly less than $\binom{120m}{3}$. A 6-design with these parameters, therefore, cannot exist.

7.14

We next give a version of Fisher's inequality for t-designs with odd t.

* *Let (P, B) be a t-(v, k, λ_t) design with $t = 2s+1$* (and $v-1 \ge k+1$). *Then $\lambda_0 \ge 2\binom{v-1}{s}$.*

Proof. Let $a \in P$. Form the derived and residual designs on $P - \{a\}$; both of these are $2s$-designs on $v-1$ points. Applying Fisher's inequality (for even t) to each one of these leads to $\lambda_0 \ge \binom{v-1}{s} + \binom{v-1}{s}$.

The lower bound on λ_0 can be sharpened with additional assumptions on the design:

* *Let (P, B) be a t-(v, k, λ_t) design with $t = 2s$* (and $v \ge k+s$). *Assume that there exists a partition of the blocks $B = B_1 \cup B_2 \cup \cdots \cup B_r$ such that (P, B_i) is a s-$(v, k, \lambda_s(i))$ design, $1 \le i \le r$. Then $\lambda_0 \ge \binom{v}{s} + r - 1$.*

Proof. With the same notation as in the proof of Fisher's inequality we recall that

$$V = V(B) = \langle \hat{\alpha} : \alpha \in B \rangle, \qquad \text{where } \hat{\alpha} = \sum_{\sigma_s \in \alpha} \sigma_s.$$

From our assumption we have

$$\sum_{\alpha \in B_i} \hat{\alpha} = \sum_{\alpha \in B_i} \sum_{\sigma_s \in \alpha} \sigma_s = \sum_{\text{all}} \sum_{\substack{\alpha \\ \sigma_s \subseteq \alpha}} \sigma_s = \lambda_s(i) \sum_{\text{all}} \sigma_s \tag{7.6}$$

[We understand Σ_{all} to mean the sum of all σ_s with $\sigma_s \in \Sigma_s(P)$.] Choose now (and fix) $\alpha_i \in B_i$, $i = 1, 2, \ldots, r$. Expression (7.6) can be written as

$$\hat{\alpha}_i = \lambda_s(i) \sum_{\text{all}} \sigma_s - \sum_{\substack{\alpha \in B_i \\ \alpha \ne \alpha_i}} \hat{\alpha}.$$

Hence

$$V = V(B) = \langle \hat{\alpha} : \alpha \in B \rangle$$

$$= \left\langle \sum_{\text{all}} \sigma_s \text{ and } \hat{\alpha}\text{: with } \alpha \in B - \{\alpha_1, \ldots, \alpha_r\} \right\rangle.$$

V is therefore spanned by $\lambda_0 - r + 1$ vectors; thus $\lambda_0 - r + 1 \geq \binom{v}{s} =$ dimension of V, as desired. This ends the proof.

A design whose blocks can be partitioned as above is often called *resolvable*. An example is design (7.5) from Section 7.7. Verify the inequality we just proved for this example.

7.15

We next investigate when equality is achieved in Fisher's inequality.

A nonnegative integer μ is called an *intersection number* for the t-design (P, B) if there exist two *distinct* blocks α and β ($\in B$) such that $|\alpha \cap \beta| = \mu$.

$*$ *Let* (P, B) *be a* t-(v, k, λ_t) *design, with* $t = 2s$ (and $v \geq k + s$). *Suppose* (P, B) *has precisely* s *distinct intersection numbers*: $\mu_s < \mu_{s-1} < \cdots < \mu_1$ $(< k = \mu_0)$. *Then* $\lambda_0 = \binom{v}{s}$.

Proof. Let W be a vector space (over the real numbers) with basis indexed by α, $\alpha \in B$; $\dim W = \lambda_0$. For $\sigma_s \in \Sigma_s(P)$ define

$$\sigma_s^* = \sum_{\substack{\alpha \\ \alpha \supseteq \sigma_s}} \alpha.$$

Consider the subspace

$$W(S) = \langle \sigma_s^* : \sigma_s \in \sum_s (P) \rangle.$$

We prove that $W(S) = W$. [Observe that all σ_s^*'s are *distinct* and there are $\binom{v}{s}$ of them. If $W(S) = W$, then the span of $W(S)$ contains at least as many vectors as a basis of W; hence $\binom{v}{s} \geq \lambda_0$—we already know that $\lambda_0 \geq \binom{v}{s}$ by Fisher's inequality. Therefore $\lambda_0 = \binom{v}{s}$.]

Fix $\alpha \in B$. We prove that $\alpha \in W(S)$. Set

$$h_i = \sum_{\substack{\beta \in B \\ |\beta \cap \alpha| = \mu_i}} \beta, \qquad 0 \leq i \leq s.$$

For consistency of notation we denote the block size k by μ_0. (Observe that $h_0 = \alpha$.) Let also

$$g_r = \sum_{\substack{\sigma_s \in \Sigma_s(P) \\ |\sigma_s \cap \alpha| = r}} \sigma_s^*, \qquad 0 \leq r \leq s.$$

The g_r's are of course in $W(S)$. We claim that

$$g_r = \sum_{i=0}^{s} \binom{\mu_i}{r}\binom{k-\mu_i}{s-r} h_i.$$

To see this fix $\beta \in B$, $|\beta \cap \alpha| = \mu_i$. We then want to count the number of σ_s such that $\sigma_s \subseteq \beta$ and $|\sigma_s \cap \alpha| = r$. There are $\binom{\mu_i}{r}\binom{k-\mu_i}{s-r}$ such choices as displayed below:

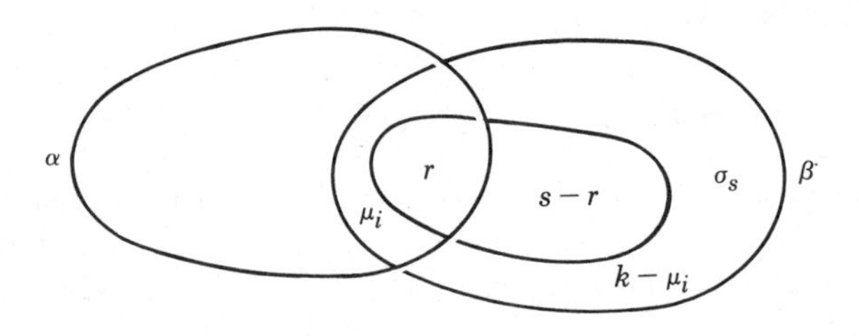

Form the matrix

$$\left(\binom{\mu_i}{r}\binom{k-\mu_i}{s-r}\right)_{\substack{0 \le r \le s \\ 0 \le i \le s}}.$$

If we show that this matrix is nonsingular, then we can solve for $h_0 = \alpha$ as a linear combination of g_r's and hence prove that $\alpha \in W(S)$; thus $W \subseteq W(S)$ and we are one.

To prove nonsingularity denote by v_r the rth row of our matrix, that is, $v_r = \left(\binom{\mu_0}{r}\binom{k-\mu_0}{s-r}, \ldots, \binom{\mu_s}{r}\binom{k-\mu_s}{s-r}\right)$. Assume $\sum_{r=0}^{s} c_r v_r = 0$ and consider the polynomial

$$f(x) = \sum_{r=0}^{s} c_r \binom{x}{r}\binom{k-x}{k-r}$$

[by $\binom{x}{r}$ we mean $[x]_r/r! = x(x-1)\cdots(x-r+1)/r!$]. The polynomial $f(x)$ has degree *at most* s. But $x = \mu_0, \mu_1, \ldots, \mu_s$ are all roots of $f(x)$ ($s+1$ of them!). Hence $f(x) \equiv 0$ for all x. Suitable choices of x give $c_r = 0$, for $0 \le r \le s$. The rows v_r of our matrix are therefore linearly independent; this establishes the nonsingularity and ends our proof.

7.16

We conclude Section 3 with a result that summarizes most of the previous ones. The reader should compare it with Proposition 7.2.

Fisher's Inequality. *Let (P, B) be a t-(v, k, λ_t) design with $t = 2s$* (and $v \geq k + s$). *Then*

(i) $\lambda_0 \geq \binom{v}{s}$.

(ii) *The number of distinct intersection numbers for (P, B) is greater than or equal to s.*

(iii) $\lambda_0 = \binom{v}{s}$ *if and only if the number of distinct intersection numbers for (P, B) equals s.*

Proof. Statement (i) is Fisher's inequality in its initial form. To prove (ii) suppose there are u distinct intersection numbers with $u < s$. (P, B) is a $2s$-design, hence also a $2u$-design. The immediately previous result now gives $\lambda_0 = \binom{v}{u}$; but Fisher's inequality assures $\lambda_0 \geq \binom{v}{s} > \binom{v}{u}$, a contradiction [because $s \leq v - k$ and $s < k$ implies $0 \leq s < v/2$; in this range $u < s$ implies $\binom{v}{u} < \binom{v}{s}$].

To establish (iii) it remains to be shown that if $\lambda_0 = \binom{v}{s}$ then there are at most s (distinct) intersection numbers for (P, B). As in the first proof of Fisher's inequality (Section 7.13) we let V be the vector space over the real numbers freely spanned by σ_s, with $\sigma_s \in \Sigma_s(P)$; dimension of $V = \binom{v}{s}$. Let also $V(B) = \langle \hat{\alpha} : \alpha \in B \rangle$ with $\hat{\alpha} = \Sigma_{\sigma_s \subseteq \alpha} \sigma_s$. We know that $V(B) = V$ (see Section 7.13). Since we assume $\lambda_0 = \binom{v}{s}$ the vectors $\{\hat{\alpha} : \alpha \in B\}$ must be in fact a *basis* for V.

Fix $\alpha \in B$. Then put $\mu_\beta = |\beta \cap \alpha|$ for $B \ni \beta \neq \alpha$. We show that μ_β is the root of a polynomial $f(x)$ (independent of α) of degree $\leq s$. To prove this let

$$m_i = \sum_{|\sigma_s \cap \alpha| = i} \sigma_s, \qquad 0 \leq i \leq s$$

and let

$$n_r = \sum_{\beta \in B} \binom{\mu_\beta}{r} \hat{\beta}, \qquad 0 \leq r \leq s.$$

We first show that $n_r = \Sigma_{i=0}^{s} c_r^i m_i$, $0 \leq r \leq s$ where $c_r^i = \Sigma_{j=0}^{i} \binom{i}{j}\binom{k-i}{r-j} \lambda_{s+r-j}$ (independent of α). Take σ_s, $|\sigma_s \cap \alpha| = i$. The coefficient of σ_s in the sum n_r is $\Sigma_{\beta, \sigma_s \subseteq \beta} \binom{\mu_\beta}{r}$, that is, it equals the number of ordered pairs (β, σ_r) such that $\sigma_s \subseteq \beta$ and $\sigma_r \subseteq \alpha \cap \beta$. (We may, however, think that this number depends on α.) Let us compute it another way: for any $\sigma_r \subseteq \alpha$ with $|\sigma_r \cap \sigma_s| = j$, the number of blocks β such that (β, σ_r) satisfy $\sigma_s \subseteq \beta$ and $\sigma_r \subseteq \alpha \cap \beta$ is λ_{s+r-j}. Thus the coefficient of σ_s in n_r is $\Sigma_{j=0}^{i} \binom{i}{j}\binom{k-i}{r-j} \lambda_{s+r-j}$, which we denote by c_r^i. It is now clear that c_r^i is independent of α. Hence $n_r = \Sigma_{i=0}^{s} c_r^i m_i$, as claimed.

The $s + 1$ vectors $n_r - c_r^s m_s$, $0 \leq r \leq s$, are contained in the span $\langle m_0, \ldots, m_{s-1} \rangle$; this span is of dimension at most s. Therefore the $s + 1$

vectors mentioned above must be linearly dependent. Let $a_0, a_1, \ldots, a_s$ be constants, not all zero, such that

$$\sum_{r=0}^{s} a_r(n_r - c_r^s m_s) = 0.$$

Or

$$\sum_{r=0}^{s} a_r \left[\sum_{\beta \in B} \binom{\mu_\beta}{r} \hat{\beta} - c_r^s \hat{\alpha} \right] = 0$$

(observe that $m_s = \Sigma_{|\sigma_s \cap \alpha| = s} \sigma_s = \Sigma_{\sigma_s \subseteq \alpha} \sigma_s = \alpha$). Since $\{\hat{\beta} : \beta \in B\}$ is a basis for V, for $\beta \neq \alpha$ the coefficient $\Sigma_{r=0}^{s} a_r \binom{\mu_\beta}{r}$ of $\hat{\beta}$ above must be 0. That is, for any $\beta \neq \alpha$ the intersection number μ_β is a root of the polynomial

$$f(x) = \sum_{r=0}^{s} a_r \binom{x}{r}$$

of degree *at most* s [recall again that $\binom{x}{r} = [x]_r / r!$]. The coefficients c_r^s are [and hence $f(x)$ is] independent of the block α; all intersection numbers are therefore roots of $f(x)$ (of which there are at most s). This ends the proof.

EXERCISES

1. Let P be the set of 16 small squares displayed below. To each small square attach the subset of the remaining 6 squares in the same row or column with it (see the figure below):

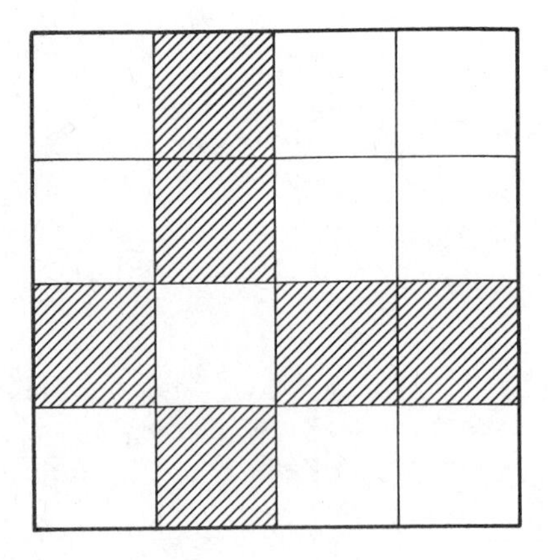

Show that the 16 subsets of size 6 thus obtained are the blocks of a symmetric 2-design.

2. Construct a Hadamard matrix of order 8 and one of order 12.

3. Suppose H is a Hadamard matrix having constant row and column sums. By taking as the ith block the index of the columns in the ith row of H in which the $+1$'s occur we obtain a symmetric 2-design. Prove this and find the parameters of the design. (Concerning the construction of such designs observe that the 4×4 Hadamard matrix $2I - J$ has constant row and column sums; this property is preserved under taking tensor products.)

4. Can a 2-(16, 10, 3) design exist? [*Hint*: Compute the determinant of NN' where N is the incidence matrix of points versus blocks.]

5. In a 2-(v, k, λ_2) design the number of blocks not disjoint from a given block α is at least

$$k(\lambda_1 - 1)^2((k-1)(\lambda_2 - 1) + \lambda_1 - 1)^{-1}.$$

Equality holds if and only if any block not disjoint from α intersects α in a constant number of points. Prove this.

6. List the 21 blocks of the projective plane $PG(2,4)$.

7. Three distinct points on the surface of an ordinary sphere determine a unique circle. Can you think of discrete analogs of this? [*Hint*: The sphere can be conveniently visualized as the field of complex numbers with a point at infinity. Given any two triples of distinct points there exists a map of the form $(az + b)(cz + d)^{-1}$, with $ad - bc \neq 0$, which sends one triple into the other. This allows a geometrical interpretation of the 3-design in (7.3).]

8. Show that a 2-(7, 3, 1) design must necessarily be $PG(2,2)$.

9. Show that a 3-(8, 4, 1) design must necessarily be $AG(3,2)$.

4 EXTENDING SYMMETRIC DESIGNS

7.17

Much of the material in the following three sections concerns symmetric 2-designs. Extending such structures means, roughly speaking, finding a larger t-design (with $t \geq 3$) that contains the initial symmetric design. The motivating examples for such a study may well have been the symmetric Hadamard 2-designs that are contained in the Hadamard 3-designs, as the derived designs at a point—see again examples (7.1) and (7.2). The formal definitions and terminology are given below.

Let (P, B) be a t-(v, k, λ_t) design. For an element a not in P let $\bar{P} = P \cup \{\alpha\}$. A $(t+1)$-design $(\bar{P}, \bar{B})$ is said to be an *extension* of (P, B) if

(P, B) is the derived design at a of $(\bar{P}, \bar{B})$. A design that admits an extension is called *extendable*.

7.18

We now prove the following result due to Cameron:

∗ *If* (P, B) *is a symmetric extendable* 2-(v, k, λ) *design* (with $k \le v - 1$), *then one of the following holds*:

(i) (P, B) *is a Hadamard 2-design with* $v = 4\lambda + 3$ *and* $k = 2\lambda + 1$.

(ii) $v = (\lambda + 2)(\lambda^2 + 4\lambda + 2)$, $k = \lambda^2 + 3\lambda + 1$.

(iii) $v = 111$, $k = 11$, $\lambda = 1$.

(iv) $v = 495$, $k = 39$, $\lambda = 3$.

Proof. Let $\bar{D} = (\bar{P}, \bar{B})$ be an extension of (P, B). Then $\bar{D}$ is a 3-$(v + 1, k + 1, \lambda_3(\bar{D}))$ design, with $\lambda_1(\bar{D}) = v$, $\lambda_2(\bar{D}) = k$, and $\lambda_3(\bar{D}) = \lambda$. For all $a \in \bar{P}$, set $\bar{D}_a = (\bar{P} - \{a\}, \{\alpha - \{a\} : a \in \alpha \in \bar{B}\})$; $\bar{D}_a = (P, B)$ for some $a \in \bar{P}$. Clearly $\bar{D}_a$ is a symmetric 2-(v, k, λ) design. So the *only* intersection number of $\bar{D}_a$ is λ (see Proposition 7.2). This implies that the only intersection numbers of $\bar{D}$ are $\lambda + 1$ and 0.

Fix $\alpha \in \bar{B}$ and define $\bar{D}_\alpha = (\bar{P} - \alpha, \{\beta \in \bar{B} : \beta \cap \alpha = \phi\})$. We claim that $\bar{D}_\alpha$ is a 2-$(v - k, k + 1, (k - \lambda)(\lambda + 1)^{-1})$ design.

Let $a, b \in \bar{P}$; $a, b \notin \alpha$. Count the cardinality of the set $\{(\beta, c) : a, b \in \beta - \alpha \text{ and } c \in \beta \cap \alpha; \beta \in \bar{B}\}$ in two different ways. We obtain (recalling that $c \in \beta \cap \alpha$ implies $|\beta \cap \alpha| = \lambda + 1$)

$$|\{\beta \in \bar{B} : a, b \in \beta \text{ and } \beta \cap \alpha \neq \phi\}|(\lambda + 1) = \lambda(k + 1).$$

It follows that $|\{\beta \in \bar{B} : a, b \in \beta \text{ and } \beta \cap \alpha = \phi\}| = \lambda_2(\bar{D}) - \lambda(k + 1)(\lambda + 1)^{-1} = k - \lambda(k + 1)(\lambda + 1)^{-1} = (k - \lambda)(\lambda + 1)^{-1} \gneq 0$ (if $k = \lambda$, then $k = v$, contrary to our assumption). This proves that $\bar{D}_\alpha$ is a 2-$(v - k, k + 1, (k - \lambda)(\lambda + 1)^{-1})$ design.

In particular, in $\bar{D}_\alpha$ the block size should not exceed the number of points, which means $v - k \ge k + 1$.

1. Suppose $v - k = k + 1$. Since (P, B) is a symmetric design we have $\lambda_0 = v$ and hence also $k = \lambda_1 = ((v - 1)/(k - 1))\lambda$; replacing $v = 2k + 1$ gives $\lambda = \frac{1}{2}(k - 1)$. Or, solving for k, $k = 2\lambda + 1$ and then $v = 4\lambda + 3$, which is (i).

2. If $v - k = k + 2$, then $(\lambda_1 =)$ $k = (v - 1)(k - 1)^{-1}\lambda = (2k + 1)(k - 1)^{-1}\lambda$, or $\lambda = k(k - 1)(2k + 1)^{-1}$, which is not an integer. Hence this case can never happen.

3. Assume $v - k > k + 2$. Then

$$\lambda_0(\overline{D}_\alpha) = \frac{(v-k)(v-k-1)}{(k+1)k}\lambda_2(\overline{D}_\alpha) = \frac{(v-k)(v-k-1)(k-\lambda)}{(k+1)k(\lambda+1)}$$

$$= \left\{\text{since } \lambda = \frac{k(k-1)}{v-1}\right\} = \frac{(v-k)^2(v-k-1)}{(k+1)(k^2-k+v-1)} \geq v-k$$

(this being Fisher's inequality applied to $\overline{D}_\alpha$). Or, equivalently,

$$v^2 - (3k+2)v - (k+1)(k^2-2k-1) \geq 0. \tag{7.7}$$

Consider the quadratic $x^2 - (3k+2)x - (k+1)(k^2-2k-1)$ with roots $x = \frac{1}{2}[(3k+2) \pm k\sqrt{4k+5}\,]$. Since $x = \frac{1}{2}[(3k+2) - k\sqrt{4k+5}\,] \leq 1$ for all integral k, the solution to (7.7) is

$$v \geq \tfrac{1}{2}\left[(3k+2) + k\sqrt{4k+5}\,\right]. \tag{7.8}$$

Put $\sqrt{4k+5} = 3 + 2\mu$, for $\mu > 0$. Then $k = \mu^2 + 3\mu + 1$ and (7.8) becomes $v \geq \mu^3 + 6\mu^2 + 10\mu + 4 = (\mu+2)(\mu^2+4\mu+2)$. Hence

$$\lambda = \frac{k(k-1)}{v-1} \leq \frac{(\mu^2+3\mu+1)(\mu^2+3\mu)}{\mu^3+6\mu^2+10\mu+3} = \frac{(\mu^2+3\mu+1)(\mu+3)\mu}{(\mu^2+3\mu+1)(\mu+3)} = \mu,$$

so that $k = \mu^2 + 3\mu + 1 \geq \lambda^2 + 3\lambda + 1$, or

$$k + 1 \geq (\lambda+1)(\lambda+2). \tag{7.9}$$

But $\lambda_0(\overline{D}) = (v+1)v/(k+1)$, which shows that $k+1$ divides $(v+1)v$. [Since $\lambda = k(k-1)/(v-1)$, we have $(v+1)v = \lambda^{-2}(k^2-k+\lambda)(k^2-k+2\lambda)$.] Therefore $k+1$ divides $(k^2-k+\lambda)(k^2-k+2\lambda)$; looking at the remainder we finally conclude that

$$k+1 \quad \text{divides} \quad 2(\lambda+1)(\lambda+2). \tag{7.10}$$

Conditions (7.9) and (7.10) restrict k very much. In fact we can only have

$$k+1 = 2(\lambda+1)(\lambda+2) \qquad \text{or} \qquad k+1 = (\lambda+1)(\lambda+2).$$

Assume $k+1 = 2(\lambda+1)(\lambda+2)$. Since $\lambda = k(k-1)/(v-1)$, λ divides $k(k-1) = (2\lambda^2+6\lambda+3)(2\lambda^2+6\lambda+2)$; looking at the remainder we conclude that λ divides 6. Hence $\lambda = 1, 2, 3$, or 6. For $\lambda = 2$ or 6, $k+1$ does not divide $v(v+1)$. Thus $\lambda = 1$ or 3, giving cases (iii) and (iv), respectively.

The case $k+1 = (\lambda+1)(\lambda+2)$ gives $\lambda = \mu$, leading to (ii); in this case $v = (\lambda+2)(\lambda^2+4\lambda+2)$ and we obtain a 2-$((\lambda+1)^2(\lambda+3), (\lambda+1)(\lambda+2), (\lambda+1))$ design for $\overline{D}_\alpha$. This ends the proof.

7.19

Consequence. *If* (P, B) *is a twice or three times extendable symmetric 2-design, then* (P, B) *is the unique* 2-(21, 5, 1) *design. There is no four times extendable symmetric 2-design.*

Proof. Let (P, B) be a symmetric twice extendable 2-(v, k, λ) design. Counting the number of blocks in the second extension [and using the fact that $\lambda_i = \binom{v-i}{k-i}\binom{k-i}{t-i}\lambda$; $i = 0, 1$] we find that $(k+1)(k+2)$ divides $v(v+1)(v+2)$. For the various possibilities for (P, B) listed in the previous result this leads to:

(i) $(2\lambda + 3)$ divides $4(\lambda + 1)(4\lambda + 3)(4\lambda + 5)$ which, upon looking at the remainder, tells us that $(2\lambda + 3)$ divides 3; this cannot happen.

(ii) $\lambda^2 + 3\lambda + 3$ divides

$$(\lambda + 1)(\lambda + 2)(\lambda^2 + 4\lambda + 2)(\lambda^2 + 5\lambda + 5)(\lambda^3 + 6\lambda^2 + 10\lambda + 6),$$

leading us to conclude that $\lambda^2 + 3\lambda + 3$ divides $\lambda + 6$ (again upon looking at the remainder). This can only happen for $\lambda = 1$. Then (P, B) is a 2-(21, 5, 1) design, the 21 point projective plane $PG(2, 4)$. There is up to isomorphism a unique such object and it is actually three times extendable [leading to the Mathieu 5-(24, 8, 1) design]. Since $4 \cdot 5 \cdot 6 \cdot 7 \cdot 8 \cdot 9$ does not divide $20 \cdot 21 \cdot 22 \cdot 23 \cdot 24 \cdot 25$ it is not four times extendable.

(iii) 13 does not divide $111 \cdot 112 \cdot 113$.

(iv) Finally, 41 does not divide $495 \cdot 496 \cdot 497$.

5 ON THE EXISTENCE OF SYMMETRIC DESIGNS

Necessary conditions for a 2-(v, k, λ_2) design to exist are that $\lambda_i = \binom{v-i}{2-i}\binom{k-i}{2-i}^{-1}\lambda_2$, $i = 0, 1$, be integers and that $\lambda_0 \geq v$ (which is Fisher's inequality). These conditions are in general far from being sufficient. Even with the additional assumption of symmetry they do not suffice. In this section we give another necessary condition for the existence of a (symmetric) 2-design. The result is known as Bruck, Ryser, and Chowla's (BRC) theorem. Since in this section we discuss 2-designs we write λ for λ_2.

7.20

The proof of the BRC theorem is based on rational congruences, Witt's cancellation law, and a result of Lagrange in number theory. We first define what is meant by rational congruence.

Two square and symmetric matrices A and B with rational entries (and of the same dimension) are called *rationally congruent* (written $A \overset{c}{=} B$) if there exists a nonsingular matrix P with rational entries such that $P'AP = B$. (The relation $\overset{c}{=}$ is an equivalence relation, and this is easy to check.)

The result of Lagrange to which we made reference is the following:

Lagrange's Result. *For any positive integer m we have $mI_4 \overset{c}{=} I_4$.*

[Here, and throughout this section, I_n denotes the $n \times n$ identity matrix. Lagrange's original result states that any positive integer m is the sum of four squares, that is, $m = a^2 + b^2 + c^2 + d^2$ with a, b, c, d nonnegative integers. But then

$$\begin{pmatrix} a & b & c & d \\ b & -a & d & -c \\ d & c & -b & -a \\ c & -d & -a & b \end{pmatrix} I_4 \begin{pmatrix} a & b & d & c \\ b & -a & c & -d \\ c & d & -b & -a \\ d & -c & -a & b \end{pmatrix} = mI_4$$

showing that $mI_4 \overset{c}{=} I_4$, as stated above.]

The other result that we need is the following.

Witt's Cancellation Law. *If* $\begin{pmatrix} A & 0 \\ 0 & B \end{pmatrix} \overset{c}{=} \begin{pmatrix} C & 0 \\ 0 & D \end{pmatrix}$ *and* $A \overset{c}{=} C$, *then* $B \overset{c}{=} D$.

The proofs of these two results are included in Appendix 3.

7.21

We now state and prove the BRC theorem:

The Theorem of Bruck, Ryser, and Chowla. *Suppose v, k, λ are natural numbers such that a symmetric 2-(v, k, λ) design exists* (and $0 \leq k + 2 \leq v$). *Then*

(i) *If v is even $k - \lambda$ must be a square.*
(ii) *If v is odd the Diophantine equation*

$$x^2 = (k - \lambda)y^2 + (-1)^{(v-1)/2}\lambda z^2$$

has a solution in integers x, y, and z, not all zero.

Proof. Let (P, B) be a symmetric 2-(v, k, λ) design. Denote by N the $v \times v$ incidence matrix (points versus blocks) of (P, B). Then $NN' = (\lambda_1 - \lambda)I_v + \lambda J$, where J is the $v \times v$ matrix with all its entries 1. By observing that $NN' - (\lambda_1 - \lambda)I_v = \lambda J$ is a matrix of rank 1 it follows immediately that the eigenvalues of NN' are $\lambda_1 - \lambda$ (of multiplicity $v - 1$) and $\lambda_1 + (v - 1)\lambda$ (of multiplicity 1); see the proof of Fisher's inequality in Section 7.12 for a more detailed derivation. We can now write

$$\det NN' = (\lambda_1 + (v - 1)\lambda)(\lambda_1 - \lambda)^{v-1} = k\lambda_1(\lambda_1 - \lambda)^{v-1} \qquad (7.11)$$

where $\det NN'$ denotes the determinant of NN'. [To explain the last sign of equality recall that $\lambda_1 = (v - 1)(k - 1)^{-1}\lambda$.]

(a) The assumption of symmetry implies $\lambda_1 = k$. Expression (7.11) is thus rewritten as

$$(\det N)^2 = \det NN' = k^2(k - \lambda)^{v-1}. \qquad (7.12)$$

Since $v-1$ is odd and since the left-hand side of (7.12) is a square it must be that each prime in the prime factorization of $k-\lambda$ occurs at an even power. Hence $k-\lambda$ is necessarily a square. This ends the proof of part (i) of the theorem.

(b) Form the $(v+1)\times(v+1)$ matrix

$$\overline{N}=\begin{bmatrix} N & \mathbf{1} \\ \mathbf{1}' & k\lambda^{-1}\end{bmatrix}$$

(with $\mathbf{1}$ the vector with all entries 1) and let

$$D=\operatorname{diag}(1,\ldots,1,-\lambda),\ E=\operatorname{diag}\big(k-\lambda,\ldots,k-\lambda,-(k-\lambda)\lambda^{-1}\big).$$

Matrices D and E are diagonal, of dimension $v+1$.

Then $NN'=(\lambda_1-\lambda)I_v-\lambda J$, $\lambda_1=(v-1)(k-1)^{-1}\lambda$, and $\lambda_1=k$ allow us to write

$$\begin{aligned}\overline{N}D\overline{N}'&=\begin{bmatrix} N & \mathbf{1} \\ \mathbf{1}' & k\lambda^{-1}\end{bmatrix}\begin{bmatrix} I_v & 0 \\ 0 & -\lambda\end{bmatrix}\begin{bmatrix} N' & \mathbf{1} \\ \mathbf{1}' & k\lambda^{-1}\end{bmatrix}\\ &=\begin{bmatrix} N & -\lambda\mathbf{1} \\ \mathbf{1}' & -k\end{bmatrix}\begin{bmatrix} N' & \mathbf{1} \\ \mathbf{1}' & k\lambda^{-1}\end{bmatrix}=\begin{bmatrix} NN'-\lambda J & 0 \\ 0 & -(k-\lambda)\lambda^{-1}\end{bmatrix}\\ &=\begin{bmatrix} (k-\lambda)I_v & 0 \\ 0 & -(k-\lambda)\lambda^{-1}\end{bmatrix}=E.\end{aligned}$$

Hence

$$D\overset{c}{=}E. \tag{7.13}$$

By Lagrange's result we know that $mI_4\overset{c}{=}I_4$. Extending this trivially to direct sums of matrices we can write

$$mI_n\overset{c}{=}I_n, \tag{7.14}$$

for any $n=0$ (modulo 4) and any positive integer m.

(b1) Let $v=1$ (modulo 4). Condition (7.13) can be rewritten as

$$\begin{aligned}I_{v-1}\oplus I_1\oplus-\lambda I_1&\overset{c}{=}(k-\lambda)I_{v-1}\oplus(k-\lambda)I_1\oplus-(k-\lambda)\lambda^{-1}I_1\\ &\overset{c}{=}\{\text{by (7.14)}\}\overset{c}{=}I_{v-1}\oplus(k-\lambda)I_1\oplus-(k-\lambda)\lambda^{-1}I_1.\end{aligned}$$

Witt's cancellation law now gives

$$\begin{pmatrix}1 & 0\\ 0 & -\lambda\end{pmatrix}\overset{c}{=}\begin{pmatrix}k-\lambda & 0\\ 0 & -(k-\lambda)\lambda^{-1}\end{pmatrix}$$

This means that there exists a nonsingular matrix $A=\begin{pmatrix}a & b\\ c & d\end{pmatrix}$ with *rational* entries such that

$$A'\begin{pmatrix}1 & 0\\ 0 & -\lambda\end{pmatrix}A=\begin{pmatrix}k-\lambda & 0\\ 0 & -(k-\lambda)\lambda^{-1}\end{pmatrix}.$$

In particular we must have $a^2 - \lambda c^2 = k - \lambda$. Multiplying by a common denominator leads us to conclude that $x^2 - \lambda z^2 = (k - \lambda)y^2$ must admit a solution in integers x, y, z, not all zero.

(b2) In case $v \equiv 3$ (modulo 4) we work with diagonal matrices of order $v + 2$ by adding an additional component $(k - \lambda)I_1$ to both D and E. Then by (7.13) and (7.14)

$$\begin{aligned} I_v \oplus (k-\lambda)I_1 \oplus -\lambda I_1 &\overset{c}{=} (k-\lambda)I_v \oplus (k-\lambda)I_1 \oplus -(k-\lambda)\lambda^{-1}I_1 \\ &\overset{c}{=} \{\text{by } (7.14)\} \overset{c}{=} I_{v+1} \oplus -(k-\lambda)\lambda^{-1}I_1 \\ &= I_v \oplus I_1 \oplus -(k-\lambda)\lambda^{-1}I_1. \end{aligned}$$

Again by Witt's cancellation law it follows that

$$\begin{pmatrix} k-\lambda & 0 \\ 0 & -\lambda \end{pmatrix} \overset{c}{=} \begin{pmatrix} 1 & 0 \\ 0 & -(k-\lambda)\lambda^{-1} \end{pmatrix}.$$

We can therefore write

$$\begin{pmatrix} a & b \\ c & d \end{pmatrix}\begin{pmatrix} k-\lambda & 0 \\ 0 & -\lambda \end{pmatrix}\begin{pmatrix} a & c \\ b & d \end{pmatrix} = \begin{pmatrix} 1 & 0 \\ 0 & -(k-\lambda)\lambda^{-1} \end{pmatrix},$$

with $\begin{pmatrix} a & b \\ c & d \end{pmatrix}$ a nonsingular matrix with *rational* entries. Equating the entries in position $(1,1)$ on both sides we obtain

$$1 = (k-\lambda)a^2 - \lambda c^2.$$

Multiplying, as before, by the common denominator of a^2 and c^2 we conclude that the Diophantine equation

$$x^2 = (k-\lambda)y^2 - \lambda z^2$$

must admit a solution in integers x, y, z, not all zero. This concludes our proof.

7.22

Relying on the BRC theorem we can conclude that certain (potential) symmetric 2-designs actually do not exist.

A 2-(22, 7, 2) *design, for example, cannot exist*. This follows from the fact that if such a design exists it must necessarily be symmetric and, since $v = 22$ is even, the BRC theorem requires that $k - \lambda = 7 - 2 = 5$ be a square. But 5 is not a square and consequently such a design does not exist.

The true strength of the BRC theorem is contained in the case with odd v. *Let us show that a potential* 2-(43, 7, 1) *design does not exist*. A design with these parameters (if it exists) must be symmetric, because $\lambda_0 = \binom{43}{2}\binom{7}{2}^{-1} \cdot 1 = 43 = v$. By the BRC theorem a necessary condition for its existence is that the equation

$$x^2 = 6y^2 - z^2 \tag{7.15}$$

admits a solution in integers x, y, z, not all zero. We now show that this equation does not have an integral nonzero solution.

Write (7.15) as $6y^2 = x^2 + z^2$. The prime 3 divides $6y^2$ and hence $x^2 + z^2$. Let r be the highest power of 3 that divides both x and z. Dividing equation (7.15) by 3^{2r} we obtain

$$\left(\frac{x}{3^r}\right)^2 + \left(\frac{z}{3^r}\right)^2 = \frac{6y^2}{3^{2r}} = 6\left(\frac{y}{3^r}\right)^2.$$

Observe that 3^r must necessarily divide y and denote $x/3^r = \bar{x}$, $z/3^r = \bar{z}$, and $y/3^r = \bar{y}$. Now (7.15) can be rewritten as

$$6\bar{y}^2 = \bar{x}^2 + \bar{z}^2. \tag{7.16}$$

Equation (7.15) admits a nonzero solution in integers if and only if equation (7.16) does; this is clear since the two equations are nonzero multiples of each other. By the way in which r was selected in (7.16) 3 divides $6\bar{y}^2$ but it does not divide $\bar{x}$ nor $\bar{y}$. Interpret now equation (7.16) over the field $GF(3)$, that is, look at it modulo 3. The numbers $\bar{x}$ and $\bar{y}$ are now nonzero elements of $GF(3)$ that satisfy

$$0 = \bar{x}^2 + \bar{y}^2. \tag{7.17}$$

Or, dividing out by $\bar{y}^2$, $(\bar{x}/y)^2 = \bar{x}^2/\bar{y}^2 = -1$, that is, we conclude that -1 is a square in $GF(3)$.

What we showed is that if a 2-(43, 7, 1) design exists, then -1 must be a square in $GF(3)$. But -1 is not a square in $GF(3)$ since $(-1)^2 = 1^2 = 1$ and $0^2 = 0$. We are now forced to conclude that a 2-(43, 7, 1) design cannot exist.

7.23 Nonexistence of Certain Projective Planes

A *projective plane* is a symmetric 2-$(v, k, 1)$ design. By letting $n = k - 1$ we can write the parameters v and k in terms of n only: $v = n^2 + n + 1$ and $k = n + 1$ (note that v is always odd, regardless of the parity of n). The number n is called the *order* of the projective plane 2-$(n^2 + n + 1, n + 1, 1)$.

If $n = 0$ or 3 (modulo 4), the BRC equation always has the solution $x = 1$, $y = 0$, $z = 1$. We therefore cannot conclude anything concerning the existence of such planes.

However, if $n = 1$ or 2 (modulo 4), the BRC equation becomes $ny^2 = x^2 + z^2$. In Section 7.22 we showed that this equation has no nonzero solution if $n = 6$. We showed, in other words, that a projective plane of order 6 does not exist. Using arguments absolutely analogous to those used in Section 7.22, and working modulo 3, 7, or 11, we conclude that projective planes of order 6, 14, 21, 22, 30, and 33 cannot exist. (It is in fact a well-known result in the theory of numbers that the equation $ny^2 = x^2 + y^2$ admits nonzero solutions in integers if and only if n is the sum of two squares. At this point it is becoming abundantly clear that knowledge about the existence of symmetric 2-designs rests fundamentally within the domain of that rich theory. We therefore direct the reader's attention to the pertinent results in the theory of numbers.)

All known projective planes have n a power of a prime. The $PG(2, n)$ are examples familiar to us (see Section 7.6). The smallest values of n for which it is not known whether a projective plane exists are: 10, 12, 15, 18, 20, 24.

7.24

A positive integer m is said to be *square free* if in the prime factorization of m all the (distinct) primes occur at power 1, that is, $m = \prod_i p_i$, with p_i distinct primes. Two positive integers a and b are called *relatively prime* [written $(a, b) = 1$] if there does not exist a prime number that divides them both. Our aim is to give a version of the BRC theorem that is easily applicable to practical situations.

Let us first consider a general Diophantine equation of the form $ax^2 + by^2 + cz^2 = 0$. Writing $a = \bar{a}A^2$, $b = \bar{b}B^2$, $c = \bar{c}C^2$ with $\bar{a}$, $\bar{b}$, and $\bar{c}$ square free, we can immediately conclude that the equation $ax^2 + by^2 + cz^2 = 0$ has nonzero integral solutions if and only if $\bar{a}x^2 + \bar{b}y^2 + \bar{c}z^2 = 0$ does.

Consider now the equation $\bar{a}x^2 + \bar{b}y^2 + \bar{c}z^2 = 0$ with $\bar{a}$, $\bar{b}$, $\bar{c}$ square free and also make the assumption that $\bar{a}$, $\bar{b}$, $\bar{c}$ are pairwise relatively prime. Let (x, y, z) be a nonzero integral solution. If p is a prime dividing $\bar{a}$, we may assume (after possibly dividing our solution through by a power of p) that p does not divide y nor z. Modulo p the equation $\bar{a}x^2 + \bar{b}y^2 + \bar{c}z^2 = 0$ becomes

$$\bar{b}y^2 = -\bar{c}z^2$$

or, multiplying through by $\bar{b}$,

$$\bar{b}^2y^2 = -\bar{b}\bar{c}z^2.$$

The last equation informs us that $-\bar{b}\bar{c}$ $(= (\bar{b}y/z)^2)$ must necessarily be a square modulo p.

We can summarize this as follows: Necessary conditions for the existence of a nonzero integral solution to $\bar{a}x^2 + \bar{b}y^2 + \bar{c}z^2 = 0$ with $\bar{a}$, $\bar{b}$, $\bar{c}$ square free and pairwise relatively prime are that, for all primes p,

1. If p divides $\bar{a}$, then $-\bar{b}\bar{c}$ is a square modulo p.
2. If p divides $\bar{b}$, then $-\bar{a}\bar{c}$ is a square modulo p.
3. If p divides $\bar{c}$, then $-\bar{a}\bar{b}$ is a square modulo p.
4. The coefficients $\bar{a}$, $\bar{b}$, and $\bar{c}$ do not all have the same sign.

(Condition 4 is obvious.)

With these remarks made let us study the BRC Diophantine equation $-x^2 + ny^2 + (-1)^{(v-1)/2}\lambda z^2$, where n stands for $k - \lambda$. First write the equation in the form $-x^2 + \bar{n}y^2 + (-1)^{(v-1)/2}\bar{\lambda}z^2$, with $\bar{n}$ and $\bar{\lambda}$ square free.

For a prime p we conclude the following:

(i) If $\bar{n}$ and $\bar{\lambda}$ are relatively prime and if p divides $\bar{n}$, then $(-1)^{(v-1)/2}\bar{\lambda}$ must be a square modulo p.

(ii) If $\bar{n}$ and $\bar{\lambda}$ are relatively prime and if p divides $\bar{\lambda}$, then n must be a square modulo p.

(iii) If p divides $\bar{n}$ and $\bar{\lambda}$, then observe that the equation

$$-x^2 + \bar{n}y^2 + (-1)^{(v-1)/2}\bar{\lambda}z^2 = 0$$

has a nonzero integral solution if and only if the equation

$$-px^2 + \frac{\bar{n}}{p}y^2 + (-1)^{(v-1)/2}\frac{\bar{\lambda}}{p}z^2 = 0$$

has such a solution. The coefficients $\bar{n}/p$ and $\bar{\lambda}/p$ may still not be relatively prime. But after eventually dividing a nonzero solution through by a power of p we can assume that p does not divide y nor z. Working again modulo p we conclude that

$$-\frac{\bar{n}}{p}y^2 = (-1)^{(v-1)/2}\frac{\bar{\lambda}}{p}z^2.$$

Multiplying through by $-\bar{n}/p$ we deduce that $(-\bar{n}/p)(-1)^{(v-1)/2}(\bar{\lambda}/p)$ must be a square modulo p.

We summarize as follows.

An Applicable Version of the BRC Theorem. *Suppose v, k, λ are natural numbers such that a symmetric 2-(v, k, λ) design exists $(0 \le k + 2 \le v)$, and v is an odd number. Denote $k - \lambda$ by n and let $\bar{n}$ and $\bar{\lambda}$ be the square free parts of n and λ. Then for every prime p the following statements are true:*

(i) *If p divides $\bar{n}$ but not $\bar{\lambda}$, then $(-1)^{(v-1)/2}\bar{\lambda}$ must be a square modulo p.*
(ii) *If p divides $\bar{\lambda}$ but not $\bar{n}$, then $\bar{n}$ must be a square modulo p.*
(iii) *If p divides both $\bar{n}$ and $\bar{\lambda}$, then $(-1)^{(v+1)/2}(\bar{n}/p)(\bar{\lambda}/p)$ must be a square modulo p.*

6 AUTOMORPHISMS OF DESIGNS

7.25

Let (P, B) be a t-(v, k, λ_t) design and g a permutation on P. Then g induces a permutation on the k-subsets of P by the "natural" action $\{x_1, \ldots, x_k\} \to \{g(x_1), \ldots, g(x_k)\}$. If g also induces a permutation on B we call g *an automorphism* of the design (P, B). More generally, a group of permutations on the points of a t-design that preserves its blocks is called an *automorphism group* of that design.

As an example, the group generated by the permutations (1 2 3 4 5 6 7), (2 7 6)(4 3 5), and (2 3 4 7)(5 6) is an automorphism group of $PG(2,2)$ as displayed in (7.2). Verify this.

The *full automorphism group* of a t-design is the group of *all* permutations on points that preserve the blocks. [It is generally difficult to find the full automorphism group of a design. Try to accomplish this for the design (7.2)!]

An automorphism group of a design has a permutation representation on the blocks of the design (in addition to its initial representation on points). It is thence natural to compare the induced action on blocks to that on points. Our first observation is the following:

$*$ *A permutation fixes all the blocks of a 2-design if and only if it fixes all the points*. (In other words the representation on blocks is faithful.)

The statement we just made is easy to justify. It is clear that the identity permutation on points fixes all the blocks; it fixes them pointwise in fact. Suppose now that a permutation on points induces the identity permutation on blocks, that is, it fixes each individual block. If x is a point, then the intersection of all blocks containing x consists of x *alone* [else we would have $\lambda_2 \geq \lambda_1$, or $\lambda_2 \geq (v-1)(k-1)^{-1}\lambda_2$, or $k \geq v$, which is a contradiction]. Since each block containing x is fixed, the intersection of these blocks (i.e., x itself) is also fixed. The permutation therefore fixes all points, as was asserted.

Another helpful observation is the following:

$*$ *The full automorphism group of a 2-(v, k, λ_2) design $(0 \leq k+2 \leq v)$ contains the whole alternating group on the v points if and only if the 2-design is a complete design.*

Proof. Denote by S_v the symmetric group on the v points; $|S_v| = v!$. It is clear that any complete design on the v points has S_v as the (full) automorphism group, and S_v contains A_v, the alternating group (of even permutations) on the v points; $|A_v| = v!/2$. Conversely, if A_v is an automorphism group and α is a block in our 2-design, then by assumption the set $\{g(\alpha) : g \in A_v\}$ consists of blocks of our 2-design. But the group A_v is $(v-2)$-transitive (thus also k-transitive) on points and hence $\{g(\alpha) : g \in A_v\}$ is a complete 2-design. This concludes the proof.

7.26

We devote this section to the proof of the following result:

$*$ *An automorphism group of a t-design $(t \geq 2)$ has at least as many orbits on the blocks of the t-design as it has on the $[t/2]$-subsets of points.*

The notation $[x]$ indicates the integral part of the fraction x.

Let (P, B) be a t-(v, k, λ_t) design and G an automorphism group of (P, B). Denote $[t/2]$ by s. The group G acts on the $\binom{v}{s}$ s-subsets of P. The resulting orbits are called *s-orbits*. For a s-subset x of P we denote by $\bar{x}$ its

s-orbit. Similarly, the orbits that G induces on B are called *block-orbits*, and we write $\bar{\alpha}$ for the block-orbit of block α. Let m be the number of s-orbits and n the number of block-orbits.

Form the $m \times n$ matrix $\bar{A} = (a_{ij})$ of s-orbits versus block-orbits with a_{ij} being the number of blocks in block-orbit j that contains a certain s-subset from s-orbit i. Observe that a_{ij} is well defined, in that it is independent of the choice of orbit representatives. Let also $\bar{B} = (b_{ij})$ be the $n \times m$ matrix of block-orbits versus s-orbits with b_{ij} being the number of s-subsets in s-orbit j contained in a block from block-orbit i; the entries b_{ij} are well defined as well.

Look now at the matrix $\bar{A}\bar{B} = (c_{ij})$. This matrix is $m \times m$. We aim to prove that $\bar{A}\bar{B}$ is *nonsingular*. The nonsingularity of $\bar{A}\bar{B}$ implies, in particular, that $\bar{A}$ has at least as many columns as it has rows, that is, $n \geq m$, which is what our result states.

Let us investigate the matrix $\bar{A}\bar{B} = (c_{ij})$ a bit more closely. The entry c_{ij} equals $\sum_{k=1}^{n} a_{ik} b_{kj}$. We next find a simpler expression for c_{ij}.

Fix a s-subset x in s-orbit i and count the cardinality of the set

$$S_x^j = \{(y, \alpha) : x \in \alpha,\ y \in \alpha,\ y \in s\text{-orbit } j, \alpha \in B\}$$

in two different ways. Let y be fixed, initially; then there are $\lambda_{|x \cup y|}$ blocks containing both s-subsets x and y, and since y runs through s-orbit j we obtain $|S_x^j| = \sum_y \lambda_{|x \cup y|}$ (the summation is over all s-subsets y in s-orbit j). Now count differently: fix α first. Then α belongs to some block-orbit, orbit k, say; we have b_{kj} choices for y and a_{ij} choices for α in block-orbit k. Summing over k leads to $|S_x^j| = \sum_{k=1}^{n} a_{ik} b_{kj}$. We therefore conclude that

$$c_{ij} = \sum_{k=1}^{n} a_{ik} b_{kj} = \sum_{y} \lambda_{|x \cup y|}.$$

The reader should understand that $c_{ij} = \sum_y \lambda_{|x \cup y|}$ (with y running over the s-subsets in s-orbit j and x being a fixed s-subset in s-orbit i) is a well-defined expression that does not depend on the particular choice of x in s-orbit i. It is this simpler expression of c_{ij} that we use to establish the nonsingularity of $\bar{A}\bar{B}$.

Define $N_s = (n_{ij})$, the $\binom{v}{s} \times \lambda_0$ incidence matrix of s-subsets versus blocks, with $n_{ij} = 1$ if the ith s-subset is contained in block j, and $n_{ij} = 0$ otherwise. Line up, moreover, the rows of N_s such that the s-subsets in s-orbit 1 come first, then those in s-orbit $2, \ldots,$ and lastly those in s-orbit m.

The matrix $N_s N_s'$ is $\binom{v}{s} \times \binom{v}{s}$ with (x, y)th entry equal to $\lambda_{|x \cup y|}$. It is a positive definite matrix and thus nonsingular (the positive definiteness is established immediately following the proof). Write $N_s N_s' = (C_{ij})$ as a partitioned matrix with C_{ij} the submatrix of $N_s N_s'$ of the ith s-orbit versus the jth s-orbit.

Observe that the row sums of C_{ij} equal $\sum_y \lambda_{|x \cup y|}$, where x is in s-orbit i and y runs over the s-subsets of s-orbit j. This common value for the row sums of C_{ij} is well defined (in the sense that it does not depend upon the choice of x in s-orbit i). We thus conclude that $\overline{A}\overline{B} = (c_{ij})$, where $c_{ij} = \Sigma_y \lambda_{|x \cup y|}$ = common value of the row sums of C_{ij}.

We end the proof by showing that the eigenvalues of $\overline{A}\overline{B}$ are necessarily among those of $N_s N_s'$. And since the eigenvalues of $N_s N_s'$ are positive we conclude that $\overline{A}\overline{B}$ is nonsingular.

Indeed, let

$$\overline{A}\overline{B}w = (c_{ij})\begin{pmatrix} w_1 \\ \vdots \\ w_m \end{pmatrix} = \mu \begin{pmatrix} w_1 \\ \vdots \\ w_m \end{pmatrix} = \mu w.$$

Then

$$N_s N_s' \begin{pmatrix} w_1 \mathbf{1} \\ \vdots \\ w_m \mathbf{1} \end{pmatrix} = (C_{ij}) \begin{pmatrix} w_1 \mathbf{1} \\ \vdots \\ w_m \mathbf{1} \end{pmatrix} = \mu \begin{pmatrix} w_1 \mathbf{1} \\ \vdots \\ w_m \mathbf{1} \end{pmatrix},$$

which shows that if μ is an eigenvalue of $\overline{A}\overline{B}$, then μ is also an eigenvalue of $N_s N_s'$. The $\mathbf{1}$ in $w_j \mathbf{1}$ is the vector with all its entries equal to 1 and of length equal to the number of columns of C_{ij}.

Proof of the Positive Definiteness of $N_s N_s'$. This proof is due to Wilson [4]. Denote by W_{ij} the $\binom{v}{i} \times \binom{v}{j}$ incidence matrix between the i-subsets and j-subsets of P. Then

$$N_s N_s' = \sum_{i=0}^{s} \lambda^i_{2s-i} W_{is}' W_{is},$$

with λ^l_m signifying the number of blocks containing m points and omitting l [see Proposition 7.1, part (b)]. Matrices $\lambda^i_{2s-i} W_{is}' W_{is}$ are nonnegative definite with one of them, $\lambda^s_s W_{ss}' W_{ss} = \lambda^s_s I$, being positive definite. Thus $N_s N_s'$ is itself positive definite as their sum. This ends the proof.

(To see that $N_s N_s'$ is indeed expressible as the sum above the reader should check that the corresponding entries on both sides are equal. This involves verification of an identity of the familiar Vandermonde type.)

7.27

By the duality between points and blocks that exists in a symmetric design result (7.18) allows us to draw the following conclusion:

∗ *An automorphism group of a symmetric design has as many point-orbits as it has block-orbits.*

In this special situation a little more can in fact be said, as was observed by Baer:

∗ *An automorphism of a symmetric design fixes precisely as many points as it does blocks.*

Indeed, an automorphism g corresponds to two permutation matrices P and Q which satisfy

$$PNQ = N,$$

where N is the (square) incidence matrix between points and blocks. Recalling that $P^{-1} = P'$ and solving for Q we obtain

$$Q = N^{-1}P'N.$$

Being conjugates of each other P' and Q have the same trace. But the traces of P and Q count the number of fixed points and fixed blocks, respectively. This ends the proof.

7.28

The relationship between the design and its automorphism group is an interesting (and sporadically a fascinating) one. If the automorphism group is large, then it usually reflects much of the structural properties of the design. Quite a number of structural characterizations through the automorphism groups are known.

The symmetric designs $PG(n, q)$ admit large automorphism groups. The blocks of $PG(n, q)$ are, by definition, the n-dimensional (vector space) subspaces of a $(n + 1)$-dimensional vector space V over $GF(q)$. The group of nonsingular linear transformations of V acts on the n-dimensional subspaces; call this group $GL(V)$. But it does not act faithfully on the projective points. We can quickly fix this small annoyance by working modulo the subgroup of scalar multiples of the identity transformation. This quotient group [which we denote by $PGL(V)$] is indeed an automorphism group of $PG(n, q)$. Up to automorphisms of the field this is the full automorphism group. We refer the reader to [7] for further reading on this classical subject.

The full automorphism group of the Mathieu 5-(12, 6, 1) design displayed in Section 7.9 is the Mathieu group M_{12} which acts 5-transitively on points and is one of the sporadic simple groups. The other 5-transitive Mathieu group, M_{24}, is the automorphism group of the Mathieu 5-(24, 8, 1) design. For more information we refer to [13].

EXERCISES

1. Go over the proof of the BRC theorem with the special case of a (potential) projective plane of order six and conclude that it cannot exist.

2. Examine what the BRC theorem tells us about a Hadamard 2-design. Do the same for a symmetric 2-$(v, k, 2)$ design.

3. Show that a 2-(21, 5, 1) design is necessarily $PG(2, 4)$.

4. Show that a (symmetric) 2-(11, 5, 2) design must necessarily be the Paley design (7.4). In addition, prove that the full automorphism group of the Paley design (7.4) has 660 elements. What is this group?

5. (Alltop.) For any positive integer k a (nontrivial) 2-design with block size k exists. [*Hint*: Find a subgraph S with k edges (a cycle, maybe) in the complete graph K_n such that $\{g(S): g \in \mathrm{Aut}(K_n)\}$ are the blocks of the nontrivial 2-design; by $\mathrm{Aut}(K_n)$ we understand the group of all the $n!$ permutations on the n vertices of K_n.]

6. Show that the full automorphism group of $PG(2,2)$ is a simple group of order 168. (A simple group is a group with no proper normal subgroups, i.e., it is like a prime number.) Find this group both as a group of 3×3 matrices and as a group of permutations on 7 points.

7. (The fundamental theorem of projective geometry.) Find the full automorphism group of $PG(n, q)$ and show that it acts transitively on noncolinear triples of points.

8. (Alltop.) Let (P, B) be a t-(v, k, λ_t) design; suppose $a \notin P$. Define

$$P_a = P \cup \{a\}$$

$$B' = \{\alpha \cup \{a\} : \alpha \in B\}$$

$$B'' = \{P - \alpha : \alpha \in B\}$$

$$B''' = \Big\{P - \sigma_k : \sigma_k \in \sum_k (P) - B\Big\}.$$

For certain sets of parameters $(t + 1)$-designs can be constructed from various combinations of B, B', B'', or B'''. Prove the following:

(a) Let (P, B) be a t-$(2k, k, \lambda_t)$ design with t even and $B'' \cap B = \phi$. Then $(P, B \cup B'')$ is a $(t + 1)$-$(2k, k, \lambda''_{t+1})$ design; $\lambda''_{t+1} = 2\lambda_t(k - t)(2k - t)^{-1}$.

(b) Let (P, B) be a t-$(2k, k, \lambda_t)$ design with t even and $B'' = B$. Then (P, B) is a $(t + 1)$-$(2k, k, \lambda_{t+1})$ design; $\lambda_{t+1} = \lambda_t(k - t)(2k - t)^{-1}$.

(c) Let (P, B) be a t-$(2k + 1, k, \lambda_t)$ design with t even. Then $(P_a, B' \cup B'')$ is a $(t + 1)$-$(2k + 2, k + 1, \lambda_t)$ design.

(d) Let (P, B) be a t-$(2k + 1, k, \lambda_t)$ design with t odd and $\lambda_0 = \frac{1}{2}\binom{2k+1}{k}$. Then $(P_a, B' \cup B''')$ is a $(t + 1)$-$(2k + 2, k + 1, \lambda_t)$ design.

7 ASSOCIATION SCHEMES

7.29

An association scheme (or scheme for short) is a set with several binary relations defined on it, which satisfy certain properties of compatibility. The association schemes were introduced by Bose in connection with the design of experiments. Over the years they turned out to be useful in the study of other combinatorial structures such as permutation groups, coding theory, and designs. Good codes and designs often arise as maximal subsets of certain association schemes.

We present the basic theory of association schemes, which centers around the Bose–Mesner algebra. Examples are then given focusing mostly on the Hamming and Johnson (or triangular) schemes. Finally, we interpret t-designs as special subsets of the Johnson scheme.

7.30 The Definition of an Association Scheme

An *association scheme with n classes* (or relations, or colors) consists of a finite set P of v points together with $n+1$ binary relations $R_0, R_1, \ldots, R_n$ that satisfy:

(i) R_0 is the identity relation, that is, $R_0 = \{(x, x) : x \in P\}$.
(ii) For every x, y in P, $(x, y) \in R_i$ for exactly one i.
(iii) Each R_i is symmetric, that is, $(x, y) \in R_i$ implies $(y, x) \in R_i$.
(iv) If $(x, y) \in R_k$, then the number of z in P such that $(x, z) \in R_i$ and $(y, z) \in R_j$, is a constant c_{ijk} depending on i, j, k but not on the particular choice of x and y in R_k.

It is often helpful to think of an association scheme as a complete graph on v points with colored edges. (Relation R_i corresponds to color i; R_0 has a somewhat degenerate meaning—it informs us that each point has color 0. Instead of coloring points one may prefer to draw a loop on top of a point; think of it as an edge and color that 0.) An edge $\{x, y\}$ is colored with color i if $(x, y) \in R_i$, and such x, y we call ith *associates*. Condition (i) states that points have color 0; (ii) tells us that each edge has a unique color; (iii) informs us that the edges are not oriented. And finally, condition (iv) is equivalent to saying that the number of triangles with a fixed base $\{x, y\}$ of color k having the edge incident with x colored i and the edge incident with y colored j is a constant c_{ijk} depending on i, j, k but not on the specific choice of the base of color k. In particular each vertex is incident with c_{ii0} edges of color i. We denote c_{ii0} by v_i. Observe that the subgraph with edges of color i (i.e., the subgraph corresponding to R_i) is regular of degree v_i. (We remind the reader that a graph is called regular of degree d if all its vertices have degree d.)

7.31 An Example—The Johnson Scheme

We give an example of a triangular scheme with two associate classes on ten points. Let the points of the scheme be the 2-subsets of $\{1, 2, 3, 4, 5\}$. Call two points $\{i, j\}$ and $\{k, l\}$ first associates if they have precisely one symbol in common, and second associates if they are disjoint.

If we place the ten points in a (symmetric) triangular array $J(5, 2)$ with diagonal entries filled by $*$ as displayed below, two points are first associated if they are in the same row or column of $J(5,2)$; they are second associates if they are not in the same row or column.

$$J(5,2) = \begin{matrix} * & \{1,2\} & \{1,3\} & \{1,4\} & \{1,5\} \\ \{1,2\} & * & \{2,3\} & \{2,4\} & \{2,5\} \\ \{1,3\} & \{2,3\} & * & \{3,4\} & \{3,5\} \\ \{1,4\} & \{2,4\} & \{3,4\} & * & \{4,5\} \\ \{1,5\} & \{2,5\} & \{3,5\} & \{4,5\} & * \end{matrix}.$$

We color the pairs of first associates red and those of second associates blue. The subgraph of blue edges that results is known as the Petersen graph (we write ij instead of $\{i, j\}$ for simplicity):

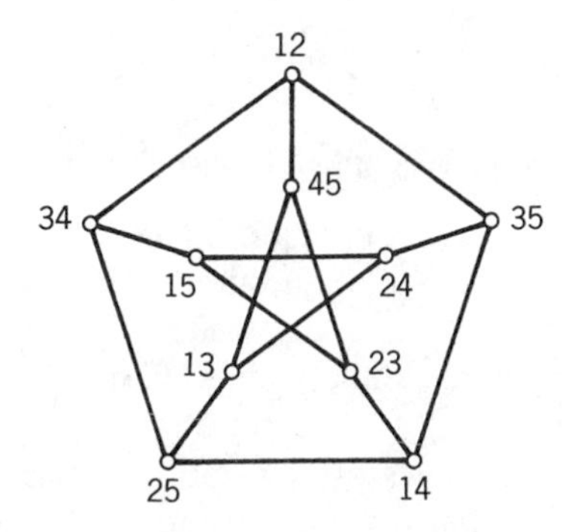

The complementary graph to this graph (in K_{10}, the complete graph on the ten points) is the subgraph of red edges.

The reader can check that $J(5,2)$ is indeed an association scheme with two associate classes. Its parameters are

$$\begin{pmatrix} c_{000} & c_{010} & c_{020} \\ & c_{110} & c_{120} \\ & & c_{220} \end{pmatrix} = \begin{pmatrix} 1 & 0 & 0 \\ & 6 & 0 \\ & & 3 \end{pmatrix}$$

$$\begin{pmatrix} c_{001} & c_{011} & c_{021} \\ & c_{111} & c_{121} \\ & & c_{221} \end{pmatrix} = \begin{pmatrix} 0 & 1 & 0 \\ & 3 & 2 \\ & & 1 \end{pmatrix} \qquad (7.19)$$

$$\begin{pmatrix} c_{002} & c_{012} & c_{022} \\ & c_{112} & c_{122} \\ & & c_{222} \end{pmatrix} = \begin{pmatrix} 0 & 0 & 1 \\ & 4 & 2 \\ & & 0 \end{pmatrix}$$

The matrices of parameters displayed above are all symmetric.

There is a general scheme of which the above example is a special case. This general scheme we denote by $J(m, n)$ and define as follows: The points of the scheme are the $\binom{m}{n}$ subsets of size n of a set with m elements; two subsets of size n are said to be ith associates if they intersect in precisely $n - i$ elements, $0 \le i \le n$. The scheme $J(m, n)$ has n associate classes. (By definition we do not count the 0th associates as an association class.) We call $J(m, n)$ the *Johnson (or triangular) scheme.*

The parameters of the scheme $J(m, n)$ are

$$c_{n-i,n-j,n-k} = \sum_{l} \binom{k}{l}\binom{n-k}{i-l}\binom{n-k}{j-l}\binom{m-2n+k}{n-i-j+l},$$

with $0 \le i, j, k \le n$. (This expression is derived by counting the number of n-subsets that intersect x in i elements and y in j elements, where x and y are a pair of n-subsets that intersect in k elements. The n-subsets in question are sorted by l, the number of elements that they have in common with $x \cap y$.)

7.32 Other Examples of Association Schemes

The association schemes are combinatorial objects of great mathematical richness and elegance. Fundamental questions of classification were addressed and partly answered by Bose and his students. Complete classification remains, however, a formidable undertaking. It has been actively researched only in the last decade or so.

We mention several examples of association schemes, the selection being motivated chiefly by the immediate connections between these schemes and other parts of combinatorics such as graph theory, finite groups, coding theory, and finite geometries.

A The Hamming Scheme H(n)

The points of the scheme are the 2^n vertices of a n-dimensional cube. The relations of association are defined as follows: Think of each vertex of the cube being a vector of n components with 0 and 1 as entries. Two vertices are called ith associates if the corresponding vectors differ in precisely i coordinates. The scheme $H(n)$ has n associate classes. It is the fundamental object of coding theory.

B The Projective Schemes

Let V be a vector space of dimension $n + 1$ over $GF(q)$. Denote by PV the associated projective geometry (see the introductory passages to Section 7.6 for the exact definition).

The points of the scheme are the m-subsets of (projective) points of PV. Two m-subsets are called ith associates if the projective points in their union span a projective subspace of PV of (projective) dimension i.

C Metric Schemes

Let G be a connected simple graph with P the set of vertices. The *distance* $d(x, y)$ between two vertices x and y is defined as the length of the shortest path joining them. The maximal distance between any two vertices is called the *diameter* of G.

The graph G is called *distance regular* if for any x and y in P with $d(x, y) = k$, the number of vertices z in P such that $d(z, x) = i$ and $d(z, y) = j$ is a constant c_{ijk} independent of the choice of x and y (so long as they are at distance k of each other).

We obtain an association scheme from a distance regular graph by calling two vertices x and y ith associates if $d(x, y) = i$. The schemes thus obtained from distance regular graphs are called *metric schemes*. (To recover the graph from the metric scheme define x and y to be adjacent if they are first associates.)

Distance regular graphs of diameter 2 are called *strongly regular*. It is easy to see that any scheme with two associate classes is metric and is obtained from a strongly regular graph.

D The Schemes Arising from Permutation Groups

Let G be a permutation group acting transitively on a set of points P. We can make G act on $P \times P$ by defining $g(x, y) = (g(x), g(y))$, for g in G and x, y in P. Let the orbits of this latter action on $P \times P$ be $R_0 = \{(x, x) : x \in P\}$, $R_1, \ldots, R_n$. The R_i's are binary relations that need not be symmetric. If they are symmetric they define an association scheme with n classes.

The Johnson and Hamming schemes correspond to special choices of the group: the symmetric group, and the symmetry group of the n-cube (of order $2^n n!$), respectively.

7.33 Relations Among the Parameters

Let c_{ijk} be the parameters of an association scheme with n classes. The reader may have already observed certain relations among the c_{ijk}'s in (7.19). In general we can say the following:

* *The parameters c_{ijk} of an association scheme with n classes satisfy*

$$c_{ijk} = c_{jik}, \qquad c_{0jk} = \begin{cases} 1 & \text{if } j = k \\ 0 & \text{otherwise} \end{cases}$$

$$v_k c_{ijk} = v_i c_{kji}$$

$$\sum_{j=0}^{n} c_{ijk} = v_i$$

$$\sum_{m=0}^{n} c_{ijm} c_{mkl} = \sum_{h=0}^{n} c_{ihl} c_{jkh}.$$

(We denote c_{ii0} by v_i.)

Proof. That $c_{ijk} = c_{jik}$ follows directly from the fact that the relation R_k is symmetric. For x and y a pair of kth associates c_{0jk} equals the number of 0th associates of x and jth associates of y. Since the only 0th associate of x is x itself, $c_{0jk} = 0$, unless $j = k$, in which case $c_{0jk} = 1$.

To see that $v_k c_{ijk} = v_i c_{kji}$ count in two ways the cardinality of the set

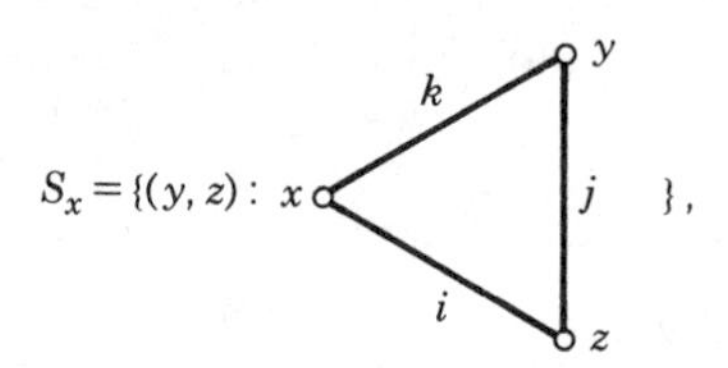

that is, S_x consists of ordered pairs (y, z) of jth associates such that (x, y) are kth associates and (x, z) are ith associates, with x a point fixed a priori.

Proving that $v_i = \sum_{j=0}^{n} c_{ijk}$ amounts to sorting out the triangles

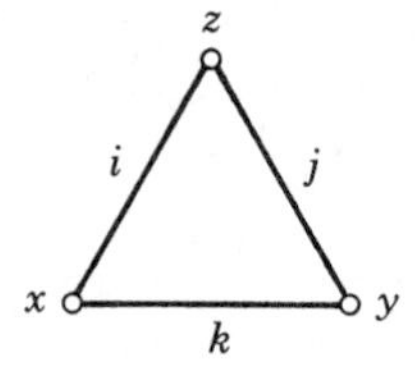

by the values of j, with x and y a fixed pair of kth associates.

The last identity follows from counting in two ways (as indicated in the figures below) the number of paths with color sequence (i, j, k) joining x and y.

In the first of the two figures fix z. For this fixed z there are c_{ijm} possible choices for u; and there actually exist c_{mkl} possibilities for z. Summing over m

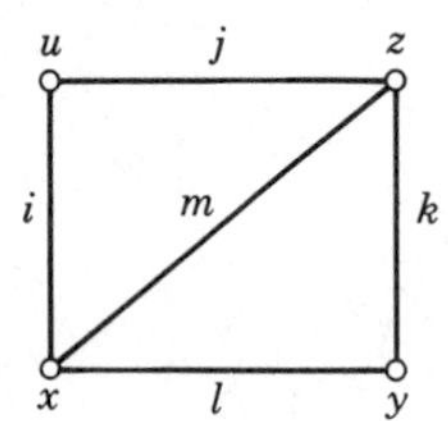

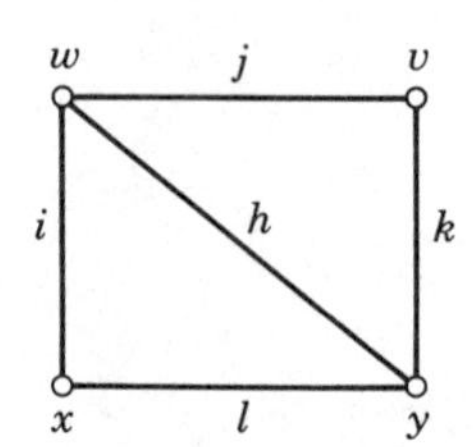

leads to $\sum_{m=0}^{n} c_{ijm}c_{mkl}$. Similar counting in the second figure gives $\sum_{h=0}^{n} c_{ihl}c_{jkh}$. We conclude that

$$\sum_{m=0}^{n} c_{ijm}c_{mkl} = \sum_{h=0}^{n} c_{ihl}c_{jkh},$$

and this ends our proof.

7.34 The Bose–Mesner Algebra

To an association scheme with n classes on a set of v points we attach an algebra of $v \times v$ matrices. Think of the association scheme as a complete graph K_v on the v vertices with edges colored with n colors. Let A_i be the $v \times v$ (points versus points) adjacency matrix of the subgraph of K_v with edges of color i, $0 \le i \le n$. To be exact, the (x, y)th entry of A_i is 1 if the edge $\{x, y\}$ is colored i, and 0 otherwise.

The defining properties of the association scheme can be readily rewritten in terms of the matrices A_i as follows:

(i) $A_0 = I$ (the identity matrix).
(ii) $\sum_{i=0}^{n} A_i = J$ (the matrix with all entries 1).
(iii) A_i is symmetric.
(iv) $A_iA_j = \sum_{k=0}^{n} c_{ijk}A_k = A_jA_i$; $0 \le i, j \le n$.

[Indeed, the (x, y)th entry of A_iA_j equals the number of paths

$$x \overset{i}{\circ\!\!-\!\!\circ}\overset{j}{\!\!-\!\!\circ}\; y$$

in K_v. This number is c_{ijk} for some k, which is what $\sum_{k=0}^{n} c_{ijk}A_k$ has for its (x, y)th entry. This explains (iv) above.]

Observe also that $A_iJ = JA_i = v_iJ$. In words this means that A_i has row and column sums equal to v_i (recall that $v_i = c_{ii0}$).

Let us consider the vector space $\mathscr{B}$ of all matrices of the form $\sum_{i=0}^{n} a_iA_i$, with a_i real numbers. From (iii) these matrices are symmetric. Condition (ii) tells us that $A_0, A_1, \ldots, A_n$ are linearly independent. We thus conclude that the A_i's form a basis for $\mathscr{B}$, and that the dimension of $\mathscr{B}$ is $n + 1$. Most importantly, (iv) informs us that $\mathscr{B}$ is closed under matrix multiplication, and that the multiplication in $\mathscr{B}$ is in fact commutative; the multiplication of $\mathscr{B}$ is, of course, associative as well (matrix multiplication is always associative). A vector space with a rule of multiplication that is associative, commutative

(distributive with respect to addition and "polite" to scalar multiplication) is called an *algebra*. We thus call $\mathscr{B}$ the *Bose–Mesner algebra* of the association scheme.

The matrices in $\mathscr{B}$ are symmetric and commute with each other. A well-known result in matrix theory tells us then that they can be simultaneously diagonalized, that is, there exists a nonsingular matrix S such that

$$S^{-1}AS = D_A$$

with D_A diagonal, for all A in $\mathscr{B}$.

The algebra $\mathscr{B}$ is now seen to be semisimple and thus admits a unique basis $J_0, J_1, \ldots, J_n$ of primitive idempotents (see [18]). These are matrices in $\mathscr{B}$ satisfying

$$\begin{aligned} J_i^2 &= J_i, \qquad 0 \le i \le n \\ J_iJ_k &= 0, \qquad i \ne k \\ \sum_{i=0}^{n} J_i &= I. \end{aligned} \tag{7.20}$$

[The matrix J (of all 1's) is in $\mathscr{B}$ and $(1/v)J$ is idempotent. *We shall therefore always choose* $J_0 = (1/v)J$.]

We now have two bases for $\mathscr{B}$: the A_i's and the J_i's. Let us relate these two bases by writing

$$A_k = \sum_{i=0}^{n} p(k,i)J_i, \qquad 0 \le k \le n$$

and

$$J_k = \sum_{i=0}^{n} v^{-1}q(k,i)A_i, \qquad 0 \le k \le n. \tag{7.21}$$

The $p(k,i)$ and $v^{-1}q(k,i)$ are real numbers. Observe, in fact, that $p(k,i)$ is an eigenvalue of A_k, in that it satisfies

$$A_kJ_i = \left(\sum_{j=0}^{n} p(k,j)J_j\right)J_i = p(k,i)J_i. \tag{7.22}$$

[The first equality sign holds by (7.21) while the second by (7.20).] Equation (7.22) also shows that (the columns of) J_i are eigenvectors of A_k, $0 \le k \le n$.

Denote by P the $(n+1) \times (n+1)$ matrix $(p(k,i))$ and by Q the matrix $(q(k,i))$. By (7.21) P and $v^{-1}Q$ are inverses of each other. Further, let μ_i denote the rank of J_i, that is, the multiplicity of the eigenvalue $p(k,i)$, as written in (7.22).

The following result holds:

* *The eigenmatrices P and Q satisfy the orthogonality conditions*

$$P'\begin{pmatrix} \mu_0 & & 0 \\ & \ddots & \\ 0 & & \mu_n \end{pmatrix} P = v \begin{pmatrix} v_0 & & 0 \\ & \ddots & \\ 0 & & v_n \end{pmatrix}$$

and

$$Q'\begin{pmatrix} v_0 & & 0 \\ & \ddots & \\ 0 & & v_n \end{pmatrix} Q = v \begin{pmatrix} \mu_0 & & 0 \\ & \ddots & \\ 0 & & \mu_n \end{pmatrix}. \tag{7.23}$$

Moreover, $p(i, m)p(j, m) = \sum_{k=0}^{n} c_{ijk} p(k, m)$. [Recall that $P = (p(k, i))$ and $Q = (q(k, i)) = vP^{-1}$.]

Proof. The eigenvalues of $A_i A_j$ are $p(i, m)p(j, m)$ with multiplicity μ_m, $0 \le m \le n$. Thus the trace of $A_i A_j$ is $\sum_{m=0}^{n} \mu_m p(i, m)p(j, m)$. But we know that

$$A_i A_j = \sum_{k=0}^{n} c_{ijk} A_k,$$

and hence the trace of $A_i A_j$ is

$$\operatorname{trace} A_i A_j = \operatorname{trace} \sum_{k=0}^{n} c_{ijk} A_k = \sum_{k=0}^{n} c_{ijk} \operatorname{trace} A_k$$

$$= c_{ij0} \operatorname{trace} A_0 = c_{ij0} v = \begin{cases} v_i v & \text{if } i = j \\ 0 & \text{if } i \ne j \end{cases}.$$

This proves $P' \operatorname{diag}(\mu_0, \ldots, \mu_n) P = v \operatorname{diag}(v_0, \ldots, v_n)$. Replacing P^{-1} by $v^{-1}Q$ gives the second equation.

To explain the last relation write

$$p(i, m)p(j, m)J_m = A_i p(j, m)J_m = A_i A_j J_m = \left(\sum_{k=0}^{n} c_{ijk} A_k\right) J_m$$

$$= \sum_{k=0}^{n} c_{ijk} A_k J_m = \left(\sum_{k=0}^{n} c_{ijk} p(k, m)\right) J_m.$$

This ends our proof.

An Algebra of Dimension n + 1 Isomorphic to $\mathscr{B}$

In an association scheme the number of relations (or colors), n, is in general much less than v. As it turns out, we can facilitate the spectral analysis of $\mathscr{B}$ by working with an algebra $\overline{\mathscr{B}}$ of $(n + 1) \times (n + 1)$ matrices, rather than with the $v \times v$ matrices of the original algebra $\mathscr{B}$.

Indeed, let $B_i = (c_{ijk})$, that is, the (j, k)th entry of B_i is c_{ijk}. Then

$$B_i B_j = \sum_{k=0}^{n} c_{ijk} B_k. \tag{7.24}$$

[That this is true can be seen as follows. The (l, m)th entry of $\sum_{k=0}^{n} c_{ijk} B_k$ is $\sum_{k=0}^{n} c_{ijk} c_{klm}$; on the other hand, the (l, m)th entry of $B_i B_j$ is $\sum_{k=0}^{n} c_{ilk} c_{jkm}$. The last of the relations among the parameters, listed in Section 7.33, allows us to write $\sum_{k=0}^{n} c_{ilk} c_{jkm} = \sum_{k=0}^{n} c_{jkm} c_{ilk} = \sum_{k=0}^{n} c_{jik} c_{klm} = \sum_{k=0}^{n} c_{ijk} c_{klm}$. This proves (7.24).]

Display (7.24) shows that the matrices B_i multiply in the same manner as the matrices A_i. Furthermore, the B_i's are linearly independent since c_{ij0} equals 0, unless $i = j$, in which case $c_{ii0} = v_i$.

Define $\overline{\mathscr{B}}$ to be the algebra of matrices $\sum_{i=0}^{n} a_i B_i$, with a_i real numbers. Note that, unlike the A_i's, the matrices B_i need not necessarily be symmetric.

Under the mapping $A_i \to B_i$, the algebras $\mathscr{B}$ and $\overline{\mathscr{B}}$ are isomorphic. (This is true since, as we said, the B_i's multiply in the same way as the A_i's). In particular A_k and B_k have the same set of eigenvalues (not with the same multiplicities, of course). Indeed, let $p(k, i)$ be an eigenvalue of A_k. Then $A_k J_i = p(k, i) J_i$, and since the isomorphic image of this equation is $B_k \bar{J}_i = p(k, i) \bar{J}_i$ we conclude that $p(k, i)$ is also an eigenvalue of B_k (we denoted by $\bar{J}_i$ the isomorphic image of J_i). In particular, this allows us to conclude that any matrix in the Bose–Mesner algebra $\mathscr{B}$ has at most $n + 1$ distinct eigenvalues.

Another happy consequence of the fact that A_k and B_k have the same set of eigenvalues is the actual computation of these; the dimension of the matrices B_k is quite small (relative to that of the A_k's) and thus the $p(k, i)$'s are easier to find as eigenvalues of the B_k's.

In the case of the Johnson scheme $J(m, n)$, for example, one finds: $p(k, i) = E(k, i)$, $q(k, i) = v_i^{-1} \mu_k E(i, k)$, where $v_i = \binom{n}{i}\binom{m-n}{i}$, $\mu_k = (m - 2k + 1)(m - k + 1)^{-1}\binom{m}{k}$ and

$$E(k, x) = \sum_{j=0}^{k} (-1)^j \binom{x}{j}\binom{n-x}{k-j}\binom{m-n-x}{k-j}, \qquad 0 \le k \le n.$$

The $E(k, x)$ are called *Eberlein* polynomials.

7.35 *t*-Designs as Subsets of Association Schemes

In this section we present a result of P. Delsarte concerning the existence of t-designs as subsets of association schemes. The language of association schemes brings into common perspective many results from coding theory, designs, and finite geometry. The reader is referred to [15] for a better understanding of this important point of view.

We work in an association scheme with point set $P = \{1, 2, \ldots, v\}$ and relations $R_0, R_1, \ldots, R_n$. Let S be a subset of P. Define

$$s_i = |R_i \cap (S \times S)|.$$

The constant s_i counts the number of ordered pairs in $S \times S$ that belong to relation R_i. Since the subsets $R_i \cap (S \times S)$ partition $S \times S$ it is clear that $\sum_{i=0}^n s_i = |S|^2$. We call the vector $(s_0, s_1, \ldots, s_n)$ the *weight distribution* of S.

We wish to interpret the weights s_i in terms of operations with the adjacency matrices A_i of the scheme. Recall that A_i is the adjacency matrix of relation R_i. To accomplish this, denote by x the $v \times 1$ indicator vector of the subset S; that is, $x' = (x_1, \ldots, x_v)$ with $x_k = 1$ if the point k belongs to S, and 0 otherwise. It then follows immediately that $s_i = x'A_i x$.

Define

$$d_i(S) = \sum_{j=0}^{n} q(i, j) s_j,$$

where $q(i, j)$ is the (i, j)th entry of the matrix Q whose inverse is $v^{-1}P$, with $P = (p(i, j))$ being the $(n + 1) \times (n + 1)$ matrix of the eigenvalues of the scheme.

A crucial observation made by Delsarte is that

$$d_i(S) = vx'J_i x, \tag{7.25}$$

where x is the indicator vector of S and J_i is the ith idempotent in the Bose–Mesner algebra $\mathscr{B}$ of the scheme. [Statement (7.25) is proved as follows: $d_i(S) = \sum_{j=0}^n q(i, j) s_j = \sum_{j=0}^n q(i, j) x'A_j x = x'(\sum_{j=0}^n \cdot q(i, j) A_j) x = vx'J_i x$. The last sign of equality is explained by (7.21).]

The idempotent J_i is a symmetric matrix that satisfies $J_i^2 - J_i = 0$. Its eigenvalues are therefore 0 or 1. We thus conclude that J_i is a nonnegative definite matrix, that is, it satisfies $y'J_i y \geq 0$ for all vectors y. With this in mind (7.25) informs us that

$$d_i(S) \geq 0, \qquad 0 \leq i \leq n.$$

Subsets S of an association scheme for which $d_i(S) = 0$, for a large number of indices i, are in a sense "extreme" and possess interesting combinatorial properties. Delsarte proved the following:

* *A nonempty subset S of the Johnson scheme $J(m, k)$ consists of the blocks of a t-(m, k, λ_t) design if and only if the vector $(d_1(S), d_2(S), \ldots, d_k(S))$ has at least t components equal to* 0.

Proof. Points of the scheme are k-subsets of a set M; $|M| = m$. For an i-subset z of M, denote by $\lambda_i(z)$ the cardinality of the set $\{\alpha : z \subset \alpha;\ \alpha \in S\}$. A two-way counting of $|\{(z, \alpha) : z \subset \alpha;\ \alpha \in S\}|$ gives

$$\sum_{z} \lambda_i(z) = \binom{k}{i} |S|.$$

Yet another two-way counting, this time of the cardinality of the set $\{(z,(\alpha,\beta)) : z \subset \alpha, z \subset \beta; \alpha, \beta \in S\}$, leads us to the following equation:

$$\sum_z \lambda_i^2(z) = \sum_{j=0}^{k} s_j \binom{j}{i}.$$

Denote by λ_i the average $\binom{m}{i}^{-1} \Sigma_z \lambda_i(z)$. The two equations we just derived allow us to write

$$\sum_z (\lambda_i(z) - \lambda_i)^2 = \left(\sum_{j=0}^{k} s_j \binom{j}{i} \right) - |S| \binom{k}{i} \lambda_i. \tag{7.26}$$

In this notation we can say that the subset S consists of the blocks of a t-(m, k, λ_t) design if and only if $\lambda_i(z) = \lambda_i$ for all $z \in \Sigma_i(M)$ and all $1 \le i \le t$ (see Proposition 7.1, part (a)). Equation (7.26) allows us now to conclude that S is a t-(m, k, λ_t) design if and only if the weights s_j satisfy the system:

$$\sum_{j=0}^{k} \binom{j}{i} s_j = |S| \binom{k}{i} \lambda_i, \qquad \text{for } 1 \le i \le t. \tag{7.27}$$

For a fixed i $(1 \le i \le t)$ observe, however, that

$$|S|^{-2} \sum_{j=0}^{k} \binom{j}{i} s_j = |S|^{-2} |S| \binom{k}{i} \lambda_i$$

$$= |S|^{-1} \binom{k}{i} \binom{m}{i}^{-1} \binom{k}{i} |S| = \binom{m}{i}^{-1} \binom{k}{i}^2. \tag{7.28}$$

The far right-hand side, that is, $\binom{m}{i}^{-1}\binom{k}{i}^2$, is seen to depend on m, k, and i only and not on the specific choice of the t-design. Select in particular the complete design consisting of all $\binom{m}{k}$ subsets of size k of M. The weight distribution of this complete design is $(v_0, v_1, \ldots, v_n)$ and thus (7.28) becomes

$$\binom{m}{k}^{-2} \sum_{j=0}^{k} \binom{j}{i} v_j = \binom{m}{i}^{-1} \binom{k}{i}^2. \tag{7.29}$$

In view of (7.28) and (7.29), system (7.27) can be written as follows:

$$|S|^{-2} \sum_{j=0}^{k} \binom{j}{i} s_j = \binom{m}{k}^{-2} \sum_{j=0}^{k} \binom{j}{i} v_j, \qquad 1 \le i \le t. \tag{7.30}$$

Let $s' = |S|^{-2}(s_0, s_1, \ldots, s_n)$, $v' = \binom{m}{k}^{-2}(v_0, v_1, \ldots, v_n)$ and let $\overline{T}$ be the $t \times (k+1)$ matrix with (i, j)th entry equal to $\binom{j}{i}$ if $i \le j$ and 0 otherwise. System (7.30) can be rewritten as

$$\overline{T}s = \overline{T}v. \tag{7.31}$$

Think of the binomial coefficient $\binom{j}{i}$ as the polynomial $\binom{x}{i}$ $(= [x]_i/i!)$ evaluated at j (by x we understand an indeterminate ready and willing to take numerical values at all times). The polynomials $\binom{x}{i}$, for $1 \le i \le t$, form a basis for the subspace of polynomials of degree at least 1 and at most t over the real numbers. The numbers $q(i, j)$ turn out also to be values of polynomials $F(i, x)$ evaluated at j. (This is a *most important* fact; the polynomials $F(i, x)$ are related to the Eberlein polynomials.) The polynomial $F(i, x)$ is of degree i and hence the $F(i, x)$, $1 \le i \le t$, also span the subspace of polynomials of degree at least 1 and at most t. Perform a change of basis by writing

$$\binom{x}{i} = \sum_{j=1}^{t} w_{ij} F(j, x), \qquad 1 \le i \le t, \tag{7.32}$$

for a $t \times t$ nonsingular matrix $W = (w_{ij})$.

Let $\bar{Q}$ be the $t \times (k+1)$ matrix with (i, j)th entry equal to $q(i, j)$. [Note that $\bar{Q}$ consists of the first t rows of the $(k+1) \times (k+1)$ matrix $Q = (q(i, j))$.] We can use the matrix W to write $\bar{T} = W\bar{Q}$ [this follows directly from (7.32)].

Observe now that $\bar{T}s = \bar{T}v$ if and only if $W\bar{Q}s = W\bar{Q}v$ if and only if $\bar{Q}s = \bar{Q}v$ (since W is nonsingular). But $\bar{Q}v = 0$ by the orthogonality relations (7.23) proved in Section 7.34. To be more exact, the ith equation of the system $\bar{Q}v = 0$ is

$$\sum_{j=0}^{k} q(i, j) v_j = 0, \qquad 1 \le i \le t.$$

Read this last equation as $\sum_{j=0}^{k} v_j q(i, j) q(0, j) = 0$ by recalling that $q(0, j) = 1$, for $0 \le j \le k$. Since $i \ne 0$ the equation is explained by the aforementioned orthogonality conditions.

To summarize, we showed that S is a t-design if and only if $\bar{T}s = \bar{T}v$ if and only if $\bar{Q}s = \bar{Q}v$ $(= 0)$ if and only if $\bar{Q}s = 0$ if and only if

$$\sum_{j=0}^{k} q(i, j) s_j = 0, \qquad 1 \le i \le t$$

if and only if

$$d_i(S) = 0, \qquad 1 \le i \le t.$$

This ends the proof.

7.36 Partial Designs

Let a set P of v points with relations $R_0, R_1, \ldots, R_n$ be an association scheme with n classes. We define the notion of a partial design, a concept less restrictive combinatorially than a 2-design. It was originally introduced by Bose in connection with statistical design of experiments.

The pair (P, B) is called a *partial design with n classes* if B is a collection of k-subsets of P (called blocks) such that points x and y appear in p_i blocks whenever $(x, y) \in R_i$, $0 \le i \le n$. [Observe, in particular, that each point occurs in p_0 blocks since $(x, x) \in R_0$.]

In case $p_1 = p_2 = \cdots = p_n$ $(= \lambda_2)$ we obtain a 2-(v, k, λ_2) design. This, however, does not happen frequently.

We can construct partial designs with two classes as follows: Let P be a set of v points. Let also G be a group that acts transitively on P and has three symmetric orbits $R_0 = \{(x, x) : x \in P\}$, R_1 and R_2 on $P \times P$. This action generates an association scheme with two associate classes on P; two points x and y are first associates if $(x, y) \in R_1$ and second associates if $(x, y) \in R_2$. Let S be *any* k-subset of P $(k \ge 2)$. We define the blocks of our partial design as

$$B = \{g(S) : g \in G\}.$$

The pair (P, B) is indeed a partial design with two classes. Suppose a point x is in blocks $\{S_i : 1 \le i \le p_0\}$. Then any other point y will be in blocks $\{g(S_i) : 1 \le i \le p_0\}$, where g is an element of G that sends x to y (we use here the transitivity of G on P). Moreover, if $(x, y) \in R_i$ and $(z, w) \in R_i$, then there exists an element g_i of G that sends (x, y) into (z, w), $i = 1, 2$. Consequently, if $\{x, y\}$ occurs in blocks $\{S_j : 1 \le j \le p_i\}$, then $\{z, w\}$ occurs in blocks $\{g_i(S_j) : 1 \le j \le p_i\}$, $i = 1, 2$. This proves that (P, B) is a partial design with two classes. [This construction can easily be remembered by a special case: Take P to be the *edges* of a complete graph on m vertices and let the group G be the whole symmetric group S_m on the m vertices. The group S_m acts transitively on edges and has three orbits on $P \times P$. One orbit is just the diagonal $R_0 = \{(x, x) : x \in P\}$; R_1 consists of all pairs of edges like

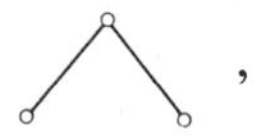

,

and R_2 consists of all pairs of parallel edges

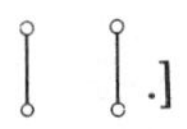

.]

Let us now introduce another concept (still due to Bose [1]). A *partial geometry* (r, k, t) is a collection of subsets (called *lines*) with the following properties: each line contains k points; each point is on r lines; two distinct points are on at most one line; given a line and a point not on it, there exist precisely t lines passing through that point and intersecting the original line.

We mention without proof a result of Bose [1]:

* *A partial geometry is a partial design with two classes. A partial design with two classes, strictly fewer blocks than points, and in which two distinct points occur in at most one block must necessarily be a partial geometry.*

We conclude our discussion on the subject of partial designs with a graphical display of a partial geometry $(r, k, t) = (3, 3, 1)$:

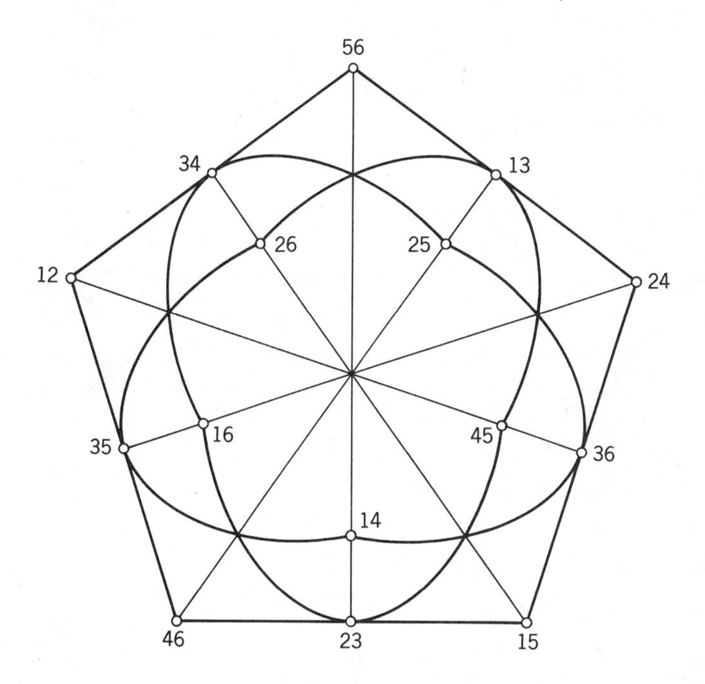

The points are labeled by the 2-subsets of the set $\{1, 2, 3, 4, 5, 6\}$. There are 15 points and 15 lines in all. Graphically we have lines of the form $\{12, 34, 56\}$, $\{13, 25, 46\}$, and "curved lines" such as $\{34, 16, 25\}$, five of each kind. To be exact, a line consists of three disjoint pairs.

EXERCISES

1. Find the 3×3 matrix of eigenvalues of an association scheme with two classes.

2. In any association scheme show that

$$p(0, j) = q(0, j) = 1, \qquad p(i, 0) = v_i, \qquad \text{and} \qquad q(i, 0) = \mu_i.$$

Show, in addition, that $|p(i, j)| \leq v_i$.

3. In the Hamming scheme $H(n)$ show that $v = 2^n$, $v_i = \binom{n}{i}$,

$$c_{ijk} = \binom{k}{2^{-1}(i - j + k)}\binom{n - k}{2^{-1}(i + j - k)},$$

if $i + j - k$ is even, and $c_{ijk} = 0$ otherwise.

4. By looking at the images of the subgraph

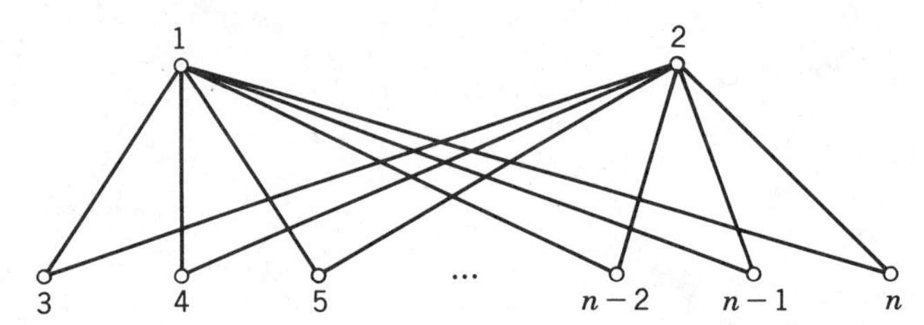

of K_n (the complete graph on n vertices) under all the $n!$ permutations on vertices we obtain a partial design. Compute its parameters.

5. (Nair.) Let N be the incidence matrix (points versus blocks) of a partial design with two classes. If the partial design has fewer blocks than points, then NN' is a singular matrix. Relying on this and looking at the algebra $\overline{\mathscr{B}}$ of 3×3 matrices (c_{ijk}) find an *explicit* relation among the parameters (c_{ijk}) of the scheme and the parameters p_i of the partial design. [*Hint*: The determinant of the 3×3 matrix (c_{ijk}) is zero.]

6. A graph like this

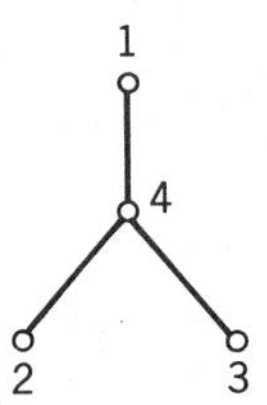

is called a 3-claw; the "missing" edges are indeed understood to be missing. Vertices 1, 2, and 3 are called the outer vertices of the 3-claw. In the Petersen graph (see Section 7.31) call three points colinear if they are the outer vertices of a 3-claw. The configuration of ten points and ten lines that results is the Desargue configuration:

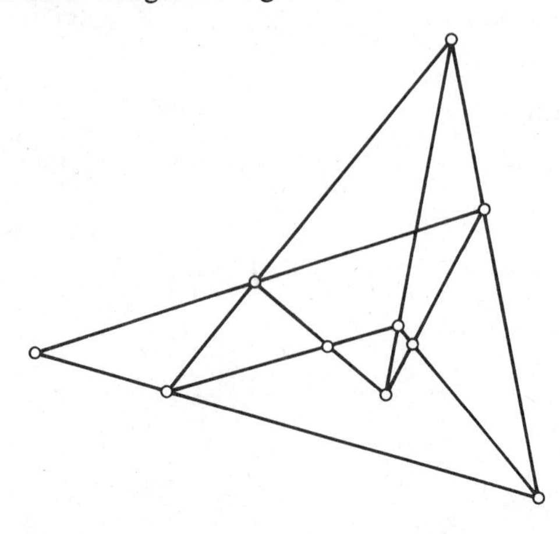

Check this. (A line is a straight line, as drawn, and it contains three points.)

7. Show that the Desargue configuration displayed above is a partial design with two classes. Find its full automorphism group.

8. Prove that the Pappus configuration

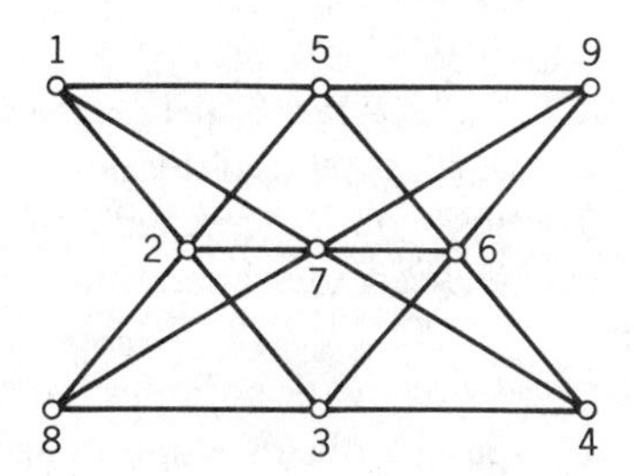

is a partial design with two classes. Compute its full automorphism group. (The blocks are the nine straight lines with three points on each, as drawn.)

9. Consider the combinatorial structure whose lines are the columns below:

2	4	1	1	4	7	1	2	3	3
10	10	10	2	5	8	4	5	6	5
6	8	9	3	6	9	7	8	9	7

Is this structure isomorphic to the Desargue configuration? Is it a partial design?

10. Show that a partial geometry (r, k, t) can exist only if

$$rk(r-1)(k-1)(k+r-t-1)^{-1}t^{-1}$$

is an integer. [*Hint*: This number is the multiplicity of an eigenvalue of the scheme.]

11. Prove that a partial design with fewer blocks than points in which a pair of distinct points appears in at most one block is necessarily a partial geometry.

NOTES

A systematic study of the subject discussed in this chapter was initiated by Fisher, Bose, and their students. The fundamental motivation stems from the planning of efficient statistical experiments.

Relatively recent contributions to t-designs of particular significance were made by Ray-Chaudhuri and Wilson [3], as well as Wilson [4]. Sections 1 and

3 are based on the results presented in these two papers. The author was exposed to this material by his former teacher, Professor Noboru Ito.

All of the methods of construction of t-designs that we present are well known and most of them are mentioned implicitly or explicitly in [5] or [6]. We highly recommend these two books, along with [7], to anyone interested in further research on the subject. The construction through hypergraphs appears to have led to the first nontrivial 6-design [16].

Section 4 consists of a result of Cameron [8]. A couple of other results on extending t-designs are found in Exercise 8 at the end of Section 6.

The well-known result of Bruck, Ryser, and Chowla is presented in Section 5. Several books contain this theorem. We recommend [9] and [6]. Much more can be said on the interplay between the design and its automorphism group than what we mention in Section 6. Except for the proof of the nonsingularity of matrix N_s, Section 7.26 is an observation due to the author. We refer to [10] for more information on the automorphism group in general, and to [7] for the automorphism groups of projective geometries in particular.

Association schemes, creations of Bose, have enjoyed a flurry of activity in recent years. Attempts at classification are vigorously pursued, influenced chiefly by Bannai and Ito [11]. Connections to designs and codes are described in Section 7.35. Recent work by Bailey et al. [12] places association schemes at the foundation of the analysis of variance.

REFERENCES

[1] R. C. Bose, Strongly regular graphs, partial geometries, and partially balanced designs, *Pacific J. Math.*, **13**, 389–419 (1963).

[2] R. C. Bose and D. M. Mesner, On linear associative algebras corresponding to association schemes of partially balanced designs, *Ann. Math. Statist.*, **30**, 21–38 (1959).

[3] D. K. Ray-Chaudhuri and R. M. Wilson, On t-designs, *Osaka J. Math.*, **12**, 737–744 (1975).

[4] R. M. Wilson, Incidence matrices of t-designs, *Lin. Alg. and Applic.*, **46**, 73–82 (1982).

[5] P. J. Cameron and J. H. van Lint, *Graphs, Codes and Designs*, LMS Lecture Note Series 43, Cambridge University Press, Cambridge, 1980.

[6] E. S. Lander, *Symmetric Designs: An Algebraic Approach*, LMS Lecture Note Series 74, Cambridge University Press, Cambridge, 1983.

[7] N. L. Biggs and A. T. White, *Permutation Groups and Combinatorial Structures*, LMS Lecture Note Series 33, Cambridge University Press, Cambridge, 1979.

[8] P. J. Cameron, Expending symmetric designs, *J. Comb. Th. (A)*, **14**, 215–220 (1973).

[9] H. J. Ryser, *Combinatorial Mathematics*, Carus Math. Monographs 14, 1963.

[10] P. Dembowski, *Finite Geometries*, Springer-Verlag, Berlin, 1968.

[11] E. Bannai and T. Ito, *Algebraic Combinatorics I: Association Schemes*, Benjamin Cummings, Menlo Park, 1984.

[12] R. Bailey, C. E. Praeger, T. P. Speed, and D. E. Taylor, *The Analysis of Variance*, manuscript, Rothamsted experimental station (private communication), 1985.

[13] F. J. MacWilliams and N. J. A. Sloane, *The Theory of Error Correcting Codes*, North-Holland, Amsterdam, 1977.

[14] H. Wielandt, *Finite Permutation Groups*, Academic Press, New York, 1964.

[15] P. Delsarte, The association schemes of coding theory, *Combinatorics*, *Part* 1, Math. Centre Tracts 55, Amsterdam, 1974.

[16] S. Magliveras and D. W. Leavitt, Simple 6-(33, 8, 36) designs from $P\Gamma L_2(32)$, preliminary report (private communication) (1984).

[17] A. Hedayat and W. D. Wallis, Hadamard matrices and their applications, *Ann. Statist.*, **6**, 1184–1238 (1978).

[18] M. Burrow, *Representation Theory of Finite Groups*, Academic Press, New York, 1965.

CHAPTER 8

Statistical Designs

The design of experiments is, however, too large a subject, and of too great importance to the general body of scientific workers, for any incidental treatment to be adequate.

The Design of Experiments, R. A. FISHER

The purpose of scientific research is to uncover laws of nature, to understand how events are interrelated, and ultimately to put to (good) use what has been learned. This almost always ends up being a long, frustrating process of many disappointments and failures. An ever present hardship is the experimental error (be it completely unexplained error, error of measurement, or error of judgement); it so easily leads us astray. Many times we try and fail, yet each time something may be learned. We slowly understand the nature of the error, its sources, its subtleties, and gradually begin to control it. Hopefully this iterative process ultimately reveals the true state of nature. Statistics is part of this scientific effort: the part that studies the nature of the error.

1 RANDOM VARIABLES

8.1

In order to discuss methods of experimentation several fundamental concepts need to be introduced. We accomplish this by way of several examples.

A chemist is interested in knowing how temperature (t) and pressure (p) affect the yield of a desired substance C. He sets the experimental apparatus at $t = 120$ and $p = 90$; obtained were 53.2 units of C. Suspecting possible fluctuations in response, he decides to repeat the experiment 19 additional times at the *same* levels $t = 120$ and $p = 90$ of temperature and pressure. The yields of C were:

53.2, 48.9, 56.1, 57, 47.3, 50.6, 51.4, 52.9, 51.2, 52.6,
49.5, 50.7, 55.2, 53.1, 48.2, 50.7, 52, 54.3, 53, 51.7.

The chemist thinks to himself: "At what seem to be identical experimental conditions, $t = 120$ and $p = 90$, I obtain *different* yields. Not widely different, thank goodness, but different still. I should thus think of the yield of C (at $t = 120$ and $p = 90$) as a variable quantity. In fact, yields between 50.5 and 53 seem more likely to come up than yields in other intervals (of the same length)."

The mental picture that emerges is as follows:

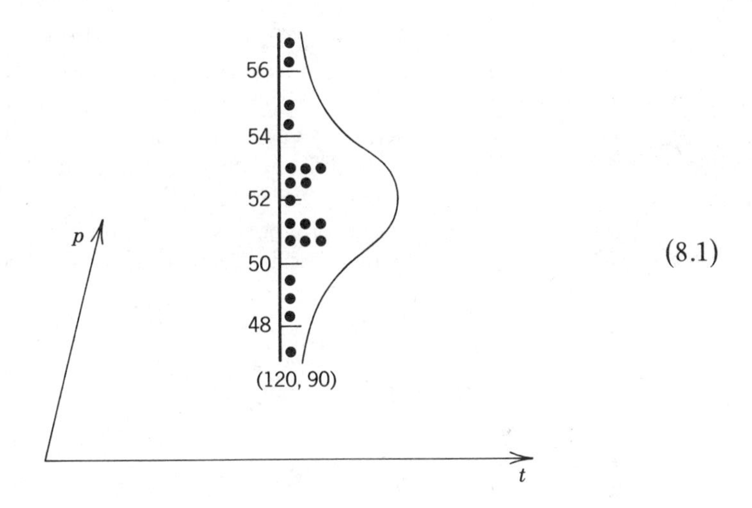

(8.1)

What the chemist observes at the point $t = 120$, $p = 90$ we call a *random variable*. Generally a random variable is a real-valued function (which takes certain values with certain probabilities).

The 20 *observations* made by the chemist are said to be a *sample* from this random variable. As is usually the case, the random variable is not completely known to the experimenter but the sample taken reveals something about it: we may empirically infer, for example, that the probability of seeing a yield of substance C between 52 and 56 is $7/20$ [see figure (8.1)].

We say that a random variable is *known* to us if we know all the values that it takes and the probabilities with which it takes those values.

Consider, as another example, rolling an ordinary fair die. The values that occur are 1, 2, 3, 4, 5, and 6, each with probability $1/6$. This random variable is thus completely known.

Now roll two ordinary fair dice and take interest in the sum of the two outcomes. The sum (call it S) is again a random variable. It is easy to describe the values that S takes: they are the integers between 2 and 12, inclusive. But it does take a little while to compute the chance with which each of these values are taken. It is clear that 12 can only happen by rolling double 6's, but 7 can occur in several ways. A bit of counting leads us to the following description of the random variable S:

2	3	4	5	6	7	8	9	10	11	12
$\frac{1}{36}$	$\frac{2}{36}$	$\frac{3}{36}$	$\frac{4}{36}$	$\frac{5}{36}$	$\frac{6}{36}$	$\frac{5}{36}$	$\frac{4}{36}$	$\frac{3}{36}$	$\frac{2}{36}$	$\frac{1}{36}$

(8.2)

On top are the values that S takes and at the bottom the corresponding probabilities. The random variable S is now completely known. (We may thus answer any probabilistic questions of interest regarding the sum S; such as, what's the chance that the sum of the two rolls is a power of two? The answer to this specific question is $\frac{1}{36} + \frac{3}{36} + \frac{5}{36}$, as the above table shows.)

Our chemist is not so lucky as to know the random variable (yield of C at $t = 120$ and $p = 90$) to any significant degree. All he really knows is the sample of 20 observations which we have listed. It seems clear that the possible values for the yield would be positive real numbers with the interval between 49 and 55 (say) carrying most of the probability. (At least this is what the sample of the 20 observations indicates.) In figure (8.1) we suggestively drew a smooth bell-shaped curve as a possible way in which the probability may be spread over the range of response. But does the true response really follow such a curve? Probably not; at any rate, no one knows for sure.

8.2

A random variable Y that takes finitely (or countably) many values $y_1, y_2, y_3, \ldots$ with probabilities $f(y_1), f(y_2), f(y_3), \ldots,$ respectively, is called a *discrete* random variable. [An example is the random variable in (8.2).]

We call a random variable *continuous* if the values that it takes include an interval (a, b) with $a < b$, and the probability is spread over the values in accordance to a (smooth) function f. [The random variable that the chemist confronts would be of this type. He does not know the shape of f but suspects that it might be bell-shaped.] We call f the *density* of the random variable.

One could conveniently think of a random variable Y as follows: to each point y (that the random variable Y takes) attach its density $f(y)$. In case Y is discrete $f(y)$ can be thought of as the *weight* of y. When Y is continuous $f(y)$ is simply the density (in a physical sense) at that point. The total weight (of all the points) is 1. [A mental image that often proves helpful is to picture a discrete random variable as a weightless but rigid string with weights pending at various points along the way. A continuous random variable can be viewed as a (possibly infinite) stick of varying density; the probability of an interval (a, b) being simply the weight of that chunk of the stick. The total weight is always 1, of course.]

8.3

Let Y be a random variable. The *mean* of Y is the number

$$\mu = \sum y f(y),$$

with the sum extending over all possible values y of Y. We often denote the mean of Y by $E(Y)$ and refer to it as the *expected value* of Y.

The *variance* of Y is

$$\sum (y - \mu)^2 f(y),$$

abbreviated usually by var(Y) or just σ^2. By *standard deviation* we understand the square root of variance; we write $sd(Y)$ or merely σ to denote the standard deviation. In general, for any function g we write

$$E(g(Y)) = \sum g(y) f(y).$$

Thus $\mu = E(Y)$ and $\sigma^2 = E((Y - \mu)^2)$. These definitions are for discrete random variables. When working with continuous ones integrals should replace the sums.

By thinking of a random variable as a string with pendant weights (or a stick), the mean becomes simply the point where one should place one's finger to keep the string with pendant weights (or the stick) in balance (i.e., it is the center of mass). The variance is somewhat less intuitive. From its definition it is clear, however, that it measures the dispersion of the weight (on the string or the stick) with reference to the center of mass. Large variance is thus associated with presence of substantial weights away from the mean; small variance, on the other hand, indicates a clustering of the weights near the mean.

8.4

Often it is necessary to work with several random variables at a time (such as the weight of a diabetic person, sugar content in the blood, and the number of dizzy spells per day, for example). In such a case it is convenient to stack up all these variables in a vector that we call a *random vector*. These random variables may in fact be dependent: high sugar content in the blood may induce more dizzy spells. Or they may behave independently, in the sense that knowledge about some of them reveals (little or) nothing about the others.

Let $\mathbf{Y}' = (Y_1, Y_2, \ldots, Y_n)$ be a random vector. The values $\mathbf{Y}$ takes form a region in the usual n-dimensional space. The probability (or weight) is spread over this region in accordance to some (known or unknown) density function f, which we call the *joint density* of $(Y_1, Y_2, \ldots, Y_n)$. One may wish to examine the (marginal) density of a subset of only k components (say $Y_1, Y_2, \ldots, Y_k$) of the vector $\mathbf{Y}$, $1 \le k < n$. Naturally this is obtained by sweeping off (or projecting) the weight from the initial n-dimensional space onto the k-dimensional subspace of $(Y_1, \ldots, Y_k)$ *parallel* to the $(n-k)$-dimensional space of $(Y_{k+1}, \ldots, Y_n)$. The weight, so reassembled, is now referred to as the *marginal density* of $(Y_1, Y_2, \ldots, Y_k)$ and denoted by $f_{Y_1, Y_2, \ldots, Y_k}$. In particular, for $k = 1$ we obtain n marginal densities, one for each Y_i, $i = 1, 2, \ldots, n$. We denote by f_{Y_i} the marginal density of Y_i.

Random variables $(Y_1, Y_2, \ldots, Y_n)$ are said to be *independent* if their joint density is equal to the product of the individual marginals, that is, if

$$f(y_1, y_2, \ldots, y_n) = f_{Y_1}(y_1) f_{Y_2}(y_2) \cdots f_{Y_n}(y_n),$$

for all $(y_1, y_2, \ldots, y_n)$. (Intuitively this says that working with the variables jointly offers no more information than working with each separately.)

Studying random variables in pairs proves to be often helpful. Let us look at (Y_1, Y_2). If it happens that high values of Y_1 induce high values of Y_2, for example, we would conclude that Y_1 and Y_2 are in some way dependent. Many forms of dependence may exist. We focus on the covariance of Y_1 and Y_2 and exploit its properties in our upcoming discussions; its definition is

$$\operatorname{cov}(Y_1, Y_2) = \sum_{y_1, y_2} (y_1 - \mu_1)(y_2 - \mu_2) f(y_1, y_2), \tag{8.3}$$

where μ_1 and μ_2 are the means of Y_1 and Y_2, respectively, and f is the joint density of (Y_1, Y_2). [Note that $\operatorname{cov}(Y_1, Y_2) = \operatorname{var}(Y_1)$, for example.] More generally, for a random vector $\mathbf{Y}^t = (Y_1, \ldots, Y_n)$ we define the $n \times n$ *covariance matrix* of $\mathbf{Y}$ as follows:

$$\operatorname{cov}(\mathbf{Y}) = \big(\operatorname{cov}(Y_i, Y_j)\big).$$

That is, $\operatorname{cov}(\mathbf{Y})$ is a $n \times n$ matrix with (i, j)th entry equal to the covariance between components Y_i and Y_j. On the diagonal we, of course, have the variances of the Y_i's.

We find it convenient to define the *expected value of a random vector* as the vector of the expected values of the components, that is,

$$E \begin{pmatrix} Y_1 \\ \vdots \\ Y_n \end{pmatrix} = \begin{pmatrix} E(Y_1) \\ \vdots \\ E(Y_n) \end{pmatrix}.$$

Going one step further, for a random matrix (Y_{ij}) we define analogously $E((Y_{ij})) = (E(Y_{ij}))$. With this handy notation we may express the covariance matrix of a random vector $\mathbf{Y}$ as follows:

$$\operatorname{cov}(\mathbf{Y}) = E\big[(\mathbf{Y} - E(\mathbf{Y}))(\mathbf{Y} - E(\mathbf{Y}))^t\big].$$

The reader should carefully examine this formula; it is used in Section 2 at a couple of important places.

It is quite easy to see that *if Y_1 and Y_2 are independent random variables, then their covariance is* 0. [Indeed, just write $f(y_1, y_2) = f_{Y_1}(y_1) f_{Y_2}(y_2)$ and separate the two sums in (8.3).] The converse is almost always false. Generally a high positive value for $\operatorname{cov}(Y_1, Y_2)$ indicates that the two variables take large values together and low values together; large negative covariance means that the two variables vary oppositely: large values on one go with low values for the other. We may summarize by saying that *the covariance measures dependence in the form of a linear trend.*

We often take a sample from a random variable in a random manner; a sample so obtained is commonly called a *random sample*. For the sake of discussion let Y_1, Y_2 be such a random sample of size 2; thus each Y_i has the same mean μ, the same variance σ^2, and they are independent random variables. It is of interest to compute the mean and variance of an arbitrary linear combination $a_1Y_1 + a_2Y_2$. Being a sum (or integral), *the operator of expected value is always linear.* This yields at once

$$E(a_1Y_1 + a_2Y_2) = a_1E(Y_1) + a_2E(Y_2)(= a_1\mu + a_2\mu).$$

In case the random variables are independent, however, the symbol of expected value behaves multiplicatively, that is

$$E(Y_1Y_2) = E(Y_1)E(Y_2).$$

[Indeed, $E(Y_1Y_2) = \Sigma_{y_1}\Sigma_{y_2}y_1y_2f(y_1, y_2) = \Sigma_{y_1}\Sigma_{y_2}y_1y_2f_{Y_1}(y_1)f_{Y_2}(y_2) = (\Sigma_{y_1}y_1f_{Y_1}(y_1))(\Sigma_{y_2}y_2f_{Y_2}(y_2)) = E(Y_1)E(Y_2)$, as stated.] Thus

$$\begin{aligned}
&\operatorname{var}(a_1Y_1 + a_2Y_2)\\
&\quad = E(a_1Y_1 + a_2Y_2 - a_1\mu - a_2\mu)^2 = E(a_1(Y_1 - \mu) + a_2(Y_2 - \mu))^2\\
&\quad = E\big(a_1^2(Y_1 - \mu)^2 + a_2^2(Y_2 - \mu)^2 + 2a_1a_2(Y_1 - \mu)(Y_2 - \mu)\big)\\
&\quad = a_1^2 \operatorname{var} Y_1 + a_2^2 \operatorname{var} Y_2 + 0,
\end{aligned}$$

since the covariance $E((Y_1 - \mu)(Y_2 - \mu))$ is zero due to independence. These observations extend directly to random samples of any size. We summarize:

∗ *If $Y_1, Y_2, \ldots, Y_n$ is a random sample, each Y_i having mean μ and variance σ^2, then*

$$E(a_1Y_1 + a_2Y_2 + \cdots + a_nY_n) = a_1\mu + a_2\mu + \cdots + a_n\mu,$$

and

$$\operatorname{var}(a_1Y_1 + a_2Y_2 + \cdots + a_nY_n) = a_1^2\sigma^2 + a_2^2\sigma^2 + \cdots + a_n^2\sigma^2.$$

Of special interest is the corollary:

$$E(\overline{Y}) = \mu \qquad \text{and} \qquad \operatorname{var}(\overline{Y}) = \frac{\sigma^2}{n}, \tag{8.4}$$

where $\overline{Y} = n^{-1}\Sigma_{i=1}^n Y_i$ is the sample mean, and the Y_i's are a random sample, each Y_i having mean μ and variance σ^2.

8.5

Several examples of random variables are now given. We first list those of the discrete kind:

Multinomial

The typical model is a *loaded die* with probability p_i to show face "i." For generality we assume that there are k faces labeled "1," "2,"..., "k." Let Y_i be the number of times face "i" turns up in n rolls. The random vector $\mathbf{Y} = (Y_1, Y_2, \ldots, Y_k)$ is called a *multinomial* random vector and its density is

$$f(y_1, y_2, \ldots, y_k) = \frac{n!}{y_1! y_2! \cdots y_k!} p_1^{y_1} p_2^{y_2} \cdots p_k^{y_k}.$$

Here $\Sigma p_i = 1$ and $\Sigma y_i = n$.

When $k = 2$ we have just two faces, so we can think of tossing a biased coin. The resulting random vector is called *binomial*. Check that the density of $\mathbf{Y}$ given above is indeed the correct one.

Hypergeometric

Think of a box with m_1 balls colored "1," m_2 balls colored "2," ..., m_k balls colored "k." Pick n balls, all in one swoop (or pick them one by one without replacing). Let Y_i be the number of balls of color "i" among the n picked. We call $\mathbf{Y} = (Y_1, Y_2, \ldots, Y_k)$ a *hypergeometric* random vector. Its density is

$$f(y_1, y_2, \ldots, y_k) = \frac{\binom{m_1}{y_1}\binom{m_2}{y_2} \cdots \binom{m_k}{y_k}}{\binom{m_1 + m_2 + \cdots + m_k}{n}}, \qquad \Sigma y_i = n.$$

(Verify this. If it seems difficult take $k = 2$.)

A few random variables of a continuous kind are now listed.

Gaussian (or Normal)

This random variable has density

$$f(y) = \frac{1}{\sigma\sqrt{2\pi}} \exp\left(-\frac{(y-\mu)^2}{2\sigma^2}\right), \qquad -\infty < y < +\infty.$$

The parameters μ and σ^2 turn out to be the mean and variance of this random variable. Simple techniques of calculus yield a bell-shaped graph for f, symmetric around μ.

The Chi-Squared Random Variable

Consider a random vector $\mathbf{X} = (X_1, X_2, \ldots, X_n)$ of n *independent* Gaussian random variables each with mean 0 and variance 1. Form $Y = \Sigma_{i=1}^n X_i^2$. The new random variable so obtained is called a *chi-squared* random variable with n degrees of freedom, written $\chi^2(n)$. It obviously takes positive values. A little work shows that Y has density

$$f(y) = c(n)y^{(n/2)-1}\exp\left(-\frac{y}{2}\right), \qquad 0 < y < \infty,$$

where $c(n)$ is a constant depending on n only.

The F Random Variable

Let V be $\chi^2(m)$ (i.e., chi-squared with m degrees of freedom) and let W be $\chi^2(n)$. Further, assume that V and W are *independent*. Then the quotient

$$Y = \frac{V/m}{W/n}$$

is said to be a F (from Fisher) random variable with m and n degrees of freedom [written $F(m, n)$]. The density of $F(m, n)$ may be shown to be

$$f(y) = c(m,n)y^{(m/2)-1}(1 + mn^{-1}y)^{-(m+n)/2}, \qquad 0 < y < \infty,$$

where $c(m, n)$ is a constant depending on m and n only.

Multivariate Gaussian

We call $\mathbf{Y}' = (Y_1, Y_2, \ldots, Y_n)$ a *multivariate Gaussian* vector if its density is

$$f(\mathbf{y}) = (2\pi)^{-n/2}|V|^{-1/2}\exp\left(-\tfrac{1}{2}(\mathbf{y} - \boldsymbol{\mu})'V^{-1}(\mathbf{y} - \boldsymbol{\mu})\right),$$

with $\mathbf{y} = (y_1, y_2, \ldots, y_n)'$ varying over the whole n-dimensional space. The parameters $\boldsymbol{\mu}$ and V are in fact the mean vector and covariance matrix of $\mathbf{Y}$. (The matrix V is positive definite.)

In closing, we suggest that the reader take the time to graph the densities of the Gaussian, χ^2, and F random variables. Then fix a point y ($y > 0$) and investigate how much area there is in the tail at the right of y for various degrees of freedom for χ^2 and F; statistical tables should be used for this task. In particular, are the tails of the chi-squares getting thinner as the degrees of freedom get larger? What about the F?

EXERCISES

1. Densities of five random variables are listed below. In each case draw a graph for the density. Eyeball in and see if in ten seconds you can come up with good guesses for the mean and standard deviation in each case. Compute the actual mean and standard deviation and check against your guess. (Sharpen your perceptions, especially that of the standard deviation.)

(a) $f(y) = \frac{1}{2}$, $-1 \le y \le 1$.

(b) $f(y) = e^{-y}$, $0 < y < \infty$.

(c) $f(y) = e^{-2}(2^y/y!)$, $y = 0, 1, 2, \ldots$.

(d) $f(y) = \pi^{-1}(1 + y^2)^{-1}$, $-\infty < y < +\infty$.

(e) $f(y) = (\frac{1}{2})^y$, $y = 1, 2, 3, \ldots$.

2. Suppose a random vector (Y_1, Y_2) has joint density $f(y_1, y_2) = 4y_1y_2$, $0 < y_1 < 1, 0 < y_2 < 1$. Find the probability that Y_1 is less than $\frac{1}{4}$ and Y_2 is greater than $\frac{1}{2}$. What is the chance that Y_1 is less than $2Y_2$?

3. The correlation coefficient of Y_1 and Y_2 is defined as follows:

$$\operatorname{corr}(Y_1, Y_2) = \frac{\operatorname{cov}(Y_1, Y_2)}{\operatorname{sd}(Y_1)\operatorname{sd}(Y_2)}.$$

(It usually is a preferred measure of linear trend to covariance.) Show that $-1 \leq \operatorname{corr}(Y_1, Y_2) \leq 1$. Describe the situations in which equality is actually attained.

4. If X is a Gaussian random variable with mean μ and variance σ^2 show that $(X - \mu)/\sigma$ is Gaussian with mean 0 and variance 1. Prove also that $((X - \mu)/\sigma)^2$ is $\chi^2(1)$.

5. Show that the sum of two independent χ^2's is again χ^2 with degrees of freedom equal to the sum of the respective degrees of freedom.

6. Prove that if $(Y_1, \ldots, Y_n)$ is multivariate Gaussian, then each Y_i is (univariate) Gaussian. The converse is false in general; construct a counterexample.

7. Which of the following would be unusual observations from a Gaussian random variable with mean -2 and variance 25: (a) 1.3 (b) -25 (c) 64.927 (d) -16.23. Explain.

8. We observe 47.39 from a χ^2 random variable. Is this observation unusual if the χ^2 has degrees of freedom equal to: (a) 2 (b) 20 (c) 51 (d) 237. To understand what is going on compute the mean and variance of $\chi^2(n)$.

2 FACTORIAL EXPERIMENTS

A chemist wishes to know how the yield of a compound of interest is affected by the temperature, pressure, and concentration at which it is produced. To detect and understand the eventual dependence of the yield on these three factors she decides to perform an experiment. Initially she chooses two settings (high and low) for each factor:

	Low (−)	High (+)
Temperature	100	140
Pressure	80	120
Concentration	15	30

This would permit her to grasp whether significant changes in the factors produce marked differences in the yield.

With two settings (− and +) for each of the three factors eight experimental conditions are possible. They are listed below:

1	2	3
−	−	−
−	−	+
−	+	−
−	+	+
+	−	−
+	−	+
+	+	−
+	+	+

(Here 1 stands for temperature, 2 for pressure, and 3 for concentration.) At each of these experimental conditions the yield would be a *random variable*. We denote the mean yield at experimental condition + − + by μ_{+-+}, the mean yield at − − + by μ_{--+}, and so on. These means are not known to the chemist. Indeed, *she would like to estimate them and see how they are affected by changes in the three factors.*

8.6 Establishing Terminology

The eight experimental conditions (or points at which the experiment is to be carried out) may conveniently be displayed as vertices of a cube:

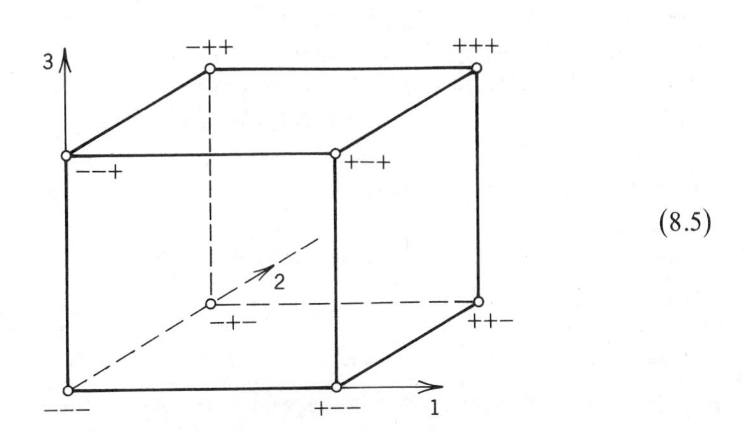

(8.5)

Main Effects

What do we understand by the effect of temperature (say) on the yield? Intuitively it is the perceived change in yield as a function of temperature only, with the other two factors averaged in. In accord with this we thus define

$$1: \tfrac{1}{4}(\mu_{+--} + \mu_{+-+} + \mu_{++-} + \mu_{+++}) - \tfrac{1}{4}(\mu_{---} + \mu_{--+} + \mu_{-+-} + \mu_{-++})$$

to be the *effect of temperature* upon the yield. [Geometrically it is the average of the four means at high temperature minus the average of the four means at low temperature; see (8.5).] For simplicity we denote this effect (effect of factor 1) simply by 1. In complete analogy effects 2 (of pressure) and 3 (of concentration) are defined. We suggest that the student write them out. Effects 1, 2, and 3 defined above are called *main effects* to distinguish them from the higher order effects that we introduce next.

Two-Factor Interactions

It may be that the difference in average yield is much greater when we move from low to high temperature at high pressure than at low pressure. If this is indeed the case we say that interaction exists between temperature and pressure. Interaction is thus a measure of nonlinear behavior in response. Detecting such behavior would be of interest, and we are led to define the *interaction* 12 between factors 1 and 2 as follows:

$$\begin{aligned} 12: \tfrac{1}{2}\Big[\Big(\tfrac{1}{2}(\mu_{+++} + \mu_{++-}) - \tfrac{1}{2}(\mu_{-++} + \mu_{-+-})\Big) \\ - \Big(\tfrac{1}{2}(\mu_{+-+} + \mu_{+--}) - \tfrac{1}{2}(\mu_{--+} + \mu_{---})\Big)\Big] \\ = \tfrac{1}{4}(\mu_{+++} + \mu_{++-} + \mu_{--+} + \mu_{---}) \\ - \tfrac{1}{4}(\mu_{-++} + \mu_{-+-} + \mu_{+-+} + \mu_{+--}). \end{aligned}$$

[Observe that the interaction between factors 1 and 2 could have been defined by interchanging the role of temperature and pressure. The resulting expression (which we would call 21) is, however, *identical* to the one given above. In symbols, $12 = 21$.] Iteractions 13 ($= 31$) and 23 ($= 32$) are defined in complete analogy. There are $\binom{3}{2} = 3$ two-factor interactions in all: 12, 13, and 23. Understanding the definition and meaning of two-factor interactions is very important. Study it in figure (8.5).

Three-Factor Interactions

Two measures of the 12 interaction are available: one at *each level* of 3 (concentration).

Restricted at the + level of factor 3 the 12 interaction would be

$$\tfrac{1}{2}(\mu_{+++} - \mu_{-++}) - \tfrac{1}{2}(\mu_{+-+} - \mu_{--+}).$$

At the low level ($-$) the 12 interaction would be

$$\tfrac{1}{2}(\mu_{++-} - \mu_{-+-}) - \tfrac{1}{2}(\mu_{+--} - \mu_{---}).$$

Half the difference between these two measures of the 12 interaction at the two levels of factor 3 is called the *three-factor interaction*. We denote it by 123.

(It can be verified that the order in which two-factor interactions are matched with the remaining factor to produce three-factor interactions is independent of order. Thus 123 = 132 = 213 = 231 = 312 = 321.) The three-factor interaction is therefore

$$\begin{aligned}123: \tfrac{1}{2}\Big[&\big(\tfrac{1}{2}(\mu_{+++} - \mu_{-++}) - \tfrac{1}{2}(\mu_{+-+} - \mu_{--+})\big)\\ &- \big(\tfrac{1}{2}(\mu_{++-} - \mu_{-+-}) - \tfrac{1}{2}(\mu_{+--} - \mu_{---})\big)\Big]\\ &= \tfrac{1}{4}(\mu_{+++} + \mu_{-+-} + \mu_{--+} + \mu_{+--})\\ &\quad - \tfrac{1}{4}(\mu_{-++} + \mu_{++-} + \mu_{+-+} + \mu_{---}).\end{aligned}$$

The chemist's objective is to accurately assess the *main effects*, *two-factor interactions*, and possibly the *three-factor interaction*. She would then form an impression of what she is up against and decide how to proceed with the investigation.

8.7

Defining the main effects and interactions might appear cumbersome. A very simple indexing rule exists, however.

Means	Effects							
	I	1	2	3	12	13	23	123
μ_{---}	+	−	−	−	+	+	+	−
μ_{--+}	+	−	−	+	+	−	−	+
μ_{-+-}	+	−	+	−	−	+	−	+
μ_{-++}	+	−	+	+	−	−	+	−
μ_{+--}	+	+	−	−	−	−	+	−
μ_{+-+}	+	+	−	+	−	+	−	−
μ_{++-}	+	+	+	−	+	−	−	+
μ_{+++}	+	+	+	+	+	+	+	+

(8.6)

We take a few minutes to explain how Table (8.6) is constructed. Column I is added for convenience (since the columns are labeled with subsets of $\{1, 2, 3\}$, which we interpret now as *effects*, we thought we would include the empty set too). Columns 1, 2, and 3 index (jointly) the vertices of the cube. The important part is how the remaining columns are obtained. They are obtained by *componentwise products*. Thus column 12 is obtained by multiplying componentwise (i.e., $- \, - = +$, $- \, + = -$, etc.) columns 1 and 2. Column 123 is, likewise, the componentwise product of columns 1, 2, and 3 (or of column 12 and 3, or of 1 and 23, or of 13 and 2), and so on. Commutativity $ij = ji$ clearly holds, and $i^2 = ii = \mathrm{I}$, for all columns i.

It will be observed that *the signs in a column* (indexed by an effect, such as 12) *produce the definition of that effect when paired up with the means* listed at left. (This is a valuable rule when it comes to writing out the various effects explicitly.) We explain in Section 8.9 how the rule is obtained. At this time the reader should simply check it for effects 1, 12, and 123, which are listed in Section 8.6.

The rule is valid for a cube in any number of dimensions (i.e., factors) so long as there are two levels on each factor.

Observe, in addition, that Table (8.6) is in fact a Hadamard matrix of order 8. Each of the seven effects may thus be identified with the two complementary blocks of the Hadamard 3-(8, 4, 1) design listed in (7.1); row indices for $+$'s form one block and row indices for $-$'s form its complement. The same is true for higher dimensional cubes.

8.8 Performing the Experiment

Our chemist decides to take 16 observations, 2 at each vertex, in order to estimate the main effects and the possible interactions. She cautiously sets out to take the observations in a *random order*, even though this may entail some additional costs and time. For reasons of convenience one would, of course, be tempted to take two (twin) observations in a row at a selected experimental condition. But at the time the twin observations are taken there may be some extraneous factors, unsuspected and unknown, that may bias the two observations (inflate the observed mean or decrease the variance, say). This is why a random order is much preferred. *Randomization allows for implicit averaging of such unsuspected effects.*

The two observations at the point $+ - +$ (say) are written as $y^{(1)}_{+-+}$ and $y^{(2)}_{+-+}$. By y_{+-+} we denote the average of the observations at $+ - +$; in this case $y_{+-+} = \frac{1}{2}(y^{(1)}_{+-+} + y^{(2)}_{+-+})$. Our estimate of the mean μ_{+-+} will be y_{+-+} (the sample mean at $+ - +$). In general, the estimate of any effect is obtained by replacing all the μ's in its expression with the corresponding y's. [The interaction 12, e.g., will be estimated by $\widehat{12}: \frac{1}{4}(y_{+++} + y_{++-} + y_{--+} + y_{---}) - \frac{1}{4}(y_{-++} + y_{-+-} + y_{+-+} + y_{+--})$.] In general we denote by $\widehat{ijk}$ the estimate (so obtained) of the effect ijk. The magnitude of these estimates then provides meaningful information about the unknown effects.

[Of major concern is the *variance*. In general the variance will vary with location (we should thus write σ^2_{-+-} for the variance of the random variable at $- + -$, whose mean we denote by μ_{-+-}). By plotting the *residuals* $y^{(i)} - y$ against the averages y one may detect certain trends in the variance. A most usual circumstance is that the variance tends to be proportional to the mean. Adjustments can usually be made, by way of *transforming* the data, such that the variance of the transformed data does not vary significantly with the mean. The transformation (or substitution) to be selected depends greatly upon the pattern displayed in the plot of residuals versus averages. Common transformations are the logarithm, square root, and inverse. We refer to [5, p. 232] for further discussion and examples.]

Modulo an eventual transformation of the data we assume that the variance is constant over the region of experimentation; denote it by σ^2. Each observation (such as $y^{(2)}_{--+}$) will thus have variance approximately σ^2. Then y_{--+}, being the average of two independent observations, has variance $\sigma^2/2$ [why?—see (8.4)]. Having two observations at each point (e.g., $+ - -$) we estimate σ^2 (which is most likely unknown) at $+ - -$ by $s^2_{+--} = \frac{1}{2}[(y^{(1)}_{+--} - y_{+--})^2 + (y^{(2)}_{+--} - y_{+--})^2]$. If the evidence is fairly strong that σ^2 is the same at each of the eight points, then we may pool our estimate and approximate σ^2 by

$$\widehat{\sigma^2} = \frac{1}{2^3}\left(\text{sum of the eight } s^2\text{'s}\right).$$

The square root of $\widehat{\sigma^2}$, which we denote by $\hat{\sigma}$, is called the (sample) *standard error* or sample standard deviation.

We have now in our possession the seven estimates $\widehat{ij}$ (symbolic writing) of the main effects and interactions, along with an estimate $\hat{\sigma}$ of the standard deviation σ of each observation. To end the discussion with a tentative conclusion we look at the data from an actual experiment:

1	2	3	Observation 1 $y^{(1)}$	Observation 2 $y^{(2)}$	Average y	Standard error $\hat{\sigma}$
−	−	−	69	71	70	
−	−	+	60	68	64	
−	+	−	60	64	62	
−	+	+	56	54	55	$\hat{\sigma} = 2$
+	−	−	84	80	82	
+	−	+	79	77	78	
+	+	−	91	95	93	
+	+	+	89	91	90	

The summary table for estimates is:

Effect	Estimate ± (Estimated standard error of estimate)
1 = temperature (T)	23.0 ± 1.0 ←
2 = pressure (P)	1.5 ± 1.0
3 = concentration (C)	−5.0 ± 1.0 ←
12 = TP	10.0 ± 1.0 ←
13 = TC	1.5 ± 1.0
23 = PC	0.0 ± 1.0
123 = TPC	0.5 ± 1.0

A few comments may be necessary as to how some of the numbers were derived. The estimates $\widehat{ij}$ were computed as discussed. For example, $\widehat{12} = 10.0$. To compute the standard error of an estimate note that each estimate is a sum of eight y's each weighted by $\pm\frac{1}{4}$. Each y has variance $\sigma^2/2$. Thus any estimate will have variance $(1/4^2)8(\sigma^2/2) = \sigma^2/4$ and standard error $\sigma/2$. We estimate σ by $\hat{\sigma} = 2$, hence the estimate for the standard error of any estimate is $\hat{\sigma}/2 = 2/2 = 1.0$. This value appears in the table above.

Tentative Conclusions

The effect of the temperature, 23.0 ± 1.0, appears to be substantial, but due to the high interaction with pressure, 10.0 ± 1.0, they must be studied *jointly*.

Concentration has negligible interactions (hardly detectable from the experimental error). We may thus conclude that increase in concentration reduces yield by about five units irrespective of the levels of temperature and pressure.

A further study should probably focus on the joint behavior of temperature and pressure, and the chemical implication of their significant interaction.

The design used to perform this experiment is called a 2^3 factorial design. By a *2^k factorial design* we understand a design based on a k-dimensional cube (i.e., a design involving k factors, each at two levels).

3 INTRODUCING THE MODEL

A convenient way to work with factorial (as well as many other kinds of) designs is to formulate *explicit formulas* for the mean of the response. By writing out such explicit representations we are creating a *model*. It is never completely clear that the model we select describes the true way that nature behaves. But models that are approximately correct are potential investigative tools. Building a good model is both a science and an art. The model should incorporate as much deterministic knowledge as possible, gathered from anyone familiar with the phenomenon of interest; some statistical safeguards should be built in and much sampling and testing for the goodness of fit of the model should be done. As a rule many tentative models are tossed out as inappropriate before a good model emerges.

8.9

We are concerned with studying the effects of k (say $k = 3$) factors, each at two levels, on the yield (of a chemical compound). A useful model to accomplish this is now formulated.

Introduce as many variables as there are factors: x_1, x_2, x_3 (say). Variable x_i takes values -1 and 1 only, the former when the ith factor is at low level, the latter when it is at high level. An experimental condition (i.e., a vertex of the cube) may now be described as a vector (x_1, x_2, x_3). The unknown mean yield (i.e., mean response) at (x_1, x_2, x_3) is written $\mu(x_1, x_2, x_3)$. We assert that *a good model to represent the mean response is*

$$\mu(x_1, x_2, x_3) = \beta_0 + \beta_1 x_1 + \beta_2 x_2 + \beta_3 x_3 + \beta_{12} x_1 x_2 + \beta_{13} x_1 x_3 + \beta_{23} x_2 x_3 + \beta_{123} x_1 x_2 x_3, \tag{8.7}$$

the β's being unknown constants. There are eight unknown μ's and eight unknown β's. It is natural to wonder what these β's are as functions of μ's. Writing (8.7) in matrix form we obtain

$$\boldsymbol{\mu} = H\boldsymbol{\beta}, \tag{8.8}$$

where $\boldsymbol{\mu}$ is the vector of the eight means listed in some fixed order [as in (8.6), say], $\boldsymbol{\beta} = (\beta_0, \beta_1, \beta_2, \beta_3, \beta_{12}, \beta_{13}, \beta_{23}, \beta_{123})'$ is the vector of unknowns, and H is the resulting coefficient matrix. It will be observed that in this case H is the full Hadamard matrix of order 8 as written in (8.6). Solving for $\boldsymbol{\beta}$ is now easy. Indeed, $H'\boldsymbol{\mu} = H'H\boldsymbol{\beta}$; thus

$$\boldsymbol{\beta} = (H'H)^{-1}H'\boldsymbol{\mu}. \tag{8.9}$$

Since $H'H = 8I$ we obtain $\boldsymbol{\beta} = \frac{1}{8}H'\boldsymbol{\mu}$. It will further be observed (and quickly checked by the reader) that $2\beta_i$ equals precisely the *main effect of factor i* (as defined in Section 8.6), $2\beta_{ij}$ equals the *interaction of factors i and j*, and $2\beta_{123}$ is precisely the *three-factor interaction*! This shows that the β's in model (8.7) have *meaning* and thus model (8.7) is a good mathematical model to work with.

REMARK. We obtained the solution $\boldsymbol{\beta} = \frac{1}{8}H'\boldsymbol{\mu}$ to system (8.8) which informs us that (up to a factor of 8) each component of $\boldsymbol{\beta}$ (i.e., each effect) is obtained by taking the inner product between a column of H and the vector $\boldsymbol{\mu}$. This is precisely the recipe we followed in Section 8.7 for computing main effects and interactions. In addition, observe that the columns of H are constructed by taking suitable *componentwise products* of columns associated with the main effects $\beta_1, \beta_2, \beta_3$; this is simply a direct consequence of the assumed model (8.7).

Revising the Definitions of Effects

At this time it seems convenient to revise our definition of effects from that given in Section 8.6 to the following more flexible one: By the *interaction of factors i, j, k,... we understand the coefficient of* $x_i x_j x_k \cdots$ *in the model with which we work*. (Thus the main effect of factor i is the coefficient of x_i, the interaction between factors i and j is the coefficient of $x_i x_j$, etc.) As we just saw, if the model is that of (8.7), the newly defined effects agree with the old ones, up to a (same) constant factor.

8.10 The Normal Equations and Optimality Criteria

If we are in the presence of k factors and wish to use a (full, or complete) factorial design, at least 2^k observations will be required: one or more at each vertex of the k-dimensional cube. The design allows estimation of interactions of *any* order. Usually most interest is manifested in *estimating main effects and two-factors interactions only*, the higher order effects being of somewhat lesser interest. (This attitude is similar to that of approximating a function by the first two powers in its Taylor series expansion.) By ignoring the higher order interactions we should be able to provide estimates for the main effects and two-factor interactions with *fewer observations* than 2^k (which increases exponentially with k). Decreasing the number of observations is of legitimate concern in many experimental situations because of the costs involved.

Throughout this section we regard as negligible the interactions of order higher than two. The rationale for doing this cannot be fully justified. In many situations, however, nature seems to show its chief intentions by way of main effects and two-factor interactions. We can at least think of this as a fair approximation to the true state of nature. With this in mind the model that suggests itself is

$$\mu_{(x_1,\ldots,x_k)} = \beta_0 + \sum_{i=1}^{k} \beta_i x_i + \sum_{\substack{i,j \\ i<j}} \beta_{ij} x_i x_j. \tag{8.10}$$

(As agreed, β_i denotes the main effect of factor i and β_{ij} the two-factor interaction between factors i and j. No higher order interactions occur in this model. The x_i's take values -1 and 1.)

We write $y_{(x_1,\ldots,x_k)}$ for an observation taken at vertex $(x_1,\ldots,x_k)$; superscripts $y^{(i)}$ are used if several observations are made at the same vertex. At $(x_1,\ldots,x_k)$ the response is a *random variable*, which we denote also by $y_{(x_1,\ldots,x_k)}$ (the same way as the observation at $(x_1,\ldots,x_k)$, an abuse of notation for the sake of convenience). The mean of $y_{(x_1,\ldots,x_k)}$ is $\mu_{(x_1,\ldots,x_k)}$, the functional form of which is expressed in (8.10).

An experiment involves the taking of n observations. In accordance with the assumed model we write

$$E\left(y_{(x_1,\ldots,x_k)}\right) = \beta_0 + \sum \beta_i x_i + \sum_{\substack{i,j \\ i<j}} \beta_{ij} x_i x_j, \tag{8.11}$$

for an observation $y_{(x_1,\ldots,x_k)}$ taken at $(x_1,\ldots,x_k)$. We assume (via an eventual transformation of the data) that the observations are uncorrelated, each with (unknown) variance σ^2. A convenient way to summarize all this is to stack up the n observations in a $n \times 1$ random vector $\mathbf{Y}$, and write (8.11) in matrix form as follows:

$$E(\mathbf{Y}) = X\boldsymbol{\beta}; \qquad \operatorname{cov} \mathbf{Y} = \sigma^2 I_{n\times n},$$

with $\boldsymbol{\beta}' = (\beta_0, \beta_1, \dots, \beta_k, \beta_{12}, \dots, \beta_{k-1,k})$ a vector with $1 + k + \binom{k}{2} = p$ components and X a $n \times p$ (observations versus effects) matrix. If the mth observation was taken at vertex $(x_1, \dots, x_k)$, the mth row of X is $(1, x_1, \dots, x_k, x_1x_2, \dots, x_{k-1}x_k)$, in accordance with (8.11). The matrix X depends on the vertices of the k-dimensional cube where the n observations are made. It is within the power of the experimenter to select these vertices in a strategically useful and clever way. The choice of locations influences the precision of the estimates of effects and affects the general understanding of how the response behaves over the entire region of study (i.e., the whole k-dimensional cube). More will be said on this important aspect at a later time.

Granted that n observations were taken (at some locations), how can we estimate the components of $\boldsymbol{\beta}$? One sensible way would be to select the $\boldsymbol{\beta}$ that minimizes the sum of squared deviations from the mean, that is, the $\boldsymbol{\beta}$ that minimizes

$$(\mathbf{Y} - E(\mathbf{Y}))'(\mathbf{Y} - E(\mathbf{Y})) = (\mathbf{Y} - X\boldsymbol{\beta})'(Y - X\boldsymbol{\beta}).$$

(Other estimates may be chosen but we stick with these.) The solution, obtained upon routine differentiation in the components of $\boldsymbol{\beta}$, is that of the linear system

$$(X'X)\boldsymbol{\beta} = X'\mathbf{Y}. \tag{8.12}$$

System (8.12) is known as the *normal equations*. If $X'X$ is invertible, the unique solution may be written as

$$\hat{\boldsymbol{\beta}} = (X'X)^{-1}X'\mathbf{Y}.$$

We call it the *least squares* estimates for the effects. Each component of the $p \times 1$ vector $\hat{\boldsymbol{\beta}}$ is therefore a linear function of the n observations. It turns out that

$$E(\hat{\boldsymbol{\beta}}) = E\left[(X'X)^{-1}X'\mathbf{Y}\right] = (X'X)^{-1}X'E(\mathbf{Y}) = (X'X)^{-1}X'X\boldsymbol{\beta} = \boldsymbol{\beta}$$

(we say that $\hat{\boldsymbol{\beta}}$ is an *unbiased estimate* of $\boldsymbol{\beta}$), and that

$$\begin{aligned}
\operatorname{cov}\hat{\boldsymbol{\beta}} &= E\left[(\hat{\boldsymbol{\beta}} - \boldsymbol{\beta})(\hat{\boldsymbol{\beta}} - \boldsymbol{\beta})'\right] \\
&= E\Big[\big((X'X)^{-1}X'\mathbf{Y} - (X'X)^{-1}X'X\beta\big) \\
&\qquad \times \big((X'X)^{-1}X'\mathbf{Y} - (X'X)^{-1}X'X\beta\big)'\Big] \\
&= (X'X)^{-1}X'E\left[(\mathbf{Y} - X\beta)(\mathbf{Y} - X\beta)'\right]X(X'X)^{-1} \\
&= (X'X)^{-1}X'(\operatorname{cov}\mathbf{Y})X(X'X)^{-1} = (X'X)^{-1}X'\sigma^2 IX(X'X)^{-1} \\
&= (X'X)^{-1}\sigma^2.
\end{aligned}$$

[Let us mention, in passing, that in case $X'X$ is singular a *generalized inverse* replaces the inverse. We denote it by $(X'X)^-$ and define it by properties $(X'X)(X'X)^{-}(X'X) = X'X$, and $(X'X)^{-}(X'X)(X'X)^{-} = (X'X)^{-}$. For any matrix $X'X$ such a generalized inverse $(X'X)^{-}$ exists. Extensive literature may be found on generalized inverses, e.g., [16].]

At the moment let us focus on what we just proved, namely that

$$\operatorname{cov} \hat{\boldsymbol{\beta}} = (X'X)^{-1}\sigma^2. \tag{8.13}$$

Formula (8.13) shows that the covariances of our estimates depend on the $n \times p$ matrix X that, in essence, describes at what vertices observations should be taken. For this reason we call X the *design matrix*. One important objective of any research worker is to select vertices in such a way as to obtain good estimates for the components of $\boldsymbol{\beta}$. What (8.13) tells us is that good estimates will be obtained if $(X'X)^{-1}$ is (in some sense) as "small" as possible. By small we might mean any one of several possible attributes such as

(i) The determinant $|X'X|^{-1}$.
(ii) The trace of $(X'X)^{-1}$.
(iii) The largest diagonal element of $(X'X)^{-1}$.
(iv) The largest eigenvalue of $(X'X)^{-1}$.
(Many others may be conceived.)

Note that trace $(X'X)^{-1}$ equals the sum of variances of the p components of $\hat{\boldsymbol{\beta}}$. The largest diagonal element of $(X'X)^{-1}$ corresponds, of course, to the largest variance among the components of $\hat{\boldsymbol{\beta}}$. Selecting the design matrix X that minimizes one of these is a sensible way of designing the experiment (provided that the model is satisfactory). In addition, the largest eigenvalue of $(X'X)^{-1}$ has statistical meaning as well: it equals the largest variance among the linear combinations $\mathbf{l}'\hat{\boldsymbol{\beta}}$, with $\mathbf{l}$ being a vector of scalars of length l. [Indeed,

$$\begin{aligned}\max_{\|\mathbf{l}\|=1} \operatorname{var} \mathbf{l}'\hat{\boldsymbol{\beta}} &= \max_{\|\mathbf{l}\|=1} \mathbf{l}'(X'X)^{-1}\mathbf{l}\sigma^2 \\ &= \left(\text{largest eigenvalue of } (X'X)^{-1}\right)\sigma^2,\end{aligned}$$

by the minimax representation of eigenvalues of a symmetric matrix; see also Section 4.23.]

It takes a bit longer to justify the rationale behind (i): minimizing the determinant of $(X'X)^{-1}$. In order to do this a distributional assumption on the response needs to be made. Specifically, if the vector $\mathbf{Y}$ of observations is multivariate Gaussian (see Section 8.5), then $\hat{\boldsymbol{\beta}}$ is also multivariate Gaussian. This is so because each component $\hat{\beta}_i$ of $\hat{\boldsymbol{\beta}}$ is a linear combination of the entries of the vector $\mathbf{Y}$ of observations, and a simple substitution shows that linear combinations of Gaussian random variables remain Gaussian. Specifically $\hat{\boldsymbol{\beta}} = (X'X)^{-1}X'\mathbf{Y}$, and thus $\hat{\boldsymbol{\beta}}$ is Gaussian with mean $\boldsymbol{\beta}$ $(= E(\hat{\boldsymbol{\beta}}))$ and covariance matrix $\operatorname{cov} \hat{\boldsymbol{\beta}} = (X'X)^{-1}\sigma^2$. It is then relatively easy to show (again via a suitable substitution—see [1, p. 54–56]) that

$$(\hat{\beta} - \beta)'\left((X'X)^{-1}\sigma^2\right)^{-1}(\hat{\beta} - \beta)$$

has a $\chi^2(p)$ distribution. In the p-dimensional space of β the inequality

$$(\hat{\beta} - \beta)'\left((X'X)^{-1}\sigma^2\right)^{-1}(\hat{\beta} - \beta) \leq c \tag{8.14}$$

describes the interior of an ellipsoid centered at $\hat{\beta}$, the shape of which depends on $(X'X)^{-1}\sigma^2$ and the size of which depends on c. For any fixed c, however, the volume of the ellipsoid in (8.14) is in fact proportional to the square root of the determinant of $(X'X)^{-1}\sigma^2$ since *its semiaxes are the square root of the eigenvalues* of $(X'X)^{-1}\sigma^2$. It is statistically desirable to have the volume of this ellipsoid as small as possible (for in that case $\hat{\beta}$ is "close" to β). This leads us to the minimization of $|X'X|^{-1}$.

Let us now summarize and introduce nomenclature as follows:

1. A design with design matrix X^* is called *D-optimal* if $|X^{*\prime}X^*| \geq |X'X|$, for all design matrices X under consideration.
2. It is called *A-optimal* (A from average trace) if

$$\operatorname{trace}(X^{*\prime}X^*)^{-1} \leq \operatorname{trace}(X'X)^{-1},$$

   for all design matrices X.
3. It is *MV-optimal* (from maximal variance, I guess) if the maximal diagonal element of $(X^{*\prime}X^*)^{-1} \leq$ the maximal diagonal element of $(X'X)^{-1}$, for all X.
4. It is *E-optimal* if the minimal eigenvalue of $X^{*\prime}X^*$ is at least as large as the minimal eigenvalue of any $X'X$, for all X under consideration.

Intuitively an optimal design (optimal in the D-, A-, MV-, or E-sense) describes an optimal way of selecting vertices on the k-dimensional cube at which observations should subsequently be made. Each one of these criteria tries to make the matrix $X'X$ as "nonsingular" as possible. This goes along well with intuition, since a "very nonsingular" $X'X$ would correspond to a choice of highly "spread out" vertices on the k-dimensional cube.

8.11 The Reduced Normal Equations

Look back at (8.11). The model expresses the expected value of an observation in terms of main effects and two-factor interactions. It may very well happen that our chief concern rests with estimating main effects (say), and interest in two-factor interactions (although these may be significant) is only secondary. For the time being we assume that this is indeed the case and we think of two-factor interactions as nuisance parameters of little interest that are, nevertheless, significant enough to demand inclusion in the model.

With such a perception of interactions the choice of design should most likely be based on maximizing precision when estimating main effects. That is, we may wish to select the design matrix X^* that leads to a minimal sum of variances of the estimates $\hat{\beta}_i$ for the main effects β_i only, $i = 1, 2, \ldots, k$. Such a design is called *A-optimal for main effects*, in agreement with our previous terminology. Similar straightforward extensions are made for D-, MV-, or E-optimal designs for main effects.

Attention being focused on main effects, it seems natural to partition the vector $\boldsymbol{\beta}$ of unknown effects as follows:

$$\boldsymbol{\beta}' = (\beta_1, \ldots, \beta_k \,\vdots\, \beta_{12}, \ldots, \beta_{k-1,k}, \beta_0) = (\boldsymbol{\beta}_m' \,\vdots\, \boldsymbol{\beta}_r'),$$

where $\boldsymbol{\beta}_m$ is the $k \times 1$ vector of main effects and $\boldsymbol{\beta}_r$ is the vector of the remaining $\binom{k}{2} + 1$ effects. (Note that β_0 is written now as the last component of $\boldsymbol{\beta}_r$. We make the corresponding switch in the design matrix X as well, reserving the last column for β_0.) This partitioning of $\boldsymbol{\beta}$ induces a partitioning of X:

$$X = [X_1 \,\vdots\, X_2],$$

with X_1 a $n \times k$ matrix and X_2 a $n \times \left[\binom{k}{2} + 1\right]$ matrix. With these notational changes the model may be written

$$E(\mathbf{Y}) = X\boldsymbol{\beta} = [X_1 \,\vdots\, X_2]\begin{pmatrix} \boldsymbol{\beta}_m \\ \hdashline \boldsymbol{\beta}_r \end{pmatrix} = X_1\boldsymbol{\beta}_m + X_2\boldsymbol{\beta}_r.$$

The normal equations of (8.12) take the form

$$\begin{bmatrix} X_1'X_1 & X_1'X_2 \\ X_2'X_1 & X_2'X_2 \end{bmatrix}\begin{pmatrix} \boldsymbol{\beta}_m \\ \hdashline \boldsymbol{\beta}_r \end{pmatrix} = \begin{pmatrix} X_1' \\ \hdashline X_2' \end{pmatrix}\mathbf{Y}. \tag{8.15}$$

Our interest is in the components of $\boldsymbol{\beta}_m$. By performing suitable row operations in (8.15) we want to produce a zero matrix in place of $X_1'X_2$ (the upper right-hand corner of the coefficient matrix); this would lead to a new system involving only the components of $\boldsymbol{\beta}_m$ as unknowns. (The process is commonly known as *Gauss–Jordan elimination*.)

The following row operations suggest themselves: multiply the bottom part of system (8.15) by $-X_1'X_2(X_2'X_2)^{-1}$ and add to the top part. What results is

$$\begin{bmatrix} X_1'X_1 - X_1'X_2(X_2'X_2)^{-1}X_2'X_1 & \mathbf{0} \\ X_2'X_1 & X_2'X_2 \end{bmatrix}\begin{pmatrix} \boldsymbol{\beta}_m \\ \hdashline \boldsymbol{\beta}_r \end{pmatrix} = \begin{pmatrix} X_1' - X_1'X_2(X_2'X_2)^{-1}X_2' \\ \hdashline X_2' \end{pmatrix}\mathbf{Y}.$$

We thus obtain the following system in the components of $\boldsymbol{\beta}_m$ only:

$$\left(X_1'X_1 - X_1'X_2(X_2'X_2)^{-1}X_2'X_1\right)\boldsymbol{\beta}_m = \left(X_1' - X_1'X_2(X_2'X_2)^{-1}X_2'\right)\mathbf{Y}.$$

Abbreviate by writing

$$C\boldsymbol{\beta}_m = Q\mathbf{Y}, \tag{8.16}$$

where $C = X_1^t X_1 - X_1^t X_2 (X_2^t X_2)^{-1} X_2^t X_1$ and $Q = X_1^t - X_1^t X_2 (X_2^t X_2)^{-1} X_2^t$. System (8.16) is known as the *reduced normal equations*. The coefficient matrix C is called *Fisher's information matrix* (or the *C-matrix*). It turns out that C is symmetric and nonnegative definite. (To see that C is nonnegative definite observe that $C = X_1^t P X_1$ and $P^2 = P$, with $P = I - X_2(X_2^t X_2)^{-1} X_2^t$.)

Solving for $\boldsymbol{\beta}_m$ in (8.16) we obtain

$$\hat{\boldsymbol{\beta}}_m = C^{-1} Q \mathbf{Y}. \tag{8.17}$$

[Observe that, in general, for a matrix of scalars A and $\mathbf{Y}$ a random vector (with mean $\boldsymbol{\mu}$) we have

$$\begin{aligned} \operatorname{cov}(A\mathbf{Y}) &= E\left[(A\mathbf{Y} - A\boldsymbol{\mu})(A\mathbf{Y} - A\boldsymbol{\mu})^t\right] \\ &= E\left[A(\mathbf{Y} - \boldsymbol{\mu})(\mathbf{Y} - \boldsymbol{\mu})^t A^t\right] \\ &= AE\left[(\mathbf{Y} - \boldsymbol{\mu})(\mathbf{Y} - \boldsymbol{\mu})^t\right] A^t \\ &= A(\operatorname{cov} \mathbf{Y}) A^t.\end{aligned}$$]

This parenthetical remark allows us to write the covariance of $\hat{\boldsymbol{\beta}}_m$ as follows:

$$\begin{aligned} \operatorname{cov} \hat{\boldsymbol{\beta}}_m &= \operatorname{cov}(C^{-1} Q \mathbf{Y}) = C^{-1} Q (\operatorname{cov} \mathbf{Y}) Q^t C^{-1} \\ &= C^{-1} Q Q^t C^{-1} \sigma^2. \end{aligned}$$

Note, however, that

$$\begin{aligned} QQ^t &= \left(X_1^t - X_1^t X_2 \left(X_2^t X_2\right)^{-1} X_2^t\right)\left(X_1 - X_2\left(X_2^t X_2\right)^{-1} X_2^t X_1\right) \\ &= \{\text{upon cancellations}\} \\ &= X_1^t X_1 - X_1^t X_2 \left(X_2^t X_2\right)^{-1} X_2^t X_1 = C. \end{aligned}$$

Thus, $\operatorname{cov} \hat{\boldsymbol{\beta}}_m = C^{-1} Q Q^t C^{-1} \sigma^2 = C^{-1} C C^{-1} \sigma^2 = C^{-1} \sigma^2$.

We have, therefore, shown that

$$\operatorname{cov} \hat{\boldsymbol{\beta}}_m = C^{-1} \sigma^2. \tag{8.18}$$

Formula (8.18) is the analogue of (8.13) with the important difference that it involves estimates of main effects only.

The ith diagonal entry of $\operatorname{cov} \hat{\boldsymbol{\beta}}_m = C^{-1}\sigma^2$ is the variance of $\hat{\beta}_i$, the estimate of the main effect β_i. Optimal designs for main effects may now be defined in ways similar to those mentioned at the end of Section 8.10. To wit, *a design is D-, A-, MV-, or E-optimal for main effects if it minimizes, respectively,* $|C^{-1}|$, *trace* C^{-1}, *the maximal diagonal element of* C^{-1}, *or the maximal eigenvalue of* C^{-1} (over all possible choices of C, or equivalently, over all choices of design matrices X).

Searching for optimal (or near optimal) designs, in any of the senses mentioned above, turns out to be a challenge in most instances. Further discussions on this subject follow in the next section.

[We owe an explanation as to the assumed nonsingularity of both $X_2'X_2$ and C—e.g., see (8.17). The information matrix, C, particularly, turns out to be *always singular* in many settings (but not under our present model (8.11)). In such instances formula (8.18) is in dire need of adjustment. A lot can be gained by working with suitable generalized inverses and we shall reexamine (8.18) more closely in Section 8.12.]

8.12 Some Optimal Factorial Designs

The Model with Main Effects Only

The simplest situation, from the viewpoint of optimality, is that in which the model includes main effects only. If such a model seems appropriate we write it as follows:

$$E\left(Y_{(x_1,\ldots,x_k)}\right) = \beta_0 + \beta_1 x_1 + \cdots + \beta_k x_k, \tag{8.19}$$

with the additional assumption that the vector $\mathbf{Y}$ of the n observations has covariance matrix $\operatorname{cov}(\mathbf{Y}) = \sigma^2 I$.

Consider an arbitrary design with design matrix X. The matrix X is $n \times (k+1)$ with entries -1 and 1. Let us quickly investigate what should X look like to be the design matrix of an optimal design (in the D-, A-, or E-sense). In ways identical to those used to derive (8.13) we obtain (under this simpler model) that

$$\hat{\boldsymbol{\beta}} = (X'X)^{-1}X'Y \qquad \text{and} \qquad \operatorname{cov}\hat{\boldsymbol{\beta}} = (X'X)^{-1}\sigma^2,$$

where $\hat{\boldsymbol{\beta}}$ is the $(k+1)\times 1$ vector of least squares estimates of the vector $\boldsymbol{\beta}$ of (main) effects.

The (i, j)th entry of $X'X$ is the inner product of columns i and j of X; hence $X'X$ has all its diagonal entries equal to n. The trace of $X'X$ is therefore $(k+1)n$, since $X'X$ is a $(k+1)\times(k+1)$ matrix. Denote by $\mu_1 \le \mu_2 \le \cdots \le \mu_{k+1}$ the eigenvalues of $X'X$. We wish to minimize

$$\operatorname{trace}(X'X)^{-1} = \sum_{i=1}^{k+1} \mu_i^{-1}\ (A\text{-optimality}),$$

$$|X'X|^{-1} = \prod_{i=1}^{k+1} \mu_i^{-1}\ (D\text{-optimality}),$$

$$\text{minimal eigenvalue of } (X'X)^{-1} = \mu_{k+1}^{-1}\ (E\text{-optimality}),$$

subject to the constraint $\Sigma_{i=1}^{k+1}\mu_i = \operatorname{trace}(X'X) = (k+1)n$. The three functions to be minimized are all symmetric in the μ_i's and the minimum therefore occurs when all the μ_i's are equal. A D-, A-, or E-optimal design is therefore one for which $X'X$ has all its eigenvalues μ_i equal [to $\operatorname{trace}(X'X)/(k+1) = (k+1)n/(k+1) = n$]. A matrix $X'X$ with all its eigenvalues equal to n must be the identity matrix. Thus an optimal design has a design matrix X with any two of its columns orthogonal. (Such a design is MV-optimal as well; the interested reader may wish to supply a proof.)

By an *optimal design* we understand a design that is optimal in the D-, A-, E-, and MV-sense.

We summarize:

> *Suppose n and k are such that design matrices with any two columns orthogonal exist* [under model (8.19)]. *A design is then optimal if and only if its design matrix has this property.* (8.20)

Two things are clear. First, for any n and any k an optimal design (in any specific sense) *exists*—since there are only finitely many design matrices to consider and some of them will, of course, minimize trace $(X'X)^{-1}$, say. Secondly, for relatively few n and k, design matrices with orthogonal columns [as written in (8.20)] will actually exist.

In this simplest of factorial models the answer to the following general question remains unknown:

> *For arbitrary n and k what are the optimal designs?* (8.21)

(By taking $n = k + 1$ to be a multiple of 4, the question reduces to that of the existence of a Hadamard matrix of order n, a problem known to be difficult.)

If n is a multiple of 4, $n \geq k + 1$, and if a Hadamard matrix H of dimension (or order) n exists, then by taking a column of all +'s and k additional columns of H one (obviously) obtains a design matrix X^* that corresponds to an optimal design. Such designs are sometimes called *Plackett–Burman* designs.

The Model with Main Effects and Two-Factor Interactions

Model (8.11), with main effects and two-factor interactions, is more comprehensive and of wider usage than that written in (8.19). Indeed, one of the reasons for introducing factorial designs is to study the interactions between factors, which model (8.11) includes. We now try to identify some optimal designs under this model.

An arbitrary design matrix X is a $n \times [1 + k + \binom{k}{2}]$ matrix, with a column of all +'s corresponding to β_0, and with the column of β_{ij} being the *componentwise product* of the column of β_i and that of β_j, in conformity to what model (8.11) prescribes.

Equal interest is manifested in each of the components of the vector $\boldsymbol{\beta}$ (which consists of β_0, main effects β_i, and interactions β_{ij}). Equation (8.13) gives the covariance of the least squares estimates for $\boldsymbol{\beta}$:

$$\operatorname{cov} \hat{\boldsymbol{\beta}} = (X'X)^{-1}\sigma^2.$$

(As before, we assume that the vector of observations $\mathbf{Y}$ has covariance $\operatorname{cov}\mathbf{Y} = \sigma^2 I$.)

The diagonal entries of $X'X$ are all equal to n and thus $X'X$ has trace $n(1 + k + \binom{k}{2})$. Arguing the same way as in the case of the model (8.19) we conclude that *an optimal design matrix should have any two columns orthogonal.* One of the columns consists of +'s only (that corresponding to β_0); the remaining $k + \binom{k}{2}$ columns, in addition to being pairwise orthogonal, must also satisfy the condition that the column corresponding to β_{ij} is the componentwise product of the columns of β_i and β_j.

Denote by I the column of β_0, by i that of β_i, and by ij (componentwise product) that of β_{ij}. It is clear that $ii = I$ for all columns, and that $ij = ji$ (i.e., this operation is commutative); review Section 8.7 if necessary. If X^* is a design matrix under model (8.11) having any two columns orthogonal (i.e., if X^* corresponds to an optimal design), then for any two distinct subsets $\{i, j\}$ and $\{k, l\}$ of columns of main effects the products ij and kl yield different (orthogonal) columns, that is, $ij \neq kl$. Or, multiplying through by ij, $I \neq ijkl$. Also, a product ij must be different from any main effect k; thus $I \neq ijk$. Clearly also $I \neq ij$, for all $i \neq j$. Consequently it takes five or more distinct columns of main effects in X^* to have their componentwise products equal to I. Because of this we say that X^* is a design of *resolution* 5. (More will be said on resolution of designs in Section 8.13, upcoming.)

In conclusion:

> *Suppose n and k are such that design matrices with any two columns orthogonal exist* [under model (8.11)]. *A design is then optimal if and only if its design matrix has this property.* [Such an optimal design is necessarily of resolution 5 (for $k \geq 5$).] (8.22)

An example of such an optimal design appears in (8.23).

Little is known (to the author) about design optimality when n and k preclude the existence of design matrices with pairwise orthogonal columns. The general answer is no less difficult than that to question (8.21).

One last remark. An optimal design as in (8.22) is jointly optimal for main effects and interactions (e.g., the A-optimal design is one that minimizes the sum $\operatorname{var}\hat{\beta}_0 + \Sigma_i \operatorname{var}\hat{\beta}_i + \Sigma_{i<j} \operatorname{var}\hat{\beta}_{ij}$). Is such a design also optimal for main effects only (i.e., does it also minimize $\Sigma_i \operatorname{var}\hat{\beta}_i$)? The answer is *yes*. [The crux of the proof lies in observing that for any design matrix $X = [X_1 \vdots X_2]$ the matrix $X_1'X_2(X_2'X_2)^{-1}X_2'X_1$ is nonnegative definite, and it becomes the $\mathbf{0}$ matrix when X is as in (8.22)—since in that case $X_1'X_2 = \mathbf{0}$. One needs to examine the consequences that this has on the eigenvalues of the information matrix C written in (8.16). We do not write out the details but summarize the consequences below.]

* *If a design as in* (8.22) *exists, it is optimal for main effects only and for interaction effects only.*

A very last remark. Similar results can be written for more general models that include main effects as well as two- and three- (or higher) factor interactions. The resulting optimal designs (with orthogonal columns) will, of course, also have an appropriately high resolution.

8.13 Resolution of Designs

We return now to the most general model for 2^k factorial experiments: that which includes interactions of any order (up to k, of course). The model contains 2^k effects in all, $\binom{k}{j}$ of which are j-factor interactions. [For $k = 3$ the model is written out fully in (8.7).]

If we take n observations the design matrix will be a $n \times 2^k$ matrix X of $+$ and $-$ signs. (As usual, $+$ abbreviates $+1$ and $-$ abbreviates -1.) The underlying model is

$$E(\mathbf{Y}) = X\boldsymbol{\beta}; \qquad \operatorname{cov} \mathbf{Y} = \sigma^2 I,$$

where $\boldsymbol{\beta}$ is the $2^k \times 1$ vector of effects (lined up as β_0, main effects, two-factor interactions, etc.). The design matrix is X, and $\mathbf{Y}$ is the $n \times 1$ vector of observations. The columns of X are indexed by the effects, the rows of X by observations. The first column of X consists of $+$'s only.

In accordance with the specified model the column corresponding to $\beta_{ijk\cdots}$ is the *componentwise product* of the columns of $\beta_i, \beta_j, \beta_k, \ldots$; therefore, if we write $i, j, k, \ldots$ for the columns of main effects $\beta_i, \beta_j, \beta_k, \ldots$, then $ijk \cdots$ signifies the column of $\beta_{ijk\cdots}$. It is thus clear that once columns $1, 2, \ldots, k$ of the main effects are selected, all the columns of the design matrix X are determined. Depending on the value of n and on the choice made for columns $1, 2, \ldots, k$, repeated columns in X are expected to occur. If columns $ij \cdots$ and $klm \cdots$ are identical or opposite in sign we say that the effects $\beta_{ij\cdots}$ and $\beta_{klm\cdots}$ are *aliased* (or *confounded*). [Aliased effects always have either identical or opposite coefficients in the model under the design matrix X and, for this reason, there is no point in distinguishing between them. We therefore attempt to work modulo such aliases.] If columns 13 and 234 are the same (or opposites) we write $13 = \pm 234$; recalling that $i^2 = I$, we may multiply throughout by 13 and obtain $1^2 3^2 = I = \pm 123^2 4$, or $I = \pm 124$ (the last writing being the one we favor and which we call a *defining relation* for the design matrix X). In general we may have several (independent) defining relations, such as $I = 123$, $I = -24$, and $I = -1345$. The independent defining relations generate a subgroup called the *subgroup of the relations*. With $I = 123$, $I = -24$, and $I = -1345$, the subgroup generated by these relations is denoted by $\langle 123, -24, -1345 \rangle$; it is a subgroup of order 2^3. In general, m independent defining relations generate a subgroup of order 2^m; denote it by W. The subgroup W partitions the set of columns of X into 2^{k-m} classes (called *cosets*) of size $|W| = 2^m$ each; two columns are in the same coset if their componentwise product belongs to W. We can easily eliminate the aliasing between the columns of X by introducing a new matrix $\tilde{X}$ whose columns are labeled by the *cosets* of W, its rows still being labeled by the n observations (as in the case of X). The matrix $\tilde{X}$ is $n \times 2^{k-m}$, if $|W| = 2^m$. When W contains all the existing relations, $\tilde{X}$ will not contain any repeated columns. We call $\tilde{X}$ the *design matrix of cosets of effects* (since the number of parameters in the model has effectively decreased from 2^k to 2^{k-m} by identifying aliases).

Special terminology exists for designs in which n, the number of observations, is a power of 2. When $n = \frac{1}{2}2^k$ we call the design a *half fraction* (of the complete design 2^k). In general a 2^{-p} *fraction* is a design with $n = 2^{k-p}$. Such fractions are called *fractional factorial* designs.

[As an example, consider the half fraction of a 2^3 design given below:

$X =$

I	1	2	3	12	13	23	123
+	−	−	−	+	+	+	−
+	−	+	+	−	−	+	−
+	+	−	+	−	+	−	−
+	+	+	−	+	−	−	−

Aliasing exists since $I = -123$, $1 = -23$, $2 = -13$, $3 = -12, \ldots$. All these relations are equivalent (obviously), so only one independent relation exists: $I = -123$. The matrix $\tilde{X}$ is:

$\tilde{X} =$

$I = -123$	$1 = -23$	$2 = -13$	$3 = -12$
+	−	−	−
+	−	+	+
+	+	−	+
+	+	+	−

For this half fraction there is no need to talk about two-factor and three-factor interactions (because they are aliased with main effects and β_0, respectively). We may thus rewrite the initial model, see (8.7), as

$$E\left(Y_{(x_1, x_2, x_3)}\right) = \beta_0 + x_1\beta_1 + x_2\beta_2 + x_3\beta_3.$$

Or, in matrix form,

$$E(\mathbf{Y}) = \tilde{X}\tilde{\boldsymbol{\beta}},$$

where $\tilde{\boldsymbol{\beta}}' = (\beta_0, \beta_1, \beta_2, \beta_3)$. Observe that we could have (equally well) kept the two-factor interactions β_{ij} in the model instead of the main effects; generally lower order effects are kept though.]

We said that a half fraction is any choice of half of the vertices of a k-dimensional cube. Such designs are useful in practice (as are 2^{-p} fractions, for that matter). A half fraction is "good" if it only aliases i-factor and j-factor interactions with the sum $i + j$ being large. (Thus in a half fraction of a 2^5 design it would be nice if $i + j$ could be 4, say; in that case main effects would be aliased with four-factor interactions and with no interactions of lower order, and two-factor interactions would be aliased with three-factor interactions but not aliased among themselves.) One way to construct such half fractions is as follows:

(i) Write out the 2^{k-1} vertices of a $(k-1)$-dimensional cube as rows.
(ii) Obtain the kth column by multiplying componentwise the initial $k-1$ columns, that is, let $k = 1\,2\,3 \cdots (k-1)$.

(The k columns so obtained will represent the main effects. Componentwise products will yield the remaining $2^k - k - 1$ columns, if necessary.) The defining relation for such a half fraction is $I = 1\,2\,3 \cdots (k-1)k$. A half fraction of a 2^5 factorial appears below:

1	2	3	4	5	12	13	$\cdots$	25	$\cdots$	45	$\cdots$
−	−	−	−	+	+	+		−		−	
−	−	−	+	−	+	+		+		−	
−	−	+	−	−	+	−		+		+	
−	−	+	+	+	+	−		−		+	
−	+	−	−	−	−	+		−		+	
−	+	−	+	+	−	+		+		+	
−	+	+	−	+	−	−		+		−	
−	+	+	+	−	−	−		−		−	
+	−	−	−	−	−	−		+		+	
+	−	−	+	+	−	−		−		+	
+	−	+	−	+	−	+		−		−	
+	−	+	+	−	−	+		+		−	
+	+	−	−	+	+	−		+		−	
+	+	−	+	−	+	−		−		−	
+	+	+	−	−	+	+		−		+	
+	+	+	+	+	+	+		+		+	

(8.23)

Note the *defining relation* $5 = 1234$. From it the confounding pattern for this half fraction may be derived. For example, we easily obtain (multiplying both sides by 125) that $12 = 345$, which tells us that two-factor interactions are confounded with three-factor interactions, and so on.

In like fashion quarter fractions, and generally 2^{-p} fractions, of a 2^k factorial design may be constructed. Specifically, to write out a "good" 2^{-p} fraction do the following:

(i) Write out the 2^{k-p} vertices of a $(k-p)$-dimensional cube.
(ii) Generate each of the remaining p columns by multiplying componentwise several of the initial columns (usually as many of them as possible,

keeping in mind that the shortest defining relation should be as long as possible).

(Before proceeding write out two different 2^{-4} fractions of a 2^7 factorial design. Examine the confounding patterns.)

We now attempt to explain the notion of resolution of a fractional factorial design with the help of the half fraction displayed in (8.23). The notion of resolution will only be defined for designs that are fractional factorials. Examination of the existing aliasing pattern in (8.23) reveals that main effects are aliased with four-factor interactions (and not with interactions of any lower order). Two-factor interactions are not aliased among themselves but are aliased with three-factor interactions; $12 = 345$, for example. We say that the design (8.23) is of resolution five; a full abbreviation for it being 2_{V}^{5-1} with the lower suffix denoting the resolution (in roman numerals).

In general a fractional factorial design is of *resolution R* if no p-factor effect is aliased with any other effect containing less than $R - p$ factors. (Thus a design of resolution three does not alias main effects with one another, but it does alias them with two-factor interactions. A design of resolution four does not alias main effects with two-factor interactions, but it does alias the two-factor interactions among themselves.) An alternate (and obviously equivalent) definition of the resolution is that *it equals the length of the shortest defining relation*; by length we mean the number of distinct main effects that occur in the defining relation. (For example, the defining relation $I = -1345$ has length four. A fractional factorial with the following three independent relations, $I = -123$, $I = 24$, $I = 1345$, is of resolution two, since the shortest defining relation is of length two.) The aliasing patterns (i.e., the cosets) follow immediately from the defining relations.

Probably *the most important property that a fractional factorial design of resolution R possesses is that it contains complete factorials in any $R - 1$ (or fewer) factors*, possibly repeated several times. [It is not hard to see that this is true, since in such a design no defining relations of length $R - 1$ (or less) exist. A defining relation is simply a relation of linear dependence among columns (written multiplicatively). Thus any $R - 1$ columns of main effects define a subspace of dimension $R - 1$, that is, a 2^{R-1} factorial design.]

For this reason it is easy to understand the practical importance of fractional factorial designs of high resolution. If k factors are studied, but it seems most plausible that only less than R of them (specific identity unknown) have significant detectable effects, a design of resolution R is the preferred choice of experiment. For if the conjecture is justified, we will have complete factorial designs in any choice of $R - 1$ factors, *including the factors of interest*. Analysis of these 2^{R-1} complete factorials would then yield further information about the $R - 1$ (or fewer) significant factors.

The notion of resolution, so far confined to fractional factorials, can be extended in several ways to arbitrary factorial designs (at two or more levels). For the time being we restrict attention to designs at two levels, as usual. An arbitrary design would be represented by a $n \times k$ matrix X with first column consisting of all $+$'s, the rest of the entries being $+$'s and $-$'s. A desirable feature, from the experimental point of view, is to select X such that complete factorials in any p factors exist. To be specific, index a vector (call it $\mathbf{R}_p$) by the $\binom{k}{p}$ available choices of p factors and let the $\{i_1, i_2, \ldots, i_p\}$th entry of $\mathbf{R}_p$ equal the number of complete factorials to be found in factors $i_1, i_2, \ldots, i_p$. The vector $\mathbf{R}_p$ is called the *resolution vector* of the design. Some of our wishes about choice of design can be conveniently expressed in terms of the resolution vector $\mathbf{R}_p$.

First, $\mathbf{R}_p$ should have all entries greater than or equal to one (if possible). Secondly, if all factors are believed (at least initially) to be equally important, then the entries of $\mathbf{R}_p$ should be as nearly equal as possible. [In fact, if accurate assessment about the relative importance of each subset of d factors exists, we can attach a vector of $\binom{k}{p}$ positive rational numbers (which sum to 1) to reflect their respective relevance and then select the entries of $\mathbf{R}_p$ approximately proportional to these numbers.] Subject to these conditions p should be *as high as possible*.

Several general questions of construction suggest themselves:

1. For a fixed value of p and a fixed resolution vector $\mathbf{R}_p$, what is the minimum number of vertices needed to obtain a design with that resolution vector? What is the design?
2. For a fixed number of factors and a fixed number of vertices to be selected, what is the design with highest possible p and highest possible resolution vector? (We leave it to the imagination of the interested reader to interpret the phrase "highest resolution vector." Many meaningful choices are possible.)

Partial answers to these problems exist. One important notion that comes into play here is that of an orthogonal array. We do not define it but refer the interested reader to the literature [17, 15, 16]. Among the main contributors to design construction are Fisher, Finney, Bose, Box, Rao, Bush, and many of their students.

This section, Section 3, acquainted the reader with factorial designs through their simplest representatives, the 2^k factorials. These designs involve k factors, each at two levels. The most general experimental situation would, of course, require l_1 levels of factor 1, l_2 of factor 2, $\ldots$, l_k of factor k. Many of the notions introduced for 2^k designs extend naturally to the general case and most easily to the case when the number of levels on each factor is a (same) power of a prime. Our main intent was to expose the reader to some of the difficulties of construction of (good) factorial designs. Such difficulties are already present in 2^k factorials so we confined our attention entirely to them. The general case is treated in Section 5.

EXERCISES

1. Estimate the main effects and interactions from the following data:

Nickel (%)	Manganese (%)	Breaking strength (ft/lb)
0	1	27
3	1	39
0	2	35
3	2	33

The object is to discover the effects of these two alloys (nickel and manganese) on the ductility of a certain product. What do the data indicate?

2. Consider a 2^{7-3} fractional factorial design (i.e., a 2^{-3} fraction of a 2^7 design). How many factors are there in this fraction? How many observations? How many levels are used for each factor? How many independent generators are there for this design? What is the cardinality of the subgroup of the relations?

3. Construct a 2^{7-2} fractional factorial design with as high a resolution as possible. What is the resolution? Write out the aliasing pattern.

4. Write out the generators of a 2^{-4} fraction of resolution five of the complete 2^{11} factorial.

5. (Fold-over designs.) The matrix

$$\begin{bmatrix} X \\ \hdashline X^f \end{bmatrix}$$

is called the fold-over matrix of the design matrix X. Here X is a $n \times k$ design matrix and X^f is a $n \times k$ matrix whose ith row is the negative of the $(n - i + 1)$th row of X. The fold-over matrix is $(2n) \times k$. Show that folding over will convert any design of resolution three into one of resolution four (or more); it will also convert a design of resolution five into one of resolution at least six.

6. (Margolin.) Show that any design of resolution four on k factors must contain at least $2k$ vertices (i.e., it must have a design matrix with at least $2k$ rows). If it has precisely $2k$ rows, then it must be a fold-over design.

7. Is the quarter fraction 2^{8-2} defined by the relations $I = 12347$ and $I = -12568$ an optimal design (among all designs on eight factors based on 2^6 observations)? By optimal we mean optimal under the model with direct effects and two-factor interactions. Based on this exercise can you make a general statement about the optimality of certain fractional factorials?

4 BLOCKING

Throughout experimental statistics the notion of "blocking" is of much significance. Even though this idea can be explained with the help of 2^k factorial designs, which we have investigated in Sections 2 and 3, we prefer to start afresh and introduce it by way of another class of statistical designs.

8.14

A manufacturer produces three brands of tennis shoes for boys. Label the brands 1, 2, 3. Interest arises in comparing the wearing qualities of the three brands. Four boys (of varying weights and habits) are talked into wearing the shoes for a time, to make the comparison possible. (Ideally one would wish to test all three brands at one time on each boy, but this seems prohibited by the bipedal form of the species.) The wear on a shoe would certainly depend on the weight and habits of the boy wearing it, in addition to the actual quality of the shoe. Giving each boy both shoes of the same brand is therefore a poor design, since we would not know if the wear was due to the quality of the shoes or to the weight of the boy. So we give them shoes of different brands, as follows (say):

	Boys			
	1	2	3	4
Left shoe	2	1	3	1
Right shoe	1	3	2	2

(8.24)

Thus boy 1 wears a shoe of brand 2 and another of brand 1, boy 2 wears brands 1 and 3, and so on. [An overly imaginative boy may still offset the experiment by hopping on one (and the same) foot all of the time, but we shall dismiss such eccentricities.]

By y_{ij} we denote the wear on shoe of brand i worn by boy j. To be exact, the notation y_{ij} stands for two separate things: it is the wear on shoe of brand i worn by boy j perceived as a *random variable*, and also the actual *observed* wear after the completion of the experiment. (This is an abuse of notation that we make consistently.) Whenever there is talk of expected values, or covariances, we treat y_{ij} as a random variable, of course. When analyzing the data y_{ij} is simply a number.

Let μ_{ij} denote the mean of (the random variable) y_{ij}. This mean is a function of the brand i of shoes as well as of the weight and playing habits of the jth boy. We say that there are two *effects* that determine μ_{ij}: a *variety effect* α_i due to the brand i of shoe and a *block effect* β_j due to the jth boy who wears it. The notions of *varieties* and *blocks* are general standard terminology. Boys represent the blocking variable in this experiment; we say that the experiment consists of comparing three varieties in four blocks of size two each [cf. (8.24)].

One may seek, or postulate, functional forms (i.e., models) for μ_{ij} in terms of the effects α_i and β_j. Common choices are $\mu_{ij} = \alpha_i + \beta_j$ or $\mu_{ij} = \alpha_i\beta_j$. (Modulo a logarithmic transformation these two models are in fact the same.) We assume that (by an eventual transformation of the data) the functional form of μ_{ij} is

$$\mu_{ij} = E(y_{ij}) = \alpha_i + \beta_j; \qquad \operatorname{cov}(y_{ij}) = \sigma^2 I. \tag{8.25}$$

[Even modulo such transformations this kind of linear behavior of μ_{ij} in α_i and β_j may not provide an adequate approximation to reality. However, we restrict attention only to situations in which model (8.25) seems appropriate.]

Our interest lies in estimating the parameters α_i (and β_j) from the observations made. As a matter of fact, *we only care about estimating the* α_i's since they represent the qualities of the three brands of shoes. Effects β_j exert significant influence upon the wear (i.e., the observations) as to be included in the model, but *the* β_j's *(as such) are of no interest to us*. (Ideally one would have wished that all β_j's were the same, for in that case there would be no need to even include them in the model. But we all know that boys differ in temperament and weight. Besides, the manufacturer wishes to test the shoes on a wide cross section of the population of boys.) We call the β_j's *nuisance parameters*. (The boys-shoes example appears in [5].)

Model (8.25) may be conveniently written in matrix form. To do so let $\mathbf{Y}$ be the vector of observations; in our case an 8×1 vector containing the eight y_{ij}'s. Observe that the entries of $\mathbf{Y}$ (the y_{ij}'s) are doubly indexed for convenience, but $\mathbf{Y}$ itself is an 8×1 vector and not a matrix. Furthermore, write model (8.25) as follows:

$$E(\mathbf{Y}) = [X_1 \vdots X_2]\begin{pmatrix} \boldsymbol{\alpha} \\ \hdashline \boldsymbol{\beta} \end{pmatrix}; \qquad \operatorname{cov}\mathbf{Y} = \sigma^2 I. \tag{8.26}$$

where the matrix $X = [X_1 \vdots X_2]$ is an indicator matrix (having exactly two ones in each row and the rest of the entries 0) and

$$\begin{pmatrix} \boldsymbol{\alpha} \\ \hdashline \boldsymbol{\beta} \end{pmatrix}$$

is the vector of unknown effects; that is,

$$(\boldsymbol{\alpha}' \vdots \boldsymbol{\beta}') = (\alpha_1\ \alpha_2\ \alpha_3 \vdots \beta_1\ \beta_2\ \beta_3\ \beta_4).$$

The rows of the matrix $X = [X_1 \vdots X_2]$ are indexed by the observations and the columns by the effects. The partitioning of X is induced by that in the vector of effects and thus X_1 has as many columns as there are α_i's and X_2 as many columns as there are β_j's. A row of $X = [X_1 \vdots X_2]$ indicates which variety and block effect is present in the observation to which that row corresponds. Thus both X_1 and X_2 have exactly one 1 in each row (the rest of the entries being zero).

The least squares estimates for the α_i's are now obtained from the *reduced normal equations*, derived in exactly the same way as was indicated in Section 8.11. To be exact, we obtain

$$C\boldsymbol{\alpha} = Q\mathbf{Y}, \tag{8.27}$$

where $C = X_1^t X_1 - X_1^t X_2 (X_2^t X_2)^{-1} X_2^t X_1$ and $Q = X_1^t - X_1^t X_2 (X_2^t X_2)^{-1} X_2^t$. We call C the Fisher information (or C-) matrix, as was mentioned in Section 8.11. In the case of design (8.24) we have $\mathbf{Y}^t = (y_{21}, y_{11}, y_{12}, y_{32}, y_{33}, y_{23}, y_{14}, y_{24})$ and

$$X = [X_1 \vdots X_2] = \left[\begin{array}{ccc|cccc} & 1 & & 1 & & & \\ 1 & & & 1 & & & \\ 1 & & & & 1 & & \\ & & 1 & & 1 & & \\ & & 1 & & & 1 & \\ & 1 & & & & 1 & \\ 1 & & & & & & 1 \\ & 1 & & & & & 1 \end{array}\right]. \tag{8.28}$$

Straightforward computation yields

$$C = \frac{1}{2}\begin{bmatrix} 3 & -2 & -1 \\ -2 & 3 & -1 \\ -1 & -1 & 2 \end{bmatrix}. \tag{8.29}$$

[Observe that $2C$ is in fact the Kirchhoff matrix of the graph obtained from design (8.24) by letting the brands of shoes be vertices and joining two vertices whenever they appear in a block of (8.24). Readers familiar with Section 6 of Chapter 2 and Section 3 of Chapter 4 are already well acquainted with properties of Kirchhoff matrices. We review them in Section 8.15.]

The matrix C has row sums zero and is therefore singular. For this reason system (8.27) can only be solved for linear functions $\mathbf{l}^t\boldsymbol{\alpha}$, with $\mathbf{l}^t$ a row vector in the row span of C. (In particular, any such $\mathbf{l}$ will necessarily have the sum of its components equal to 0.) A linear function $\mathbf{l}^t\boldsymbol{\alpha}$, with $\mathbf{l}^t$ in the row span of C, is called an *estimable function* (of $\boldsymbol{\alpha}$). Singularity of matrix (8.29) prohibits solutions for individual components of the vector $\boldsymbol{\alpha}$, but it allows estimation of linear functions of the form $\alpha_i - \alpha_j$, $i \neq j$, for example. Such pairwise differences, called *elementary contrasts*, allow us to compare the relative magnitudes of the variety effects α_i (and subsequently form an opinion as to which brand of shoes wears least).

The coefficient matrix C is defined in terms of X_1 and X_2, where $X = [X_1 \vdots X_2]$ is the *design matrix*; see (8.27). As a consequence, the whole estimation process for the α_i's depends on the design matrix X and therefore on the actual design that is used—our discussion having been based on design (8.24), for concreteness. The general question of efficient design centers around finding the design with information matrix C "as large as possible." A detailed examination of this matrix follows.

8.15

Design (8.24) arose from the need of comparing three varieties in four blocks of size two. We assume, more generally, that v varieties are to be compared via b blocks of size k each. A *design* will be any $k \times b$ array d with entries from the set $P = \{1, 2, \dots, v\}$ (of varieties). The columns of d are the blocks. Denote by $\Omega_{v,b,k}$ the collection of all such designs.

Let $d \in \Omega_{v,b,k}$. By y_{ij} we denote an observation on variety i that belongs to the jth block of d. We assume that

$$E(y_{ij}) = \alpha_i + \beta_j,$$

for unknown parameters α_i and β_j. The parameter α_i is called the *effect of variety* i, and β_j is called the *effect of block* j. The α_i's are subject to estimation from the observations. Let $\mathbf{Y}$ be the $(kb) \times 1$ vector of observations (for simplicity we assume that exactly one observation is taken at each of the kb entries of d). The model we assume may be written in matrix form as follows:

$$E(\mathbf{Y}) = [X_1 \vdots X_2]\begin{pmatrix} \boldsymbol{\alpha} \\ \hline \boldsymbol{\beta} \end{pmatrix}, \qquad \operatorname{cov}(\mathbf{Y}) = \sigma^2 I.$$

Matrices X_1 and X_2 depend on the design d. If the mth row of $X = [X_1 \vdots X_2]$ corresponds to observation y_{ij}, then the mth row of X_1 has a 1 in column i (which corresponds to α_i) and zero everywhere else in that row; X_2 has a 1 in the jth column of its mth row and zero everywhere else in that row. Therefore X_1 and X_2 are indicator matrices for the variety and block effects, respectively. The reduced normal equations for variety effects are

$$C_d \boldsymbol{\alpha} = Q_d \mathbf{Y},$$

where

$$C_d = X_1^t X_1 - X_1^t X_2 (X_2^t X_2)^{-1} X_2^t X_1$$

and

$$Q_d = X_1^t - X_1^t X_2 (X_2^t X_2)^{-1} X_2^t. \tag{8.30}$$

[Observe that $Q_d Q_d^t = C_d$; see also (8.18).]

The Fisher information matrix C_d can be computed explicitly (for any design d). Indeed, $X_1^t X_1 = \operatorname{diag}(r_1, \dots, r_v)$, a diagonal matrix with ith diagonal entry r_i equal to the number of times variety i appears throughout the design d; we call r_i the *replication number* of variety i. Clearly $X_2^t X_2 = kI$, a $b \times b$ matrix. Also, $X_1^t X_2$ has as (i, j)th entry the inner product between column i of X_1 and column j of X_2; this inner product is simply the number of times that variety i appears in block j of d (we denote this number by n_{ij}). Therefore $X_1^t X_2 = (n_{ij})$, a $v \times b$ matrix. The matrix C_d can now be written as

$$C_d = \operatorname{diag}(r_1, \dots, r_v) - k^{-1}(n_{ij})(n_{ij})^t. \tag{8.31}$$

Observe that the matrix (n_{ij}) has all its column sums equal to k, and its ith row sum equal to $\sum_j n_{ij} = r_i$, $1 \le i \le v$. The matrix (n_{ij}) is called the *incidence matrix* between varieties and blocks. (A design d for which the n_{ij}'s are 0 or 1 only is called *binary*.) [Verify expression (8.31) for the design (8.24) and compare it to (8.29). Calculate also $X_1^t X_2$ for matrices X_1 and X_2 in (8.28).]

We wish to give a still more convenient interpretation to the matrix C_d than that written in (8.31). Look at kC_d (since it has integral entries). For $i \neq j$ the product $n_{il}n_{jl}$ counts the number of joint occurrences of varieties i and j in block l. Thus $\Sigma_l n_{il}n_{jl}$ equals the total number of joint occurrences of i and j in all the blocks. But if we write $kC_d = (c_{ij})$, then [cf. (8.31)] $c_{ij} = \Sigma_l n_{il}n_{jl}$, for $i \neq j$. This suggests that we attach to the design d a graph $G(d)$ as follows: the vertices of $G(d)$ are the v varieties of $P = \{1, 2, \ldots, v\}$; place an edge between vertices i and j whenever i and j occur together in a block of d. Generally i and j occur together in several blocks, possibly more than once in each, and thus $G(d)$ is in general a graph with multiple edges. Clearly for $i \neq j$ the number of edges between i and j is equal to the number of joint occurrences of i and j in all the blocks. Thus $G(d)$ has $\Sigma_l n_{il}n_{jl} = -c_{ij}$ edges between vertices i and j; $i \neq j$. We may now conclude that the (i, j)th entry of kC_d equals the number of edges between vertices i and j in the graph $G(d)$. The next thing to note is that the row sums of kC_d are zero; indeed, by (8.31),

$$\begin{aligned}
c_{ii} + \sum_{\substack{j \\ j \neq i}} c_{ij} &= \left(kr_i - \sum_l n_{il}n_{il}\right) - \sum_{\substack{j \\ j \neq i}} \sum_l n_{il}n_{jl} \\
&= kr_i - \sum_j \sum_l n_{il}n_{jl} = kr_i - \sum_l \sum_j n_{il}n_{jl} \\
&= kr_i - \sum_l n_{il} \sum_j n_{jl} = kr_i - \sum_l n_{il}k \\
&= kr_i - k\sum_l n_{il} = kr_i - kr_i = 0.
\end{aligned}$$

Thus $c_{ii} = -\Sigma_j c_{ij}$, $j \neq i$, which says that the ith diagonal element of C_d is equal to the negative of the sum of the off-diagonal elements in the ith row. We summarize as follows:

* *The matrix kC_d has row sums zero and its (i, j)th entry equals the negative of the number of edges between vertices i and j in the graph $G(d)$, $i \neq j$.*

In accordance with established terminology we call kC_d the *Kirchhoff matrix* of the graph $G(d)$; see Section 2.15 and Section 4.17. [The Kirchhoff matrix of $G(d)$ is therefore k times the Fisher information matrix of the design d.] In Section 2.15 we called attention to the following properties of the matrix kC_d (all these properties are true for any positive multiple pC_d, and in particular for C_d itself):

1. $C_d\mathbf{1} = 0$ (where $\mathbf{1}$ is the vector with all entries 1).
2. If C_d has rank $v - 1$, then all cofactors of C_d are equal and nonzero. (8.32)

3. C_d is nonnegative definite.
4. C_d has rank $v - 1$ if and only if $G(d)$ is a connected graph.

8.16

There are many designs in $\Omega_{v,b,k}$. Which one should we use for experimentation? One sensible approach (among many other sensible approaches) is to select the design in accordance with some attractive features that its information matrix should possess. We attempt to identify specific features of C_d that may be of interest. Choices will be made by identifying designs that minimize the average variance of the estimates of $\widehat{\alpha_i - \alpha_j}$, $i \neq j$ (or other related statistical functions).

To begin with, let us look at a vector $\widehat{P'\alpha}$ of least squares estimates of $P'\alpha$, with P' a $m \times v$ matrix having its rows in the row span of C_d. We may write $P = C_d R$ for some $v \times m$ matrix R (since P' has rows in the row span of C_d; recall that C_d is a symmetric nonnegative definite matrix). Therefore

$$\begin{aligned}\operatorname{cov}\widehat{P'\alpha} &= \operatorname{cov}\widehat{R'C_d\alpha} = \operatorname{cov}\big(R'Q_d\mathbf{Y}\big)\\ &= R'Q_d\operatorname{cov}(\mathbf{Y})Q_d'R = R'Q_dQ_d'R\sigma^2\\ &= R'C_dR\sigma^2.\end{aligned}$$

[The reader should refer back to (8.30) and (8.18) in order to understand this computation. In particular, recall that $C_d\alpha = Q_d\mathbf{Y}$ and $Q_dQ_d' = C_d$.] The fact that $\operatorname{cov}\widehat{P'\alpha} = R'C_dR\sigma^2$ may look puzzling, in that it seems to depend on the choice of R. Since C_d is always singular under this model there may be many different matrices R that satisfy $P = C_dR$. But any two such matrices R_1 and R_2 (say) differ by a matrix K in the kernel of C_d; therefore $R_2'C_dR_2 = (R_1 + K)'C_d(R_1 + K) = R_1'C_dR_1$, since $K'C_d$ and C_dK are both zero matrices. Therefore

$$\operatorname{cov}\widehat{P'\alpha} = R'C_dR\sigma^2$$

is *well defined*. We may thus further write

$$\operatorname{cov}\widehat{P'\alpha} = R'C_dR\sigma^2 = R'C_dC_d^-C_dR\sigma^2 = P'C_d^-P\sigma^2.$$

[This last formula is somewhat more appealing in that it has only P and C_d in it. But it actually involves C_d^-, a generalized inverse of C_d. We just finished saying that $\operatorname{cov}\widehat{P'\alpha}$ is well defined; consequently, and in spite of its looks, $P'C_d^-P$ *does not* depend upon the choice of the generalized inverse C_d^-. (A generalized inverse of matrix B is a matrix B^- that satisfies $BB^-B = B$ and $B^-BB^- = B^-$. Such matrices B^- always exist.)] We proved that

$$\operatorname{cov}\widehat{P'\alpha} = P'C_d^-P\sigma^2. \tag{8.33}$$

Expression (8.33) is well defined and not dependent upon the choice of the generalized inverse C_d^-.

We select a C_d^- that will prove convenient to work with. By (8.22) the vector $\mathbf{1}$ is in the kernel of C_d. In terms of eigenvalues and eigenvectors the matrix C_d may be written as follows:

$$C_d = 0\mathbf{1}\mathbf{1}^t + \mu_{d1}w_1w_1^t + \cdots + \mu_{d,v-1}w_{v-1}w_{v-1}^t,$$

where $0 = \mu_{d0} \le \mu_{d1} \le \cdots \le \mu_{d,v-1}$ are the eigenvalues and $\mathbf{1}, w_1, \ldots, w_{v-1}$ are the eigenvectors of C_d. [These eigenvectors are chosen to form an orthonormal basis of the v-dimensional (real) vector space on which C_d operates; thus $w_i^t\mathbf{1} = 0$, $w_i^tw_i = 1$, for all i, and $w_i^tw_j = 0$, for all $i \ne j$.] Define a matrix C_d^- as follows:

$$C_d^- = 0\mathbf{1}\mathbf{1}^t + \mu_{d1}^{-1}w_1w_1^t + \cdots + \mu_{d,v-1}^{-1}w_{v-1}w_{v-1}^t. \tag{8.34}$$

[A word of caution here: the kernel of C_d may be more than one-dimensional; let us say it is p-dimensional. Then $\mu_{d1} = \mu_{d2} = \cdots = \mu_{d,p-1} = 0$. In such a case we make the convention that $\mu_{d1}^{-1} = \mu_{d2}^{-1} = \cdots = \mu_{d,p-1}^{-1} = 0$ in (8.34). Designs with $\mu_{d1} = 0$ have information matrix of rank at most $v - 2$ and (by (8.32), part 4) their graph is disconnected. Such designs are in fact of no statistical interest.] It is straightforward to verify, using the orthonormality of the w_i's, that $C_dC_d^-C_d = C_d$ and $C_d^-C_dC_d^- = C_d^-$. In conclusion, C_d^-, *as defined in* (8.34), *is a generalized inverse of* C_d. Moreover, (8.34) also shows that *the eigenvalues of* C_d^- *are* $0, \mu_{d1}^{-1}, \mu_{d2}^{-1}, \ldots, \mu_{d,v-1}^{-1}$. Observe also that $C_d^-\mathbf{1} = 0$, since $w_i^t\mathbf{1} = 0$ for all $1 \le i \le v - 1$. This informs us that C_d^- *has rows sums zero*; *in particular the sum of entries in* C_d^- *is zero.*

Write $C_d = (c_{ij})$ and $C_d^- = (c^{ij})$; from now on whenever we write C_d^- we mean the C_d^- in (8.34). Let P in (8.33) be the vector with 1 in position i, -1 in position j $(j \ne i)$, and 0 elsewhere. Then $P^t\boldsymbol{\alpha} = \alpha_i - \alpha_j$, and by (8.33),

$$\operatorname{cov}\widehat{P^t\boldsymbol{\alpha}} = \operatorname{var}\left(\widehat{\alpha_i - \alpha_j}\right) = P^tC_d^-P\sigma^2 = (c^{ii} + c^{jj} - 2c^{ij})\sigma^2.$$

Furthermore,

$$\begin{aligned}\sum_{\substack{i,j\\i<j}} \operatorname{var}\left(\widehat{\alpha_i - \alpha_j}\right) &= \sigma^2\left(\sum_{\substack{i,j\\i<j}} (c^{ii} + c^{jj} - 2c^{ij})\right)\\ &= \{\text{since the sum of entries in } C_d^- \text{ is zero}\}\\ &= \sigma^2(v-2)\operatorname{trace} C_d^- = \sigma^2(v-2)\sum_{i=1}^{v-1}\mu_{di}^{-1}.\end{aligned}$$

A mode of selection for d now suggests itself. Pick that d in $\Omega_{v,b,k}$ that minimizes the average variance of the estimates of the elementary contrasts:

$$\binom{v}{2}^{-1}\sum_{i<j}\operatorname{var}\left(\widehat{\alpha_i - \alpha_j}\right) = \sigma^2(v-2)\binom{v}{2}^{-1}\sum_{i=1}^{v-1}\mu_{di}^{-1}.$$

We may therefore define as follows:

* *A design d^* in $\Omega_{v,b,k}$ that minimizes $\sum_{i=1}^{v-1}\mu_{di}^{-1}$ over all d in $\Omega_{v,b,k}$ is called A-optimal* ("A" stands for average variance).

Instead of looking at the average variance of $\widehat{\alpha_i - \alpha_j}$ one could look at the maximum variance

$$\max_{\substack{i,j\\ i<j}} \operatorname{var}\left(\widehat{\alpha_i - \alpha_j}\right)$$

and pick a design that would minimize that. Such a design is called *MV-optimal.*

Yet another possibility is to look at the set of all normalized contrasts $\widehat{\mathbf{l}'\boldsymbol{\alpha}}$ (with $\mathbf{l}'\mathbf{l} = 1$ and $\mathbf{l}'\mathbf{1} = 0$), compute

$$\max_{\|\mathbf{l}\|=1} \operatorname{var} \widehat{\mathbf{l}'\boldsymbol{\alpha}}\left(= \max_{\|l\|=1} \left(\mathbf{l}'C_d^{-}\mathbf{l}\right)\sigma^2\right),$$

and select a design that minimizes this expression. By the minimax representation of eigenvalues of a nonnegative definite matrix the expression to be minimized is simply the largest eigenvalue of C_d^{-}, that is, μ_{d1}^{-1}. Minimizing μ_{d1}^{-1} is clearly the same as maximizing μ_{d1}. [Designs d with $\mu_{d1} = 0$ will actually not even be considered since they have disconnected graph $G(d)$.] We thus define as follows:

* *A design d^* in $\Omega_{v,b,k}$ that maximizes μ_{d1} over all d in $\Omega_{v,b,k}$ is called E-optimal* ("E" from eigenvalue). [As one can see from above, an E-optimal design is a very conservative kind of design in the sense that it guards well against the worst possible situation.]

Finally, a determinantal criterion arises by taking in (8.33) a $(v-1) \times v$ matrix P' with orthonormal rows, each row orthogonal to $\mathbf{1}'$. Then

$$\operatorname{cov}\widehat{P'\boldsymbol{\alpha}} = P'C_d^{-}P\sigma^2$$

is the $(v-1) \times (v-1)$ covariance matrix of $\widehat{P'\boldsymbol{\alpha}}$. One would then select the design that minimizes $|P'C_d^{-}P|$. When the vector of observations $\mathbf{Y}$ is multivariate Gaussian, then $\widehat{P'\boldsymbol{\alpha}}$ will also be multivariate Gaussian, and $|P'C_d^{-}P|$ will be proportional to the square of the volume of the (standardized) confidence ellipsoid for the vector $P'\boldsymbol{\alpha}$; see (8.14) for details. [By standardized we simply mean $c = 1$ in (8.14).] An additional pleasant feature that makes itself noticed is that $|P'C_d^{-}P| = \prod_{i=1}^{v-1}\mu_{d_i}^{-1}$, irrespective of the choice of P. This is indeed so, for the following stronger statement is true: the eigenvalues of $P'C_d^{-}P$ are $\mu_{d_1}^{-1}, \ldots, \mu_{d,v-1}^{-1}$. [To verify this let w be an eigenvector of C_d^{-} (orthogonal to $\mathbf{1}$) with eigenvalue μ. We may write $C_d^{-}w = \mu w$. Since the matrix P has $v-1$ orthonormal columns, and since each column is orthogonal to $\mathbf{1}$, these columns form an orthonormal basis for the subspace orthogonal to $\mathbf{1}$. In particular we may write $w = Py$, for some $(v-1) \times 1$ vector y. We therefore have the following chain of implications: $C_d^- w = \mu w$ implies $C_d^- Py = \mu Py$ implies $P^t C_d^- Py = \mu P^t Py = \mu Iy = \mu y$. This shows that μ is also an eigenvalue of $P^t C_d^- P$.] A particularly useful example of a matrix P^t as above is obtained from the following matrix (by dividing each row vector to its length):

$$\begin{bmatrix} 1 & -1 & & & & 0 \\ 1 & 1 & -2 & & & \\ 1 & 1 & 1 & -3 & & \\ & \vdots & & & & \\ 1 & 1 & 1 & 1 & \cdots & -(v-1) \end{bmatrix}.$$

Other interesting choices are possible.

We may now define as follows:

$*$ *A design d^* in $\Omega_{v,b,k}$ is called D-optimal if it maximizes $\prod_{i=1}^{v-1}\mu_{di}$ over all d in $\Omega_{v,b,k}$.*

For special values of v, b, and k the optimal (D-, A-, or E-) designs are combinatorial structures of remarkable beauty and elegance. We have met some of them in Chapter 7, though they were introduced there on purely combinatorial grounds (we are alluding to the 2-designs and partial designs). Sections 8.17 and 8.18 aim to explain how combinatorial structures with interesting symmetries arise out of this process of spectral extremization (via the optimality criteria introduced above). But first an important remark.

A Critique of Optimality

Our selection process for a good design is based entirely upon the information matrix; this matrix being, in turn, derived from assuming an *additive* model of response. However, questions of design usually arise *before* observations are taken and without any observations available it is nearly impossible to assess the form of response. This seems like a vicious circle. To break the circle one has to take some observations somewhere. If one wants to take the chance that the response (or a transformation thereof) is linear, an optimal design of the kind described above may prove suitable. Indeed, in certain experimental settings it may be known (from deterministic considerations such as differential equations or certain physical laws) that a transformation of the response is approximately linear. When this is the case considerations of optimal design (based on the information matrix alone) become quite useful.

On the other hand, when nothing is known about the form of the response much caution should be exercised. Blind assumption of linearity and subsequent adoption of a D-optimal design (say) could easily border on the idiotic! Consider this: you are allowed to choose ten points in the interval $[-1, 1]$ to guess the unknown function $\mu(x)$ (defined on this interval). When you choose a point x_0 you will observe $\mu(x_0)$, with small error (say). How should you select the ten points? The thinking pattern that we hope you will refrain from is the following: "Well, $\mu(x)$ is probably a line, so it can be determined by just two distinct points, intuitively the further away from each other the better. So I will take them at the endpoints, five of them at -1 and five of them at $+1$." [When $\mu(x)$ is a line this would indeed be the D-optimal choice.] If you actually did this and $\mu(x)$ is far from being a line you can see how much potential information was lost. We refer to Box and his students [6] for a description of techniques to minimize the bias (which is what the true concern here is).

8.17 Examples of Optimal Designs

Interesting geometric configurations arise out of the process of spectral extremization described in Section 8.16, the general underlying principle being that extreme spectral behavior often implies geometric symmetry.

To recapitulate and generalize, we start out with a triple of natural numbers (v, b, k). Let $\Omega_{v,b,k}$ denote the set of all $k \times b$ arrays with entries from the set $P = \{1, 2, \ldots, v\}$. The elements of P (which correspond to varieties in Section 8.16) are now called *points*. An element d of $\Omega_{v,b,k}$ is called an *(experimental) design* or simply an *array*. To d we attach a graph $G(d)$ obtained by taking as vertices the points of P and joining two vertices whenever they occur in a column of d. The columns of d are called *blocks*. A matrix is in turn associated to the graph $G(d)$. We denote it by C_d and let its (i, j)th entry be equal to the negative of the number of edges between vertices i and j in C_d $(i \neq j)$; the ith diagonal entry of C_d is equal to the negative of the sum of the off-diagonal entries in the ith row. [Observe that C_d as defined here is in fact equal to kC_d, where C_d is the Fisher information matrix of (8.30). The matrix C_d, as defined here, is called the *Kirchhoff matrix* of $G(d)$.] Let, further, $0 = \mu_{d0} \leq \mu_{d1} \leq \cdots \leq \mu_{d,v-1}$ be the eigenvalues of the Kirchhoff matrix C_d. Denote by $\mu(d)$ the vector of $v-1$ eigenvalues $(\mu_{d1}, \ldots, \mu_{d,v-1})'$. Select a symmetric, concave, and increasing function ϕ that takes real values and acts upon vectors of length $v-1$ [like $\mu(d)$]. (By increasing we mean increasing in each component.) Examples of such functions ϕ are

$$\phi(\mu(d)) = \mu_{d1},$$

$$\phi(\mu(d)) = \prod_{i=1}^{v-1} \mu_{di},$$

and

$$\phi(\mu(d)) = -\sum_{i=1}^{v-1} \mu_{di}^{-1}.$$

The former is attached to E-optimality, the one in the middle to D-optimality, and the last one to A-optimality. We may thus generally say:

* *The design d^* in $\Omega_{v,b,k}$ is ϕ-optimal if $\phi(d^*) \geq \phi(d)$, for all d in $\Omega_{v,b,k}$.* We call ϕ an *optimality criterion*.

Observe that for any triple (v, b, k) of natural numbers and any optimality criterion ϕ, there exist ϕ-optimal arrays in $\Omega_{v,b,k}$, this being so simply because $\Omega_{v,b,k}$ is a finite set and because any two arrays can be compared with respect to the ϕ-criterion. For nice choices of (v, b, k) the ϕ-optimal arrays turn out to be combinatorial structures that we met before.

Perhaps the simplest example that illustrates well the interplay between spectral extremization and combinatorial symmetry is provided by 2-designs. Think of the blocks of a 2-design as columns of a $k \times b$ array (the constant b equals λ_0, i.e., it denotes the number of blocks). A 2-design can therefore be interpreted as a $k \times b$ array; and, as we are about to see, it is an array with remarkable spectral and combinatorial properties:

* *Suppose $\Omega_{v,b,k}$ contains 2-designs. If d^* is a 2-design, then d^* is ϕ-optimal for all optimality criteria ϕ. Conversely, if d^* is ϕ-optimal for some strictly increasing and strictly concave optimality criterion ϕ, then d^* is necessarily a 2-design.*

Proof. Let $d^* \in \Omega_{v,b,k}$ be a 2-design. Denote by λ the number of blocks containing two distinct points, by r the number of blocks containing one point, and by b the total number of blocks of d^* (we usually denote these by λ_2, λ_1, and λ_0, respectively, but we make an exception here). The information matrix of d^* is $C_{d*} = v\lambda(I - (1/v)J)$, as readily follows from the definition of an information matrix. The spectrum of C_{d*} is $\mu(d^*) = (v\lambda, v\lambda, \dots, v\lambda)$, apart from the zero eigenvalue associated with the eigenvector $\mathbf{1}$ [see part 1 in (8.32)]. This is a particularly nice spectral structure.

Let d be any array in $\Omega_{v,b,k}$ with information matrix $C_d = (c_{ij})$. Since the row sums of C_d are zero, we have trace $C_d = -2\Sigma_{i,j,i<j} c_{ij}$. It is therefore easy to see that each edge of the graph $G(d)$ contributes 2 to the trace of C_d, unless the edge is a loop, in which case the contribution is 0. Hence

$$\operatorname{trace} C_d \le 2\,(\text{number of edges of } G(d)) = 2\binom{k}{2}b. \tag{8.35}$$

The last sign of equality is a consequence of the fact that each column of d generates $\binom{k}{2}$ edges in $G(d)$ (loops included). Note that the above inequality becomes equality if and only if $G(d)$ has no loops—in which case we call d a *binary array*. A 2-design d^* is a binary array, and thus (8.35) allows us to write

$$\operatorname{trace} C_d \le \operatorname{trace} C_{d*} = 2\binom{k}{2}b. \tag{8.36}$$

Let $\lambda_d = (v-1)^{-1}$ trace C_d and think of λ_d as the average of the entries of the spectrum $\mu(d)$. We can now write

$$\phi(\mu(d)) \le \phi(\lambda_d, \lambda_d, \dots, \lambda_d) \le \phi(v\lambda, \dots, v\lambda) = \phi(\mu(d^*)),$$

for any optimality criterion ϕ. The first inequality sign follows by the concavity and symmetry of ϕ; the second is a consequence of (8.36) and the monotonicity of ϕ. This shows that a 2-design is ϕ-optimal.

Assume, conversely, that $\phi(\mu(d)) \leq \phi(\mu(d^*))$, for some strictly increasing and strictly concave optimality criterion ϕ, some array d^*, and all d in $\Omega_{v,b,k}$. We show that d^* must be a 2-design. To begin with, such d^* must be a binary array. [Otherwise $G(d^*)$ has loops and thus trace $C_{d*} <$ trace $C_{\bar{d}} = 2\binom{k}{2}b$, for some 2-design $\bar{d}$ in $\Omega_{v,b,k}$; this in turn implies $\phi(\mu(d^*)) \leq \phi(\lambda^*, \ldots, \lambda^*) < \phi(v\lambda, \ldots, v\lambda) = \phi(\mu(\bar{d}))$, since ϕ is strictly increasing (λ^* denotes $(v-1)^{-1}$ trace C_{d*}). This contradicts our assumption.] Furthermore, $\mu(d^*)$ has all its entries equal to $\lambda^* = (v-1)^{-1}$trace C_{d*}. [This follows from the assumption that ϕ is strictly concave. For if $\mu(d^*)$ does not have all entries equal to λ^*, then $\phi(\mu(d^*)) < \phi(\lambda^*, \ldots, \lambda^*) \leq \phi(v\lambda, \ldots, v\lambda) = \phi(\mu(\bar{d}))$, a contradiction.] Thus trace $C_{d*} =$ trace $C_{\bar{d}} = 2\binom{k}{2}b$ where $\bar{d}$ is a 2-(v, k, λ) design. This implies $\lambda^* = v\lambda$. Hence

$$\mu(d^*) = (v\lambda, \ldots, v\lambda).$$

Look now at the matrix $C_{d*} - v\lambda I$. Since $\mu(d^*) = (v\lambda, \ldots, v\lambda)$, the matrix $C_{d*} - v\lambda I$ has rank 1. Its nonzero eigenvalue is $-v\lambda$ with eigenvector $\mathbf{1}$, the vector of all 1's. This implies that $C_{d*} - v\lambda I = -v\lambda(v^{-1/2}\mathbf{1})(v^{-1/2}\mathbf{1}') = -\lambda\mathbf{1}\mathbf{1}' = -\lambda J$. We conclude that $C_{d*} = v\lambda(I - (1/v)J)$. In words, this tells us that any two distinct points occur in λ columns of d^*. We showed that d^* is a 2-design. This ends our proof.

Optimal designs are known in other settings as well. We mention a result of Cheng [8]. A definition is needed first: we call a graph G *generalized bipartite* if its vertices can be partitioned into two sets of equal size such that two vertices are joined by λ_1 edges whenever they belong to the same set, and joined by λ_2 edges whenever they belong to different sets. An array d^* in $\Omega_{v,b,k}$ is called *bipartite* with parameters λ_1 and λ_2 if $G(d^*)$ is a generalized bipartite graph. (The usual nomenclature for such d^* is group divisible.) Cheng's result is the following:

* *Suppose $\Omega_{v,b,k}$ contains a bipartite design d^* with parameters λ_1 and $\lambda_2 = \lambda_1 + 1$. Then d^* is D-, A-, and E-optimal.* (Cheng showed that such d^* is ϕ-optimal for a very large class of criteria ϕ.)

We end Section 8.17 by stating a result concerning the E-optimality of certain partial designs. Assume that (v, b, k) are such that $\Omega_{v,b,k}$ contains a partial design with two classes in which any two distinct points occur in at most one block. It was proved in [9] that:

* *A partial design with two classes in which any pair of distinct points appears in at most one block is E-optimal.*

8.18 E-Optimality and Line Graphs

For a given triple of positive integers (v, b, k) let us denote bkv^{-1} by r, and $k(k-1)v^{-1}(v-1)^{-1}b$ by λ. The numbers r and λ so defined are fractions, in general, and we let $[r]$ and $[\lambda]$ be their respective integral parts. For any binary array d in $\Omega_{v,b,k}$, r signifies the average number of occurrences of a point throughout the array d and λ is the average number of edges between two distinct vertices in $G(d)$.

The task of finding E-optimal arrays for an arbitrary triple (v, b, k) is a difficult one. We simplify by assuming that (v, b, k) is such that r is an integer and look only at binary arrays in which a pair of points occurs in either $[\lambda]$ or $[\lambda] + 1$ columns. Denote the collection of such arrays by $\Omega^*_{v,b,k}$. An array d^* in $\Omega^*_{v,b,k}$ that satisfies $\mu_{d*1} \geq \mu_{d1}$, for all d in $\Omega^*_{v,b,k}$, is called E-optimal over $\Omega^*_{v,b,k}$. (Often, but not always, an E-optimal array belongs in fact to $\Omega^*_{v,b,k}$. E-optimality in $\Omega_{v,b,k}$ and E-optimality over $\Omega^*_{v,b,k}$ are only related problems with no implication going either way, in general.)

To an array d in $\Omega^*_{v,b,k}$ we associate a simple graph $H(d)$ whose (vertex versus vertex) adjacency matrix A_d is defined by

$$A_d = C_d + ([\lambda] + 1)J - (r(k-1) + [\lambda] + 1)I.$$

The matrix A_d is clearly symmetric, has 0 on the diagonal (since the diagonal entries of C_d are $r(k-1)$), and off-diagonal entries 0 or 1 (since C_d has off-diagonal entries $-([\lambda] + 1)$ or $-[\lambda]$). Observe that $A_d\mathbf{1} = r(k-1)\mathbf{1}$, which shows that $H(d)$ is a regular graph of degree $r(k-1)$. Denote by $\tilde{\mu}_{d1} \leq \tilde{\mu}_{d2} \leq \cdots \leq \tilde{\mu}_{d,v} = r(k-1)$ the eigenvalues of A_d, and note that an array d^* in $\Omega^*_{v,b,k}$ satisfies

$$\mu_{d*1} \geq \mu_{d1} \quad \text{if and only if} \quad \tilde{\mu}_{d*1} \geq \tilde{\mu}_{d1}.$$

This shows that an array d^* is E-optimal over $\Omega^*_{v,b,k}$ if and only if $\tilde{\mu}_{d*1} \geq \tilde{\mu}_{d1}$, for all d in $\Omega^*_{v,b,k}$.

This suggests studying the following related (but not equivalent) problem in graph theory: *Among all simple and regular graphs G on v vertices and of degree s find the graph G^* with maximal minimal eigenvalue of the adjacency matrix.* [If such a graph G^* equals $H(d^*)$ for some d^* in $\Omega^*_{v,b,k}$, then d^* is E-optimal over $\Omega^*_{v,b,k}$.]

Our interest thus focuses on regular graphs with a large minimal eigenvalue of the adjacency matrix. Let us denote by $A(G)$ the adjacency matrix of a simple graph G and by $\mu_1(G)$ the minimal eigenvalue of $A(G)$. If G has at least one edge, then $\mu_1(G) \leq -1$ [to see this look at the 2×2 principal minor of $A(G)$ corresponding to the two vertices of that edge and compute its eigenvalues]. Values of -2 or more for $\mu_1(G)$ are therefore quite high.

It is not hard to think of simple graphs G with $\mu_1(G) \geq -2$. One large family of such graphs is obtained by starting out with any simple graph H and constructing from it a new graph $L(H)$, called the *line graph* of H, as follows: The vertices of $L(H)$ are the edges of H with two vertices of $L(H)$ being joined by an edge if and only if the two corresponding edges of H intersect. If M is the incidence matrix of vertices of H versus the edges of H, then the adjacency matrix of $L(H)$ is

$$A(L(H)) = M'M - 2I.$$

Since $M'M$ is nonnegative definite this informs us at once that $\mu_1(L(H)) \geq -2$.

Another such graph is the simple graph on $v = 2m$ vertices whose complement consists of m parallel edges. A graph like this is suggestively called a *cocktail party* graph, and denoted by $CP(m)$. It is easy to check that $CP(m)$ is not a line graph, but still $\mu_1(CP(m)) \geq -2$.

In a successful effort to classify the simple graphs G with $\mu_1(G) \geq -2$ Hoffmann [19] defines the notion of a generalized line graph. Let H be a simple graph with vertices labeled $1, 2, \dots, k$. A *generalized line graph* $L(H; m_1, \dots, m_k)$ is the graph obtained from $L(H)$ by adjoining k disjoint cocktail party graphs $CP(m_i)$, $1 \leq i \leq k$, where every vertex of the ith cocktail party graph is adjacent to every vertex of $L(H)$ that corresponds to an edge of H containing i. An example is:

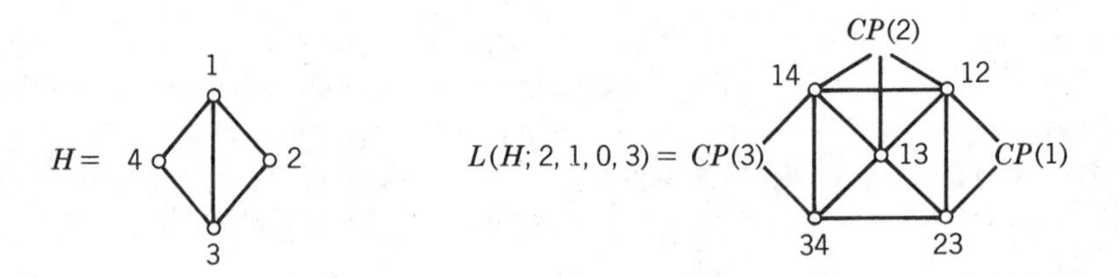

Peterson's graph shows, however, that a graph with $\mu_1(G) \geq -2$ need not necessarily be a generalized line graph (see Section 7.31). It is, on the other hand, remarkable that *apart from a finite number of exceptional graphs, a graph has minimal eigenvalue* -2 *or more if and only if it is a generalized line graph.*

The exceptional graphs are best understood by working with root systems of semisimple Lie algebras (which are simply systems of lines through the origin in the n-dimensional space R^n at angles of 60° or 90°). To introduce the root systems denote by $\{e_i : 1 \leq i \leq n\}$ an orthonormal basis of R^n. By a line through the origin we understand all the scalar multiples of a nonzero vector. Such a vector is said to generate the line.

The *root system* A_n is the set of $\binom{n+1}{2}$ lines generated by $e_i - e_j$, for all $i \neq j$, where $\{e_i\}$ is an orthonormal basis of R^{n+1}. [The vectors $e_i - e_j$ $(i \neq j)$ generate a subspace of dimension n in R^{n+1}, orthogonal to the vector $\mathbf{1}$, with all $n + 1$ entries equal to 1.]

The *root system* D_n is the set of $n(n-1)$ lines generated by $\pm e_i \pm e_j$, for $i \neq j$, where $\{e_i\}$ is an orthonormal basis of R^n.

The (*exceptional*) *root system* E_8 is the set of 120 lines in R^8 generated by the vectors $2e_i$ and $\pm e_i \pm e_j \pm e_k \pm e_l$, with the sets of indices $\{i, j, k, l\}$ being the 14 blocks of the Hadamard 3-(8, 4, 1) design [see (7.1)].

In [7] Cameron et al. show the following:

* *If G is a connected simple graph with n vertices and* $\mu_1(G) \geq -2$, *then*

$$2I + A(G) = M'M,$$

where the matrix M can be selected to satisfy one of the following three properties:

(a) *M is* $(n+1) \times n$ *with vectors of length* $\sqrt{2}$ *along some lines of* A_n *as columns.*
(b) *M is* $n \times n$ *with vectors of length* $\sqrt{2}$ *along some lines of* D_n *as columns.*
(c) *M is* $8 \times n$ *with vectors of length* $\sqrt{2}$ *along some lines of* E_8 *as columns.*

It is not too hard to show that if M satisfies (a) or (b), then G is a generalized line graph, necessarily. The fact that (c) can occur explains the existence of exceptional graphs (such as Petersen's graph and possibly many, but only finitely many, others).

We summarize as follows:

Hoffmann's Theorem. *A simple and connected graph G satisfies* $\mu_1(G) \geq -2$ *if and only if G is either a generalized line graph or G is represented by a subset of lines in* E_8.

[A connected *regular* simple graph G with $\mu_1(G) \geq -2$ must be either a line graph, cocktail party graph, or represented by a subset of lines in E_8—this follows at once upon examining what a regular generalized graph must be.]

The results in graph theory that we just mentioned give insight into what the E-optimal arrays over $\Omega^*_{v,b,k}$ are. Motivated by the above we call an array d^* in $\Omega^*_{v,b,k}$ of the *Lie type* if the (regular) graph $H(d^*)$ is a line graph, cocktail party graph, or represented by lines of the root system E_8.

It readily follows that:

* *If* (v, b, k) *are such that* bkv^{-1} *is an integer and* $\Omega^*_{v,b,k}$ *contains arrays of the Lie type, then an E-optimal array over* $\Omega^*_{v,b,k}$ *must necessarily be of the Lie type.*

For certain (rather scarce) sets of triples (v, b, k), arrays of the Lie type are known to exist but (to the author's knowledge) no systematic study of these has been undertaken.

A general question that translates directly into issues of statistical efficiency is the following: *Find a function f of v, b, and k such that* $f(v, b, k) \leq \mu_{d*1}$ *and* $\mu_{d*1} - f(v, b, k)$ *is uniformly small, for* d^* *E-optimal in* $\Omega_{v,b,k}$.

EXERCISES

1. Find the D-, A-, and E-optimal designs in $\Omega_{3,3,2}$.

2. Show that any matrix (square or rectangular) has a generalized inverse. [*Hint*: Think modulo the kernel of the matrix.]

3. Show that under the model $E(y_{ij}) = \alpha_i + \beta_j$, with $\text{cov}(y_{ij}) = \sigma^2 I$, all pairwise differences $\alpha_i - \alpha_j$ are estimable if and only if the information matrix has kernel of dimension precisely 1.

4. Let A be a matrix of full column rank. Set $B = A(A^tA)^{-1}A^t$. Prove that:

(a) The matrix B is symmetric, nonnegative definite.

(b) $B^2 = B$.

(c) Eigenvalues of B are 0 and 1 only.

(d) Trace B = rank B.
What is the geometrical meaning of B as a linear map? What does it do? Show that $I - B$ satisfies (a) through (d) also. How are B and $I - B$ geometrically related as linear maps?

5. Find the D-, A-, and E-optimal designs in $\Omega_{3,b,2}$ with b arbitrary.

6. What is the D- and E-optimal design in $\Omega_{10,15,2}$? [*Hint*: Petersen's graph could be a choice.]

7. Let $d \in \Omega_{v,b,k}$ and δ be the design d to which one additional block was added; thus $\delta \in \Omega_{v,b+1,k}$. Show that $C_\delta \geq C_d$. Prove that equality takes place if and only if all the points (i.e., varieties) in the added block are the same.

8. Identify the designs in $\Omega_{v,b,k}$ whose information matrix has maximal trace. (Identify those with trace of the information matrix being zero.)

9. Can you think of some way of designing an experiment (of comparing varieties in blocks) that would adequately guard against possible nonadditivity in the variety and block effects? Write an essay.

5 MIXED FACTORIALS

8.19

Situations in which k factors are believed to influence a response would generally require the selection of a_i levels of interest on the ith factor. [In Sections 1 through 3 we focused entirely on the case with all a_i's equal to 2 (high and low).] Let us denote by A_i the set of levels selected on factor i. The set of all possible combinations of levels is then described by the Cartesian product $A = A_1 \times A_2 \times \cdots \times A_n$. There are $|A| = \prod_{i=1}^{k}|A_i| = \prod_{i=1}^{k} a_i$ points in A (which we may either call combinations of levels or experimental conditions); for simplicity, as well as geometrical appeal, we simply call them *points*.

A *factorial design* d is a selection (possibly with repetitions) of points of A. When the number of points selected divides $|A|$, d is called a *fractional factorial*. The set A itself, when viewed as a design, is known as a *full* (or *complete*) *factorial*; it is a *mixed factorial* when not all a_i's are the same. In most situations a design d will not have repeated points and we may therefore think of d as being simply a subset of A.

One of the main objectives when designing experiments is to pick a design d^* with *as few points as possible*, which would yield *as much information as possible* about the response (upon analyzing the observations made at the points in d^*). At this stage the notion of information remains vague; many useful formulations could be made, however. We choose one (to be described below) that has to do with maximizing the number of points on several "projections."

At the initial stages of planning an experiment decisions must be made as to which factors significantly effect the response. Some such factors will usually be known; others we may have doubts about. It seems preferable, however, to include some of the factors on which we are in doubt (thus increase dimensionality) rather than to omit altogether factors that may turn out to be really important. We thus end up selecting k factors in all, of which at most p will be really significant for explaining the response; $p \leq k$. By choosing a_i levels on factor i we obtain $\prod_{i=1}^{k} a_i$ points at which observations could be taken. The number of points is surely larger than 2^k (for we select at least two levels on each factor) and thus it increases exponentially fast with k. For this reason the a_i's should be kept as small as possible. Each tentative observation entails costs (of setting the apparatus, of filing the data), which suggests—at least to those who incur the expense—that as few observations as possible should be taken in order to determine the truly relevant factors and their impact on the response. The pressure is now on selecting the design; we should take observations only at points that are "as strategically placed as possible." Since at most p factors (specific identity unknown) are affecting the response, it would make sense to take observations at a set d^* of points that projects as full (and possibly repeated) factorials in *any* subset of p (out of the k) factors. In case $k - p$ or more factors indeed turn out to be inert, we would have at least one complete factorial design in the remaining p or fewer significant factors, no matter which specific p factors these turn out to be.

One critical aspect in all this is the actual value of p, which is not known. Initially we may take p as large as possible, subject to the budgetary constraints, which limit the number of observations to be taken. (If we want full factorials in any p factors, and $a_1, a_2, \ldots, a_p$ are the p largest a_i's, we obviously need at least $a_1 a_2 \cdots a_p$ observations. When compared to the available budget this allows us to decide whether p needs to be increased or decreased.) Studying the response would usually involve designing several experiments in a row. The rule of thumb (as expressed by Box) is that less than

40% of the budget for the entire study should be spent on the first design. The first stage of the study will thus have to be conducted under such budgetary restrictions. If the conclusions from the first stage appear ambiguous (as to the relative importance of some factors or to the kind of impact they have upon the response), the subsequent stages of the study should aim at clearing such questions up. Especially at the first few (first and second, say) stages of the study, selecting designs that project as full factorials on any p factors, whatever p may be, proves informative. The actual construction of such designs is discussed in Sections 8.20 and 8.21, upcoming.

8.20

Having thus motivated the selection of designs we formulate the following general combinatorial problem:

> *Given k sets $A_1, A_2, \ldots, A_k$, find a set d^* of the Cartesian product $A_1 \times A_2 \times \cdots \times A_k$ with as few points as possible whose projections onto any p of the k sets is equal to the Cartesian product of those p sets.* (8.37)

[Let $|A_i| = a_i \geq 2$, for $1 \leq i \leq k$. We denote by $m(a_1, \ldots, a_k; p)$ the number of points in d^*; it is clear that $m(a_1, \ldots, a_k; p)$ is an increasing function in each of its arguments. A lower bound on $m(a_1, \ldots, a_k; p)$ is obviously the product of the p largest a_i's. As we see, for small values of k and p this bound is actually equal to $m(a_1, \ldots, a_k; p)$.]

Problem (8.37) will now be examined in some special cases. To begin with, let $k = 3$ and let $|A_i| = 2$, for $i = 1, 2, 3$. Without loss $A_1 = A_2 = A_3 = \{1, 2\}$. Then $A = A_1 \times A_2 \times A_3$ consists of eight points, the vertices of a three-dimensional cube. The eight vertices are (x, y, z), with $x, y, z \in \{1, 2\}$.

If $p = 1$, then d^* must have at least two points. It is easy to see that $(1, 1, 1)$ and $(2, 2, 2)$ will suffice for the task; the two points project onto each of the three A_i's. (Clearly the two points at the ends of any one of the four cross-diagonals would do just as well, but all these choices are essentially the same.)

If $p = 2$ we must have four or more vertices in d^*. It turns out that $m(2, 2, 2; 2) = 4$, and d^* consists of the points circled below:

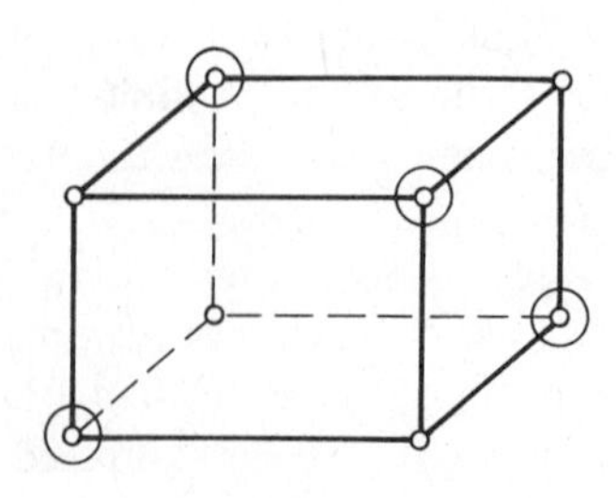

It can be seen that the three projections of d^* are indeed the required sets $A_1 \times A_2$, $A_1 \times A_3$, and $A_2 \times A_3$. Note that the coordinates of the points of d^* may be obtained by *superimposing* the three "squares":

$$\begin{matrix} 1 & 1 \\ 2 & 2 \end{matrix}, \quad \begin{matrix} 1 & 2 \\ 1 & 2 \end{matrix}, \quad \begin{matrix} 1 & 2 \\ 2 & 1 \end{matrix}.$$

They are (1, 1, 1), (1, 2, 2), (2, 1, 2), (2, 2, 1). The design d^* is again unique, up to rotations and/or reflections of the cube.

More generally, let $A_1 = A_2 = A_3$ ($= \{1, 2, \ldots, a\}$); thus $k = 3$, and each A_i has a elements. If $p = 1$, then we may take as d^* the set of a points $(x, x, \ldots, x)$ on the cross-diagonal of $A_1 \times A_2 \times A_3$.

For $p = 2$ we have $m(a, a, a; 2) = a^2$, and a d^* as we propose may be obtained by superimposing the three squares:

$$\begin{matrix} 1 & 1 & 1 & \cdots & 1 \\ 2 & 2 & 2 & \cdots & 2 \\ 3 & 3 & 3 & \cdots & 3 \\ & & & \cdots & \\ a & a & a & \cdots & a \end{matrix} \quad \begin{matrix} 1 & 2 & 3 & \cdots & a \\ 1 & 2 & 3 & \cdots & a \\ 1 & 2 & 3 & \cdots & a \\ & & & \cdots & \\ 1 & 2 & 3 & \cdots & a \end{matrix} \quad \begin{matrix} 1 & 2 & 3 & \cdots & a \\ 2 & 3 & 4 & \cdots & 1 \\ 3 & 4 & 5 & \cdots & 2 \\ & & & \cdots & \\ a & 1 & 2 & \cdots & (a-1) \end{matrix}.$$

Any two squares, when superimposed, produce all the pairs (i, j), with $i, j \in \{1, 2, \ldots, a\}$. The relevant property of the third square is that each symbol occurs exactly once in each row and each column. (One convenient way to have this happen is to cyclically expand the first column, which is what we did above.) This leads us to define as follows: A square arrangement of a symbols $\{1, 2, \ldots, a\}$ such that each symbol occurs exactly once in each row and each column is called a *latin square*. We call two latin squares *essentially the same* if one can be obtained from the other upon permutations of rows and columns; otherwise they are said to be *essentially different*.

In the setting $|A_1| = |A_2| = |A_3| = a$, any design d^* that projects as a full factorial in any two of the three factors is clearly uniquely associated with a latin square of size a. Two such designs d^* are said to be *essentially different* if the corresponding latin squares are such.

[To specialize just for an instant, let $a = 3$; we are thus looking at a three-dimensional cube with three points on each side, and thus 27 points in all. What are the essentially different designs d^* that project as full factorials (with a points) on any pair of coordinate planes? There are exactly *two* of them, listed below:

$$\begin{matrix} 1 & 1 & 1 \\ 2 & 2 & 2 \\ 3 & 3 & 3 \end{matrix} \quad \begin{matrix} 1 & 2 & 3 \\ 1 & 2 & 3 \\ 1 & 2 & 3 \end{matrix} \quad \begin{matrix} 1 & 2 & 3 \\ 2 & 3 & 1 \\ 3 & 1 & 2 \end{matrix}$$

and

$$\begin{matrix} 1 & 1 & 1 \\ 2 & 2 & 2 \\ 3 & 3 & 3 \end{matrix} \quad \begin{matrix} 1 & 2 & 3 \\ 1 & 2 & 3 \\ 1 & 2 & 3 \end{matrix} \quad \begin{matrix} 1 & 3 & 2 \\ 3 & 2 & 1 \\ 2 & 1 & 3 \end{matrix}.$$

The reason for having only two is that there are just two essentially different latin squares of size 3; the last ones listed in the above displays of the two designs. Geometrically, the two designs look like this:

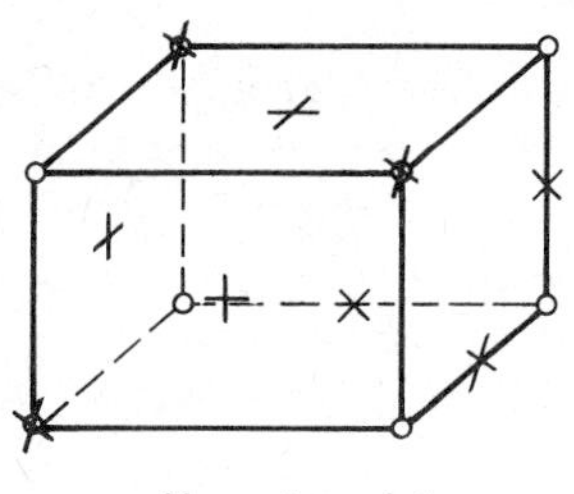

No center point

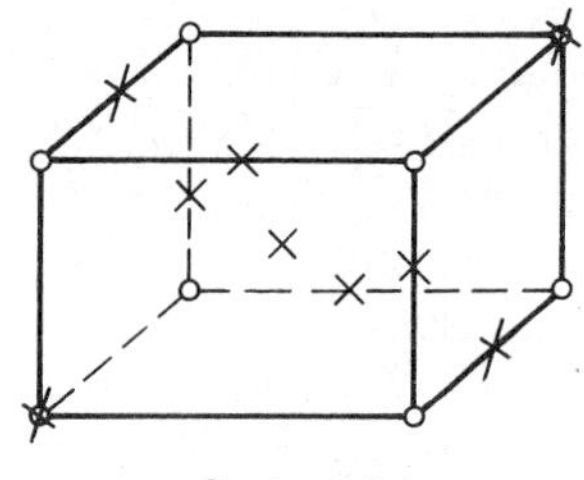

Center point

The exact number of essentially different latin squares of size a remains unknown. To the author's knowledge many bounds exist, but an exact formula is not available.]

We really should not close Section 8.20 without discussing the four-dimensional case. This case has received a lot of attention in the statistical as well as combinatorial literature because it raises the question of existence of orthogonal latin squares. We have four sets A_i, each of cardinality a; thus $k = 4$. When $p = 1$, d^* can again be the main cross-diagonal (x, x, x, x), with $x \in A_i$ (we take $A_i = \{1, 2, \ldots, a\}$, $i = 1, 2, 3, 4$). It clearly projects fully on each A_i.

Let $p = 2$. A full two-dimensional projection contains a^2 points; thus d^* will have at least that many points. A celebrated result of Bose, Shrikhande, and Parker informs us that a design d^* with a^2 points exists for all a, except when $a = 2$ and $a = 6$. One obtains such a design d^* by superimposing four squares selected in such a way that any two superimposed produce all the a^2 pairs (x, y), with $x, y \in \{1, 2, \ldots, a\}$. An example for $a = 3$ is

$$\begin{matrix} 1 & 1 & 1 \\ 2 & 2 & 2 \\ 3 & 3 & 3 \end{matrix} \qquad \begin{matrix} 1 & 2 & 3 \\ 1 & 2 & 3 \\ 1 & 2 & 3 \end{matrix} \qquad \begin{matrix} 1 & 2 & 3 \\ 2 & 3 & 1 \\ 3 & 1 & 2 \end{matrix} \qquad \begin{matrix} 1 & 2 & 3 \\ 3 & 1 & 2 \\ 2 & 3 & 1 \end{matrix} \tag{8.38}$$

It is clear that if a design d^* with a^2 points projecting fully on any two A_i's exists, it *must* be constructed as in (8.38). The relationship between the last two latin squares in (8.38) suggests that we make the following definition: Two latin squares of size a are said to be *orthogonal* if their superimposition yields all the a^2 pairs (x, y), with x and y elements of $\{1, 2, \ldots, a\}$. [The last two latin squares in (8.38) are orthogonal.]

Existence of a design d^* (on four factors, each at a levels) that projects fully on any two factors is therefore equivalent to the existence of two orthogonal latin squares of size a. Bose, Shrikhande, and Parker [4] showed that orthogonal latin squares of any size different from 2 and 6 exist. (For 2 and 6 they in fact do not exist.) When the size a is a power of a prime, orthogonal latin squares exist in profusion. The difficult cases are of the form $a = 4t + 2$ (such as 2, 6, 10, etc.). Euler conjectured that no pair of orthogonal latin squares exists for numbers of this form. His conjecture was disproved, as we said, the only critical values being 2 and 6. The most difficult case in constructing orthogonal latin squares turned out to be 10. We do not include the general proof of Bose, Shrikhande, and Parker's result; however, two orthogonal latin squares of size 10 are displayed below:

$$\begin{array}{cccccccccc}
00 & 67 & 58 & 49 & 91 & 83 & 75 & 12 & 24 & 36 \\
76 & 11 & 07 & 68 & 59 & 92 & 84 & 23 & 35 & 40 \\
85 & 70 & 22 & 17 & 08 & 69 & 93 & 34 & 46 & 51 \\
94 & 86 & 71 & 33 & 27 & 18 & 09 & 45 & 50 & 62 \\
19 & 95 & 80 & 72 & 44 & 37 & 28 & 56 & 61 & 03 \\
38 & 29 & 96 & 81 & 73 & 55 & 47 & 60 & 02 & 14 \\
57 & 48 & 39 & 90 & 82 & 74 & 66 & 01 & 13 & 25 \\
21 & 32 & 43 & 54 & 65 & 06 & 10 & 77 & 88 & 99 \\
42 & 53 & 64 & 05 & 16 & 20 & 31 & 89 & 97 & 78 \\
63 & 04 & 15 & 26 & 30 & 41 & 52 & 98 & 79 & 87
\end{array}.$$

To summarize: when $a \neq 2, 6$ we have $m(a, a, a, a; 2) = a^2$; as to the exceptions, $m(2, 2, 2, 2; 2) = 6$, and $36 < m(6, 6, 6, 6; 2) \leq 42$. In the case $a = 2$, the six points written as columns are

$$\begin{array}{cccccc}
1 & 1 & 2 & 2 & 1 & 2 \\
1 & 2 & 1 & 2 & 1 & 2 \\
1 & 2 & 2 & 1 & 2 & 1 \\
2 & 1 & 1 & 2 & 2 & 1
\end{array}.$$

When $p = 3$ (and $k = 4$) the value of $m(a, a, a, a; 3)$ is not known in general. The lower bound on this number is a^3 and sometimes we find designs $d*$ with as few as a^3 points; one such case exists when $a = 3$, and it is illustrated below. We write the points of the four-dimensional cube as columns:

$$\begin{array}{ccccccccccccccccccccccccccc}
1 & 2 & 3 & 1 & 2 & 3 & 1 & 2 & 3 & 1 & 2 & 3 & 1 & 2 & 3 & 1 & 2 & 3 & 1 & 2 & 3 & 1 & 2 & 3 & 1 & 2 & 3 \\
1 & 2 & 3 & 2 & 3 & 1 & 3 & 1 & 2 & 2 & 3 & 1 & 3 & 1 & 2 & 1 & 2 & 3 & 3 & 1 & 2 & 1 & 2 & 3 & 2 & 3 & 1 \\
1 & 2 & 3 & 3 & 1 & 2 & 2 & 3 & 1 & 2 & 3 & 1 & 1 & 2 & 3 & 3 & 1 & 2 & 3 & 1 & 2 & 2 & 3 & 1 & 1 & 2 & 3 \\
1 & 1 & 1 & 1 & 1 & 1 & 1 & 1 & 1 & 2 & 2 & 2 & 2 & 2 & 2 & 2 & 2 & 2 & 3 & 3 & 3 & 3 & 3 & 3 & 3 & 3 & 3
\end{array}. \quad (8.39)$$

It can be verified that these 27 points project fully on any three of the four available factors; in fact, each point in the image of the projection has exactly one preimage among the 27 points. [To check these properties one has to verify that in *any* three rows of (8.39) all the 27 vectors $(x, y, z)^t$, with $x, y, z \in \{1, 2, 3\}$, occur.]

The design displayed in (8.39) is an example of an orthogonal array. Motivated by statistical concerns similar to those presented in this section Rao introduced this notion in 1947 (see [17]). We recall his definition: A $k \times m$

array with entries from the set $\{1, 2, \ldots, a\}$ is called an *orthogonal array* if any p rows of the array contain all of the a^p possible column vectors, each with the same multiplicity. [Geometrically, therefore, an orthogonal array is a set of m points of the k-dimensional cube (with a points on each coordinate axis) that projects as a full factorial the same number of times on any choice of p-coordinate axes.] Much work on orthogonal arrays is available. We direct the interested reader to the original work of Rao [17] and to [15] for many additional references. The close connection between orthogonal arrays and the problem raised in (8.37) is self-evident. Two differences will be observed, however: in (8.37) we take an interest in k-dimensional product spaces that are not necessarily cubes and wish to find a set of minimal cardinality that projects at least once as a full factorial onto any choice of p-coordinate axes.

Let us remark, in closing, that when $|A_i| = q$ (a power of a prime), $i = 1, 2, \ldots, k$, we can easily construct a design d^* that projects fully and exactly once upon any choice of $k - 1$ A_i's. Indeed, the Cartesian product $A_1 \times A_2 \times \cdots \times A_k$ can in this case be made into a vector space of dimension k over $GF(q)$, the ith coordinate axis corresponding to A_i. Obtain d^* as the subspace of vectors orthogonal to the vector $\mathbf{1}$ (with all entries 1); this subspace has dimension $k - 1$ and will sometimes contain the vector $\mathbf{1}$ itself. More explicitly,

$$d^* = \left\{ (x_1, x_2, \ldots, x_k) : \sum_{i=1}^{k} x_i = 0 \right\}.$$

[To check its properties on projections, any vector of the form $(y_1, y_2, \ldots, y_{k-1}, 0)$, for example, is the projection of the vector $(y_1, y_2, \ldots, y_{k-1}, y)$ in d^*, where y is the unique vector verifying $(\sum_{i=1}^{k-1} y_i) + y = 0$.] This shows that $m(q, q, \ldots, q; k-1) = q^{k-1}$ if we have k q's in $m(q, q, \ldots, q; k-1)$. Design (8.39) is of this kind, with $k = 4$ and $q = 3$.

8.21

Most of Section 8.20 was devoted to situations in which the sets A_i in (8.37) were of the same cardinality. While examining these cases several combinatorial notions suggested themselves, such as orthogonal latin squares and orthogonal arrays. We devote this section to the mixed (or asymmetrical) cases; these are cases in which not all the A_i's have the same cardinality. General constructions for two and three dimensions are given.

Let $k = 2$ with $A_1 = \{1, 2, \ldots, a\}$ and $A_2 = \{1, 2, \ldots, b\}$; without loss we assume $a \geq b$. We seek d^* of minimal cardinality that projects fully onto each A_i. Its cardinality must obviously be a or more, and a d^* with exactly a points can be obtained by taking a *zigzag of cross-diagonals* in $A_1 \times A_2$, for example,

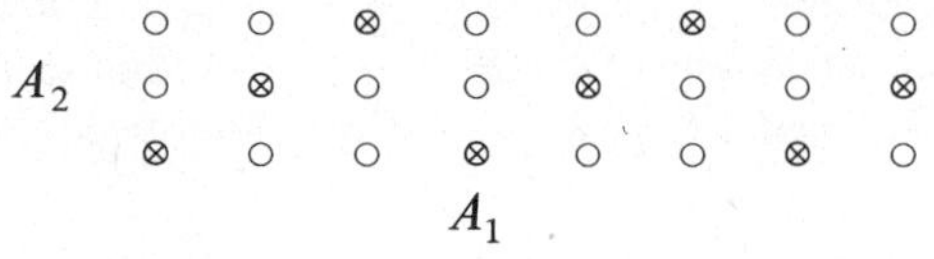

	○	○	⊗	○	○	⊗	○	○
A_2	○	⊗	○	○	⊗	○	○	⊗
	⊗	○	○	⊗	○	○	⊗	○
				A_1				

$(a = 8, b = 3)$

(the points of d^* being visibly marked). Generally we may read the coordinates of d^* as columns below:

$$\begin{array}{cccccccccccc}
1 & 2 & 3 & 4 & 5 & 6 & 7 & 8 & 9 & \cdots & a & \\
1 & 2 & \cdots & & b & 1 & 2 & \cdots & & b & & \cdots
\end{array}$$

Let now $k = 3$. Take $A_1 = \{1, 2, \ldots, a\}$, $A_2 = \{1, 2, \ldots, b\}$, $A_3 = \{1, 2, \ldots, c\}$, with $a \geq b \geq c$. If $p = 1$, that is, if we wish to project fully on each one of the three sets, then the d^* we seek [still in accord with (8.37)] is a *zigzag of cross-diagonals*, which may be written as

$$\begin{array}{cccccccccccc}
1 & 2 & 3 & 4 & 5 & 6 & 7 & 8 & 9 & \cdots & a & \\
1 & 2 & \cdots & & b & 1 & 2 & \cdots & b & & \cdots & \\
1 & \cdots & c & & 1 & \cdots & c & 1 & \cdots & & c & \cdots
\end{array} .$$

(Columns indicate points of d^* in $A_1 \times A_2 \times A_3$.)

If $p = 2$, and thus we wish to project fully on any pair of factors, d^* will have ab points that are obtained by superimposing the following three rectangles (each with a rows and b columns):

$$\begin{array}{ccccccccccccccccccc}
1 & 1 & 1 & \cdots & 1 & & 1 & 2 & 3 & \cdots & b & & 1 & 2 & \cdots & c & 1 & 2 & \cdots & c & 1 & \cdots \\
2 & 2 & 2 & \cdots & 2 & & 1 & 2 & 3 & \cdots & b & & 2 & 3 & \cdots \\
3 & 3 & 3 & \cdots & 3 & & 1 & 2 & 3 & \cdots & b & & \vdots & \vdots \\
 & & & & & & & & & & & & c & 1 & \cdots \\
 & & & \cdots & & & & & & \cdots \\
 & & & & & & & & & & & & 1 & 2 & \cdots \\
 & & & \cdots & & & & & & \cdots \\
 & & & & & & & & & & & & & & & \cdots \\
a & a & a & \cdots & a & & 1 & 2 & 3 & \cdots & b & & & & & \cdots
\end{array}$$

When the dimension exceeds three things get complicated. These designs are useful for practical experiments, especially for a relatively small (say ≤ 6) number of factors and not too large a number of levels (say ≤ 7). It would be handy to have tables of such d^*'s in the practical range just described.

Problem (8.37) remains largely unsolved.

EXERCISES

1. What is $m(4, 3, 2; 2)$? Display the design.

2. Give full details on the construction of an orthogonal array on four factors, with four levels on each factor, projecting once fully on any three factors. How many points are involved?

3. Show that $m(2, 2, 2, 2; 2) = 6$.

4. What is $m(6, 6, 6, 6; 2)$?

5. Find a design d^* with a minimal number of points that projects fully on any two of the four sets A_i, $i = 1, 2, 3, 4$. Take $|A_1| = 5$, $|A_2| = 4$, and $|A_3| = |A_4| = 3$.

6. Write an essay describing why it would be practically useful to have a design with as few points as possible that projects fully onto some "good fractions" only (as opposed to full factorials) when the number of levels on the factors get to be large.

NOTES

This chapter presented some basic aspects of statistical planning, and described several combinatorial problems that arise from them. For the most part we chose a motivational approach, attempting to explain how we are led to these problems rather than treating the problems themselves in painstaking detail.

To those interested in the actual task of designing experiments we recommend the book by Box, Hunter, and Hunter [5]. The reader will find in it many interesting insights and actual uses of factorial experiments. Along the same lines one can benefit from the text by John [13]. The treatise [11] is devoted entirely to factorial designs.

Over the years Raghavarao's book provided much information to students and researchers interested in the combinatorial side of statistical design. In it the reader will find a proof of the falsity of Euler's conjecture by Bose, Shrikhande, and Parker, the construction of projective planes from pairwise orthogonal latin squares, and many other interesting results that we had to omit.

New research trends in factorial and block designs will be found in [2], [12], and [18]. In particular, research initiated by Srivastava [18] focuses on the so-called search designs.

REFERENCES

[1] T. W. Anderson, *An Introduction to Multivariate Statistical Analysis*, Wiley, New York, 1958.

[2] R. A. Bailey, A unified approach to design of experiments, *J. Royal Stat. Soc. (A)*, **144**, 214–223 (1981).

[3] R. C. Bose, Mathematical theory of symmetrical factorial design, *Sankhyā*, **8**, 107–166 (1947).

[4] R. C. Bose, S. S. Shrinkhande, and E. T. Parker, Further results on the construction of mutually orthogonal latin squares and the falsity of Euler's conjecture, *Can. J. Math.*, **12**, 189–203 (1960).

[5] G. E. P. Box, G. W. Hunter, and J. S. Hunter, *Statistics for Experimenters*, Wiley, New York, 1978.

[6] G. E. P. Box, and J. S. Hunter, Multifactor experimental designs for exploring response surfaces, *Ann. Math. Statist.*, **28**, 195–241 (1957).

[7] P. J. Cameron, J. M. Goethals, J. J. Seidel, and E. E. Shult, Line graphs, root systems, and elliptic geometry, *J. Algebra*, **43**, 305–327 (1976).

[8] C.-S. Cheng, Optimality of certain asymmetrical experimental designs, *Ann. Statist.*, **6**, 1239–1261 (1978).

[9] G. M. Constantine, On the *E*-optimality of PBIB designs with a small number of blocks, *Ann. Statist.*, **3**, 1027–1031 (1982).

[10] R. A. Fisher, The theory of confounding in factorial experiments in relation to the theory of groups, *Ann. Eugenics*, **11**, 341–353 (1942).

[11] B. L. Raktoe, A. Hedayat, and W. T. Federer, *Factorial Designs*, Wiley, New York, 1981.

[12] M. Jacroux, On the *E*-optimality on regular graph designs, *J. Royal Statist. Soc.* (*B*), **42**, 205–209 (1980).

[13] P. W. M. John, *Statistical Design and Analysis of Experiments*, Macmillan, New York, 1971.

[14] J. C. Kiefer, Construction and optimality of generalized Youden designs, in *A Survey of Statistical Design and Linear Models* (J. N. Srivastava, Ed.), North Holland, Amsterdam, 1975, pp. 333–353.

[15] D. Raghavarao, *Constructions and Combinatorial Problems in Design of Experiments*, Wiley, New York, 1971.

[16] C. R. Rao, *Linear Statistical Inference and its Applications*, 2nd ed., Wiley, New York, 1973.

[17] C. R. Rao, Factorial experiments derivable from combinatorial arrangements of arrays, *J. Royal Statist. Soc.* (*B*), **9**, 128–139 (1947).

[18] J. N. Srivastava, Designs for searching non-negligible effects, in *A Survey of Statistical Design and Linear Models* (J. N. Srivastava, Ed.), North Holland, Amsterdam, 1975, pp. 507–519.

[19] A. J. Hoffman, $-1 - \sqrt{2}$? In *Combinatorial Structures and Their Applications* (R. Guy, Ed.), Gordon and Breach, New York, 1970, pp. 173–176.

CHAPTER 9

Möbius Inversion

Order leads to all virtues!
But what leads to order?

Aphorismen, G.C. LICHTENBERG

The general process of inversion that we study in this chapter is known as Möbius inversion. It is a "natural" process that occurs in many counting problems. In discrete settings it links (roughly speaking) the often readily available information on certain kinds of subsets of a set to knowledge about the "points" of that set. The points are usually the fundamental objects for which we seek information. Once we possess the information we need about the points, we can fully control the phenomenon under study.

As a simple example, consider rolling a fair die three times. Denote by X_i the number of spots that appear on the ith roll, $i = 1, 2, 3$. Let $Y = \max(X_i)$, that is, Y is the maximum of the three outcomes. Find $P(Y = y)$, the probability that Y equals y, $y = 1, \ldots, 6$.

While $P(Y = y)$ (the information about points) may be a bit difficult to compute, $P(Y \leq y)$ is easy to obtain: $P(Y \leq y) = P(\max(X_i) \leq y) = \prod_{i=1}^{3} P(X_i \leq y) = (P(X_1 \leq y))^3 = (y/6)^3$. So we easily gained access to the information $P(Y \leq y) = (y/6)^3$ about subsets of the form $Y \leq y$. Möbius inversion is particularly simple in this case and it transfers the information from these subsets to the points themselves in the following obvious way: $P(Y = y) = P(Y \leq y) - P(Y \leq y - 1) = (y/6)^3 - ((y - 1)/6)^3$, $y = 1, \ldots, 6$. This last step is not always easy to perform. More often than not it involves intricate and ingenious constructions; but the spirit of the inversion process remains the same. [Once we possess information about the fundamental objects, in this case points, we can essentially answer any probabilistic questions that may arise, such as: What is the chance that Y is odd, i.e., $Y \in \{1, 3, 5\}$? Since $\{1, 3, 5\} = \{1\} \cup \{3\} \cup \{5\}$ (a disjoint union of points), we have $P(Y \in \{1,3,5\}) = P(Y=1) + P(Y=3) + P(Y=5) = (\frac{1}{6})^3 - (\frac{0}{6})^3 + (\frac{3}{6})^3 - (\frac{2}{6})^3 + (\frac{5}{6})^3 - (\frac{4}{6})^3$. But any subset is the disjoint union of the points it contains and we can therefore compute its probability as above. By knowing the probability of points we now fully control this phenomenon.]

The example we just discussed, along with many other counting problems in discrete settings, resembles the well-known result in calculus: If $g(x) = \int_0^x f(t)\,dt$, then $f(x) = (d/dx)g(x)$. The function f may often be difficult to compute directly, while g may be known or possible to obtain by some method with little effort. In such a case the process of differentiation yields f easily. For example, if $g(x) = x^2 + 3x$ $(= \int_0^x f(t)\,dt)$, then $f(x) = (d/dx)g(x) = 2x + 3$.

When extended properly to discrete settings this process of "differentiation" (or information at a point) becomes a powerful counting machine that provides answers to many intricate combinatorial problems. We find, for example, the number of automorphisms of an Abelian group essentially by visualizing automorphisms as "derivatives" of homomorphisms. Counting homomorphisms turns out to be much easier. In general there is a process of inversion hidden here that we call Möbius inversion.

The terminology is inspired by classical work in the theory of numbers related to the distribution of primes. The Riemann zeta function $\hat{\zeta}(s) = \sum_{n=1}^{\infty} 1/n^s$ (expressed here as a Dirichlet series) is intimately connected to the distribution of primes. Its inverse, that is, the function $\hat{\mu}$ that satisfies $\hat{\zeta}(s)\hat{\mu}(s) = 1$, can be also expressed as a Dirichlet series: $\hat{\mu}(s) = \sum_{n=1}^{\infty} \mu(n)/n^s$, where $\mu(n) = (-1)^k$ if n is a product of k distinct primes, and 0 otherwise. (Multiplying the two series $\hat{\zeta}$ and $\hat{\mu}$ in the usual way, by convolution, we indeed obtain $\hat{\zeta}\hat{\mu} = \hat{\mu}\hat{\zeta} = 1$.) The function $\hat{\mu}$ is called the Möbius function. It gives a lot of information about primes; for example, the fact that $\hat{\mu}(1) = 0$ is equivalent to the prime number theorem which states that there are approximately $x(\ln x)^{-1}$ primes less than x.

The aforementioned study of the multiplicative structure of the natural numbers has as its basis the partial order of divisibility. In this chapter we extend the notions of zeta and Möbius functions to any partially ordered set (cf. Weisner, Hall, Delsarte, Rota, and others). Many examples are given from probability, set theory, number theory, finite-dimensional vector spaces over finite fields, partitions of a set, p-groups, and nilpotent groups. Basic results by Weisner and Hall are included as well. In Section 2 we learn counting methods using Möbius inversion. The well-known sieve formula is there established, along with numerous other classical examples in counting. In general the emphasis is on concrete results. The theoretical development clusters around the major results and is presented with a minimum amount of notation. We conclude the chapter with a theorem of Delsarte which counts the number of subgroups of a given isomorphism type in an Abelian p-group, and its extension to counting surjective homomorphisms between non-Abelian p-groups regular in the sense of Phillip Hall.

1 THE MÖBIUS FUNCTION

9.1

Let the ordered pair $(X, \leq)$ consist of a set X with a relation $\leq$ of *partial order*. By definition $\leq$ satisfies

$$x \leq x,$$
$$x \leq y \quad \text{and} \quad y \leq x \text{ implies } x = y,$$
$$x \leq y \quad \text{and} \quad y \leq z \text{ implies } x \leq z,$$

for $x, y, z \in X$. We assume that there exists a unique element 0 in X satisfying $0 \leq x$ for all $x \in X$. The ordered pair $(X, \leq)$ is called a *partially ordered set*.

For x, y in X define the *segment* $[x, y]$ as the set of all u in X satisfying $x \leq u \leq y$. A partially ordered set is called *locally finite* if for all x, y in X the segment $[x, y]$ is a finite set. Throughout this chapter, unless we specify otherwise, all the partially ordered sets are assumed (and those in the examples simply are) locally finite.

It is convenient to extend the notation as follows: We write

$$x < y \quad \text{if } x \leq y \quad \text{and} \quad x \neq y,$$
$$x \geq y \quad \text{if } y \leq x,$$

and by $x \nleq y$ we understand the negation of $x \leq y$.

A good model to occasionally check general definitions (or theorems) against is $(N, |)$, the set of natural numbers with divisibility as relation of partial order. This partially ordered set is locally finite. The segment $[3, 18] = \{u \in N: 3|u|18\} = \{3, 6, 9, 18\}$, for example. On the other hand, the segment $[4, 30]$ is the empty set.

9.2

With a partially ordered set $(X, \leq)$ we associate a set of functions

$$A = \{\alpha: X \times X \to R,$$

$$\alpha \textit{ a function satisfying}$$
$$\alpha(x, y) \neq 0 \textit{ if } x = y \textit{ and } \alpha(x, y) = 0 \textit{ if } x \nleq y\}.$$

Define a multiplication, called *convolution*, on A by

$$(\alpha \cdot \beta)(x, y) = \sum_{x \leq u \leq y} \alpha(x, u)\beta(u, y).$$

(The operation of convolution on A is motivated by the multiplication of formal Dirichlet series; example (b) below makes this connection more explicit.)

It is well worth observing that:

* *The set A endowed with multiplication by convolution is a group* (called the group of *arithmetic functions*).

Convolution is an associative operation. The identity element of A is

$$\delta(x, y) = \begin{cases} 1 & \text{if } x = y \\ 0 & \text{if } x \neq y \end{cases}.$$

By α^{-1} we denote the *inverse* of α in A, and α^{-1} is defined inductively as follows:

$$\text{for } y = x \qquad \alpha^{-1}(x, x) = \frac{1}{\alpha(x, x)},$$

$$\text{for } y > x \qquad \alpha^{-1}(x, y) = \frac{-1}{\alpha(y, y)} \sum_{x \leq u < y} \alpha^{-1}(x, u)\alpha(u, y)$$

(note the strict inequality in the summation index, crucial to the inductive definition),

$$\text{for } y \ngtr x \qquad \alpha^{-1}(x, y) = 0.$$

[Let us check that α^{-1}, as defined above, is the inverse of α. Indeed, $(\alpha^{-1}\alpha)(x, x) = (1/\alpha(x, x))\alpha(x, x) = 1$, and if $x < y$, then

$$\begin{aligned}(\alpha^{-1}\alpha)(x, y) &= \sum_{x \leq u \leq y} \alpha^{-1}(x, u)\alpha(u, y) \\ &= \sum_{x \leq u < y} \alpha^{-1}(x, u)\alpha(u, y) + \alpha^{-1}(x, y)\alpha(y, y) \\ &= -\alpha^{-1}(x, y)\alpha(y, y) + \alpha^{-1}(x, y)\alpha(y, y) = 0.\end{aligned}$$

Thus $\alpha^{-1}\alpha = \delta$ and similarly $\alpha\alpha^{-1} = \delta$.]

The most "natural" function $\zeta \in A$ defined by

$$\zeta(x, y) = \begin{cases} 1 & \text{if } x \leq y \\ 0 & \text{otherwise} \end{cases}$$

is called the *Riemann zeta function* of the partially ordered set X.

We denote by μ the inverse of ζ in A and call it the *Möbius function* of X. By its definition

$$\mu(x, x) = 1, \quad \text{and for } y > x,\ \mu(x, y) = -\Sigma_{x \leq u < y}\mu(x, u).$$

(We often write the defining properties of μ as $\mu(x, x) = 1$, and for $y > x$ we implicitly define $\mu(x, y)$ by $\Sigma_{x \leq u \leq y}\mu(x, u) = 0$, obviously equivalent to the above.)

9.3

We now state and prove the major result of this chapter:

Theorem 9.1 (Möbius Inversion Formula). *Let f and g be real-valued functions defined on the partially ordered set $(X, \leq)$. If*

$$g(x) = \sum_{0 \leq u \leq x} f(u),$$

then

$$f(x) = \sum_{0 \leq u \leq x} g(u)\mu(u, x).$$

(The function f should in some sense be thought of as the "derivative" of g.)

Proof.

$$\begin{aligned}
\sum_{0 \leq u \leq x} g(u)\mu(u, x) &= \sum_{0 \leq u \leq x} \Big(\sum_{0 \leq z \leq u} f(z) \Big) \mu(u, x) \\
&= \sum_{0 \leq u \leq x} \Big(\sum_{z \in X} f(z)\zeta(z, u) \Big) \mu(u, x) \\
&= \{\text{interchanging the order of summation}\} \\
&= \sum_{z \in X} f(z) \sum_{0 \leq u \leq x} \zeta(z, u)\mu(u, x) \\
&= \sum_{z \in X} f(z) \sum_{z \leq u \leq x} \zeta(z, u)\mu(u, x) \\
&= \sum_{z \in X} f(z)\delta(z, x) = f(x),
\end{aligned}$$

which ends the proof.

An easy extension yields the more general Möbius inversion formula:

$$g(x) = \sum_{0 \leq u \leq x} f(u)\alpha(u, x) \quad \text{if and only if} \quad f(x) = \sum_{0 \leq u \leq x} g(u)\alpha^{-1}(u, x)$$

where f and g are functions on X and $\alpha, \beta \in A$.

The proof is just as before,

$$\begin{aligned}
&\sum_{0 \leq u \leq x} f(u)\alpha(u, x) \\
&\quad = \sum_{0 \leq u \leq x} \Big(\sum_{0 \leq z \leq u} g(z)\alpha^{-1}(z, u) \Big) \alpha(u, x) \\
&\quad = \sum_{0 \leq z \leq x} g(z) \sum_{0 \leq u \leq x} \alpha^{-1}(z, u)\alpha(u, x) \\
&\quad = \sum_{0 \leq z \leq x} g(z) \sum_{z \leq u \leq x} \alpha^{-1}(z, u)\alpha(u, x) \\
&\quad = \sum_{0 \leq z \leq x} g(z)\delta(z, x) = g(x).
\end{aligned}$$

Conversely, $\Sigma_{0 \leq u \leq x} g(u)\alpha^{-1}(u, x)$ equals $f(x)$ in the same manner.

9.4

We can sometimes formulate what we just said in terms of matrices. Let X be a countable set and suppose that the elements of X can be lined up as $0, x_1, x_2, x_3, \ldots$ so that $x_1 < x_j$ implies $i < j$ (this can always be done for finite sets). To an arithmetic function α we associate an upper triangular matrix A (rows and columns indexed by $0, x_1, x_2, \ldots$) whose (i, j)th entry equals $\alpha(x_i, x_j)$. This allows us to embed the group of arithmetic functions into the group on nonsingular upper triangular matrices with usual matrix multiplication as group operation. Moreover, if $\alpha \to \overline{A}$, $\beta \to \overline{B}$, then $\alpha\beta \to \overline{A}\,\overline{B}$. With $\zeta \to \overline{Z}$, $\mu \to \overline{M}$ ($\overline{Z}\overline{M} = \overline{M}\overline{Z} = I$), $f(x) = (f(0), f(x_1), f(x_2), \ldots)$, and $g(x) = (g(0), g(x_1), (x_2), \ldots)$, the Möbius formula reads: *If* $g(x) = f(u)\overline{Z}$, *then* $f(x) = g(u)\overline{M}$.

2. MÖBIUS INVERSION ON SPECIAL PARTIALLY ORDERED SETS

This section is devoted to the computation of the Möbius functions for partially ordered sets of special importance. These partially ordered sets (such as the subsets of a set, subspaces of a vector space, partitions of a set, subgroups of a group, and others) occur naturally in many nontrivial counting problems. We have the opportunity to use these Möbius functions and obtain answers to several intricate counting problems, especially in the last section.

Example 1. The simplest example, perhaps, is the set of natural numbers (with zero), that is, $X = N \cup \{0\}$, with $\leq$ the usual "less than or equal" order. The Riemann zeta function is in this case $\zeta(m, n) = 1$ if $m \leq n$, and $\zeta(m, n) = 0$ otherwise. We compute the Möbius function μ as follows. For $m \in N \cup \{0\}$ by definition $\mu(m, m) = 1$. Inductively, $\mu(m, m) + \mu(m, m + 1) = 0$, giving $\mu(m, m + 1) = -1$. It now easily follows from the definition of μ that $\mu(m, n) = 0$ for $n \geq m + 2$. Hence

$$\mu(m, n) = \begin{cases} 1 & \text{if } n = m \\ -1 & \text{if } n = m + 1 . \\ 0 & \text{otherwise} \end{cases}$$

In matrix notation

$$\zeta \to \overline{Z} = \begin{pmatrix} 1 & 1 & 1 & 1 & 1 & \cdots \\ & 1 & 1 & 1 & 1 & \cdots \\ & & 1 & 1 & 1 & \cdots \\ & & & 1 & 1 & \cdots \\ 0 & & & & 1 & \cdots \end{pmatrix};$$

$$\mu \to \overline{M} = \begin{pmatrix} 1 & -1 & 0 & 0 & 0 & \cdots \\ & 1 & -1 & 0 & 0 & \cdots \\ & & 1 & -1 & 0 & \cdots \\ & & & 1 & -1 & \cdots \\ 0 & & & & 1 & \cdots \end{pmatrix}.$$

With $\overline{Z}$ and $\overline{M}$ as above the Möbius inversion formula (i.e., Theorem 9.1) reads: If $(g(n)) = (f(n))\bar{Z}$, then $(f(n)) = (g(n))\bar{M}$. Or, written in terms of sums,

$$\textit{If } g(n) = \sum_{k=0}^{n} f(k) \textit{ then } f(n) = g(n) - g(n-1).$$

This is the fundamental theorem of calculus for discrete spaces. In the theory of probability $g(n)$ plays the role of the distribution function and $f(n)$ is the probability density. The example with the die that we gave in the introduction to this chapter is an illustration of Möbius inversion on this partially ordered set.

Example 2. Let X be the set of positive integers (i.e., natural numbers). For d and n two positive integers write $d \leq n$ if $d|n$ (d divides n).

The function

$$\zeta(d, n) = \begin{cases} 1 & \text{if } d|n \\ 0 & \text{otherwise} \end{cases}$$

is the Riemann zeta function.

Let us compute $\mu(d, n)$. Visualize geometrically the segment $[d, n]$ and in it the hypercube as drawn below:

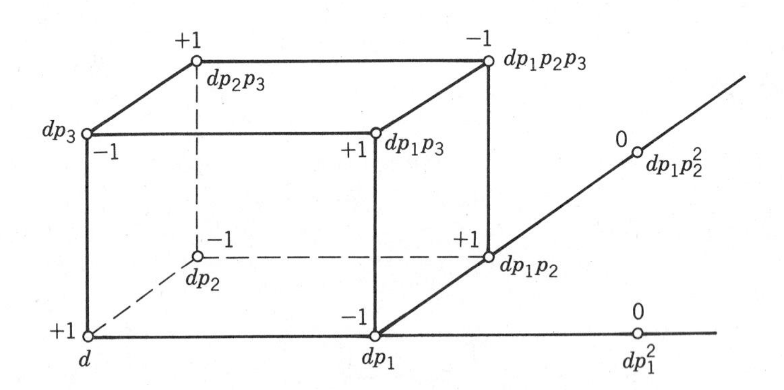

with p_i's distinct primes. If we label the vertices of this hypercube by vectors with entries 0 and 1 (d corresponding to the vertex with all entries 0), then the Möbius function assigns (by its definition) a $+1$ to a vertex with an even (or zero) number of 1's and -1 to a vertex with an odd number of 1's. Outside the hypercube the Möbius function has value zero. The values of μ are written next to each vertex. The reader should carefully study the details on the figure above.

We summarize:

* *The Möbius function of the set of natural numbers with divisibility as partial order is* $\mu(d, n) = (-1)^k$, *if* $n = dp_1p_2 \cdots p_k$ (*for* p_i*'s distinct primes*), *and is* 0 *otherwise.*

[Observe, in addition, that $\mu(d, n) = \mu(1, n/d)$ if d divides n. Number theorists habitually use $\mu(n/d)$ for $\mu(1, n/d)$.] The Möbius inversion formula reads, therefore:

$$\textit{If } g(n) = \sum_{\substack{d \\ d|n}} f(d) \textit{ then } f(n) = \sum_{\substack{d \\ d|n}} g(d)\mu(d, n),$$

with $\mu(d, n)$ as above.

As we mentioned, the general terminology that we adopted originated with work in number theory. We present now a succinct explanation. Let X still be the set of positive integers with divisibility as relation of partial order. For x, y in X look at the functions, $f \in A$ (the group of arithmetic functions) of the form $f(x, y) = F(y/x)$, with y/x integral. The product FG of two such functions is

$$FG\left(\frac{y}{x}\right) = fg(x, y) = \sum_{\substack{u \\ x|u|y}} f(x, u)g(u, y) = \sum_{\substack{u \\ x|u|y}} F\left(\frac{u}{x}\right)G\left(\frac{y}{u}\right),$$

or substituting m for y/x, d for u/x, and n for y/u,

$$FG(m) = \sum_{\substack{d \\ dn=m}} F(d)G(n). \tag{9.1}$$

In this notation $1 = \zeta(x, y) = \zeta(y/x)$ and $\mu(x, y) = \mu(y/x)$ (the Möbius function found above).

To a function F we associate its *formal Dirichlet series* $\hat{F}(s) = \sum_{n=1}^{\infty} F(n)/n^s$. Then $\hat{\zeta}(s) = \sum_{n=1}^{\infty} 1/n^s$ becomes the well-known Riemann zeta function in number theory. Observe that the coefficient of $1/m^s$ in the product $\hat{F}(s)\hat{G}(s)$ of two Dirichlet series is given by the convolution formula (9.1) (check this quickly). We have

$$\hat{\zeta}^{-1}(s) = \sum_{n=1}^{\infty} \frac{\mu(n)}{n^s} = \hat{\mu}(s).$$

Indeed, the coefficient of $1/m^s$ in $\hat{\zeta}(s)\hat{\mu}(s)$ is

$$\sum_{\substack{d \\ dn=m}} \zeta(d)\mu(n) = \sum_{\substack{u \\ x|u|y}} \zeta(x, u)\mu(u, y)$$

$$= \begin{cases} 1 & \text{if } x = y \\ 0 & \text{otherwise} \end{cases} = \begin{cases} 1 & \text{if } m = 1 \\ 0 & \text{otherwise} \end{cases}.$$

This justifies the general terminology.

Example 3. Let X be the set of all subsets of a finite set. The partial order on X is the usual set inclusion $\subseteq$. For subsets S and T the Riemann zeta function is $\zeta(S, T) = 1$ if $S \subseteq T$, and $\zeta(S, T) = 0$ otherwise.

Let us compute the Möbius function $\mu(S, T)$ (for $S \subseteq T$). The figure below indicates how to begin the inductive process:

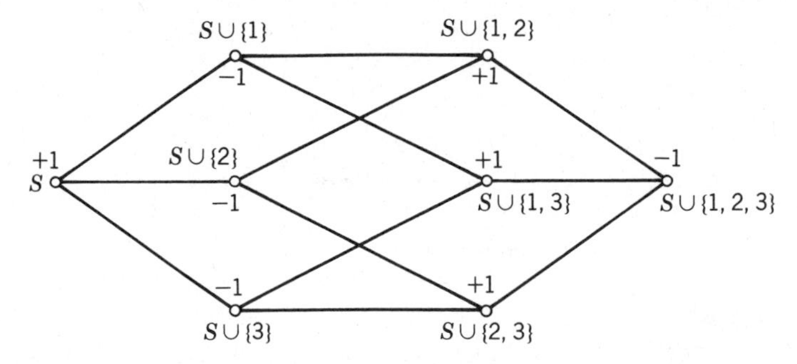

The figure suggests that $\mu(S, T) = (-1)^{|T|-|S|}$. An easy inductive argument allows us to verify this. Indeed, assume that $\mu(S, T) = (-1)^k$ if $|T - S| = k$, for $0 \leq k \leq n - 1$. Then, for $|T - S| = n$, we may write

$$\begin{aligned} \mu(S, T) &= -\left[1 + (-1)\binom{n}{1} + 1 \cdot \binom{n}{2} + \cdots + (-1)^{n-1}\binom{n}{n-1}\right] \\ &= -\left[(1-1)^n - (-1)^n\right] = (-1)^n \\ &= (-1)^{|T-S|} = (-1)^{|T|-|S|}, \end{aligned}$$

as desired. We thus conclude that:

* *The Möbius function for the subsets of a set is*

$$\mu(S, T) = (-1)^{|T|-|S|}$$

if $S \subseteq T$, *and is* 0 *otherwise.*

Whence the Möbius inversion formula:

$$\textit{If } g(T) = \sum_{S \subseteq T} f(S) \textit{ then } F(T) = \sum_{S \subseteq T} (-1)^{|T|-|S|} g(S).$$

This inversion formula takes a familiar form when f (and g) are functions of the cardinality of subsets only. In such a case we may write $f(S) = b_k$ and $g(S) = a_k$, for subsets S of cardinality k. Sorting by cardinality, the above Möbius inversion formula becomes: If

$$a_n = g(T) = \sum_{S \subseteq T} f(S) = \sum_{k=0}^{n} \binom{n}{k} b_k,$$

then

$$b_n = f(T) = \sum_{S \subseteq T} g(S)(-1)^{|T|-|S|} = \sum_{k=0}^{n} \binom{n}{k} (-1)^{n-k} a_k$$

(and vice versa), in which the reader will recognize the familiar binomial inversion mentioned in Section 3.3.

Example 4. Let X be the set of subspaces of a vector space V of dimension n over $GF(q)$, the finite field with q elements (with q a power of a prime). The partial order $S \leq T$ means that subspace S is included in subspace T, that is, S is a subspace of T. The Riemann zeta function is $\zeta(S,T) = 1$ if $S \leq T$ and 0 otherwise. We compute the Möbius function of $(X, \leq)$.

It helps to know the exact number of subspaces of dimension k in V. The vector space V contains q^n vectors. There are $(q^n - 1)(q^n - q)(q^n - q^2) \cdots (q^n - q^{k-1})$ ordered selections of k linearly independent vectors in V. (Indeed, we have $q^n - 1$ choices for the first nonzero vector, $q^n - q$ for the second, ..., $q^n - q^{k-1}$ for the kth. Note that at step r we subtract from q^n the q^{r-1} vectors in the subspace generated by our first $r - 1$ selections of linearly independent vectors.) But each subspace of dimension k is spanned by $(q^k - 1)(q^k - q) \cdots (q^k - q^{k-1})$ such ordered selections. (The counting here is again done as above.) We therefore conclude that:

* *A vector space of dimension n over $GF(q)$ has*

$$\frac{(q^n - 1)(q^n - q) \cdots (q^n - q^{k-1})}{(q^k - 1)(q^k - q) \cdots (q^k - q^{k-1})}$$

distinct subspaces of dimension k.

Upon cancellation, we find that the number of subspaces of dimension k in V is $\begin{bmatrix} n \\ k \end{bmatrix}(q)$, where

$$\begin{bmatrix} n \\ k \end{bmatrix}(x) = \frac{(x^n - 1)(x^{n-1} - 1) \cdots (x^{n-k+1} - 1)}{(x^k - 1)(x^{k-1} - 1) \cdots (x - 1)} \quad \text{(notation)}.$$

The $\begin{bmatrix} n \\ k \end{bmatrix}(x)$ are called Gaussian polynomials. (It might not look like it, but they are indeed polynomials.) A simple recurrence relation exists, and it can easily be checked from the above definition of $\begin{bmatrix} n \\ k \end{bmatrix}(x)$. From it, we inductively deduce that $\begin{bmatrix} n \\ k \end{bmatrix}(x)$ are polynomials. The recurrence is

$$\begin{bmatrix} n \\ k \end{bmatrix}(x) = \begin{bmatrix} n-1 \\ k \end{bmatrix}(x) + x^{n-k}\begin{bmatrix} n-1 \\ k-1 \end{bmatrix}(x).$$

(Observe that for all n, $\begin{bmatrix} n \\ n \end{bmatrix}(x) = 1$; for convenience we take $\begin{bmatrix} n \\ k \end{bmatrix}(x) = 0$ for $k > n$ and $k < 0$.) The reader is referred to Section 6 of Chapter 3 for more information on Gaussian polynomials.

We now address ourselves to the task of computing the Möbius function for the set of subspaces of the vector space V. The dimension of V over the field $GF(q)$ is n. We begin by observing that $\mu(0,0) = 1$, with 0 signifying the zero subspace. In general we denote by $\langle v_1, v_2, \ldots \rangle$ the subspace of V generated by vectors $v_1, v_2, \ldots$. By the definition of μ, for any one-dimensional subspace $\langle v \rangle$ we have $\mu(0,0) + \mu(0, \langle v \rangle) = 0$, which gives $\mu(0, \langle v \rangle) = -1$. For a two-dimensional subspace $\langle v_1, v_2 \rangle$ of V we may similarly write

$$1 + (-1)\,(\text{the number of one-dimensional subspaces of } \langle v_1, v_2 \rangle) + \mu(0, \langle v_1, v_2 \rangle) = 0.$$

The subspace $\langle v_1, v_2\rangle$, being of dimension 2, contains $\left[{2 \atop 1}\right](q) = q + 1$ subspaces of dimension 1. The above computation thus yields

$$\mu(0, \langle v_1, v_2\rangle) = -[1 + (-1)(q+1)] = q.$$

To develop a feel for the general pattern that emerges we suggest that the reader calculate in similar ways $\mu(0, W)$ for W a subspace of dimension 3 and 4. It will then be conjectured that

$$\mu(0, W) = (-1)^k q^{\binom{k}{2}},$$

for W a subspace of dimension k. Assume this is true for subspaces of dimension k, with $0 \le k \le m$. Inductively we prove that it is also true for subspaces of dimension $m + 1$.

Indeed, for a subspace W of dimension $m + 1$ we have

$$\begin{aligned}
-\mu(0, W) &= \sum_{k=0}^{m} (-1)^k q^{\binom{k}{2}} \left[{m+1 \atop k}\right](q) \\
&= \left\{\text{by the recurrence for the } \left[{n \atop k}\right]\text{'s}\right\} \\
&= \sum_{k=0}^{m} (-1)^k q^{\binom{k}{2}} \left(\left[{m \atop k}\right](q) + q^{m-k+1}\left[{m \atop k-1}\right](q)\right) \\
&= (-1)^m q^{\binom{m}{2}} + \sum_{k=0}^{m-1} (-1)^k q^{\binom{k}{2}} \left[{m \atop k}\right](q) \\
&\quad + \sum_{k=0}^{m} (-1)^k q^{\binom{k}{2}} q^{m-k+1} \left[{m \atop k-1}\right](q) \\
&= \{\text{by the inductive assumption, and by writing } s \text{ for } k-1\} \\
&= (-1)^m q^{\binom{n}{2}} - (-1)^m q^{\binom{m}{2}} \\
&\quad + \sum_{s=0}^{m-1} (-1)^{s+1} q^{\binom{s+1}{2}} q^{m-s} \left[{m \atop s}\right](q) \\
&= -q^m \sum_{s=0}^{m-1} (-1)^s q^{\binom{s}{2}} \left[{m \atop s}\right](q) \\
&= \{\text{by the inductive assumption}\} \\
&= q^m (-1)^m q^{\binom{m}{2}} = (-1)^m q^{\binom{m+1}{2}}.
\end{aligned}$$

We have thus shown that

$$\mu(0, W) = (-1)^m q^{\binom{m}{2}},$$

for a subspace W of dimension m in V.

To compute $\mu(S, T)$ for two subspaces $S \leq T$ of V, we first remind the reader of the bijection that exists between subspaces W, with $S \leq W \leq T$, and subspaces of T/S (the bijection is simply $W \to W/S$). This observation allows us to write

$$\mu(S, T) = \mu(0, T/S) = (-1)^{t-s} q^{\binom{t-s}{2}},$$

where s is the dimension of S and t is that of T. The factor space T/S is of dimension $t - s$, which explains the last sign of equality above.

In conclusion:

* *The Möbius function on the subspaces of a vector space is* $\mu(S, T) = (-1)^k q^{\binom{k}{2}}$ *if S is of index q^k in T, and is* 0 *otherwise.*

We can now write the Möbius inversion formula of for the subspaces of a vector space as follows:

$$\textit{If } g(T) = \sum_{0 \leq S \leq T} f(S) \textit{ then}$$

$$f(T) = \sum_{0 \leq S \leq T} g(S)(-1)^{t-s} q^{\binom{t-s}{2}},$$

with S of dimension s and T of dimension t.

Recalling that the number of subspaces of dimension k in a vector space of dimension n over $GF(q)$ is $\left[{n \atop k}\right](q)$ and observing that the Möbius function is a function of dimension only, we can write the above Möbius inversion formula as

$$a_n = \sum_{k=0}^{n} \left[{n \atop k}\right](q) b_k \quad \text{iff} \quad b_n = \sum_{k=0}^{n} \left[{n \atop k}\right](q)(-1)^{n-k} q^{\binom{n-k}{2}} a_k,$$

for sequences of scalars (a_n) and (b_n). Many of the inversion formulas in Chapter 3 can be obtained in analogous ways, by working with suitably chosen partially ordered sets.

9.5 The Möbius Function of a Product

Let $(X_1, \leq); \zeta_1, \mu_1$ and $(X_2, \leq); \zeta_2, \mu_2$ be two partially ordered sets, with their respective zeta and Möbius functions. The $\leq$ signs are in general different, of course, but we retain the same symbol. We define a partial order

$\leq$ on $X_1 \times X_2$ as follows:

$$(x_1, x_2) \leq (y_1, y_2) \qquad \text{if } x_1 \leq y_1 \quad \text{and} \quad x_2 \leq y_2.$$

With this partial order we obtain $(X_1 \times X_2, \leq)$; ζ, μ. By definition $\zeta = \zeta_1\zeta_2$, that is, $\zeta((x_1, x_2), (y_1, y_2)) = \zeta_1(x_1, x_2)\zeta_2(y_1, y_2)$. We assert that also $\mu = \mu_1\mu_2$. Indeed,

$$\begin{aligned}
&\zeta\mu_1\mu_2((x_1, x_2), (y_1, y_2)) \\
&\quad = \sum_{(x_1, x_2) \leq (u_1, u_2) \leq (y_1, y_2)} \zeta_1(x_1, u_1)\zeta_2(x_2, u_2)\mu_1(u_1, y_1)\mu_2(u_2, y_2) \\
&\quad = \sum_{x_1 \leq u_1 \leq y_1} \zeta_1(x_1, u_1)\mu_1(u_1, y_1) \sum_{x_2 \leq u_2 \leq y_2} \zeta_2(x_2, u_2)\mu_2(u_2, y_2) \\
&\quad = \delta_{X_1}\delta_{X_2} = \delta_{X_1 \times X_2},
\end{aligned}$$

as desired. This proves:

Theorem 9.2. *The Möbius function of a product equals the product of the Möbius functions.*

The theorem we just proved is useful in finding the Möbius function for the partitions of a set. This is what we attempt in our next example.

Example 5. Let X be the list of all partitions of a set with n elements; $|X| = B_n$ (the Bell number). The partial order on X is as follows: For P and Q in X we write $P \leq Q$ if each class of P is union of classes of Q. We say that Q is a *refinement* of P. As an example,

$$(P =)\ 12/345 \leq 1/2/35/4\ (= Q).$$

We denote by $\hat{0}$ the partition $1, 2, \ldots, n$ (with 1 class only) and by $\hat{1}$ the partition $1/2/\cdots/n$ with n classes. For all P in X we have $\hat{0} \leq P$ and $P \leq \hat{1}$.

Let P_0 be a *fixed* partition with m classes. One can label, for convenience, the classes of P_0 by $\bar{1}, \bar{2}, \ldots, \bar{m}$ and then $P \leq P_0$ simply means that P is a partition of $\{\bar{1}, \bar{2}, \ldots, \bar{m}\}$ (this is an important thought process in what we do next). For any real number x, look at $\sum_{P,\, P \leq P_0} [x]_k$, where k denotes the number of classes of P. By visualizing P $(\leq P_0)$ as a partition of $\{\bar{1}, \bar{2}, \ldots, \bar{m}\}$ we can write

$$\sum_{\substack{P \\ P \leq P_0}} [x]_k = \sum_{k=1}^{m} S_m^k [x]_k = x^m,$$

the last equality being the well-known polynomial identity for the Stirling numbers S_m^k (see Section 1.7(c), if necessary). By the Möbius inversion formula:

$$x^m = \sum_{\substack{P \\ P \leq P_0}} [x]_k \qquad \text{implies} \qquad [x]_m = \sum_{\substack{P \\ P \leq P_0}} x^k \mu(P, P_0) \tag{9.2}$$

(for $x = 1, 2, 3, \ldots$, say; but then the above equality holds for all real numbers x, because the two polynomials take the same values on all positive integers. Statement (9.2) becomes therefore a polynomial identity). By Section 1.8(c) we know that

$$[x]_m = \sum_{k=1}^{m} s_m^k x^k. \tag{9.3}$$

From (9.2) and (9.3) we obtain

$$\sum_{\substack{P \\ P \leq P_0}} x^k \mu(P, P_0) = \sum_{k=1}^{m} s_m^k x^k.$$

Comparing the coefficients of x^k on both sides gives

$$\sum_{\substack{P \\ P \leq P_0}} \mu(P, P_0) = s_m^k \tag{9.4}$$

(P has exactly k classes).

Taking $P_0 = \hat{1}$ and $k = 1$ in (9.4) we obtain

$$\mu(\hat{0}, \hat{1}) = s_n^1 = (-1)^{n-1}(n-1)!.$$

We now proceed in computing $\mu(P, Q)$, for two partitions $P \leq Q$. Since $P \leq Q$, P is the union of classes of Q; let r_i be the number of classes of P made up of exactly i classes of Q, $1 \leq i \leq t$. (To visualize this better we may, as above, label the classes of Q by $\{\bar{1}, \bar{2}, \ldots, \bar{t}\}$; P is then assumed to be of type $1^{r_1} 2^{r_2} \cdots t^{r_t}$, with $\sum_{i=1}^{t} i r_i = t$.) Inspection of the segment $[P, Q]$, that is, $\{R\colon P \leq R \leq Q\}$, reveals that it is in fact the *direct product* of r_1 copies of the partitions of a set with one element, r_2 copies of the partitions of a set with two elements, $\ldots$, r_t copies of the partitions of a set with t elements. By Theorem 9.2 we can now write

$$\mu(P, Q) = \prod_{i=1}^{t} \left((-1)^{i-1}(i-1)!\right)^{r_i} = (-1)^{\sum_i^t (i-1) r_i} \prod_{i=1}^{t} \left((i-1)!\right)^{r_i}.$$

We obtained, therefore (recalling that $\sum_{i=1}^{t} i r_i = t$)

$$\mu(P, Q) = (-1)^{\sum_{i=1}^{t}(r_i - 1)} \prod_{i=1}^{t} \left((i-1)!\right)^{r_i},$$

where r_i is the number of classes of P consisting of the union of exactly i classes of Q. The Möbius inversion reads:

$$g(Q) = \sum_{P \leq Q} f(P) \qquad \textit{implies} \qquad f(Q) = \sum_{P \leq Q} g(P) \mu(P, Q)$$

with $\mu(P, Q)$ as above.

3 RESULTS OF WEISNER AND HALL

9.6

For α and β arithmetic functions, we now define their *sum* as follows: $(\alpha + \beta)(x, y) = \alpha(x, y) + \beta(x, y)$. We also define $\hat{0}$ and $\hat{1}$ as elements that satisfy $\hat{0} \leq x$ and $x \leq \hat{1}$, for all x in the partially ordered set. Such elements need not always exist, but whenever they do they are necessarily unique.

A *chain of length k* (or a chain with k edges) between x and y is a set of elements $u_1, u_1, \ldots, u_{k-1}$ satisfying

$$x < u_1 < u_2 < \cdots < u_{k-1} < y.$$

Assume $x < y$ and look at $(\zeta - \delta)^m(x, y)$. With ζ the zeta function and δ the identity, for $m = 1$ we have $(\zeta - \delta)(x, y) = 1 =$ number of chains of length 1 between x and y. When $m = 2$ we obtain $(\zeta - \delta)^2(x, y) = \sum_{x \leq u_1 \leq y}(\zeta - \delta)(x, u_1)(\zeta - \delta)(u_1, y) =$ number of chains of length 2 between x and y. Similarly,

$$(\zeta - \delta)^m(x, y) = \textit{number of chains of length } m \textit{ between } x \text{ and } y.$$

Theorem 9.3 (P. Hall). *Let X be a partially ordered set containing $\hat{0}$ and $\hat{1}$ ($\hat{0} \neq \hat{1}$). Then $\mu(\hat{0}, \hat{1}) = -c_1 + c_2 - c_3 + \cdots + (-1)^n c_n$, where c_k equals the number of chains of length k between $\hat{0}$ and $\hat{1}$, and n is the length of a longest chain between $\hat{0}$ and $\hat{1}$.*

Proof. Indeed,

$$\begin{aligned}
\mu(\hat{0}, \hat{1}) \\
&= \zeta^{-1}(\hat{0}, \hat{1}) = [\delta + (\zeta - \delta)]^{-1}(\hat{0}, \hat{1}) \\
&= \left[\delta - (\zeta - \delta) + (\zeta - \delta)^2 - (\zeta - \delta)^3 + \cdots\right](\hat{0}, \hat{1}) \\
&= \left[\delta - (\zeta - \delta) + (\zeta - \delta)^2 - (\zeta - \delta)^3 + \cdots + (-1)^n(\zeta - \delta)^n\right](\hat{0}, \hat{1}) \\
&= -c_1 + c_2 - c_3 + \cdots + (-1)^n c_n.
\end{aligned}$$

The last equality follows from the above interpretation of $(\zeta - \delta)^m(x, y)$. From the definition of n we have $(\zeta - \delta)^m(\hat{0}, \hat{1}) = 0$ for all $m > n$. This ends the proof.

9.7

If for two elements x and y of a partially ordered set there exists an element z that satisfies: (i) $z \leq x$ and $z \leq y$, and (ii) for any w that satisfies $w \leq x$ and $w \leq y$ we also have $w \leq z$, then we call z the *meet* of x and y (notation: $x \wedge y$). If an element z exists and satisfies: (i) $x \leq z$, $y \leq z$, and (ii) $x \leq w$ and $y \leq w$ implies $z \leq w$, then we call z the *join* of x and y (notation: $x \vee y$). Note that *the meet and join are unique whenever they exist.*

We call $(L, \leq)$ a *lattice* if each pair of elements in L has a meet and a join. For the remainder of Section 3 we reserve the notation $(L, \leq)$ for a lattice with $\hat{0}$ and $\hat{1}$; $\hat{0}$ and $\hat{1}$ are defined by: $\hat{0} \leq x$ for all $x \in L$, and $x \leq \hat{1}$ for all $x \in L$. For a lattice $(L, \leq)$ we define its *dual lattice* $(L, \leq^*)$ as the lattice with the same elements as L and with $\leq^*$ defined by: $x \leq^* y$ if and only if $x \geq y$ in L. The Möbius function for $(L, \leq^*)$ is $\mu^*(x, y) = \mu(y, x)$ as may easily be verified. The join (respectively meet) of x and y in $(L, \leq^*)$ is the meet (respectively joint) of x and y in $(L, \leq)$. Any segment $[x, y]$ of a lattice is itself a lattice (with x as its $\hat{0}$ and y as its $\hat{1}$). A finite lattice L always has $\hat{0}$ and $\hat{1}$: $\hat{0} = \bigwedge_{z \in L} z$ and $\hat{1} = \bigvee_{z \in L} z$.

In a lattice (and in any partially ordered set, actually) we say that *a covers b* if $b < a$ and there is no c for which we can write $b < c < a$. An *x-atom* is an element y that covers x. An element y in a lattice is said to be the *join of x-atoms* if $y = y_1 \vee y_2 \vee \cdots \vee y_n$ and each y_i is an x-atom. Element y is said to be an *x-coatom* if x covers y. We say that y is the *meet of x-coatoms* if $y = y_1 \wedge y_2 \wedge \cdots \wedge y_n$, for y_i's x-coatoms.

A finite partially ordered set may often be conveniently displayed by way of a *Hasse diagram*: Draw a vertex to represent 0, and work your way "up" by placing an edge between x and y whenever y covers x.

Theorem 9.4 (P. Hall). *Let $(L, \leq)$ be a lattice with Möbius function μ. Then*:

(a) $\mu(x, x) = 1$ *and if $x \neq y$ then $\mu(x, y) = 0$, unless y is a join of x-atoms.*
(b) *Dually, $\mu(x, x) = 1$ and if $x \neq y$, then $\mu(y, x) = 0$, unless y is a meet of x-coatoms.*

Proof. (a) By definition $\mu(x, x) = 1$. Let z be an x-atom. Then $\mu(x, x) + \mu(x, z) = 0$, and hence $\mu(x, z) = -1$. We now proceed (by induction) as follows: Assume that the theorem holds for all u such that $x \leq u < y$ and then prove it for y.

Assume y is *not* a join of x-atoms. Let $z_1, z_2, \ldots, z_n$ be *all* the x-atoms less than y and put $z = z_1 \vee z_2 \vee \cdots \vee z_n$. Then $z < y$. From the definition of z it is also clear that no w that satisfies $z < w < y$ can be a join of x-atoms. By the inductive assumption $\mu(x, w) = 0$ for all such w, and therefore

$$0 = \sum_{x \leq u \leq y} \mu(x, y) = \mu(x, y) + \sum_{x \leq u < y} \mu(x, u)$$

$$= \mu(x, y) + \sum_{x \leq u \leq z} \mu(x, u) = \mu(x, y) + 0 = \mu(x, y).$$

The fact that $\mu(x, w) = 0$ is used in explaining the third equality sign above.

Part (b) is dual to (a) and is proved in the same way by interchanging the notions of meet and join. This ends the proof.

The computation of $\mu(d, n)$ for the lattice of natural numbers partially ordered by divisibility (see Example 2) becomes more transparent if one uses

Theorem 9.4. It is not unlikely that Theorem 9.4 was inspired by this special case, and it may easily be remembered by recalling this case.

9.8

We continue our assortment of examples by computing the Möbius function of the lattice of subgroups of a p-group.

Example 6. A finite group G is called a *p-group* if $|G| = p^n$, for p a prime number. We now compute the Möbius function of a p-group (to be exact, the Möbius function of the lattice of subgroups of a p-group partially ordered by inclusion on subgroups). The case in which the p-group G is *elementary Abelian* (i.e., a vector space of finite dimension over Z_p) has been examined in Example 4 of Section 2 where we established that

$$\mu(H, G) = \mu(0, G/H) = (-1)^k p^{\binom{k}{2}},$$

with H a subgroup of index p^k in G.

In order to find the Möbius function of a general p-group we need to become sufficiently well acquainted with the lattice of its subgroups. The reader is referred to Section 4 of Appendix 1 for a list of helpful results.

Let G be a p-group of order p^n. Perhaps the single most important thing about such a group is that *its center $Z(G)$ is nontrivial, that is,* $|Z(G)| \geq p$. (The *center* consists, by definition, of elements z that commute with all elements of G.) It is easy to show that $Z(G) \triangleleft G$, that is, the center is normal in G. The nontriviality of $Z(G)$ allows opportunity for much inductive work by passing to quotient groups. All books on finite groups prove the nontriviality of $Z(G)$, in particular [8]; we take this result for granted here.

A subgroup M of G is called *maximal* if no proper subgroup of G contains M properly. The maximal subgroups are the coatoms of G. We assert that a maximal subgroup M is normal in G and $|M| = p^{n-1}$. [That $|M| = p^{n-1}$ can be proved by induction on n working in $G/\langle z \rangle$, $z \in Z(G)$; $z^p = 1$. By representing G on the left cosets of M we obtain a homomorphism from G into the symmetric group on p letters. The kernel of this homomorphism is precisely M since p divides $p!$ and no higher power of p does. Hence M is normal in G.]

Denote by $\Phi(G)$ the intersection of all maximal subgroups of G. The subgroup $\Phi(G)$ is normal in G and is called the *Frattini* (or *principal*) *subgroup* of G. For a maximal subgroup M denote by $\bar{G}$ the factor group G/M and by $\bar{g}$ the image of g in $\bar{G}$. Since $M \triangleleft G$ and $|M| = p^{n-1}$ it follows that $\bar{G} \cong Z_p$. Hence $\bar{1} = \bar{g}^p = \overline{g^p}$ and therefore $g^p \in M$, for all g in G. Since $[G, G]$, the commutator subgroup of G, is the smallest normal subgroup of G with Abelian quotient group, it follows that also $[G, G] \leq M$. This being true of an arbitrary maximal subgroup M, we conclude that $[G, G] \leq \Phi(G)$ and that $g^p \in \Phi(G)$, for all g in G. The quotient group $G/\Phi(G)$ is therefore Abelian and has all its elements of order p; it is therefore an elementary

Abelian group. To summarize:

The meet (or intersection) of all coatoms of G is the Frattini subgroup $\Phi(G)$, and $G/\Phi(G)$ is an elementary Abelian p-group. (9.5)

The subgroups of G form a lattice, with the meet of two subgroups being their intersection, and the join being the subgroup that they generate. By Hall's result, Theorem 9.4(b), we now conclude that $\mu(H,G) = 0$ unless H is a meet of coatoms. In other words, $\mu(H,G) = 0$, unless $\Phi \leq H$ (since Φ is the meet of all coatoms). A subgroup H that contains Φ is called a *major subgroup* of G. It is clear that major subgroups are normal in G, since G/Φ is Abelian (and thus has all its subgroups normal). For a major subgroup H the quotient G/H is an elementary Abelian group (since $G/H \cong \overline{G}/\overline{H}$, with $\overline{G} = G/\Phi$ and $\overline{H} = H/\Phi$, and since (9.5) informs us that $\overline{G}$ is itself elementary Abelian). Thus

$$\mu(H,G) = \mu(0, G/H) = (-1)^k p^{\binom{k}{2}},$$

if H is a major subgroup of index p^k in G. We have therefore proved the following result.

Proposition 9.1. *The Möbius function of the lattice of subgroups of a finite p-group G is $\mu(K,G) = 0$, unless K is a major subgroup of index p^k in G, in which case*

$$\mu(K,G) = (-1)^k p^{\binom{k}{2}}.$$

9.9

A theorem due to Weisner proves quite helpful when computing the Möbius function of a lattice. We first state and prove it, and then illustrate how it can be used in the lattice of subspaces of a vector space.

Theorem 9.5 (Weisner). *Let $(L, \leq)$ be a finite lattice with $\hat{0} \neq \hat{1}$. Then:*

(i) *For any a in L $(a \neq \hat{0})$*

$$\sum_{\substack{x \\ x \vee a = \hat{1}}} \mu(\hat{0}, x) = 0.$$

(ii) *For any a in L $(a \neq \hat{1})$*

$$\sum_{\substack{x \\ x \wedge a = \hat{0}}} \mu(x, \hat{1}) = 0.$$

Proof. Let $a \neq \hat{0}$ be fixed. For $u \geq a$ define

$$f(u) = \sum_{\substack{x \\ x \vee a = u}} \mu(\hat{0}, x),$$

and for all other elements u in L define $f(u) = 0$. We prove that in fact $f(u) = 0$ for *all* u in L. In particular $f(\hat{1}) = 0$, which is what statement (i) asserts.

The proof is by induction on $l(\hat{0}, u)$, the maximum length of a chain from $\hat{0}$ to u. If $l(\hat{0}, u) = 1$, then $f(u) = 0$ unless $l(\hat{0}, a) = 1$ and $u = a$, in which case still

$$f(u) = \mu(\hat{0}, \hat{0}) + \mu(\hat{0}, u) = 0,$$

by the definition of the Möbius function μ.

Assume that for $l(\hat{0}, u) \le n - 1$ we have $f(u) = 0$. Then for u with $l(\hat{0}, u) = n$ we can write

$$0 = \sum_{\hat{0} \le x \le u} \mu(\hat{0}, x) = \sum_{a \le y \le u} \sum_{\substack{x \\ x \vee a = y}} \mu(\hat{0}, x)$$

$$= \sum_{a \le y \le u} f(y) = f(u) + \sum_{a \le y < u} f(y).$$

(The second sign of equality is explained by sorting the elements x in the interval $[\hat{0}, u]$ by the joins y that they make with a.) But each term $f(y)$ in the last sum is zero, by the inductive assumption, since for y such that $a \le y < u$ we have $l(\hat{0}, y) \le n - 1$. It follows that $f(u) = 0$. By taking $u = \hat{1}$ we obtain (i).

Statement (ii) is the dual of (i) and can be proved similarly by replacing joins with meets. This ends the proof of Weisner's theorem.

The computation of the Möbius function for the lattice of subspaces of a vector space (Example 4) and for the lattice of partitions (Example 5) becomes easier if Weisner's result is used. In the former case, for instance, let μ_n denote $\mu(0, V)$ for a vector space V of dimension n over $GF(q)$. Let a be a fixed subspace of V of dimension $n - 1$. We look at subspaces x of V such that $x \wedge a = 0$. Such x is either 0 or a one-dimensional subspace outside of a. There exist $q^n - 1 - (q^{n-1} - 1) = (q - 1)q^{n-1}$ nonzero vectors (and hence q^{n-1} one-dimensional subspaces) outside a. Weisner's theorem gives

$$0 = \sum_{\substack{x \\ x \wedge a = 0}} \mu(x, V) = \mu(0, V) + \sum_{\substack{x \\ x \wedge a = 0 \\ x \ne 0}} \mu(x, V) = \mu_n + q^{n-1}\mu_{n-1},$$

which establishes the recurrence $\mu_n = -q^{n-1}\mu_{n-1}$. [We used the fact that $\mu(x,V) = \mu_{n-1}$, by working in the quotient space $V/\langle x \rangle$.] Since $\mu_1 = -1$ the recurrence inductively leads to

$$\mu_n = \mu(0, V) = (-1)^n q^{\binom{n}{2}},$$

the familiar expression for the Möbius function of this lattice.

9.10

We now share with the reader the interesting combinatorial description of the Möbius function of a lattice due to Weisner.

Let $(L, \le)$ be a lattice with Möbius function μ. For any segment $[x, y]$ with $x < y$ denote by $P[x, y]$ the *set* of elements in $[x, y]$ possessing property

P. We assume that P satisfies the following three conditions:

$$\begin{aligned}
&\text{(i) } y \notin P[x, y]. \\
&\text{(ii) } P[x, y] \neq \varnothing, \text{ for all } [x, y] \text{ with } x < y. \\
&\text{(iii) } [u, y] \subseteq [x, y] \text{ implies } P[u, y] = [u, y] \cap P[x, y].
\end{aligned} \tag{9.6}$$

Soon we see actual examples of such properties P.

For $x < y$ define a nonnegative integer $Q_k(x, y)$ by setting

$$\begin{aligned}
Q_k(x, y) = |\{&\text{subsets with } k \text{ elements } z_1, z_2, \ldots, z_k \\
&\text{from } [x, y] \text{ having property } P \text{ and} \\
&\text{satisfying } z_1 \wedge z_1 \wedge \cdots \wedge z_k = x\}|.
\end{aligned}$$

In addition, define a function $\bar{\mu}$ on $L \times L$ as follows:

$$\begin{aligned}
\bar{\mu}(x, x) &= 1, \qquad \text{for all } x \text{ in } L \\
\bar{\mu}(x, y) &= \sum_{k \geq 1} Q_k(x, y), \qquad \text{for } x < y \\
\bar{\mu}(x, y) &= 0, \qquad \text{otherwise.}
\end{aligned}$$

(The reader will observe that the sum in the definition of $\bar{\mu}$ is actually finite, for any segment contains only finitely many elements.)

Theorem 9.6 (Weisner). *The function $\bar{\mu}$ defined above is the Möbius function of the lattice $(L, \leq)$.*

Proof. We prove that $\bar{\mu}$ (as defined above) satisfies $\zeta\bar{\mu} = \delta$, where ζ is the Riemann zeta function and δ is the identity function of the group of arithmetic functions. Since the Möbius function μ of $(L, \leq)$ satisfies by definition $\zeta\mu = \delta$, we conclude that $\bar{\mu} = \mu$, by the uniqueness of the inverse (see Section 9.2, if necessary).

What needs to be shown is that $(\zeta\bar{\mu})(x, y) = 0$, if $x < y$. Indeed,

$$\begin{aligned}
(\zeta\bar{\mu})(x, y) &= \sum_{x \leq u \leq y} \zeta(x, u)\bar{\mu}(u, y) = \sum_{x \leq u \leq y} \bar{\mu}(u, y) \\
&= \bar{\mu}(y, y) + \sum_{x \leq u < y} \bar{\mu}(u, y) \\
&= 1 + \sum_{x \leq u < y} \sum_{k \geq 1} (-1)^k Q_k(u, y) \\
&= 1 + \sum_{k \geq 1} (-1)^k \sum_{x \leq u < y} Q_k(u, y) \\
&= 1 + \sum_{k \geq 1} (-1)^k \sum_{x \leq u < y} |\{\{z_1, z_2, \ldots, z_k\} \subseteq [u, y] \\
&\qquad \text{with property } P \text{ and satisfying } z_1 \wedge z_2 \wedge \cdots \wedge z_k = u\}| \\
&= 1 + \sum_{k \geq 1} (-1)^k |\{\{z_1, z_2, \ldots, z_k\} \subseteq [x, y] \text{ with property } P\}| \\
&= 1 + \sum_{k \geq 1} (-1)^k \binom{N}{k} = (1 - 1)^N = 0,
\end{aligned}$$

where $N(\geq 1)$ denotes the total number of elements in $[x, y]$ with property P. To explain the fourth equality sign from the end up we use all three defining characteristics of property P displayed in (9.6). It is particularly easy to see how (i) is used when $k = 1$, for example. This shows that $\zeta\bar{\mu} = \delta$ and we may conclude that $\bar{\mu}$ is indeed the Möbius function of $(L, \leq)$. The proof is now complete.

On the lattice of subgroups of a p-group G we can define a property P as follows: for $x < y$ let $P[x, y]$ be the set of subgroups of G containing x and maximal in y. This P satisfies conditions (i), (ii), and (iii) listed in (9.6). Moreover, if G is not elementary Abelian the intersection of any number of maximal subgroups in G contains the nontrivial Frattini subgroup of G, and hence is not the identity of the group; see (9.5). This translates directly into $Q_k(\hat{0}, \hat{1}) = 0$ for all $k \geq 1$. Theorem 9.6 leads us now to conclude that $\mu(\hat{0}, \hat{1}) = \Sigma_{k \geq 1} Q_k(\hat{0}, \hat{1}) = 0$, with $\hat{1} = G$, which is in agreement with Proposition 9.1.

A result closely related to Theorem 9.6 is due to Rota [5].

Proposition 9.2 (Rota). *Let $(L, \leq)$ be a finite lattice, with $\hat{0} \neq \hat{1}$. If M is a subset of L containing all the coatoms and not containing $\hat{1}$, then*

$$\mu(\hat{0}, \hat{1}) = \sum_{k \geq 1} (-1)^k Q_k,$$

where Q_k denotes the number of k-subsets of M with meet $\hat{0}$. (We remind the reader that a coatom is an element of the lattice covered by $\hat{1}$.)

Proof. The initial condition $\mu(\hat{1}, \hat{1}) = 1$ and the equation

$$\sum_{x \leq u \leq \hat{1}} \mu(u, \hat{1}) = 0$$

determine uniquely the values $\mu(x, \hat{1})$ of the Möbius function μ of $(L, \leq)$, for all $x < \hat{1}$.

Define now $\bar{\mu}$ by setting $\bar{\mu}(\hat{1}, \hat{1}) = 1$, and for $x < \hat{1}$ let

$$\bar{\mu}(x, \hat{1}) = \sum_{k \geq 1} (-1)^k Q_k(x, \hat{1}),$$

where $Q_k(x, \hat{1})$ denotes the number of k-subsets of M contained in the segment $[x, \hat{1}]$ and having meet x. We assert that for $x < 1$ the function $\bar{\mu}$ verifies the equation

$$\sum_{x \leq u \leq \hat{1}} \bar{\mu}(x, \hat{1}) = 0$$

(the same equation that μ satisfies). This allows us to conclude that $\bar{\mu}(x, \hat{1}) = \mu(x, \hat{1})$, for all $x \leq \hat{1}$. In particular $\bar{\mu}(\hat{0}, \hat{1}) = \mu(\hat{0}, \hat{1})$, which is what we seek to prove.

Indeed,

$$\begin{aligned}
\sum_{x \le u \le \hat{1}} \bar{\mu}(x, \hat{1}) &= \bar{\mu}(\hat{1}, \hat{1}) + \sum_{x \le u < \hat{1}} \bar{\mu}(u, \hat{1}) \\
&= 1 + \sum_{x \le u < \hat{1}} \sum_{k \ge 1} (-1)^k Q_k(u, \hat{1}) \\
&= 1 + \sum_{k \ge 1} (-1)^k \sum_{x \le u < \hat{1}} Q_k(u, \hat{1}) \\
&= 1 + \sum_{k \ge 1} (-1)^k \sum_{x \le u < \hat{1}} |\{k\text{-subsets of } M \\
&\qquad\qquad \text{in } [u, \hat{1}] \text{ with meet } u\}| \\
&= 1 + \sum_{k \ge 1} (-1)^k |\{k\text{-subsets of } M \text{ in } [x, \hat{1}]\}| \\
&= 1 + \sum_{k \ge 1} (-1)^k \binom{N}{k} = (1 - 1)^N = 0,
\end{aligned}$$

as stated. By N we denote the total number of elements of M in $[x, \hat{1}]$.
The proposition is proved by taking $x = \hat{0}$ and observing that $Q_k(\hat{0}, \hat{1}) = Q_k$.

EXERCISES

1. For simplicity we write 123 for the set $\{1, 2, 3\}$. Draw the Hasse diagrams for the sets ordered by inclusion given below:

(a) $\{1, 12, 13, 1234, 1235, 12345\}$.

(b) $\{1, 12, 13, 123, 124, 134, 1234\}$.

(c) $\{1, 12, 13, 19, 124, 126, 135, 134, 1268, 1267, 1356\}$.

(d) $\{1, 12, 14, 13, 124, 145, 134, 1246, 1456, 1346, 123456\}$.

Which of these four partially ordered sets are lattices? Calculate the Möbius function in each case.

2. Consider the group $Z_2 \oplus Z_2$ written as a permutation group on four letters: $\{(1), (12), (34), (12)(34)\}$. Display the lattice of its subgroups. Compute the Möbius function of this lattice. Is it in agreement with Example 4 of Section 2?

3. The *dihedral group* D_8 (of order 8) consists of eight permutations on the four vertices of a square which preserve its four edges. Find the eight permutations in question. Determine the lattice of subgroups of D_8 (there are ten subgroups in all). Compute the Möbius function of this lattice. [Verify Proposition 9.1 by computing the Frattini subgroup of D_8 and identifying the major subgroups.]

4. Find the Möbius function for the lattice of subgroups of the following groups:

(a) S_3, the symmetric group on three letters; $|S_3| = 6$.

(b) A_4, the (alternating) group of even permutations on four letters; $|A_4| = 12$.

(c) A_5, the alternating group on five letters; $|A_5| = 60$.

5. Show that the operations $\wedge$ and $\vee$ of meet and join in a lattice obey the following rules:

(a) $(x \vee y) \vee z = x \vee (y \vee z)$

(b) $x \vee x = x$

(c) $x \vee y = y \vee x$

(d) $x \vee (x \wedge y) = x$,

along with the dual rules obtained by replacing $\vee$ by $\wedge$ in (a) through (d). Conversely, any set with two operations $\wedge$ and $\vee$ that obey rules (a) through (d) and their duals is necessarily a lattice. [*Hint*: Write $x \leq y$ if $y = x \vee y$.]

6. Consider the set of all partitions of a finite set. For two partitions P and Q write $P \leq Q$ if each class of P is union of classes of Q. Convince yourself that the set of partitions with this partial order forms a lattice. Describe the meet and join.

7. Let f be a real-valued function defined on the set of partitions of a finite set and assume that it has the following property: $f(P) = a_k$, for all partitions P with k classes. Partially order the partitions by refinement (see Example 4 in Section 2) and write out the Möbius inversion formula for such a function f.

8. We say that a permutation on n letters has cycle type $(\lambda_1, \lambda_2, \ldots, \lambda_n)$, with $\lambda_1 \geq \lambda_2 \geq \cdots \geq \lambda_n \geq 0$ integers satisfying $\Sigma_{i=1}^{n} \lambda_i = n$, if its first cycle is of length λ_1, its second cycle is of length $\lambda_2, \ldots,$ its nth cycle is of length λ_n. Some of the λ_i's may be 0. For example, the permutation (324)(15)(67) on seven letters has cycle type $(3, 2, 2, 0, 0, 0, 0)$. Consider the set X_n of all cycle types on n letters. Partially order X_n as follows: For two cycle types $(\lambda_1, \ldots, \lambda_n)$ and $(\mu_1, \ldots, \mu_n)$ write $(\lambda_1, \ldots, \lambda_n) \leq (\mu_1, \ldots, \mu_n)$ if $\Sigma_{i=1}^{k} \lambda_i \leq \Sigma_{i=1}^{k} \mu_i$, for $k = 1, 2, \ldots, n$. (Observe that for $k = n$ we always have equality.) Find the Möbius function for $(X_n, \leq)$. Do so at least for $n = 7$ and $n = 8$.

9. Let L be a finite lattice and $\{A, B, C\}$ a partition of L such that if $x \in A$ and $y \leq x$, then $y \in A$; on the other hand if $x \in C$ and $x \leq y$, then

$y \in C$. Show that

$$1 + \sum_{x \in A} \sum_{y \in C} \mu(x, y) = \sum_{x, y \in B} \mu(x, y),$$

where μ is the Möbius function of L.

10. Let C be a set of pairwise incomparable elements of the lattice L such that every maximal chain contains an element of C. Denote by q_k the number of those k-tuples from C whose join is 1 and whose meet is 0. Show that $\mu(0,1) = \sum_{k=0}^{\infty}(-1)^k q_k$, where μ is the Möbius function of L.

4 COUNTING WITH THE MÖBIUS FUNCTION

9.11

Many problems in counting can be solved through the process of Möbius inversion. In Section 2 we computed the Möbius function for the lattice of subsets of a set, Example 3, and that of subspaces of a vector space over a finite field, Example 4. Relying on this information we now illustrate how one can use the Möbius function as a computational device to obtain answers to a large variety of problems in combinatorics.

Two Examples

Example 1. (Le Problème des Rencontres.) Let T be a subset of $\{1, 2, \ldots, n\}$. *How many permutations leave every point in T fixed and fix no other point outside T*? That is, how many permutations fix *exactly* T pointwise?

Let $f(t)$ denote the number of permutations that fix *exactly* T pointwise. We want to compute $f(t)$. Observe, however, that $\sum_{S, S \supseteq T} f(S)$ is much easier to calculate. Indeed, $\sum_{S, S \supseteq T} f(S)$ equals the number of permutations that leave *at least* T fixed pointwise. But so long as T is fixed pointwise every permutation on the points outside T is counted in this sum. Hence

$$\sum_{\substack{S \\ S \supseteq T}} f(S) = (n - |T|)! \; (= g(T), \text{say}).$$

The Möbius inversion formula gives us at once

$$f(T) = \sum_{\substack{S \\ S \supseteq T}} g(S)\mu(S, T) = \sum_{\substack{S \\ S \supseteq T}} (n - |S|)!(-1)^{|S|-|T|}.$$

[Let us explain how we think here: we let $(L, \leq)$ be the lattice of subsets of $\{1, 2, \ldots, n\}$ with $S \leq T$ if and only if $S \supseteq T$; denote by μ its Möbius function. Then $(L, \supseteq)$ is the lattice *dual* to the lattice of subsets of $\{1, 2, \ldots, n\}$ with the usual order $P \subseteq R$ and Möbius function μ^*, say. By the definition of μ in terms of μ^* for $S \supseteq T$ we have $\mu(S, T) = \mu^*(T, S) = (-1)^{|S|-|T|}$. The actual computation of μ^* was carried out in Example 3 of Section 2.]

We can make our last expression look nicer by taking into account the special structure of the lattice of subsets. Denote $|T| = t$ and $|S| = k$. Then

$$f(T) = \sum_{\substack{S \\ S \supseteq T}} (n - |S|)!(-1)^{|S|-|T|} = \sum_{k=t}^{n} \binom{n-t}{k-t}(n-k)!(-1)^{k-t},$$

which gives the solution to "le problème des rencontres."

[This problem is often formulated for $T = \varnothing$, the empty set. The answer in that case becomes $f(\varnothing) = n!\sum_{k=0}^{n}(-1)^k/k!$. In probabilistic terms, if each permutation on n points carries equal probability, then the chance of picking a permutation with no fixed points is $\sum_{k=0}^{n}(-1)^k/k! \cong 1/e$, for reasonably large n, say $n \geq 10$.]

We may also answer the following: *How many permutations of* $\{1, 2, \ldots, n\}$ *have exactly* t *fixed points*? And the answer is, of course, $\sum_{T, |T|=t} f(T)$. A rewriting gives

$$\sum_{\substack{T \\ |T|=t}} f(T) = \binom{n}{t} \sum_{k=t}^{n} \binom{n-t}{k-t}(n-k)!(-1)^{k-t},$$

since $f(T)$ is only a function of $|T| = t$.

Example 2. (Counting Certain Kinds of Subspaces.) Let V be an n-dimensional vector space over $GF(q)$ and let W be a fixed subspace of dimension l. *Show that the number of k-dimensional subspaces of V that intersect W in a specific subspace of dimension t is given by*

$$\sum_{s=t}^{l} \begin{bmatrix} l-t \\ s-t \end{bmatrix}(q) \begin{bmatrix} n-s \\ k-s \end{bmatrix}(q)(-1)^{s-t} q^{\binom{s-t}{2}}.$$

Find also the number of k-dimensional subspaces of V that intersect W in a subspace of dimension (exactly) t.

To solve this problem, let $T \leq W$ be a subspace of W of dimension t. Define

$f(T)$ = number of k-dimensional subspaces of V that intersect W in *exactly* T.

Observe that

$$\sum_{\substack{S \\ W \geq S \geq T}} f(S) = \text{number of } k\text{-dimensional subspaces of } V \text{ whose intersection with } W \text{ contains } T$$

$$= \text{number of } k\text{-dimensional subspaces of } V \text{ that contain } T$$

$$= \begin{bmatrix} n-t \\ k-t \end{bmatrix}(q) = g(T), \text{ say.}$$

[The $\begin{bmatrix} m \\ j \end{bmatrix}(x)$ are Gaussian polynomials.] Möbius inversion formula now gives

$$f(T) = \sum_{\substack{S \\ W \geq S \geq T}} g(S)\mu(S,T) = \sum_{\substack{S \\ W \geq S \geq T}} \begin{bmatrix} n - |S| \\ k - |S| \end{bmatrix}(q)(-1)^{|S|-t} q^{\binom{|S|-t}{2}}$$

$$= \sum_{s=t}^{l} \begin{bmatrix} l - t \\ s - t \end{bmatrix}(q) \begin{bmatrix} n - s \\ k - s \end{bmatrix}(q)(-1)^{s-t} q^{\binom{s-t}{2}},$$

where s is the dimension of the subspace S of W.

[An explanatory note is perhaps due here as well. We are in actuality working with the lattice dual to that of the subspaces of V with usual subspace inclusion, $\leq$, as partial order. Its Möbius function was computed in Example 4 of Section 2.]

Note again that, as in Example 1 in this Section 9.11, $f(T)$ is only a function of the dimension t of T. It follows that the number of k-dimensional subspaces of V that intersect W in a subspace of dimension precisely t is

$$\sum_{\substack{T \\ \dim T = t}} f(T) = \begin{bmatrix} l \\ t \end{bmatrix}(q) \sum_{s=t}^{l} \begin{bmatrix} l - t \\ s - t \end{bmatrix}(q) \begin{bmatrix} n - s \\ k - s \end{bmatrix}(q)(-1)^{s-t} q^{\binom{s-t}{2}}.$$

9.12

To understand the counting techniques that follow, we remind the reader of a fundamental convention: When we list the elements of a set we always list them without repetitions, that is, each element is listed precisely once. It is apparent, therefore, that the number of elements in the union of two sets equals the sum of the number of elements in each, minus the number of common elements; the minus occurs because of the convention we just invoked. Counting the cardinality of the union in terms of cardinalities of intersections is known as the principle of inclusion-exclusion. We generalize this principle roughly as follows. Suppose that finite sets A_i are defined by saying that an element belongs to A_i if it has property "i." The sets A_i may in general overlap a lot. If often happens that we become interested in the cardinality of the set of elements with *precisely* p of these properties. This problem leads to expressions known as sieve formulas. We derive these formulas by using Möbius inversion on the lattice of subsets of a set. Similar formulas may be established on other partially ordered sets in analogous ways.

Sieve Formulas

Let X be a finite set and $m\colon X \to [0, \infty)$ a function. Call $m(x)$ the *measure of* $x \in X$. For $A \subseteq X$ define

$$m(A) = \sum_{x \in A} m(x) \qquad \text{if} A \neq \varnothing,$$

and

$$m(A) = 0 \qquad \text{if } A = \varnothing.$$

Call $m(A)$ the *measure of* A. [In most cases we look at $m(A)$ will be $|A|$, the cardinality of A.]

Let $\{A_i\}$ be subsets of X, $i \in \{1, 2, \ldots, q\} = Q$. Establish notation as follows:

$$\underline{m}(K) = m\Big(\bigcap_{i \in K} A_i\Big) \qquad \text{if } \varnothing \neq K \subseteq Q,$$

$$\underline{m}(K) = m(X) \qquad \text{if } K = \varnothing,$$

and

$$\overline{m}(K) = m\Big(\bigcup_{i \in K} A_i\Big) \qquad \text{if } \varnothing \neq K \subseteq Q,$$

$$\overline{m}(K) = 0 \qquad \text{if } K = \varnothing.$$

For $0 \leq p \leq q$ define also

$$E_q^p = m(\{x \in X\colon x \text{ belongs to } \textit{exactly } p \text{ subsets } A_i\}).$$

By its definition E_q^p is (of course) a function of p and q, but it is primarily a function of the subsets A_i themselves. We do not indicate this in the notation chiefly for reasons of esthetic appearance.

The figure below is a graphical attempt to describe the elements that occur in the definition of E_q^p:

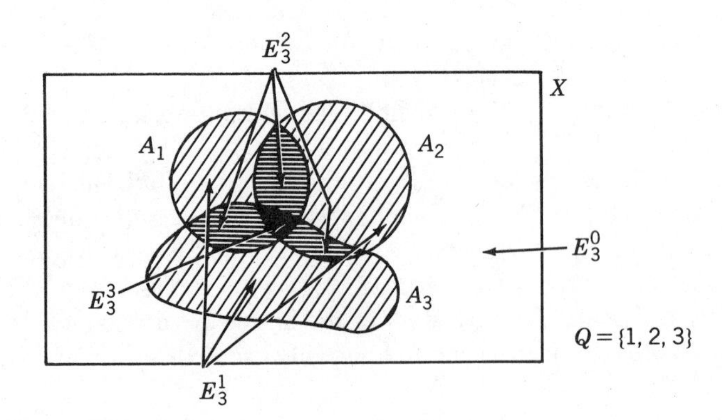

Our aim is to calculate E_q^p *in terms of measures of intersections*. We find such an expression for E_q^p using Möbius inversion.

Let $f(P) = m(\{x\colon x \text{ belongs to } \textit{exactly} \text{ the } A_i\text{'s with } i \in P\})$. The word "exactly" is of crucial importance here. We first want to find $f(P)$. But observe that

$$\sum_{\substack{K \\ K \supseteq P}} f(K) = \sum_{\substack{K \\ K \supseteq P}} m(\{x\colon x \text{ belongs to exactly the } A_i\text{'s with } i \in K\})$$

$$= m\Big(\bigcap_{i \in P} A_i\Big) = \underline{m}(P).$$

By Möbius' inversion formula we then also have

$$f(P) = \sum_{\substack{K \\ K \supseteq P}} \underline{m}(K)\mu(K, P) = \sum_{\substack{K \\ K \supseteq P}} \underline{m}(K)(-1)^{|K|-|P|}.$$

Moreover, observe that

$$E_q^p = m(\{x\colon x \text{ belongs to } \textit{exactly } p\ A_i\text{'s}\})$$

$$= \sum_{\substack{P \\ |P|=p}} f(P) = \sum_{\substack{P \\ |P|=p}} \sum_{\substack{K \\ K \supseteq P}} \underline{m}(K)(-1)^{|K|-|P|}.$$

This is the *sieve formula* for the lattice of subsets. Analogous formulas can be derived in other lattices. [Example 2 in Section 9.11 was an instance of a sieve formula on the lattice of subspaces of a vector space.]

We now simplify the above expression for E_q^p by interchanging the order of summation. The sums involved in the expression of E_q^p are over the set of pairs $\{(P, K)\colon |P| = p \text{ and } K \supseteq P\}$. By first fixing K, the second coordinate, and summing over P we explain the third equality sign below:

$$E_q^p = \sum_{\substack{P \\ |P|=p}} \sum_{\substack{K \\ K \supseteq P}} \underline{m}(K)(-1)^{|K|-|P|}$$

$$= \sum_{\substack{\text{pairs } (P, K) \\ \text{as above}}} \underline{m}(K)(-1)^{|K|-|P|}$$

$$= \sum_{\substack{K \\ |K| \geq p}} \sum_{\substack{P \\ K \supseteq P \\ |P|=p}} \underline{m}(K)(-1)^{|K|-|P|}$$

$$= \sum_{\substack{K \\ |K| \geq p}} \binom{|K|}{p}(-1)^{|K|-|P|}\underline{m}(K)$$

$$= \sum_{k=p}^{q} \sum_{\substack{K \\ |K|=k}} \binom{k}{p}(-1)^{k-p}\underline{m}(K)$$

$$= \sum_{k=p}^{q} \binom{k}{p}(-1)^{k-p} \sum_{\substack{K \\ |K|=k}} \underline{m}(K).$$

We have proved that

$$E_q^p = \sum_{k=p}^{q} \binom{k}{p} (-1)^{k-p} \sum_{\substack{K \\ |K|=k}} \underline{m}(K),$$

an expression known as *the sieve formula.*

The useful case with $p = 0$ describes the measure of elements that are in none of our sets and is known as *Sylvester's formula*:

$$E_q^0 = \sum_{k=0}^{q} (-1)^k \sum_{\substack{K \\ |K|=k}} \underline{m}(K).$$

Since E_q^0 is the measure of the elements that are in the complement of the union $\bigcup_{i=1}^{q} A_i$, we obtain

$$\overline{m}(Q) = m(X) - \sum_{k=0}^{q} (-1)^k \sum_{\substack{K \\ |K|=k}} \underline{m}(K) = \sum_{k=1}^{q} (-1)^{k+1} \sum_{\substack{K \\ |K|=k}} \underline{m}(K),$$

popularly known as the *principle of inclusion-exclusion.*

9.13

When counting, the sieve formulas we just proved arise quite often. We describe now several such instances. It is truly instructive to derive the results in all these examples by directly applying the process of Möbius inversion. In each instance this can be achieved by specializing the arguments in our derivation of the sieve formula.

More Examples

In what follows we denote by $\bar{A}$ the complement of the subset A in X, that is, the elements that are in X but not in A.

Example 3. (Sieve of Eratosthenes.) *We wish to construct the prime numbers up to n knowing those* $\leq \sqrt{n}$, *and to count how many prime numbers there are between* $\sqrt{n}$ *and* n. Let $p_1, p_2, \ldots, p_q$ be the list of primes $\leq \sqrt{n}$. Set $X = \{2, 3, \ldots, n\}$ and let $A_i = \{x \in X\colon\ p_i | x\}$, $i \in Q = \{1, 2, \ldots, q\}$. Observe that $\bar{A}_1 \cap \bar{A}_2 \cap \cdots \cap \bar{A}_q$ is the set of primes between $\sqrt{n}$ and n. (Indeed, let $m \in X$ not be a prime. Denote by p_m the smallest prime dividing m. Then $m \geq p_m^2$, hence $p_m \leq \sqrt{m}$ $(\leq \sqrt{n})$, and hence m is in one of our A_i's. The only numbers that are not in any of the A_i's are therefore the primes between $\sqrt{n}$ and n.)

This shows that we can obtain the list of primes between $\sqrt{n}$ and n by crossing out (or sieving out) all multiples of p_1 (up to n), then all multiples of p_2 (up to n), and so on. The numbers that are left are the ones we seek. An example with $n = 25$ is given below:

1	(2)	(3)	4	(5)
6	(7)	8	9	10
(11)	12	(13)	14	15
16	(17)	18	(19)	20
21	22	(23)	24	25

This is the classical sieve method for constructing primes due to Eratosthenes. Our general terminology is, of course, inspired by it.

As to counting the number of primes between $\sqrt{n}$ and n, by Sylvester's formula we obtain

$$|\bar{A}_1 \cap \cdots \cap \bar{A}_q| = \sum_{k=0}^{q} (-1)^k \sum_{\substack{K \subseteq Q \\ |K|=k}} m(K)$$

$$= \left\{ \text{here } \underline{m}(K) = \left| \bigcap_{i \in K} A_i \right| = \left[\frac{n}{\prod_{i \in K} p_i} \right] \right.$$

$$\left. \text{and } [y] \text{ denotes the integral part of } y \right\}$$

$$= \sum_{k=0}^{q} (-1)^k \sum_{\substack{K \subseteq Q \\ |K|=k}} \left[\frac{n}{\prod_{i \in K} p_i} \right].$$

9.14

The example that follows is another well-known application of the sieve formula. It provides us with a closed form expression for the Eulerian function ϕ.

Example 4. (The Eulerian Function) *Given n, find $\phi(n)$, the number of positive integers less than n that are relatively prime to n.*

Let $n = p_1^{\alpha_1} p_2^{\alpha_2} \cdots p_q^{\alpha_q}$ be the prime factorization of n. Let also $X = \{1, 2, \ldots, n\}$ and $A_i = \{x \in X\colon p_i | x\}$, with $i \in Q = \{1, 2, \ldots, q\}$. Then

$$\phi(n) = |\bar{A}_1 \cap \bar{A}_2 \cap \cdots \cap \bar{A}_q| = \{\text{by Sylvester's formula}\}$$

$$= \sum_{k=0}^{q} (-1)^k \sum_{\substack{K \subseteq Q \\ |K|=k}} \underline{m}(K) = \left\{ \text{here } \underline{m}(K) = \left| \bigcap_{i \in K} A_i \right| = \frac{n}{\prod_{i \in K} p_i} \right\}$$

$$= \sum_{k=0}^{q} (-1)^k \sum_{\substack{K \subseteq Q \\ |K|=k}} \frac{n}{\prod_{i \in K} p_i} = n\left(1 - \frac{1}{p_1}\right)\left(1 - \frac{1}{p_2}\right) \cdots \left(1 - \frac{1}{p_q}\right).$$

To see this last step, recall that $(x - a_1)(x - a_2) \cdots (x - a_q) = x^q - (\Sigma a_i)x^{q-1} + (\Sigma_{i<j} a_i a_j)x^{q-2} - \cdots \pm a_1 a_2 \cdots a_q x^0$. Let $x = 1$ and $a_i = 1/p_i$ to explain the above.

9.15

Our next example is somewhat more intricate, yet well worth musing over. It is french in origin and quite likely inspired by certain kinds of recreational activities.

Example 5. (Problème des Ménages.) We begin with a couple of lemmas.

Lemma 9.1. *The number of subsets of size k of $\{1, 2, 3, \ldots, n\}$ that contain no two consecutive numbers is* $\binom{n-k+1}{k}$.

Proof. Represent a k-subset by a vector with n entries (labeled $1, 2, \ldots, n$—in this order) having k 1's and $n - k$ 0's; for example, with $n = 6$ and $k = 3$ we have

$$\{2, 4, 5\} \leftrightarrow (0\ 1\ 0\ 1\ 1\ 0).$$

We should now count the number of vectors containing precisely k 1's, no two of which are next to each other. Consider $n - k$ digits of 0, labeled from 1 to $n - k$, and interpose k digits of 1 in such a way that no two of them are next to each other. Each digit of 1 can be characterized by the label given to the 0 immediately preceeding it (to an eventual digit of 1 in the first position we attach a label of $*$). We now see that the problem is the same as that of selecting k distinct labels from the set $\{*, 1, 2, \ldots, n - k\}$. And therefore the answer is $\binom{n-k+1}{k}$.

Lemma 9.2. *Assume* $1 \le 2 \le 3 \le \cdots \le n \le 1$ (cyclic order). *The number of subsets of size k of $\{1, 2, \ldots, n\}$, endowed with this cyclic order, that contain no two consecutive numbers is* $n(n-k)^{-1}\binom{n-k}{k}$.

Proof. The k-subsets that contain n cannot contain either $n - 1$ or 1, and by Lemma 1 there are $\binom{(n-3)-(k-1)+1}{k-1}$ of them. Those not containing n number $\binom{(n-1)-k+1}{k}$. A total of $\binom{n-k-1}{k-1} + \binom{n-k}{k} = n(n-k)^{-1}\binom{n-k}{k}$. This ends the proof of Lemma 9.2.

Consider now the following counting problem known as "Le problème des ménages":

Find the number of ways of seating n husbands (labeled $1, 2, \ldots, n$) and their respective wives (labeled $\bar{1}, \bar{2}, \ldots, \bar{n}$) at a circular table in such a way that each man has a woman seated on either side of him, neither being his wife.

Such a seating arrangement can be described by a bijection f: seat man 1; seat to his right woman $f(1)$; seat man 2; seat to his right woman $f(2)$; and so on. (Note that the counting is done modulo cyclic shifts around the table.)

Let A_{2i-1} ($i \in \{1,2,\dots,n\}$) be the set of bijections f satisfying $f(i) = \bar{i}$. For $i \neq n$ let A_{2i} be the set of bijections f satisfying $f(i) = \overline{i+1}$; for $i = n$, A_{2n} denotes the set of bijections f with $f(n) = \bar{1}$. We want the number T_{2n}^p of bijections that belong to exactly p of the A_i's; $i \in \{1,2,\dots,2n\} = Q$. (The problem asks in fact for T_{2n}^0 only.) By the sieve formula,

$$T_{2n}^p = \sum_{k=p}^{2n} (-1)^{k-p}\binom{k}{p} \sum_{\substack{K\subseteq Q\\|K|=k}} \underline{m}(K).$$

But

$$\underline{m}(K) = \left|\bigcap_{i\in K} A_i\right| = \begin{cases} (n-k)! & \text{if } K \text{ does not contain two consecutive integers from the cyclic sequence } (1,2,\dots,2n,1) \\ 0 & \text{otherwise} \end{cases}$$

For example, $|A_1| = (n-1)!$, $|A_1 \cap A_2| = 0$, $|A_1 \cap A_4| = (n-2)!$, and so on.

By Lemma 9.2 we know that the number of subsets K, with $|K| = k$, that contain no two consecutive numbers from the cyclic ordering $(1,2,\dots,2n,1)$ is $2n(2n-k)^{-1}\binom{2n-k}{k}$. Thus

$$\sum_{\substack{K\subseteq Q\\|K|=k}} \underline{m}(K) = \frac{2n}{2n-k}\binom{2n-k}{k}(n-k)!.$$

This gives

$$T_{2n}^p = \sum_{k=p}^{2n} (-1)^{k-p}\binom{k}{p}\frac{2n}{2n-k}\binom{2n-k}{k}(n-k)!.$$

When $p = 0$ we obtain the answer to "Le problème des ménages":

$$T_{2n}^0 = \sum_{k=0}^{2n} (-1)^k \frac{2n}{2n-k}\binom{2n-k}{k}(n-k)!.$$

[It may be helpful to recall, when evaluating these formulas, that $(-m)! = 0$ for $m = 1,2,\dots$.]

9.16

Graph theory is a great source of entertaining problems in counting and enumeration. Chapter 6 describes systematic techniques to approach these sorts of problems. Here we restrict attention to a result on graph colorations due to Whitney.

Example 6. (Counting Proper Colorations). We call G a *simple graph* if between any two distinct vertices of G there is at most one edge (no loops being allowed).

Let G be a simple graph. If we color each vertex of G with one of λ available colors we obtain a *coloration* of G. A coloration is said to be *proper* if whenever two vertices are joined by an edge they have different colors. (The smallest number of colors needed to properly color a graph is called the *chromatic number* of the graph.) Label by $1, 2, \ldots, n$ the vertices of G. Let $M_G(\lambda)$ denote the number of proper colorations of G that can be obtained with at most λ colors. Denote by $B(s, k)$ the number of subgraphs of G with k edges and s nontrivial connected components. (A *subgraph* of G is a graph using *all* the vertices but only some of the edges of G. A connected component is called nontrivial if it contains *at least one edge*.)

Whitney showed that $M_G(\lambda)$ is a polynomial in λ. Specifically:

∗ *The explicit form of the polynomial M_G is*

$$M_G(\lambda) = \sum_{k=0}^{q} (-1)^k \sum_{s=1}^{k} B(s, k)\lambda^s,$$

where q is the number of edges in G.

We now prove this result. Let X be the set of all colorations of G (proper or not); $|X| = \lambda^n$. For vertices i and j joined by an edge we define

$$A_{ij} = \text{colorations of } G \text{ that assign the same color to } i \text{ and } j$$

($A_{ij} \subseteq X$, and we have q such A_{ij}'s—one for each edge of G). Let $Q = \{\{i, j\} : \text{with } \{i, j\} \text{ an edge of } G\}$, $|Q| = q$. We thus wish to compute E_q^0, the number of colorations that are not in any of the A_{ij}'s. Using Sylvester's formula

$$M_G(\lambda) = E_q^0 = \sum_{k=0}^{q} (-1)^k \sum_{\substack{|K|=k \\ K \subseteq Q}} \underline{m}(K) = \sum_{K} (-1)^{|K|} \underline{m}(K),$$

with $\underline{m}(K) = |\bigcap_{\{i,j\} \in K} A_{ij}|$. But what is $\underline{m}(K)$, actually? Let $|K| = k$ and suppose K has $c(K)$ nontrivial connected components. Note that

$$\bigcap_{\{i,j\} \in K} A_{ij} = \text{colorations of } G \text{ that assign the same color to } i \text{ and } j, \text{ with } \{i, j\} \in K.$$

The edges in K are edges of a subgraph. It is clear that all vertices in a nontrivial connected component of K must have the same color (by sliding from one to the next on edges). Thus $\underline{m}(K) = |\bigcap_{\{i,j\} \in K} A_{ij}|$ counts the number of ways to color the $c(K)$ nontrivial connected components of K with λ colors (all points in a connected component of K having the same color). This number is $\lambda^{c(K)}$, since we have $c(K)$ objects to color and at most λ colors to use. We can now rewrite

$$M_G(\lambda) = E_q^0 = \sum_K (-1)^{|K|}\underline{m}(K) = \sum_K (-1)^{|K|}\lambda^{c(K)}$$

$$= \sum_{k=0}^{q} (-1)^k \sum_{s=1}^{k} B(s,k)\lambda^s.$$

The last equality emerges upon sorting out the terms in the sum by the possible values of $c(K)$ (which we index by s). This proves Whitney's result.

5. SURJECTIVE MORPHISMS

In 1948 Delsarte generalized the classical Möbius function in number theory to answer questions regarding the automorphisms of Abelian groups. As is well known (and quite easy to prove), a cyclic group of order n has $\phi(n)$ automorphisms, where ϕ is the Eulerian function that we discussed in Example (4) in Section 9.14. More generally now, if x is any finite Abelian group can we find the number of automorphisms of x? In addition, we may also ask: How many subgroups of a given isomorphism type does x contain? Möbius inversion is the central tool that leads to explicit formulas for both these questions.

The reader is reminded of the basic results on finite Abelian groups in our Section 9.17 of preliminaries. Delsarte's result is then presented. The proof is followed by an example and a detailed justification of some technical points that occur in the main proof.

An extension of these results to non-Abelian p-groups regular in the sense of P. Hall concludes the chapter.

9.17 Preliminaries

Letters x, y, and d denote Abelian groups. The group of complex numbers of modulus 1 (the *circle group*) is written as T.

Let x be an Abelian group. The set of all homomorphisms from x into T is denoted by $\mathrm{Hom}(x, T)$. For $h \in \mathrm{Hom}(x, T)$ we denote by $\langle g, h\rangle$ the image of the element $g \in x$ under the homomorphism h. For $h_1, h_2 \in \mathrm{Hom}(x, T)$ we let the product $h_1h_2 \in \mathrm{Hom}(x, T)$ be defined by $\langle g, h_1h_2\rangle = \langle g, h_1\rangle\langle g, h_2\rangle$. With this rule of multiplication $\mathrm{Hom}(x, T)$ becomes a group; an Abelian group, in fact, as can easily be checked. We denote the group $\mathrm{Hom}(x, T)$ by x^* and call it the *dual* of x.

[It is easy to see that *an Abelian group is the direct sum of its Sylow p-subgroups*, and it is a fundamental result that *an Abelian p-group is the direct sum of cyclic groups* (this decomposition being unique up to reorderings of the cyclic factors). *If* $x = \oplus x_i$ *and* $y = \oplus_j y_j$ *then, again, it is known that* $\mathrm{Hom}(x, y) \cong \oplus_i \oplus_j \mathrm{Hom}(x_i, y_j)$ (The symbol $\cong$ indicates isomorphism). By

Z_{p^m} we denote the cyclic group of order p^m. We refer the reader to Appendix 1 for a review of results in group theory.]

Let

$$x \cong Z_{p^{n_1}} \oplus \cdots \oplus Z_{p^{n_r}}$$

be an Abelian p-group. Then

$$\mathrm{Hom}(x, T) \cong \bigoplus_{i=1}^{r} \mathrm{Hom}(Z_{p^{n_i}}, T).$$

Let g be a generator of $Z_{p^{n_i}}$. If $w \in T$ is a homomorphic image of g, then $|w|$ divides $|g| = p^{n_i}$. There are precisely p^{n_i} such choices for $w \in T$, each being identified with a different homomorphism. Hence

$$\mathrm{Hom}(Z_{p^{n_i}}, T) \cong Z_{p^{n_i}}.$$

By using the above mentioned behavior of Hom with $\oplus$, we establish: $x^* \cong x$. The same arguments yield in fact the following more general conclusion:

The dual x^* of an Abelian group x is isomorphic to x.

By writing $\langle x, x^* \rangle$ we can visualize (not only x^* as the homomorphism group of x into T, but also) x as the group of homomorphisms of x^* into T, that is, we have a "natural" way of interpreting x as x^{**}.

Let d be a subgroup of x (we write $d \leq x$). Look at $d^* = \{h \in x^* : \langle d, h \rangle = 1\}$ (by $\langle d, h \rangle = 1$ we mean $\langle g, h \rangle = 1$ for all $g \in d$). The set d^* is a subgroup of x^*. We now prove:

Lemma 9.3. *The mapping $d \leftrightarrow d^*$ is a bijection and $d^* \cong x/d$.*

Proof. An element h of d^* defines a homomorphism $x/d \to T$ by $\langle gd, h \rangle = \langle g, h \rangle$; it is well defined since $\langle d, h \rangle = 1$. This gives a map from d^* into $(x/d)^*$. This map is in fact a homomorphism; clearly $\langle gd, h_1 h_2 \rangle = \langle gd, h_1 \rangle \langle gd, h_2 \rangle$, for all $g \in x$. Further, it is both injective and surjective. Indeed, $\langle gd, h \rangle = 1$ for all $g \in x$ implies $h = 1$ (showing injectivity); for $z \in (x/d)^*$ we define

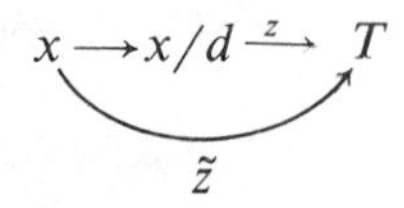

$\tilde{z}$ being the composition map as indicated above. But $\tilde{z} \in d^*$ (since $\langle d, \tilde{z} \rangle = 1$, as the above diagram indicates) and it is mapped into $z \in (x/d)^*$ (showing surjectivity). This proves that $d^* \cong (x/d)^* \cong x/d$, fulfilling one of our objectives.

We now show that $d^{**} = d$. First, it is clear that $d \leq d^{**}$. Indeed, let $g \in d$; then for all $h \in d^*$, $\langle g, h \rangle = 1$; hence $\langle g, d^* \rangle = 1$, or $g \in d^{**}$. We just finished proving that $d^* \cong x/d$; hence $|d^*| = |x/d| = |x|\,|d|^{-1}$. Similarly (working with x^* instead of x) we have $|d^{**}| = |x^*/d^*| =$

$|x^*|\,|d^*|^{-1} = |x|(|x|\,|d|^{-1})^{-1} = |d|$. But $d \le d^{**}$ and $|d^{**}| = |d|$ implies $d^{**} = d$. The map $d \to d^*$ is seen now to be a bijection: d determines uniquely and unambiguously d^*; d^*, in turn, determines d uniquely (as $d^{**} = d$). This proves Lemma 9.3.

9.18 Delsarte's Result

Let x be an Abelian p-group and y a subgroup of x; we write $y \le x$. By Y we denote the collection (or class) of all subgroups of x isomorphic to y. (We sometimes conveniently understand by Y an arbitrary representative of this class.) Y is called the *isomorphism type* of y. By $Y \le X$ we mean that X admits a subgroup of isomorphism type Y. We write $x \cong y$ to express the fact that x is isomorphic to y. By $\mathrm{Hom}(x, y)$ we denote the set of all homomorphisms from x to y; $\mathrm{Aut}\, x$ denotes the set of all automorphisms of x.

For x and y Abelian p-groups (y not necessarily a subgroup of x) we introduce the following notation:

$$|\mathrm{Hom}(x, y)| = H(x, y)$$
$$|\{e : e \le y \text{ such that } e \cong x\}| = h(x, y)$$
$$|\mathrm{Aut}\, e| = \phi(e).$$

(It is clear that H, h, and ϕ are actually functions of type, so we can unambiguously use capital letters as their arguments. It is in fact important that one be fully aware of this fact before proceeding.)

Let $d \le x$ and $e \le y$ and look at $f \in \mathrm{Hom}(x, y)$ such that $\ker f = d$ and $\mathrm{im} f = e$. The figure below summarizes the relevant information:

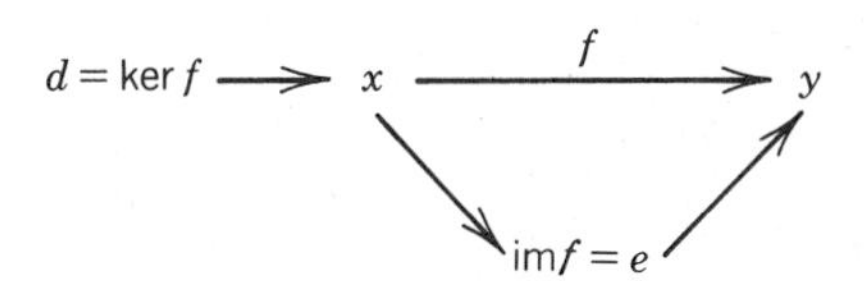

It is well known that $e \cong x/d$. From this diagram it is apparent that

$$|\{f : \ker f = d \text{ and } \mathrm{im} f = e\}| = |\mathrm{Aut}\, e| = \phi(e)$$

and, restricting $\ker f = d$ only,

$$|\{f : \ker f = d\}| = \phi(e)h(e, y) = \phi\left(\frac{x}{d}\right)h\left(\frac{x}{d}, y\right), \tag{9.7}$$

where $e \cong x/d$.

Summing (9.7) over all subgroups d, $1 \le d \le x$, we obtain

$$H(x, y) = \sum_{1 \le d \le x} \phi\left(\frac{x}{d}\right)h\left(\frac{x}{d}, y\right). \tag{9.8}$$

By the bijection that exists between the subgroups of x and the quotient groups of x (exhibited in Lemma 9.3), we can write (9.8) as

$$H(x, y) = \sum_{1 \le d \le x} \phi\left(\frac{x}{d}\right) h\left(\frac{x}{d}, y\right) = \sum_{1 \le d \le x} \phi(d) h(d, y).$$

To summarize:

$$H(x, y) = \sum_{1 \le d \le x} \phi(d) h(d, y). \tag{9.9}$$

Our task is to find explicit expressions for $\phi(x)$ *and* $h(x, y)$. [It turns out that $H(x, y)$ is not difficult to find explicitly. Möbius inversion will then yield $\phi(x)$.]

In (9.9) let y be *fixed* and think of $H(x, y)$ and $h(x, y)$ as functions of x only; denote temporarily $H(x, y)$ by $H(x)$ and $h(x, y)$ by $h(x)$. Expression (9.9) can be rewritten as

$$H(x) = \sum_{1 \le d \le x} \phi(d) h(d) = \sum_{1 \le d \le x} (\phi h)(d),$$

where ϕh is a function on the lattice of subgroups d of x, defined by $(\phi h)(d) = \phi(d) h(d)$. By Möbius inversion formula $H(x) = \sum_{1 \le d \le x} (\phi h)(d)$ implies

$$(\phi h)(x) = \sum_{1 \le d \le x} H(d) \mu(d, x), \tag{9.10}$$

with μ the Möbius function on the lattice of subgroups of x. Resuming standard notation, (9.10) becomes

$$\phi(x) h(x, y) = \sum_{1 \le d \le x} H(d, y) \mu(d, x). \tag{9.11}$$

Special instances of (9.11) are:

(a) *If* $y < x$ $(y \ne x)$ *then* $h(x, y) = 0$ *and* (9.11) *becomes*

$$\sum_{1 \le d \le x} H(d, y) \mu(d, x) = 0. \tag{9.12}$$

(b) *If* $x \cong y$, *then* $h(x, y) = 1$ *and* (9.11) *gives*

$$\sum_{1 \le d \le x} H(d, x) \mu(d, x) = \phi(x). \tag{9.13}$$

Let D be the isomorphism type of d and Y that of y. By grouping terms we can rewrite (9.12) as

$$\sum_{1 \le D \le X} H(D, Y) \left(\sum_{\substack{d \\ d \cong D}} \mu(d, x) \right) = 0.$$

Denoting $\sum_{d,\, d \cong D} \mu(d, x)$ by $M(D)$, (9.11) can be rewritten as

$$\phi(X) h(X, Y) = \sum_{1 \le D \le X} H(D, Y) M(D). \tag{9.14}$$

Expressions (9.12) and (9.13) become, respectively:

(a) *If* $Y < X$ $(Y \neq X)$, *then*

$$\sum_{1 \leq D \leq X} H(D, Y) M(D) = 0. \tag{9.15}$$

(b) *If* $X = Y$, *then*

$$\sum_{1 \leq D \leq X} H(D, X) M(D) = \phi(X). \tag{9.16}$$

[Note that $M(x) = \mu(x, x) = 1$.]

[From here on we proceed as follows: First we establish an explicit formula for $H(D, Y)$. We then let Y range over all proper isomorphism types of subgroups of X, and treat (9.15) as a system of linear equations with $M(D)$'s as unknowns and $H(D, Y)$'s as known coefficients. Since $M(X) = 1$ system (9.15) can be written as

$$\sum_{\substack{1 \leq D \leq X \\ D \neq X}} H(D, Y) M(D) = -H(X, Y).$$

We show that the coefficient matrix $(H(D, Y))$, with both D and Y ranging independently over all proper isomorphism types of subgroups of X, is nonsingular. Hence the $M(D)$'s are uniquely determined by (9.15). To find $\phi(X)$ we essentially substitute the $M(D)$'s and $H(D, X)$ in (9.16). We then substitute $\phi(X)$, $M(D)$, and $H(D, Y)$ in (9.14) to obtain $h(X, Y)$.]

Let us find $H(x, y)$ for two Abelian p-groups x and y. Assume

$$x \cong Z_{p^{m_1}} \oplus \cdots \oplus Z_{p^{m_k}}, \qquad m_i \geq 1.$$

Let $x_1 = \{g \in x\colon \text{order of } g \text{ divides } p\}$; $x_1 \leq x$. Let $|x_1| = p^{r_1}$ ($r_1 = k$, in fact). Repeat this process in x/x_1, that is, look at $(x/x_1)_1$ and denote its order by p^{r_2}. Continuing this process we associate to x a sequence of nonnegative integers ending in 0's $(r_1, r_2, r_3, \ldots, 0, 0, \ldots)$, which satisfies $r_1 \geq r_2 \geq r_3 \geq \cdots$ and which we call the *signature* of x. Observe that, in fact, $r_n = |\{i : m_i \geq n\}|$. Conversely, a given signature $r_1 \geq r_2 \geq r_3 \geq \cdots$ determines uniquely the isomorphism type of an Abelian p-group as

$$(Z_p)^{r_1 - r_2} \oplus (Z_{p^2})^{r_2 - r_3} \oplus \cdots \oplus (Z_{p^l})^{r_l - 0},$$

where $r_n = 0$ for $n \geq l + 1$. [$(Z_{p^2})^{r_2 - r_3}$ means $Z_{p^2} \oplus \cdots \oplus Z_{p^2}$, a direct sum of $r_2 - r_3$ factors.] Denote by $s(x)$ the signature of x. [If, e.g., $x \cong Z_p \oplus Z_{p^2} \oplus Z_{p^2} \oplus Z_{p^3}$, then $s(x) = (4, 3, 1, 0, \ldots)$; conversely, $(4, 3, 1, 0, \ldots)$ leads uniquely to $(Z_p)^{4-3} \oplus (Z_{p^2})^{3-1} \oplus (Z_{p^3})^{1-0} = Z_p \oplus Z_{p^2} \oplus Z_{p^2} \oplus Z_{p^3} \cong x$.]

For a direct sum $X \oplus Y$ we have $s(X \oplus Y) = s(X) + s(Y)$ with usual componentwise addition of sequences (this is very easy to check). We also denote by $\langle s(X), s(Y) \rangle$ the usual inner product of two sequences, that is, if $s(X) = (r_1, r_2, \ldots)$ and $s(Y) = (s_1, s_2, \ldots)$, then $\langle s(X), s(Y) \rangle = \sum_{i=1}^{\infty} r_i s_i$.

Now let x be cyclic, $x \cong Z_{p^m}$, and let y also be cyclic, $y \cong Z_{p^n}$. Then we claim that $H(x, y) = |\text{Hom}(x, y)| = p^{\min(m, n)} = p^{\langle s(x), s(X)\rangle}$. [Indeed, a homomorphism is determined by what it does to a generator g of x. If $m \le n$ and h is the image of g, then $|h|$ divides $|g| = p^m$. However, $y \cong Z_{p^n}$ contains precisely one subgroup isomorphic to Z_{p^m} and h can be any one of its p^m elements. Each such choice determines uniquely a homomorphism; hence $H(x, y) = p^m$. If $n \le m$, then h can be anything in Z_n, and hence there are p^n choices for h; thus $H(x, y) = p^n$. In any case $H(x, y) = p^{\min(m, n)}$.] Observe that $s(x) = s(Z_{p^m}) = (1, 1, \ldots, 1, 0, \ldots)$ with m 1's, and $s(y) = s(Z_{p^n}) = (1, 1, \ldots, 1, 0, \ldots)$ with n 1's. Hence $\langle s(x), s(y)\rangle = \min(m, n)$ and therefore $H(x, y) = p^{\langle s(x), s(y)\rangle}$, as claimed (for x and y cyclic).

Let now x and y be arbitrary Abelian p-groups. Write $x = \oplus_i x_i$ and $y = \oplus_j y_j$ with x_i and y_j cyclic groups. Then it is well known that

$$\text{Hom}\left(\bigoplus_i x_i, \bigoplus_j y_j\right) \cong \bigoplus_i \bigoplus_j \text{Hom}(x_i, y_j),$$

and we obtain

$$\begin{aligned} H(x, y) &= |\text{Hom}(\oplus_i x_i, \oplus_j y_j)| = |\oplus_i \oplus_j \text{Hom}(x_i, y_j)| \\ &= \prod_i \prod_j p^{\langle s(x_i), s(y_j)\rangle} = p^{\langle s(x), s(y)\rangle} = p^{\langle s(X), s(Y)\rangle}. \end{aligned}$$

We have thus shown that

$$H(D, Y) = p^{\langle s(D), s(Y)\rangle}. \tag{9.17}$$

Looking back at (9.15) and recalling that $M(X) = 1$, we can rewrite (9.15) as follows:

$$\sum_{\substack{1 \le D \le X \\ D \ne X}} H(D, Y) M(D) = -H(X, Y), \qquad 1 \le Y \le X \; (Y \ne X),$$

or, using (9.17),

$$\sum_{\substack{1 \le D \le X \\ D \ne X}} p^{\langle s(D), S(Y)\rangle} M(D) = -p^{\langle s(X), s(Y)\rangle}, \qquad 1 \le Y \le X \; (Y \ne X). \tag{9.18}$$

Regard (9.18) as a (square) system of equations with $M(D)$'s as unknowns. The coefficient matrix $(p^{\langle s(D), s(Y)\rangle})$ is *nonsingular*. We establish its nonsingularity in detail immediately following Delsarte's main result. System (9.18) has therefore a unique solution. To summarize:

* *System* (9.18) *determines uniquely the unknowns* $M(D)$*'s.*

Let X have signature $(r_1, \ldots, r_k, 0, \ldots)$. Then for $D \le X$ we can write $s(D) = (i_1, \ldots, i_k, 0, \ldots)$, with $i_1 \le r_1, \ldots, \; i_k \le r_k, \ldots$. Conversely, for *any* sequence of nonincreasing nonnegative integers ending in zeros $(i_1, \ldots, i_k, 0, \ldots)$ and satisfying $i_1 \le r_1, \ldots, \; i_k \le r_k, \ldots$ (along with $i_{l-1} - i_l \le r_{l-1} - r_l$, $2 \le l \le k$), there exists a subgroup $D \le X$ with signature $(i_1, \ldots, i_k, 0, \ldots)$. This is easy to see by recalling how one constructs the group from a signature. Denote $M(D)$ by $M(i_1, \ldots, i_k, 0, \ldots)$, where $(i_1, \ldots, i_k,$

$0, \ldots) = s(D)$, and define the polynomial

$$F_X(Z_1, \ldots, Z_k) = \sum_{(i_1, \ldots, i_k, 0, \ldots)} M(i_1, \ldots, i_k, 0, \ldots) Z_1^{i_1} Z_2^{i_2} \cdots Z_k^{i_k}$$

with Z_i's indeterminates (the sum extends over the signatures of all isomorphism types of subgroups of X, including X itself). F_X is a monic polynomial, since $1 = M(X) = M(r_1, \ldots, r_k, 0, \ldots)$. Observe also that for Y *any* p-group with signature $s(Y) = (j_1, \ldots, j_k, 0, \ldots)$ we have

$$p^{\langle s(D), s(Y) \rangle} = Z_1^{i_1} Z_2^{i_2} \cdots Z_k^{i_k},$$

with $Z_1 = p^{j_1}$, $Z_2 = p^{j_2}, \ldots, Z_k = p^{j_k}$. System (9.18) can now be written as

$$F_X(p^{j_1}, p^{j_2}, \ldots, p^{j_k}) = 0, \tag{9.19}$$

for all signatures $(j_1, \ldots, j_k, 0, \ldots)$ of $Y \leq X$ (and $Y \neq X$). [Comparing with (9.18), in (9.19) the right-hand side is 0 simply because we switched the free term $-p^{\langle s(X), s(Y) \rangle}$ in (9.18) to the left side of the equality sign.]

By now one ought to be motivated to look at the polynomial

$$G_X(Z_1, \ldots, Z_k) = Z_1^{r_2} Z_2^{r_3} \cdots Z_{k-1}^{r_k} \left[\prod_{i_1 = r_2}^{r_1 - 1} (Z_1 - p^{i_1}) \right]$$
$$\times \left[\prod_{i_2 = r_3}^{r_2 - 1} (Z_2 - p^{i_2}) \right] \cdots \left[\prod_{i_k = 0}^{r_k - 1} (Z_k - p^{i_k}) \right].$$

(The first parenthesis is of degree $r_1 - r_2$ in Z_1; the second is of degree $r_2 - r_3$ in Z_2; the last of degree $r_k - 0$ in Z_k.) It is easy to see that G_X is a monic polynomial. In general, a typical monomial in the expansion of G_X is of the form $Z_1^{j_1} Z_2^{j_2} \cdots Z_k^{j_k}$ with

$$r_2 \leq j_1 \leq r_1, \quad r_3 \leq j_2 \leq r_2, \quad \ldots, \quad r_k \leq j_{k-1} \leq r_{k-1}, \quad 0 \leq j_k \leq r_k. \tag{9.20}$$

The sequence $(j_1, j_2, \ldots, j_k)$ obviously satisfies $j_1 \geq j_2 \geq \cdots \geq j_k \geq 0$ and $j_1 \leq r_1$, $j_2 \leq r_2, \ldots, j_k \leq r_k$; one can therefore visualize the (trivially extended) sequence $(j_1, \ldots, j_k, 0, \ldots)$ as the signature of a subgroup Y of X. We claim that G_X does in fact satisfy the same conditions that F_X satisfies in (9.19).

This will be verified if $G_X(p^{j_1}, \ldots, p^{j_k}) = 0$ for $(j_1, \ldots, j_k, 0, \ldots)$ any sequence that can be interpreted as the signature of a subgroup Y of X (and $Y \neq X$). But such a sequence $(j_1, \ldots, j_k, 0, \ldots)$, being visualized as the signature of some $Y \leq X$ *and* $Y \neq X$, satisfies at least one of the following inequalities:

$$r_2 \leq j_1 \leq r_1 - 1, \quad r_3 \leq j_2 \leq r_2 - 1, \quad \ldots, \quad 0 \leq j_k \leq r_k - 1.$$

(This may become more apparent if one reads these inequalities from right to left, starting with $0 \leq j_k \leq r_k - 1$, etc.) It follows, by examining the factors in parenthesis in the definition of G_X, that at least one of these factors will become 0 when computing $G_X(p^{j_1}, \ldots, p^{j_k})$. Thus G_X verifies the same condition that F_X verifies in (9.19).

Since (9.19) is just a rewriting of system (9.18), the coefficients of G_X satisfy (9.18) and are hence *uniquely* determined [as our $M(D)$'s]. Therefore $F_X = G_X$, as polynomials. The polynomial F_X is now explicitly determined. We proved that

$$F_X(Z_1,\ldots,Z_k) = Z_1^{r_2}Z_2^{r_3}\cdots Z_{k-1}^{r_k}\left[\prod_{i_1=r_2}^{r_1-1}(Z_1-p^{i_1})\right]\left[\prod_{i_2=r_3}^{r_2-1}(Z_2-p^{i_2})\right]$$
$$\times\cdots\left[\prod_{i_k=0}^{r_k-1}(Z_k-p^{i_k})\right].$$

[Relations (9.20) indicate that in the expansion of G_X only those signatures that satisfy (9.20) appear as exponents of Z's. In F_X *all* signatures appear but those that do not satisfy (9.20) lead to monomials in Z's with coefficient 0.]

By (9.17) $\phi(X) = \Sigma_{1\le D\le X}H(D,X)M(D) = \Sigma_{1\le D\le X}p^{\langle s(D),\,s(X)\rangle}M(D) = F_X(p^{r_1},\ldots,p^{r_k})$ [see the original definition of F_X in terms of $M(D)$'s to understand the last equality]. Also, from (9.14),

$$\phi(X)h(X,Y) = \sum_{1\le D\le X} H(D,Y)M(D) = \sum_{1\le D\le X} p^{\langle s(D),\,s(Y)\rangle}M(D)$$
$$= F_X(p^{s_1},p^{s_2},\ldots,p^{s_k}),$$

with $(s_1,\ldots,s_k,0,\ldots) = s(Y)$. We know $\phi(X)$ now, so

$$h(X,Y) = \frac{F_X(p^{s_1},\ldots,p^{s_k})}{F_X(p^{r_1},\ldots,p^{r_k})},$$

which is an *explicit* expression for $h(X,Y)$ the number of subgroups of Y isomorphic to X; $h(X,Y)$ is in fact a polynomial (which generalizes the Gaussian polynomials).

In summary, we now formally state Delsarte's result:

Delsarte's Theorem. *For Abelian p-groups x and y with signatures* $s(x) = (r_1,\ldots,r_k,0,\ldots)$ *and* $s(y) = (s_1,\ldots,s_l,0,\ldots)$, $r_1\le s_1,\ \ldots,\ r_k\le s_k$, *we have the following*:

(a) *The number of automorphisms of x equals* $F_X(p^{r_1},\ldots,p^{r_k})$.

(b) *The number of subgroups of y isomorphic to x equals*

$$\frac{F_X(p^{s_1},\ldots,p^{s_k})}{F_X(p^{r_1},\ldots,p^{r_k})},$$

where

$$F_X(Z_1,\ldots,Z_k) = Z_1^{r_2}Z_2^{r_3}\cdots Z_{k-1}^{r_k}$$
$$\left[\prod_{i_1=r_2}^{r_1-1}(Z_1-p^{i_1})\right]\left[\prod_{i_2=r_3}^{r_2-1}(Z_2-p^{i_2})\right]$$
$$\times\cdots\left[\prod_{i_k=0}^{r_k-1}(Z_k-p^{i_k})\right].$$

(c) *The number of homomorphisms from x to y equals* $p^{\langle s(x),\,s(y)\rangle}$.

REMARK. The result extends to arbitrary (finite) Abelian groups through the standard theory: An Abelian group is the direct sum of its Sylow p-subgroups. The theory developed so far extends canonically from the Sylow p-subgroup to the entire group.

An Example. Let $X \cong Z_p \oplus Z_p \oplus Z_{p^3}$ with $s(x) = (3, 1, 1, 0, \ldots)$ and $y \cong Z_{p^2} \oplus Z_{p^3} \oplus Z_{p^3} \oplus Z_{p^5} \oplus Z_{p^7}$ with $s(y) = (5, 5, 4, 2, 2, 1, 1, 0, \ldots)$. Then $F_X(Z_1, Z_2, Z_3) = Z_1 Z_2 (Z_1 - p)(Z_1 - p^2)(Z_3 - 1)$, and

$$\phi(x) = p^3 p(p^3 - p)(p^3 - p^2)(p - 1) = p^7(p^2 - 1)(p - 1)^2$$

$$h(x, y) = \frac{p^5 \cdot p^5(p^5 - p)(p^5 - p^2)(p^4 - 1)}{p^7(p^2 - 1)(p - 1)^2}$$

Also, upon some computation,

$$\phi(y) = F_Y(p^5, p^5, p^4, p^2, p^2, p, p) = p^{69}(p - 1)^4(p^2 - 1)$$

and

$$H(x, y) = p^{\langle s(x), s(y) \rangle} = p^{3 \cdot 5 + 1 \cdot 5 + 1 \cdot 4} = p^{24}.$$

9.19 Establishing the Nonsingularity of the Coefficient Matrix $C = (p^{\langle s(D), s(Y) \rangle})$.

Let the signature of x be $(r_1, r_2, \ldots, r_k, 0, \ldots)$. As we already pointed out, there is a bijection $[Y \leftrightarrow s(Y)$, actually$]$ between the isomorphism types of subgroups of x and sequences $(s_1, s_2, \ldots, s_k, 0, \ldots)$ satisfying: $s_1 \geq s_2 \geq \cdots \geq s_k \geq \cdots$ *and* $s_1 \leq r_1,\ s_2 \leq r_2,\ \ldots,\ s_k \leq r_k,\ \ldots$. Denote the coefficient matrix $(p^{\langle s(D), s(Y) \rangle})$ by C. Its rows and columns are labeled by the proper (isomorphism types of) subgroups of x; think, instead, of its rows and columns being labeled by the *corresponding signature* sequences. A block of C is a set of signatures $(s, s_2, \ldots, s_k, 0, \ldots)$ with $s_2, \ldots, s_k$ *fixed* and s varying, $s_2 \leq s \leq r_1$. The block $(s, s_2, \ldots, s_k, 0, \ldots)$ is said to be larger or of the same magnitude as the block $(l, l_2, \ldots, l_k, 0, \ldots)$ if $l_2 \geq s_2$. Arrange the rows and columns of C by blocks in their order of magnitude (largest to smallest—blocks of same magnitude can be arranged among themselves in any order).

Write now $C = (C_{ij})$ as a partitioned matrix with $C_{ij} = (p^{\langle s(D), s(Y) \rangle})$, with $s(D)$ running through block i and $s(Y)$ running through block j.

The strategy of proof is as follows: Find an explicit form for the submatrix

$$\begin{bmatrix} C_{11} & C_{1i} \\ C_{i1} & C_{ii} \end{bmatrix}, \qquad i > 1.$$

Observe that C_{11} is a *nonsingular* matrix of Vandermonde type. Show that C_{1i}

can be made the zero matrix by column operations. Hence reduce

$$\begin{bmatrix} C_{11} & C_{1i} \\ C_{i1} & C_{ii} \end{bmatrix} \text{ to } \begin{bmatrix} C_{11} & 0 \\ C_{i1} & \tilde{C}_{ii} \end{bmatrix}.$$

Show that $\tilde{C}_{ii}$ is again of nonsingular Vandermonde type. Repeat the process to reduce C to a matrix with blocks of nonsingular Vandermonde type on the diagonal and 0's on top of these blocks. The reduced matrix has determinant equal to that of C; its determinant equals the product of the determinants of the diagonal blocks of Vandermonde type. Each such block has nonzero determinant; hence C has also nonzero determinant. The matrix C is therefore nonsingular.

Let block 1 be $(s, s_2, \ldots, s_k, 0, \ldots)$, $s_2 \leq s \leq r_1$ and let block 2 be $(l, l_2, \ldots, l_k, 0, \ldots)$, $l_2 \leq l \leq r_1$ (with $l_2 \geq s_2$). The signatures in block 1 and 2 (written as columns and truncated at the kth entry for simplicity) are, respectively,

$$\begin{bmatrix} s_2 & s_2+1 & s_2+2 & \cdots & r_1 \\ s_2 & s_2 & s_2 & \cdots & s_2 \\ s_3 & s_3 & s_3 & \cdots & s_3 \\ s_4 & s_4 & s_4 & \cdots & s_4 \\ \vdots & \vdots & \vdots & & \vdots \end{bmatrix}$$

and

$$\begin{bmatrix} l_2 & l_2+1 & l_2+2 & \cdots & r_1 \\ l_2 & l_2 & l_2 & \cdots & l_2 \\ l_3 & l_3 & l_3 & \cdots & l_3 \\ l_4 & l_4 & l_4 & \cdots & l_4 \\ \vdots & \vdots & \vdots & & \vdots \end{bmatrix}$$

Let $\alpha = \sum_{i=2}^{r_1} s_i^2$, $\gamma = \sum_{i=2}^{r_1} l_i^2$, and $\beta = \sum_{i=2}^{r_1} s_i l_i$. (Note that the index of summation starts at 2.) Look at the matrix $\begin{bmatrix} C_{11} & C_{12} \\ C_{21} & C_{22} \end{bmatrix}$. We have

$$\begin{array}{c} \qquad\quad \begin{array}{cccccc} p^{s_2^2} & p_2^{s_2(s_2+1)} & \cdots & p^{s_2 l_2} & \cdots & p^{s_2 r_1} \end{array} \\ C_{11} = p^{\alpha} \begin{bmatrix} 1 & 1 & \cdots & 1 & \cdots & 1 \\ p^{s_2} & p^{s_2+1} & \cdots & p^{l_2} & \cdots & p^{r_1} \\ p^{2s_2} & p^{2(s_2+1)} & \cdots & p^{2l_2} & \cdots & p^{2r_1} \\ p^{3s_2} & p^{3(s_2+1)} & \cdots & p^{3l_2} & \cdots & p^{3r_1} \\ \vdots & \vdots & & \vdots & & \vdots \end{bmatrix} \end{array}$$

$$C_{12} = p^{\beta}\left[\begin{array}{cccc} p^{s_2 l_2} & p^{s_2(l_2+1)} & \cdots & p^{s_2 r_1} \\ \hline 1 & 1 & \cdots & 1 \\ p^{l_2} & p^{l_2+1} & \cdots & p^{r_1} \\ p^{2l_2} & p^{2(l_2+1)} & \cdots & p^{2r_1} \\ p^{3l_2} & p^{3(l_2+1)} & \cdots & p^{3r_1} \\ \vdots & \vdots & & \vdots \end{array}\right]$$

$$C_{21} = p^{\beta}\left[\begin{array}{cccccc} p^{l_2 s_2} & p^{l_2(s_2+1)} & \cdots & p^{l_2^2} & \cdots & p^{l_2 r_1} \\ \hline 1 & 1 & \cdots & 1 & \cdots & 1 \\ p^{s_2} & p^{s_2+1} & \cdots & p^{l_2} & \cdots & p^{r_1} \\ p^{2s_2} & p^{2(s_2+1)} & \cdots & p^{2l_2} & \cdots & p^{2r_1} \\ p^{3s_2} & p^{3(s_2+1)} & \cdots & p^{3l_2} & \cdots & p^{3r_1} \\ \vdots & \vdots & & \vdots & & \vdots \end{array}\right]$$

$$C_{22} = p^{\gamma}\left[\begin{array}{cccc} p^{l_2^2} & p^{l_2(l_2+1)} & \cdots & p^{l_2 r_1} \\ \hline 1 & 1 & \cdots & 1 \\ p^{l_2} & p^{l_2+1} & \cdots & p^{r_1} \\ p^{2l_2} & p^{2(l_2+1)} & \cdots & p^{2r_1} \\ p^{3l_2} & p^{3(l_2+1)} & \cdots & p^{3r_1} \\ \vdots & \vdots & & \vdots \end{array}\right].$$

What do we mean by a matrix with a bar and some numbers on top? We mean that each entry in the jth column is multiplied by the number that lies on top of that jth column, that is,

$$3^2\left[\begin{array}{ccc} 2 & 3 & 5 \\ \hline 3 & 2 & 1 \\ 2 & 4 & 2 \\ 1 & 2 & 3 \end{array}\right] = 3^2\begin{bmatrix} 3 & 2 & 1 \\ 2 & 4 & 2 \\ 1 & 2 & 3 \end{bmatrix}\begin{bmatrix} 2 & 0 & 0 \\ 0 & 3 & 0 \\ 0 & 0 & 5 \end{bmatrix}.$$

Since $l_2 \geq s_2$ each column of C_{12} appears also as a column in C_{11} (with the same number on top, as displayed above). Multiplying the $(l+i)$th column of C_{11} by $p^{\beta-\alpha}$ and subtracting it from the ith column of C_{12}, we reduce C_{12} to the zero matrix $(1 \leq i \leq r_1 - l_2 + 1)$. This process changes C_{22} into $(p^{\gamma} - p^{2\beta-\alpha})C_{22} = p^{-\alpha}(p^{\gamma+\alpha} - p^{2\beta})C_{22}$ $(= \tilde{C}_{22}$, say). We thus reduced by column operations the matrix

$$\begin{bmatrix} C_{11} & C_{12} \\ C_{21} & C_{22} \end{bmatrix} \quad \text{to} \quad \begin{bmatrix} C_{11} & 0 \\ C_{21} & \tilde{C}_{22} \end{bmatrix}.$$

Observe that $\tilde{C}_{22} \neq 0$ since $C_{22} \neq 0$ and since $p^{\gamma+\alpha} - p^{2\beta} \neq 0$ [In fact $p^{\gamma+\alpha} - p^{2\beta} > 0$, since $\gamma + \alpha - 2\beta = \Sigma_{i=2}^{r_1}(l_i^2 + s_i^2 - 2l_i s_i) = \Sigma_{i=2}^{r_1}(l_i - s_i)^2 \geq 0$;

$\gamma + \alpha - 2\beta$ is strictly positive since the two blocks in question are not identical, i.e., $l_i \neq s_i$ for *some* i, $2 \leq i \leq r_1$].

An $n \times n$ matrix V of the form (v_j^{i-1}) with v_j constants, $1 \leq i, j \leq n$, is called a *Vandermonde matrix*. The determinant of V is $\pm\prod_{i<j}(v_i - v_j)$. The matrix V is hence nonsingular, unless $v_i = v_j$ for some $i \neq j$. We call an $n \times n$ matrix W of *Vandermonde type* if

$$W = a \begin{array}{c} \begin{array}{cccc} b_1 & b_2 & \cdots & b_n \end{array} \\ \hline \left[\begin{array}{cccc} 1 & 1 & \cdots & 1 \\ v_1 & v_2 & \cdots & v_n \\ v_1^2 & v_2^2 & \cdots & v_n^2 \\ v_1^3 & v_2^3 & \cdots & v_n^3 \\ \vdots & \vdots & & \vdots \end{array}\right] \end{array}$$

with $a \neq 0$ and $b_i \neq 0$, for all $1 \leq i \leq n$.

Observe that such W has determinant $\pm ab_1 \cdots b_n\prod_{i<j}(v_i - v_j)$. It is therefore nonsingular, unless $v_i = v_j$ for some $i \neq j$. Both C_{11} and C_{22} are therefore *nonsingular* matrices of Vandermonde type (as a quick glance now shows). So is $\tilde{C}_{22}$. Hence $\begin{bmatrix} C_{11} & 0 \\ C_{21} & \tilde{C}_{22} \end{bmatrix}$ is nonsingular and so is the original matrix $\begin{bmatrix} C_{11} & C_{12} \\ C_{21} & C_{22} \end{bmatrix}$. We now repeat the process to make all $C_{ij} = 0$, for $i < j$. The matrix C will be reduced to a lower block-triangular matrix (as mentioned before) with *nonsingular* diagonal blocks of Vandermonde type. This proves the nonsingularity of C and completes the proof of Delsarte's result.

9.20 Extensions to Regular p-Groups

In the pages to follow we attempt to obtain explicit formulas for the number of surjective homomorphisms between certain kinds of (non-Abelian) p-groups regular in the sense of Phillip Hall [1]. Our results generalize those of Delsarte (for the Abelian case) to which we have devoted the first part of Section 5. It will be fortunate if the reader is already familiar with the relevant parts of Hall's fundamental contribution to the theory of groups of prime power order [1, Section 4.5].

1. Regular p-groups possess *uniqueness bases* (definitions follow shortly) just like Abelian groups, and this renders them particularly appealing from a combinatorial viewpoint. Although non-Abelian in general, their structure appears to be more tractable due in large part to the existence of such a uniqueness basis (abbreviated as UB). The UB allows in many instances an explicit expression for the order of the automorphism group in terms of the *type* of the group and the type of its Frattini subgroup. An immediate consequence is a formula for the number of nonequivalent minimal bases that such a group possesses. This latter point was addressed only briefly and in rather general terms in [1, Section 1.3], but a more extensive study was subsequently undertaken [2].

The case of Abelian p-groups has been investigated by Delsarte [3]. He found polynomials that give the order of the automorphism group, as well as the number of subgroups of a given type. An important role in his proof is played by the dual group (i.e., the group of homomorphisms into the circle group). In the Abelian case the dual is isomorphic to the original group, thus offering the convenience of interpreting quotients of the original group as subgroups of the dual. When moving to non-Abelian cases such suitabilities no longer exist. The results can nevertheless be extended to large classes of regular p-groups by devising ways to avoid reference to dual groups. A principal factor that allows the extension is the existing similarity between the lattice of *major subgroups* in the Abelian and certain regular cases. (A subgroup is called major, the reader will recall, if it contains the intersection of all maximal subgroups.) The polynomials that emerge are more general than those of Delsarte, though the two classes coincide for Abelian p-groups. In principle one expects a more general form for these polynomials because, unlike in the Abelian situation, the type of a regular p-group does not uniquely determine the type of its principal (or Frattini) subgroup. The freedom of choice for the latter accounts for the greater generality.

2. Let G be a p-group. An important characteristic subgroup of G is the *principal* (or Frattini) subgroup, which we denote by ϕ. It is usually defined as the intersection of all maximal subgroups of G, a subgroup being maximal if it is of index p in G. The principal subgroup is also the smallest normal subgroup ϕ (with inclusion as partial order) such that G/ϕ is Abelian with all its nonidentity elements of order p. The quotient group G/ϕ is therefore isomorphic to a vector space over the field of p elements. We denote by d the dimension of this vector space and note that $|G/\phi| = p^d$. Another convenient way to visualize the subgroup ϕ is by observing that

$$\phi = \langle x^p, [G,G] : x \in G \rangle, \tag{9.21}$$

which, in words, says that ϕ is generated by the commutator subgroup $[G, G]$ of G, along with all the pth powers. That these three ways of seeing ϕ are equivalent is elementary.

A subgroup of G is called *major* if it contains the principal subgroup ϕ. Since G/ϕ is Abelian, all major subgroups are normal in G.

A *basis* of G is an (ordered) sequence $x_1, x_2, \ldots, x_r$ of distinct elements that generate G. We call a basis *minimal* if no proper subsequence of its elements generates G.

The basis theorem of Burnside informs us that

> *Any minimal basis (or MB) of G contains exactly d elements, where d is the dimension of the vector space G/ϕ.* (9.22)

Consequently, any basis of G with d elements is necessarily an MB.

The number of MB's is easy to calculate since it mainly involves counting the number of vector space bases for G/ϕ.

A *p-group of order p^n has*

$$p^{d(n-d)}(p^d-1)(p^d-p)\cdots(p^d-p^{d-1}) \tag{9.23}$$

distinct MB's.

(We emphasize that two MB's are considered the same only if they consist of the same elements listed in the same order.) Expression (9.23) appears in [1, p. 36].

Interest arises, however, not in counting the number of MB's, but rather in determining the number of MB's essentially different. For it is evident that if $x_1, \ldots, x_d$ is an MB of G, its image through an automorphism f of G $fx_1, \ldots, fx_d$ is an MB of the same kind. To be precise, we observe that the automorphism group of G acts on the set of MB's (as indicated above) and *we ask for the number of orbits generated by this action*. This number we denote by E, and call it the Eulerian invariant of the p-group (cf. [2, pp. 135–137]), the nomenclature being motivated by the Eulerian function of arithmetic.

It is clear that nothing but the identity automorphism fixes an MB, and therefore that each orbit is of cardinality A, where A denotes the number of automorphisms of G. For this reason the totality of MB's expressed in (9.23) falls into E classes, each containing A elements. Hence we may write

$$EA = p^{d(n-d)}(p^d-1)(p^d-p)\cdots(p^d-p^{d-1}). \tag{9.24}$$

Equation (9.24) shows that finding the number of essentially different MB's is equivalent to determining the cardinality of the automorphism group.

3. Following Hall [1], a p-group G is called *regular* if, given any positive integer α and any pair of elements x and y of G, it is always possible to find elements $s_3, s_4, \ldots, s_r$ all belonging to the commutator subgroup of $\langle x, y\rangle$ (the subgroup generated by x and y) and satisfying the equation

$$(xy)^{p^\alpha} = x^{p^\alpha}y^{p^\alpha}s_3^{p^\alpha}s_4^{p^\alpha}\cdots s_r^{p^\alpha}.$$

The elements s_i and the number r depend in general on α, x, and y.

Apart from Abelian p-groups, every p-group whose lower central series is of length less than p is regular, and thus p-groups of order p^n with $p \geq n$ are regular. Subgroups and quotient groups of regular p-groups remain regular.

A sequence $x_1, x_2, \ldots, x_r$ of elements of a group G of orders $n_1, n_2, \ldots, n_r$, respectively (with $n_i > 1$, for all i), is said to be a *uniqueness basis* (or UB) of G if each element x of G can be expressed in one and only one way in the form

$$x = x_1^{m_1}x_2^{m_2}\cdots x_r^{m_r},$$

with $0 \leq m_i < n_i$ $(i = 1, 2, \ldots, r)$.

It was shown in [1, Section 4.5] that

$$\textit{Every regular p-group has at least one UB.} \tag{9.25}$$

More has in fact been shown: Let G be a regular p-group of order p^n. Denote a UB of G by $x_1, \ldots, x_r$. We write nonincreasingly the orders of these base elements as $p^{\mu_1}, \ldots, p^{\mu_r}$. (This does not necessarily mean that x_i is of order p^{μ_i}.) The vector $\mu = (\mu_1, \ldots, \mu_r)$ is called the *type* of the group G. It turns out that any two UB's of G have r elements and yield the same μ. The number r and the vector μ are therefore invariants of the group. The type does not determine the group to within an isomorphism, but it does so in the Abelian case. We can write, however, $\mu_1 + \cdots + \mu_r = n$.

Two important characteristic subgroups of a regular p-group are denoted by $\mho_\alpha$ and Ω_β. The subgroup $\mho_\alpha$ consists of all p^αth powers of elements of G, while Ω_β contains all elements whose order divides p^β. It is known that $|\Omega_\alpha| = |G/\mho_\alpha|$, and that $|G/\mho_1| = p^r$, where r equals the number of elements in a UB of G. An immediate consequence of this fact is that

$$\text{In a regular } p\text{-group an UB is a MB if and only if } [G, G] \leq \mho_1. \tag{9.26}$$

For two vectors $t = (t_1, \ldots, t_r)$ and $u = (u_1, \ldots, u_r)$, with their components listed nonincreasingly, we write $t \leq u$ if $t_i \leq u_i$, for all i. By $\mathbf{1}$ we denote the vector with all entries 1.

It is known that $H \leq G$ implies $h \leq \mu$, where h denotes the type of the subgroup H. We take special interest in the type ω of the subgroup $\mho_1$ and prove that $\omega = \mu - \mathbf{1}$. The order of $\mho_1$ is known to depend only on the type μ of G, and hence we can compute it in the Abelian group of type μ. This remark easily yields $|\mho_1| = \prod_{i=1}^r p^{\mu_i - 1} = p^{n-r}$. To derive the type of $\mho_1$, let $y_i = x_i^p$ and observe that $y_1, \ldots, y_r$ is a UB for $\mho_1$. (Some of the y_i's may in fact be 1, so that the actual cardinality of a UB of $\mho_1$ could be less than r.) Indeed, the elements $\prod_{i=1}^r y_i^{\alpha_i}$ $(1 \leq \alpha_i \leq p^{\mu_i - 1})$ are all distinct (since the x_i's form a UB of G) and $\prod_{i=1}^r p^{\mu_i - 1} = p^{n-r}$ in number. Every element of $\mho_1$ occurs therefore among them. For this reason they form a UB of $\mho_1$. And since the orders of the y_i's could be found among the numbers $p^{\mu_1 - 1}, \ldots, p^{\mu_r - 1}$ the type of $\mho_1$ is $\mu - \mathbf{1}$.

We observed in (9.21) that the principal subgroup ϕ is equal to $\langle \mho_1, [G, G] \rangle$. In particular, if we denote by ν the type of ϕ, the inclusions $\mho_1 \leq \phi < G$ allow us to conclude that $\mu - \mathbf{1} \leq \nu < \mu$.

In the Abelian case $\nu = \mu - \mathbf{1}$ (since $\phi = \mho_1$). By (9.26) and (9.22) we may easily characterize this situation:

$$\begin{aligned} \nu = \mu - \mathbf{1} \quad & \text{if and only if } [G, G] \leq \Omega_1, \\ & \text{if and only if an UB of } G \text{ is also a MB}, \\ & \text{if and only if an UB of } G \text{ has } d \text{ elements} \\ & \left(\text{where } p^d = |G/\phi|\right). \end{aligned} \tag{9.27}$$

The number of surjective homomorphisms from H to G depends, as we see, on the type of the principal subgroup ϕ of G, in addition to the dependence upon the types of the initially given groups H and G.

4. Along with the type of a regular p-group we find particularly useful the "dual" notion of *signature*. Readers familiar with the theory of partitions will recognize the signature as the conjugate partition to the partition that the type describes. The two notions are connected graphically through the matrix known as the *figure* of the group, cf. [1, p. 80]. (This latter notion corresponds in the theory of partitions to the Ferrer diagram.) The type is the vector of column sums of the figure, while the signature is the vector of row sums. Given either the type or the signature, the other is easily derived by writing out the figure as indicated below:

$$\begin{array}{cccc|c} 1 & 1 & 1 & 1 & 4 \\ 1 & 1 & 1 & 0 & 3 \\ 1 & 0 & 0 & 0 & 1 \\ \hline 3 & 2 & 2 & 1 & \end{array} \quad \text{(signature)}$$

(type)

The existing bijection between types (t) and signatures (s) is denoted by $t \leftrightarrow s$.

Formally, the components of t and s are related as follows: s_i is equal to the number of t_j's greater than or equal to i, and t_i is equal to the number of s_j's greater than or equal to i.

In the context of regular p-groups the signature has important group theoretical meaning. For the most part, however, we need only rely on its formal description given above. The relationship described below between the types and signatures of two groups, say $\mu \leftrightarrow s$ and $\lambda \leftrightarrow u$, will be of help in the next section:

$$\sum_k \sum_i \min(\mu_k, \lambda_i) = \langle s, u \rangle, \tag{9.28}$$

where $\langle s, u \rangle = \Sigma_j s_j u_j$. In equation (9.28) and throughout the next section we conveniently interpret the types and signatures as vectors of indefinite lengths ending in zeros. By $\min(\mu_k, \lambda_i)$ we understand the minimum of the two numbers.

Relation (9.28) becomes clear upon observing that, more generally, the sum of the m^2 inner products between the columns of two $m \times m$ matrices equals the inner product of their row sums. Verifying this last statement is a straightforward task.

5. For H and G p-groups, $\mathrm{Hom}(H, G)$ denotes the number of homomorphisms from H to G, and $\mathrm{Surj}(H, G)$ denotes the number of homomorphisms that are surjective. (This is a departure from the notation we used in the first part of this section, but it proves equally effective.)

A homomorphism being obviously surjective on its image allows us to write

$$\mathrm{Hom}(H, G) = \sum_{1 \le K \le G} \mathrm{Surj}(H, K).$$

Möbius inversion now yields

$$\operatorname{Surj}(H,G) = \sum_{1 \le K \le G} \operatorname{Hom}(H,K)\mu(K,G). \qquad (9.29)$$

By Proposition 9.2 $\mu(K,G) = 0$, unless K is a major subgroup of G. And if K is major of index p^k in G, then $\mu(K,G) = (-1)^k p^{k(k-1)/2}$.

Equation (9.29) thus becomes

$$\operatorname{Surj}(H,G) = \sum_{\phi \le K \le G} \operatorname{Hom}(H,K)\mu(K,G), \qquad (9.30)$$

where ϕ is the principal subgroup of G.

6. Let us assume now that H is regular and that it admits a UB that is also an MB. [We characterized this assumption in (9.27).] The group K will be any regular p-group. An expression for $\operatorname{Hom}(H, K)$ in terms of the signatures of the two groups can easily be found.

Fix a UB $x_1, \dots, x_s$ of H. A homomorphism f from H to K is unambiguously defined if we specify the image $f(x_i)$ of each x_i, subject only to the condition that the order of $f(x_i)$ divides that of x_i. An element of order p^α of the above UB of H can therefore be mapped into any element of $\Omega_\alpha(K)$, the subgroup of K consisting of all elements whose order divides p^α. To find $|\Omega_\alpha(K)|$, select first a UB $y_1, \dots, y_r$ of K. An element $z = z_1 z_2 \cdots z_r$ (with z_i a power of y_i) belongs to $\Omega_\alpha(K)$ if and only if all the z_i's do; see [1, p. 91]. If we denote by $\langle y_i \rangle$ the cyclic group of order p^{m_i} generated by y_i, then there are $p^{\min(\alpha, m_i)}$ elements of $\langle y_i \rangle$ whose order divides p^α. In terms of the type $(t_1, \dots, t_r)$ of K we can therefore write

$$|\Omega_\alpha(K)| = p^{\min(\alpha, t_1)} \dots p^{\min(\alpha, t_r)}.$$

(Note that the vectors of the m_i's and t_i's are the same upon a permutation of entries.)

Denoting by $(v_1, \dots, v_s)$ the type of H, we can write as follows:

$$\operatorname{Hom}(H,K) = \prod_{i=1}^{s} |\Omega_{v_i}(K)| = \prod_{i=1}^{s} \prod_{j=1}^{r} p^{\min(v_i, t_j)}$$

$$= p^{\sum_{i=1}^{s} \sum_{j=1}^{r} \min(v_i, t_j)} = p^{\langle h, k \rangle},$$

with h and k the signatures of H and K, respectively. The last sign of equality is a consequence of (9.28).

In conclusion, if K is any regular p-group and H is a regular p-group in which a UB is also an MB, then

$$\operatorname{Hom}(H,K) = p^{\langle h, k \rangle}, \qquad (9.31)$$

where h and k are the *signatures* of H and K, respectively.

7. Equation (9.31) suggests that we sort the terms of (9.30) by signature. We thus let $\overline{K}$ be a representative of the class of all subgroups of G with signature the same as K. In particular this class contains all the subgroups of

G isomorphic to K, but is generally larger. The new appearance assumed by (9.30) is

$$\operatorname{Surj}(H, G) = \sum_{\phi \le \overline{K} \le G} \operatorname{Hom}(\overline{H}, \overline{K}) M(\overline{K}, G),$$

where

$$M(\overline{K}, G) = \sum_{\substack{\phi \le K \le G \\ (K \text{ has same} \\ \text{signature as } \overline{K})}} \mu(K, G).$$

Or, more explicitly still,

$$\operatorname{Surj}(H, G) = \sum_{f \le k \le g} p^{\langle h, k \rangle} M(k, g). \tag{9.32}$$

Expression (9.32) should be interpreted with care. The index of summation, k, runs over all signatures of *representatives of existing classes of major subgroups* $\overline{K}$ of G. By f, h, and g we understand the signatures of ϕ, H, and G, respectively. Lastly, the notation $M(k, g)$ stands for $M(\overline{K}, G)$. Note, in particular, that $M(g, g) = M(\overline{G}, G) = \mu(G, G) = 1$.

8. Special cases of (9.32) are important in determining the numbers $M(k, g)$. Note that these numbers *do not depend* upon the choice of H. In particular, selecting H Abelian with the same signature as a *proper* major subgroup of G, we evidently have

$$0 = \operatorname{Surj}(H, G) = \sum_{f \le k \le g} p^{\langle h, k \rangle} M(k, g). \tag{9.33}$$

Allowing H to run over the Abelian groups of the same signature as the classes of *proper* major subgroups of G, we obtain the system

$$\sum_{f \le k < g} p^{\langle h, k \rangle} M(k, g) = -p^{\langle h, g \rangle} \tag{9.34}$$

for $f \le h < g$. The equations of system (9.34) follow from (9.33) upon moving to the right-hand side the (last) term $p^{\langle h, g \rangle} M(g, g)$. We treat $M(k, g)$ as unknowns, and thus the *square* matrix $(p^{\langle h, k \rangle})$ is the coefficient matrix of the system. Being a principal minor of the matrix C shown positive definite in Section 9.19 the above matrix is *nonsingular*. System (9.34) thus *uniquely* determines the unknowns $M(k, g)$.

Substituting the values of the $M(k, g)$'s in (9.32) we obtain, in principle at least, a solution for $\operatorname{Surj}(H, G)$. We attempt to find more explicit expressions, which in many cases take the form of polynomials similar to those in [3].

9. Define the polynomial

$$P_G(Z_1, \ldots, Z_r) = \sum_{\substack{g \\ f \le k \le g}} M(k, g) Z_1^{k_1} Z_2^{k_2} \cdots Z_r^{k_r}, \tag{9.35}$$

where $k = (k_1, \ldots, k_r)$ runs over the signatures of all *classes* of major subgroups of G [the definition being motivated by equations (9.33) and (9.34)].

By moving $-p^{\langle h, g\rangle}$ to the left-hand side, system (14) takes the form:

$$P(p^{h_1}, \ldots, p^{h_r}) = 0, \tag{9.36}$$

for $h = (h_1, \ldots, h_r)$ running over the signatures of all classes of *proper* major subgroups of G (these signatures being in fact attached to Abelian p-groups, as we have mentioned).

The polynomial P_G is [in view of (9.34)—or (9.36)] characterized by the following three properties:

(i) It is a monic polynomial.
(ii) Its monomials are precisely

$$Z_1^{k_1} Z_2^{k_2} \cdots Z_r^{k_r}$$

where $k = (k_1, \ldots, k_r)$ runs over the signatures of all existing classes of major subgroups of G.
(iii) It satisfies equation (9.36).

During the next section we need be reminded that

A polynomial that possesses properties (*i*), (*ii*), *and* (*iii*) *is necessarily the polynomial that appears in* (9.35). (9.37)

10. *The results.* We are in a position to examine the consequences of the work done so far.

Combining the contents of (9.32), (9.34), and (9.35) we conclude the following:

$$\mathrm{Surj}(H, G) = P_G(p^{h_1}, p^{h_2}, \ldots, p^{h_r}),$$

and

$$\mathrm{Aut}(H) = \mathrm{Surj}(H, H) = P_H(p^{h_1}, \ldots, p^{h_s}). \tag{9.38}$$

In these formulas G is any regular p-group, and H is a regular p-group in which a UB is also an MB.

The signature of H is $(h_1, \ldots, h_s)$. We conveniently extend indefinitely this vector with zeros, as explained in the sentence following (9.28), and write $h = (h_1, \ldots, h_s, 0, 0, \ldots)$. When we substitute $p^{h_1}, \ldots, p^{h_r}$ in $\mathrm{Surj}(H, G)$ of (9.38), $h_1, \ldots, h_r$ are the first r components of the infinite vector h.

The polynomial that occurs in (9.38) is

$$P_W(Z_1, \ldots, Z_m) = \sum_{\substack{k \\ v \le k \le w}} M(k, w) Z_1^{k_1} Z_2^{k_2} \cdots Z_m^{k_m}, \tag{9.39}$$

where w is the signature of the regular p-group W, and v is the signature of the principal subgroup of W. The index k runs over the signatures of all classes of major subgroups of W. System (9.34) yields the coefficients $M(k, w)$ of P_W.

In certain cases the regular p-group G has major subgroups of every possible signature k, $f \le k \le g$. (The vectors g and f are the signatures of G and its principal subgroup ϕ, respectively.) Whenever this is the case, the polynomial P_G written in (9.35) takes the form

$$P_G(Z_1, \ldots, Z_r) = Z_1^{f_1} Z_2^{f_2} \cdots Z_r^{f_r} \prod_{i_1 = f_1}^{g_1 - 1} (Z_1 - p^{i_1}) \cdots \prod_{i_r = f_r}^{g_r - 1} (Z_r - p^{i_r}). \quad (9.40)$$

One can easily verify that this is indeed true, by observing that the polynomial defined in (9.40) conforms to the three requirements mentioned in (9.37).

Consequently,

∗ *If H is a regular p-group in which an UB is also a MB, and G is a regular p-group that has major subgroups of every possible signature, then*

$$\mathrm{Surj}(H, G) = P_G(p^{h_1}, p^{h_2}, \ldots, p^{h_r}), \quad (9.41)$$

where P_G appears explicitly in (9.40).

Statement (9.41) is the explicit formula mentioned at the beginning of Section 9.20, and is our main result.

11. The Eulerian invariant E of a regular p-group G of the kind mentioned below is

$$E = \frac{p^{d(n-d)}(p^d - 1)(p^d - p) \cdots (p^d - p^{d-1})}{P_G(p^{g_1}, p^{g_2}, \ldots)}.$$

This formula follows from (9.24) and (9.41). It is valid for regular p-groups G of signature $(g_1, g_2, \ldots)$ that possess an UB with d elements and major subgroups of every signature. The number d is the dimension of the vector space G/ϕ, and p^n is the order of G. In all these groups the principal subgroup ϕ has signature $(g_2, g_3, \ldots)$, as (9.27) implies.

12. We now turn to Abelian p-groups and observe that in their case the signature g of the group determines uniquely the signature f of the principal subgroup. In fact $f_i = g_{i+1}$, for all i, that is, $f = (g_2, g_3, \ldots, g_r, 0, \ldots)$. The polynomials written in (9.40) become in this instance those of Delsarte mentioned at the end of Section 9.18.

Counting the number of subgroups of an Abelian group G of a given type is the same as counting the number of injective homomorphisms from an Abelian group H of that type to G. By the existing "duality" in the Abelian case, we conclude that the number of injective homomorphisms from H to G equals the number of surjective homomorphisms from G to G/H. Equation (9.41) is therefore directly applicable. Delsarte expresses the answer as a quotient of two polynomials (see the end of Section 9.18), but easy cancellations show that it reduces to the polynomial written in (9.41).

13. The results can be extended in a canonical way to nilpotent groups whose Sylow subgroups are regular p-groups of the kind we consider. Generally they allow a better understanding of the size of normalizers in a general finite group, since the normalizer is part of the automorphism group.

EXERCISES

1. Use Möbius inversion to prove that $\sum_{s=p}^{n}(-1)^{s-p}\binom{n-p}{s-p}2^{n-s} = 1$. [*Hint:* Let $T = \{1, 2, \ldots, n\}$. For $P \subseteq T$ let $f(P)$ denote the number of sequences of length n formed with letters a and b such that a occurs in positions i only, with $i \in P$. Clearly $f(P) = 1$, for all P. On the other hand, if $g(P) = \sum_{S \supseteq P} f(S)$ = number of sequences of length n with a's in positions i with $i \in P$ and possibly other positions, then $g(P) = 2^{n-p}$. Invert to obtain the other expression for $f(P)$.]

2. Use Möbius inversion on the lattice of subsets of a set to prove Stirling's formula:

$$S_m^n = \frac{1}{n!}\sum_{k=0}^{n}\binom{n}{k}(-1)^{n-k}k^m,$$

with S_m^n the Stirling number of the second kind. [*Hint:* Express the number of functions from A to B in terms of surjections onto subsets of B. Invert.]

3. How many spanning (or generating) subsets are there for a vector space of finite dimension over a finite field? [*Hint:* Let $f(U)$ = number of spanning subsets of U. Compute first $g(U) = \sum_{T \leq U} f(T)$ and then invert.]

4. Show that $\sum_{n=1}^{\infty} x^n(1 - x^n)^{-1}\mu(1, n) = x$, with μ the Möbius function on the lattice of natural numbers with divisibility as relation of partial order.

5. Suppose we have beads of n colors. How many distinct necklaces with r beads can we make ($r \leq n$)? (Two necklaces are considered identical if one can be obtained from the other by a planar rotation. A necklace may contain more than one bead of the same color.) [*Sketch of proof.* To a necklace, such as

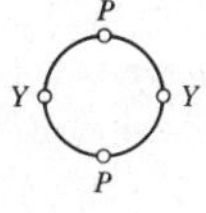

where Y = yellow and P = purple, we associate its *strings* as follows. Cut the necklace open at some point, obtain a string, and consider the set of strings obtained by cyclic permutations of the beads on the original string. In our case:

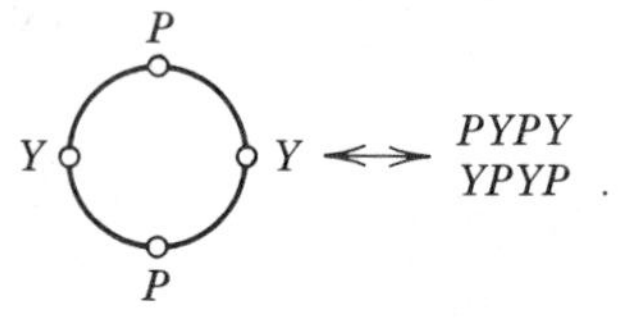

P

Y Y $\longleftrightarrow$ PYPY YPYP.

P

The set of strings of a necklace is called the *orbit* of the necklace. Observe that the cardinality of an orbit always divides r. Furthermore, since $r \leq n$, for each divisor d of r there exists a necklace with orbit of cardinality d. Let $M(d)$ = number of necklaces with orbit of cardinality d. We want to compute $\Sigma_{d|r} M(d)$. But observe that $\Sigma_{d|r} dM(d)$ = number of all strings with r beads of at most n colors $= n^r$. Möbius inversion gives now $dM(d)$ and hence also $M(d)$.]

6. Each of n men checks a hat and an umbrella when entering a restaurant. When leaving they are handed back a hat and an umbrella at random. What is the chance that precisely k men get back their own hats and umbrellas? What is the chance that precisely k men get back either their own hat or their own umbrella (or both)? [*Hint:* Looks like a sieve method should take care of this, doesn't it?]

7. Determine the number of connected simple graphs on n (labeled) vertices. [*Hint:* Work on the lattice of partitions of the n vertices, with refinement as partial order. Note that there are $2^{\binom{k}{2}}$ graphs in all on k vertices. Use Example 4 of Section 2 to invert.]

8. (*The sieve formula for finite vector spaces.*) Identify (by coloring in red, say) subsets $C_1, C_2, \ldots, C_m$ of a vector space V of dimension n over $GF(q)$. For a subspace S of V denote by $f(S)$ the number of red subsets (i.e., the number of C_i's) contained in S but not in any of the proper subspaces of S. By $g(S)$ denote the total number of red subsets in S. Express f in terms of g. [*Note:* It is easy to express g in terms of f.]

9. Let G be a group of order p^n. Prove the following "Anzahl" theorems for p-groups:

(a) The number of subgroups of G of order p^m is congruent to 1 modulo p, for $0 \leq m \leq n$.

(b) If $p > 2$ and G is not cyclic, then the number of cyclic subgroups of G of order p^m is congruent to 0 modulo p, for $1 < m < n$.

(c) If $p > 2$ and G is not cyclic, then the number of subgroups of G of order p^m is congruent to $1 + p$ modulo p^2, for $0 < m < n$.

(d) The number of subgroups of index p^k in G is congruent to $\begin{bmatrix} d \\ k \end{bmatrix}(p)$ modulo p^{d-k+1}, where d denotes the dimension of the vector space G/ϕ (with ϕ the Frattini subgroup) and $\begin{bmatrix} d \\ k \end{bmatrix}(x)$ is the Gaussian polynomial. [*Hint:* Use the sieve formula for vector spaces; see Exercise 8.]

10. Let $(L, \leq)$ be a finite lattice; $L = \{x_1, \ldots, x_n\}$. Let f be a function from L into the real numbers and let g be defined by $g(x) = \sum_{y \leq x} f(y)$. Denote $g(x_i \wedge x_j)$ by g_{ij}. Show that

$$\det(g_{ij}) = f(x_1) \cdots f(x_n),$$

where $\det(g_{ij})$ denotes the determinant of the $n \times n$ matrix (g_{ij}). [*Hint:* Define

$$f_{ij} = \begin{cases} f(x_i) & \text{if } i = j \\ 0 & \text{otherwise} \end{cases}.$$

What is the relationship between (f_{ij}) and (g_{ij})?]

11. Evaluate the determinant

$$\begin{vmatrix} (1,1) & (1,2) & \cdots & (1,n) \\ (2,1) & (2,2) & \cdots & (2,n) \\ & & \vdots & \\ (n,1) & (n,2) & \cdots & (n,n) \end{vmatrix}$$

where (i, j) denotes the greatest common divisor of i and j. [*Hint:* See Exercise 10.]

12. Let T be a tree. Partially order T by selecting as 0 one of its vertices of degree 1 and writing $x_i \leq x_j$ for two vertices x_i and x_j if the branch from 0 to x_j contains x_i. Denote by d_{ij} the length of the chain between x_i and x_j. Show that $\det(d_{ij}) = -(-2)^{n-2}(n-1)$.

13. Use Delsarte's result to compute the number of automorphisms of the group

$$Z_2 \oplus Z_4 \oplus Z_4 \oplus Z_8 \oplus Z_9 \oplus Z_{27}.$$

14. Find the number of subgroups isomorphic to $Z_2 \oplus Z_2$ in $Z_2 \oplus Z_2 \oplus Z_4 \oplus Z_{16} \oplus Z_{16}$.

15. Find the Delsarte polynomials F_X for all Abelian groups X of order p^3 and p^4.

16. Show that for an Abelian group X of order p^n the Delsarte polynomial F_X has degree n.

17. Write down the explicit form of $\phi(X)$ and $h(X, Y)$ for X a subgroup of the cyclic group $Y \cong Z_{p^n}$. Do the same for $Y \cong (Z_p)^n$.

18. Show that a group of order p^n with $n < p$ is necessarily regular (in the sense of P. Hall [1]).

19. Is the group of upper triangular $n \times n$ matrices with entries from $GF(q)$ and 1's on the diagonal always a regular group in the sense of Phillip Hall?

20. What is the type of the group in the previous exercise?

NOTES

Möbius inversion occupies a prominent place in the theory of numbers and it dates back to the works of Dedekind. Surprisingly perhaps, its potential as a general principle of enumeration was not realized until the 1930s. Hall formulated in 1933 an enumeration principle that involves the Möbius function of the lattice of subgroups of a p-group (see [1, Theorem 1.4]). In 1935 Weisner [4] realized that the inversion in question is one of significant generality and may be performed in quite abstract systems which he called hierarchies. We hope to have extracted the essence of his abstract results in the two theorems attributed to him in Section 3, for the notation in [4] is a bit cumbersome. A clear exposition of the abstract Möbius inversion appeared in another of Hall's papers [2] a few months later. It is the notation in [2] that we have followed, and it is there that Hall's results of Section 3 are found. Three years later Delsarte rediscovered Möbius inversion on the lattice of subgroups of an Abelian group. His contribution was the subject of the first part of Section 5. The extension of his results to p-groups regular in the sense of Phillip Hall is due to the author [7]. Section 9.19 is author's joint work with Kulkarni.

During the past twenty years systematic studies of partially ordered sets were undertaken by Rota and his students. Rota's paper [5] has generated a great deal of interest and activity in this area. It is a pleasure to note how such influences continue to greatly benefit the lovely subject to discrete mathematics.

REFERENCES

[1] P. Hall, A contribution to the theory of groups of prime-power order, *Proc. LMS*, **36**, 29–95 (1933).

[2] P. Hall, The Eulerian functions of a group, *Quart. J. Math. Oxford Series*, **7**, 134–151 (1936).

[3] S. Delsarte, Fonctions de Möbius sur les groupes Abeliens finis, *Ann. Math.*, **49**, 600–609 (1948).

[4] L. Weisner, Abstract theory of inversion of finite series, *Trans. AMS*, **38**, 474–484 (1935).

[5] G. C. Rota, On the foundations of combinatorial theory I. Theory of Möbius functions, *Z. Wahrscheinlichkeitstheorie*, **2**, 340–368 (1964).

[6] G. Constantine and R. S. Kulkarni, On a result of S. Delsarte, *Proc. AMS*, **92**, 149–152 (1984).

[7] G. Constantine, Möbius functions on regular p-groups of prime-power order, submitted for publication.

[8] J. J. Rotman, *The Theory of Groups*, Allyn and Bacon, Boston, 1973.

APPENDIX 1

Finite Groups

We mention definitions and list basic results. The material contained in here is usually covered in a one-semester course on finite groups. For individual reading we suggest [1], [2], and (if possible) the first five chapters of [3].

All the groups that occur are assumed finite. The notation $\doteq$ should be read "equal by definition." We mark by an asterisk the more relevant results and definitions.

1 NOTATION AND DEFINITIONS

1.1

A *binary operation* on a set G is a rule that associates to each pair of elements x and y of G another element of G denoted by xy. The operation is said to be *associative* if $(xy)z = x(yz)$, for all x, y, and z in G. It is called *commutative* if $xy = yx$ for all x and y in G. (If G, e.g., consists of the six permutations on three letters, we may select composition of permutations as a binary operation on G. This is an associative operation, but not a commutative one. Examples of commutative operations are addition of integers and multiplication of integers. These two operations are associative as well.)

*1.2

A *group* is a set G with an associative binary operation [sending (x, y) to xy] that contains an element called the *identity* (and denoted by 1) satisfying

$$x1 = 1x = x, \qquad \text{for all } x \text{ in } G$$

and such that for each x in G there exists x^{-1} in G (called the *inverse* of x) that satisfies

$$xx^{-1} = 1 = x^{-1}x.$$

The group G is *finite* if it has finitely many elements.

*1.3

A nonempty subset H of G is called a *subgroup* of G if H is a group under the binary operation of G.

1.4

$H \leq G$ means H is a subgroup of G. $H < G$ means $H \leq G$ and $H \neq G$. $X \subseteq G$ means X is a subset of G, and $|X|$ means the number of elements in X.

1.5

Let X and Y be nonempty subsets of G. Then $XY \doteq \{xy : x \in X, y \in Y\}$. If $x \in G$, then $xY \doteq \{x\}Y$. Since multiplication of subsets is associative, we may disregard parentheses. Let $g \in G$. Then $X^g \doteq g^{-1}Xg$. If $x \in G$, then $x^g \doteq g^{-1}xg$.

Let $H \leq G$. The subset gH is a (*right*) *coset* of H in G. By $|G: H|$ we denote the number of (right) cosets of H in G. (Note that the cosets of H partition G into $|G: H|$ classes of size $|H|$ each.)

*1.6

A subgroup H of G is called *normal* if $H^g = H$, for all g in G. By $H \triangleleft G$ we indicate that H is normal in G. If $H \triangleleft G$, then the set of cosets $G/H \doteq \{gH : g \in G\}$ is a group with operation defined by $(gH)(hH) = (gh)H$. We call G/H the *quotient group* of G over H.

*1.7

Suppose G and H are groups. A mapping T: $G \to H$ is called a *homomorphism* if $T(xy) = T(x)T(y)$ for all $x, y \in G$. We define the kernel of T by $\ker T \doteq \{g \in G : T(g) = 1_H\}$ and observe that $\ker T \triangleleft G$. A homomorphism that is both injective (or one-to-one) and surjective (or onto) is called an *isomorphism*. An isomorphism of G onto G is called an *automorphism*. $\mathrm{Aut}(G) \doteq$ {automorphisms of G}. The notation $G \cong H$ informs us that there is an isomorphism between G and H.

1.8

Let S be a set. A bijection $f: S \to S$ is called a *permutation*. Let $\Sigma_S =$ {permutations on S}. If $f \in \Sigma_S$ and $\alpha \in S$, then α^f denotes the image of α under f. The set Σ_S forms a group under composition of mappings, called the *symmetric group* on S. If $S = \{1, 2, \ldots, n\}$, we write Σ_n for Σ_S.

*1.9

We say that the group G *acts on the set* S if there is a homomorphism $T: G \to \Sigma_S$. For α in S and g in G we set $\alpha^g \doteq \alpha^{T(g)}$. Then for $h \in G$ we have $\alpha^{gh} = \alpha^{T(gh)} = \alpha^{T(g)T(h)} = (\alpha^{T(g)})^{T(h)} = (\alpha^g)^h$. If $\ker T = 1$, we say that

G acts *faithfully*. Let $\alpha \in S$. Define a subgroup $G_\alpha \doteq \{g \in G : \alpha^g = \alpha\}$ and call it the *stabilizer* of α in G.

For $\alpha \in S$ set $\alpha^G = \{\alpha^g : g \in G\}$ and call this subset of S the *orbit* of α under the action of G. The set S can be written as disjoint union of orbits. Also $|\alpha^G| = |G : G_\alpha|$. Thus $|S| = \Sigma_i |G : G_{\alpha_i}|$, where α_i is a representative from orbit i.

1.10

Suppose $X \subseteq G$, $X \neq \varnothing$. We call $N_G(X) \doteq \{g \in G : X^g = X\}$ the *normalizer* of X in G. We call $C_G(X) \doteq \{g \in G : gx = xg, \text{ for all } x \in X\}$ the *centralizer* of X in G. Let also $Z(G) \doteq C_G(G) = \{g \in G : gx = xg \text{ for all } x \text{ in } G\}$, and call it the *center* of G.

1.11

Suppose $X \subseteq G$. Then $\langle X \rangle \doteq \bigcap_{H \leq G,\, X \subseteq H} H$, and $\langle X \rangle$ is called the *subgroup of G generated by X*. We write $\langle x : x \text{ has property } P \rangle$ for $\langle \{x : x \text{ has property } P\} \rangle$.

1.12

Let $X \subseteq G$. Then $X^G \doteq \{g^{-1}Xg : g \in G\}$ and is called the *set of conjugates of* X or the *conjugacy class containing* X (in G). The group $\langle X^G \rangle$ is called the *normal closure* of X in G. (Note that $\langle X^G \rangle \triangleleft G$.)

1.13

Suppose $\psi \in \text{Aut}(G)$ and $X \subseteq G$. We say X is ψ-*invariant* if $\psi(X) = X$. If $\Gamma \subseteq \text{Aut}(G)$, we say X is Γ-*invariant* if $\psi(X) = X$, for all $\psi \in \Gamma$. We say X is *characteristic in G* if X is ψ-invariant for all $\psi \in \text{Aut}(G)$.

*1.14

Let $x, y \in G$. Then $[x, y] \doteq x^{-1}y^{-1}xy$ and is called the *commutator* of x and y. If $X, Y \subseteq G$, we set $[X, Y] \doteq \langle [x, y] : x \in X, y \in Y \rangle$. We set $G^{(1)} = [G, G]$ and $G^{(n+1)} = [G^{(n)}, G^{(n)}]$. The subgroup $G^{(1)}$ (which we actually denote by G') is called the *commutator* subgroup of G.

*1.15

Let p be a prime. A group P whose order is a power of p is called a *p-group*. For a p-group P we define $\phi(P) = \langle P', x^p : x \in P \rangle$, the *Frattini subgroup* of P. When $\phi(P) = 1$ we say that P is *elementary*.

1.16

Let p^a divide $|G|$, but p^{a+1} not divide $|G|$. Then the set of *Sylow p-subgroups* of G, $\mathrm{Syl}_p(G)$, is defined by $\mathrm{Syl}_p(G) \doteq \{P \leq G : |P| = p^a\}$.

1.17

Suppose $A, B \leq G$. We say that G is a *direct sum* of A and B (written $G = A \oplus B$) if $G = AB$ and $[A, B] = 1$.

1.18

Let A and B be groups. We say the group G is an *extension* of A by B if there exists $H \lhd G$ such that $H \cong A$ and $G/H \cong B$. We say that G *splits over* H if there exists $K \leq G$ such that $G = HK$ and $H \cap K = 1$.

1.19

The group G *acts* on the group H if there exists a homomorphism of G into $\mathrm{Aut}(H)$. If $H \lhd G$ and the homomorphism sends g into the automorphism $h \to g^{-1}hg$, we say that G *acts by conjugation* on H.

1.20

A group G is called *Abelian* if $xy = yx$ for all $x, y \in G$.

1.21

A group is called *simple* if it has no proper normal subgroups (i.e., if $H \lhd G$, then $H = G$ or $H = 1$).

2 BASIC RESULTS

2.1

If $H \leq G$, then $|H|$ divides $|G|$ (Lagrange).

2.2

If $H, K \leq G$, then $|HK| = |H|\,|K|/|H \cap K|$.

*2.3

For x in G set $x^G = \{g^{-1}xg : g \in G\}$. Then

(a) $|x^G| = |G : C_G(x)|$.

(b) There exist $x_1, \ldots, x_r \in G$ such that G is a disjoint union of $x_1^G, \ldots, x_r^G$. Thus $|G| = \sum_{i=1}^{r} |G : C_G(x_i)|$ (the "class equation"). (We call x_i^G the ith *orbit*, and the x_i's *orbit representatives*.)

2.4

Suppose H is a nonempty subset of G. Then $H \leq G$ if and only if $HH = H$.

2.5

The commutator subgroup G' and the center $Z(G)$ are normal subgroups of G. (They are in fact characteristic in G.)

***2.6**

The group G is Abelian if and only if $G' = 1$ [if and only if $Z(G) = G$].

2.7

A group C is called *cyclic* if $C = \langle c \rangle$, for some element $c \in C$. If $G/Z(G)$ is cyclic, then G is Abelian.

2.8

Suppose $H \leq G$ and set $H^G = \{g^{-1}Hg : g \in G\}$. Then $|H^G| = |G : N_G(H)|$.

2.9

Suppose $H \leq G$. Then $N_G(H)/C_G(H)$ is isomorphic to a subgroup of $\mathrm{Aut}(H)$.

2.10

Suppose $E \triangleleft G$. Define a mapping T from G to $\mathrm{Aut}(E)$ by $T(g) : l \to g^{-1}lg$; $l \in E$. Then T is a homomorphism and $\ker T = C_G(E)$, a normal subgroup of G.

2.11

Let G be an Abelian group. By G^* we denote the set of all homomorphisms into the unit circle (of complex numbers of modulus 1). Define a binary operation on G^* by $(gh)(x) = g(x)h(x)$ for all x in G, the product on the right-hand side being the ordinary multiplication of complex numbers. With this multiplication G^* also becomes an Abelian group. In fact G^* and G are isomorphic.

For $H \leq G$ let $H^* = \{h \in G^* : h(x) = 1, \text{ for all } x \text{ in } H\}$. We call H^* the *annihilator* of H. Then $H^* \leq G^*$, and $H^* \cong G/H$.

3 NORMAL SUBGROUPS AND FACTOR GROUPS

3.1

If $H, K \triangleleft G$, then $HK \triangleleft G$ and $H \cap K \triangleleft G$.

3.2

If $H \leq G$ and $K \triangleleft G$, then $HK \leq G$ and $(H \cap K) \triangleleft H$.

*3.3

Suppose T is a homomorphism of G *onto* $\overline{G}$. Then:

(a) $G/\ker T \cong \overline{G}$ (first isomorphism theorem).
(b) $H \triangleleft G$ implies $T(H) \triangleleft \overline{G}$.
(c) $\overline{H} \triangleleft \overline{G}$ implies $T^{-1}(\overline{H}) \triangleleft G$, where $T^{-1}(\overline{H}) \doteq \{h \in G : T(h) \in \overline{H}\}$.
(d) There exists a bijection between the normal subgroups of $\overline{G}$ and the normal subgroups of G containing the kernel $\ker T$. The bijection is $H \leftrightarrow T(H)$, $H \geq \ker T$.

3.4

If $H \leq G$ and $K \triangleleft G$, then $HK/K \cong H/(H \cap K)$ (second isomorphism theorem).

3.5

If $H, K \triangleleft G$ and $K \leq H$, then $G/H \cong (G/K)/(H/K)$ (third isomorphism theorem).

3.6

There exists a sequence $G = G_0 \geq G_1 \geq \cdots \geq G_n = 1$ of subgroups such that each $G_{i+1} \triangleleft G_i$ and G_i/G_{i+1} is a simple group.

3.7

Suppose $H \leq G$ and set $\mathrm{Core}_G(H) \doteq \langle K \leq H : K \triangleleft G \rangle$. Then $|G/\mathrm{Core}_G(H)|$ divides $|G : H|!$.

4 SYLOW SUBGROUPS

*4.1 Sylow's Theorem

Let p be a prime. Then the following hold:

(a) $\mathrm{Syl}_p(G) \neq \varnothing$.

(b) If $P, Q \in \mathrm{Syl}_p(G)$, then $P = g^{-1}Qg$ for some g in G.
(c) $|\mathrm{Syl}_p(G)| = |G: N_G(P)|$ for any $P \in \mathrm{Syl}_p(G)$, and $|\mathrm{Syl}_p(G)| \equiv 1 \pmod{p}$.
(d) If H is a p-subgroup of G, then $H \leq P$ for some $P \in \mathrm{Syl}_p(G)$.

4.2

Suppose G is Abelian and let $P_1, \ldots, P_r$ be Sylow subgroups of G for a complete set $p_1, \ldots, p_r$ of distinct prime divisors of $|G|$. Then $G = P_1 \oplus \cdots \oplus P_r$.

*4.3

Suppose for any prime p and $P \in \mathrm{Syl}_p(G)$ that $P \triangleleft G$. Then $G = P_1 \oplus \cdots \oplus P_r$, where $P_1, \ldots, P_r$ are the Sylow subgroups of G.

4.4

If $P \in \mathrm{Syl}_p(G)$ and $H \triangleleft G$, then $P \cap H \in \mathrm{Syl}_p(H)$.

4.5 The Frattini Argument

If $H \triangleleft G$ and $P \in \mathrm{Syl}_p(H)$, then $G = HN_G(P)$.

4.6

Suppose T is a homomorphism of G *onto* $\overline{G}$. If $P \in \mathrm{Syl}_p(G)$, then $T(P) \in \mathrm{Syl}_p(\overline{G})$.

5 GROUPS OF PRIME POWER ORDER

Let p be a prime number. A *p-group* is a group whose number of elements is a power of p.

*5.1

Suppose G is a p-group with more than one element. Then:

(a) $Z(G) \neq 1$.
(b) $1 \neq H \triangleleft G$ implies $H \cap Z(G) \neq 1$.
(c) $G' < G$ (note *strict* inclusion).
(d) $H < G$ implies $H < N_G(H)$ (note *strict* inclusions).
(e) Every subgroup of index p in G is normal.

(f) Every proper subgroup of G is contained in some subgroup of index p.
(g) A group of order p^2 is necessarily Abelian.
(h) Every normal subgroup of G of index p^2 contains the commutator group G'.

*5.2

A subgroup is called *maximal* if it is of index p in the p-group G. By $\phi(G)$ we denote the intersection of all maximal subgroups. The subgroup $\phi(G)$ is normal in G and is called the *Frattini* (*or principal*) subgroup of G. The subgroup $\phi(G)$ has the following properties:

(a) $\phi(G) = \langle G', x^p : x \in G \rangle$.
(b) $G/\phi(G)$ is Abelian with all its elements (except identity) or order p. The subgroup $\phi(G)$ is the smallest normal subgroup of G with this property.
(c) If $G = \langle X, \phi(G) \rangle$ for some subset X of G, then $G = \langle X \rangle$. [This is why $\phi(G)$ is sometimes called the subgroup of nongenerators of G.]

*5.3

If G is a p-group and $|G/\phi(G)| = p^d$, then d is the minimal number of elements required to generate G.

*5.4

Suppose P is an Abelian p-group. Let A be a cyclic subgroup of maximal order in P. Then we can find $B \leq P$ such that $P = A \oplus B$.

*5.5

By Z_p we mean a cyclic group of order p. An elementary Abelian p-group is a direct sum of Z_p's. Suppose E is an elementary Abelian p-group. Then there is a "natural" interpretation of E as a vector space over the field $GF(p)$ of p elements. If $|E| = p^n$, then the dimension of E over $GF(p)$ is n. Finally, $\mathrm{Aut}(E) \cong \{n \times n$ invertible matrices with coefficients in $GF(p)\}$.

REFERENCES

[1] W. R. Scott, *Group Theory*, Prentice-Hall, Englewood Cliffs, NJ, 1964.
[2] J. J. Rotman, *The Theory of Groups*, Allyn and Bacon, Boston, 1973.
[3] D. Gorenstein, *Finite Groups*, Harper and Row, New York, 1968.

APPENDIX 2

Finite Fields, Vector Spaces, and Finite Geometries

A number of results on the topics listed in the title play a central role in the construction of block designs; they also form the backbone for interesting computations with the Möbius function. We list the definitions and mark with an asterisk the facts that are needed (implicitly or explicitly) throughout the text. Suggested readings are appropriate selections from [1], [2], and [3].

1 FINITE FIELDS

*1.1

A set F with two binary operations, one denoted by $+$ and the other by $\cdot$ (each being commutative and associative) is said to be a *field* if $(F, +)$ is an Abelian group (with 0 as identity), if $(F - \{0\}, \cdot)$ is an Abelian group (with 1 as identity), and if the two operations are related by the law of *distributivity*:

$$(a + b)c = ac + bc.$$

A field is called *finite* if it has finitely many elements.

1.2

In a finite field F we can add, subtract, multiply, and divide (except that division by 0 is not allowed), and remain at all times *within* the finite set F. [That a *finite* set can allow such great computational diversity is fascinating indeed. For this reason the elements of F should be thought of as (magical) "numbers," in a way. We use them in this capacity when we consider vector spaces over finite fields.]

1.3

The set $\{0, 1, \ldots, p - 1\}$ with p a *prime* number forms a field with addition and multiplication done modulo p. (The sum of x and y modulo p is the remainder upon dividing $x + y$ to p. Multiplication modulo p is analogously defined.) We denote this field by $GF(p)$. The capital "G" honors Evariste Galois, the french mathematician who discovered the finite fields.

1.4

A polynomial $P(x)$ [with coefficients in $GF(p)$] is called *irreducible* over $GF(p)$ if it has no root in $GF(p)$.

***1.5**

If $P(x)$ is irreducible over $GF(p)$ and has degree m, then the set of all polynomials in x of degree $m - 1$ or less [and coefficients from $GF(p)$] is a *finite field* with addition and multiplication modulo the polynomial $P(x)$. This field has p^m elements and we denote it by $GF(q)$, where $q = p^m$.

***1.6**

A field with q elements exists if and only if q is a power of a prime.

***1.7**

Up to field isomorphisms, there is a *unique* field with q elements [which we denote by $GF(q)$].

1.8

$GF(q) - \{0\}$ is a *cyclic* group with respect to multiplication. [The additive structure of $GF(q)$ is an elementary Abelian group, that is, $(Z_p)^m$, where $q = p^m$.]

1.9

Every element of $GF(q)$ is a root of the polynomial $x^q - x$.

1.10

$GF(p^r)$ contains a subfield (isomorphic to) $GF(p^s)$ if and only if s divides r. Moreover, an element β of $GF(p^r)$ belongs to $GF(p^s)$ if and only if $\beta^{p^s} = \beta$.

1.11

An *automorphism* σ of $GF(q)$ (with $q = p^m$) is a bijection on $GF(q)$ that *fixes* the subfield $GF(p)$ elementwise and satisfies

$$\sigma(\alpha + \beta) = \sigma(\alpha) + \sigma(\beta),$$

and

$$\sigma(\alpha\beta) = \sigma(\alpha)\sigma(\beta),$$

for all α and β in $GF(q)$.

The set of automorphisms of $GF(q)$ forms a group (under composition of functions) called the *automorphism group* of $GF(q)$, and is written as Aut $GF(q)$.

1.12

If $|F| = p^m$, then Aut F is a cyclic group of order m. Aut F is in fact generated by

$$\sigma\colon \alpha \to \alpha^p,$$

where α is a nonzero element of the field F that *generates* the multiplicative (cyclic) group $F - \{0\}$.

2 VECTOR SPACES

*2.1

An Abelian group V (with addition as group operation, denoted by $+$) and with an operation of scalar multiplication by elements of a field F is called a *vector space* if the two operations satisfy

$$\alpha(v + w) = \alpha v + \alpha w$$

$$(\alpha + \beta)v = \alpha v + \beta v$$

$$(\alpha\beta)v = \alpha(\beta v)$$

$$1v = v$$

for all $v, w \in V$ and all $\alpha, \beta \in F$ (also $1 \in F$).

The elements of V are called *vectors*. The identity with respect to addition is called the *zero vector* and is denoted by 0 (this is not the same as the scalar 0 in the field F).

2.2

A subset W of vectors of V is called a *vector subspace* if W is itself a vector space (with the operations defined on V). By $W \leq V$ we denote the fact that W is a subspace of V.

2.3

For vectors $v_1, v_2, \ldots, v_n$ in V we call the vector $\sum_{i=1}^{n} \alpha_i v_i$ a *linear combination* of the v_i's (here $\alpha_i \in F$). The *subspace spanned* by $v_1, \ldots, v_n$ is the set of all linear combinations of $v_1, \ldots, v_n$; this subspace we denote by $\langle v_1, v_2, \ldots, v_n \rangle$.

2.4

A set of vectors $v_1, \ldots, v_n$ are called *dependent* if there are scalars $\alpha_1, \ldots, \alpha_n$ not all zero such that $\sum_{i=1}^{n} \alpha_i v_i = 0$. A set of vectors $v_1, \ldots, v_n$ are *independent* if they are not dependent. (Alternatively, $v_1, \ldots, v_n$ are called independent if $\sum_{i=1}^{n} \alpha_i v_i = 0$ implies $\alpha_i = 0$, for all $1 \le i \le n$.)

2.5

A *basis* of V is a set of independent vectors whose span is V.

*2.6

Let V be a vector space over the field F. Then:

(a) V has a basis.
(b) Any two bases of V contain the same number of vectors.
(c) Given a basis, each vector can be written in a unique way as a linear combination of the vectors in that basis.

2.7

The number of vectors in a basis of V is called the *dimension* of V (and we denote it by $\dim V$).

*2.8

A vector space of dimension n over $GF(q)$ contains q^n vectors.

*2.9

By the sum $S + T$ of subspaces S and T we mean the set of vectors $x + y$, with x in S and y in T. (The intersection $S \cap T$ is the usual set intersection.)

Both $S + T$ and $S \cap T$ are subspaces, and

$$\dim(S + T) = \dim S + \dim T - \dim(S \cap T).$$

2.10

Let V be a vector space over the field F. An *inner product* (denoted by $(\,,\,)$) is a function defined on pairs of vectors in V with values in F, such that:

1. $(x, y) = (y, x)$.
2. $(\alpha x + \beta y, z) = \alpha(x, z) + \beta(y, z)$.
3. If $(x, y) = 0$ for all y in V, then $x = 0$.

2.11

Vectors x and y are said to be *orthogonal* if $(x, y) = 0$. If S is a subspace, then we define the subspace

$$S^{\perp} = \{x : (x, y) = 0, \text{ for all } y \text{ in } S\},$$

and call $S^{\perp}$ the *orthogonal subspace* of S.

2.12

For S and T subspaces of V we have the following properties:

(a) $(S + T)^{\perp} = S^{\perp} \cap T^{\perp}$.
(b) $\dim S + \dim S^{\perp} = \dim V$.
(c) $(S^{\perp})^{\perp} = S$.

3 FINITE GEOMETRIES

*3.1

A *finite projective geometry* is a finite set of *points* and a collection of subsets of the points, called *lines*, that satisfy axioms (a) through (d):

(a) There is a unique line passing through any two distinct points.
(b) Every line contains at least three points.
(c) If the distinct lines (l) and (m) intersect in 1, and if 2, 3 are points on (l) not equal to 1, and if 4 and 5 are points on (m) not equal to 1, then the line through 2 and 4 intersects the line through 3 and 5. (See figure below.)

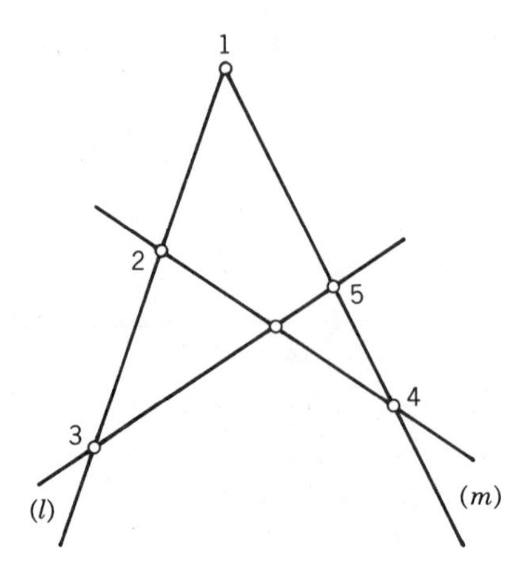

(d) For any point there are at least two lines not containing it, and for any line there are at least two points not on it.

3.2

A *subspace* of the projective geometry is a subset of points with the property that together with two distinct points it contains all points on the line determined by the two points.

A *hyperplane* is a maximal proper subspace. (By maximal we mean maximal with respect to inclusion on subspaces and by proper we mean that it is not the whole set of points.)

*3.3

A *finite affine geometry* is obtained by deleting the points of a fixed hyperplane (called the hyperplane at infinity) from the subspaces of a finite projective geometry. The resulting subsets are called the *subspaces* of the affine geometry. (In particular when removing the points of the hyperplane from the lines of the projective geometry we obtain the *lines* of the affine geometry.)

*3.4

Let V be a vector space of dimension $n + 1$ over $GF(q)$. We can construct a projective geometry as follows: The points are the one-dimensional subspaces of V, and the lines are the subspaces of dimension 2 of V. We denote the hyperplanes of this geometry (and also the geometry itself) by $PG(n, q)$.

*3.5

Let V be a vector space of dimension n over $GF(q)$. An affine geometry is obtained by letting the points be the vectors in V, and letting the lines of the geometry be the subspaces of dimension 1, along with their cosets. We denote this geometry by $AG(n, q)$. [$AG(n, q)$ can be obtained from $PG(n, q)$ by deleting the points of a hyperplane of $PG(n, q)$.]

3.6

A nonempty set T of points in a projective geometry is called *independent* if for every point x the smallest subspace containing $T - \{x\}$ does not contain x. [Interpret this geometric notion in algebraic terms in $PG(n, q)$.]

If the largest independent set of points in a subspace has cardinality m, then $m - 1$ is defined to be the (*projective*) *dimension* of that subspace. This defines, in particular, the dimension of the projective geometry itself by letting the subspace consist of all points of the geometry.

Similar definitions hold for an affine geometry.

3.7

$PG(n, q)$ has projective dimension n, and $AG(n, q)$ has affine dimension n.

*3.8 The First Fundamental Theorem of Projective Geometry

If $n \geq 3$, then a finite projective geometry of projective dimension n is a $PG(n, q)$ for some prime power q, and an affine geometry of affine dimension n is an $AG(n, q)$ for some prime power q.

*3.9 The Principle of Duality

Any statement about $PG(n, q)$ remains true if we make the interchanges

$$\begin{aligned} \text{point} &\leftrightarrow \text{hyperplane}, \\ PG(r, q) &\leftrightarrow PG(n - r - 1, q), \\ \text{intersect} &\leftrightarrow \text{span}, \\ \text{contained in} &\leftrightarrow \text{contained by}. \end{aligned}$$

[Algebraically, the underlying vector space V over $GF(q)$ can be endowed with an inner product. Then duality corresponds to orthogonality. If V has dimension $n + 1$, then there is a natural bijection

$$\langle x \rangle \leftrightarrow \langle x \rangle^{\perp}$$

that maps the one-dimensional subspaces into their (n-dimensional) orthogonal complements. In the geometry this induces the duality between points and hyperplanes.]

3.10

A *collineation* of a projective geometry is a permutation on points that maps lines into lines. It follows that every subspace is mapped into a subspace of the same dimension.

The set of all collineations of $PG(n, q)$ forms a group, called the *collineation group* (or the *automorphism group*) of $PG(n, q)$.

3.11

With p a prime, let $q = p^s$ and recall that the automorphism group of $GF(q)$ is cyclic of order s, and generated by $\sigma_p : \beta \to \beta^p$ [$\beta \in GF(q)$].

Let V be an $(n + 1)$-dimensional vector space over $GF(q)$. Any nonsingular linear transformation induces a collineation of $PG(n, q)$. Moreover, if T is such a linear transformation, then T and λT [$0 \neq \lambda \in GF(q)$] induce the same collineation. In addition, the field automorphism σ_p induces also a collineation.

The nonsingular transformations (modulo the scalar multiples) and σ_p generate a group that we denote by $P\Gamma L(n + 1, q)$.

*3.12 The Second Fundamental Theorem of Projective Geometry

The collineation group of $PG(n, q)$ is $P\Gamma L(n+1, q)$.

$$|P\Gamma L(n+1, q)| = \frac{s}{q-1} \prod_{i=1}^{n} (q^{n+1} - q^i),$$

where $q = p^s$, for p a prime.

3.13

The *collineation* (or *automorphism*) *group* of $AG(n, q)$ is defined similarly, that is, it is the group of all permutation on the points of the affine geometry that preserves its lines.

3.14

The automorphism group of $AG(n, q)$ is the subgroup of $P\Gamma L(n+1, q)$ that fixes (setwise) the hyperplane at infinity; its order is

$$s \prod_{i=1}^{n-1} (q^n - q^i),$$

where $q = p^s$, with p a prime.

3.15

A projective geometry of dimension 2 is called a *projective plane*. Unlike in dimensions 3 or more, a projective plane need not be a $PG(2, q)$.

A projective plane has $n^2 + n + 1$ points ($n \geq 2$) with $n + 1$ points on each line, and a unique line passing through any two distinct points (this being in fact a characterization of a projective plane).

Yet another equivalent definition of a projective plane (commonly used) is as follows. A projective plane is a set of points satisfying the following axioms:

(a) There is a single line containing any two distinct points.
(b) Any two lines intersect in one point.
(c) There exist four points no three of which lie on a line.

REFERENCES

[1] F. J. MacWilliams and N. J. A. Sloane, *The Theory of Error-Correcting Codes*, North-Holland, Amsterdam, 1977.

[2] V. Pless, *Introduction to the Theory of Error-Correcting Codes*, Wiley, New York, 1982.

[3] E. Snapper and R. Troyer, *Metric Affine Geometry*, Academic Press, New York, 1971.

APENDIX 3

The Four Squares Theorem and Witt's Cancellation Law

These two classical results are used to prove Bruck, Ryser, and Chowla's theorem, presented in Section 5 of Chapter 7. As references we recommend [1], [2], and [3].

1 THE FOUR SQUARES THEOREM

This theorem informs us as follows:

Lagrange's Four Squares Theorem. *Every positive integer is the sum of four squares.*
[We allow 0's as well as repetitions among the four squares, i.e., $5 = 2^2 + 1^2 + 0^2 + 0^2$, or $9 = 2^2 + 2^2 + 1^2 + 0^2$.]

The proof is in three "steps."

Step 1 (Euler's Identity).

$$\begin{aligned}(x_1^2 + x_2^2 + x_3^2 + x_4^2)(y_1^2 + y_2^2 + y_3^2 + y_4^2) \\ = (x_1y_1 + x_2y_2 + x_3y_3 + x_4y_4)^2 + (x_1y_2 - x_2y_1 + x_3y_4 - x_4y_3)^2 \\ + (x_1y_3 - x_2y_4 - x_3y_1 + x_4y_2)^2 + (x_1y_4 + x_2y_3 - x_3y_2 - x_4y_1).\end{aligned} \quad (1)$$

[This shows that if both a and b are sums of four squares, then so is the product ab. It thus suffices to prove the theorem for prime numbers, since every positive integer factors into primes.]

To prove (1), set

$$H = \begin{bmatrix} y_1 & y_2 & y_3 & y_4 \\ y_2 & -y_1 & y_4 & -y_3 \\ y_3 & -y_4 & -y_1 & y_2 \\ y_4 & y_3 & -y_2 & -y_1 \end{bmatrix}.$$

Then $H^tH = (y_1^2 + y_2^2 + y_3^2 + y_4^2)I$. By defining $z = Hx$, we can write

$$z_1^2 + z_2^2 + z_3^2 + z_4^2 = z^tz = (Hx)^tHx = x^tH^tHx$$
$$= (x_1^2 + x_2^2 + x_3^2 + x_4^2)(y_1^2 + y_2^2 + y_3^2 + y_4^2),$$

which shows that the product of two integers, each the sum of four squares, is itself a sum of four squares. The explicit expressions of the components of z are written in (1).

Step 2. For p a prime ($p \neq 2$) there exists an integer m, $1 \leq m < p$, such that mp is the sum of four squares.

To see this, let $P = \{0^2, 1^2, 2^2, \ldots, ((p-1)/2)^2\}$ and $Q = \{-1 - 0^2, -1 - 1^2, -1 - 2^2, \ldots, -1 - ((p-1)/2)^2\}$. Observe that all entries in P are distinct modulo p and thus $|P| = (p+1)/2$; similarly $|Q| = (p+1)/2$. There are p residue classes modulo p, hence $p \geq |P \cup Q| = |P| + |Q| - |P \cap Q|$, which informs us that $|P \cap Q| \geq 1$. For some x^2 in P, and $-1 - y^2$ in Q, we can therefore write $x^2 + y^2 + 1 \equiv 0 \pmod{p}$. Then $mp = x^2 + y^2 + 1^2 + 0^2$, and

$$1 \leq m \leq \frac{1}{p}(x^2 + y^2 + 1) \leq \frac{1}{p}\left[\left(\frac{p-1}{2}\right)^2 + \left(\frac{p-1}{2}\right)^2 + 1\right]$$
$$\leq \frac{1}{p}\left(\frac{p^2}{2} + 1\right) < p.$$

Step 3. The integer m in Step 2 equals 1. [Step 2 allows us now to conclude that any prime number is a sum of four squares; by Step 1 this concludes the proof of the theorem.]

Let m be the *smallest* integer such that $mp = x_1^2 + x_2^2 + x_3^2 + x_4^2$ (such m exists by Step 2). We prove that $m = 1$; first we assume that m is even and derive a contradiction to its minimality; then we assume m odd and $m \geq 3$, and contradict minimality once again.

Indeed, if m is even so is $mp = x_1^2 + x_2^2 + x_3^2 + x_4^2$, and hence either none, two, or four of the x_i's are even. [If not all four x_i's are odd let x_1 and x_2 be even (without loss of generality).] Then in all three cases $x_1 - x_2$, $x_1 + x_2$, $x_3 - x_4$, and $x_3 + x_4$ are even, and

$$\left(\frac{x_1 - x_2}{2}\right)^2 + \left(\frac{x_1 + x_2}{2}\right)^2 + \left(\frac{x_3 - x_4}{2}\right)^2 + \left(\frac{x_3 + x_4}{2}\right)^2 = \frac{m}{2}p,$$

thus contradicting the minimality of m.

Assume now that m is odd and $3 \leq m < p$. Define numbers y_i, $i = 1, 2, 3, 4$, as follows:

$$y_i \equiv x_i \pmod{m}, \qquad -\frac{m-1}{2} \leq y_i \leq \frac{m-1}{2}. \tag{2}$$

Then $y_1^2 + y_2^2 + y_3^2 + y_4^2 \equiv x_1^2 + x_2^2 + x_3^2 + x_4^2 \equiv 0 \pmod{m}$, since $x_1^2 + x_2^2 + x_3^2 + x_4^2 = mp$. We can therefore write

$$y_1^2 + y_2^2 + y_3^2 + y_4^2 = mn, \qquad \text{and} \qquad 0 \le n \le \frac{1}{m}4\left(\frac{m-1}{2}\right)^2 < m.$$

Observe that n is strictly positive. [Else we would have $y_1 = y_2 = y_3 = y_4 = 0$, and thus $x_1 \equiv x_2 \equiv x_3 \equiv x_4 \pmod{m}$ by (2). This would imply $mp = x_1^2 + x_2^2 + x_3^2 + x_4^2 \equiv 0 \pmod{m^2}$, which cannot happen since $3 \le m < p$.]

Now we can write

$$\begin{aligned} m^2np = mpmn &= \left(x_1^2 + x_2^2 + x_3^2 + x_4^2\right)\left(y_1^2 + y_2^2 + y_3^2 + y_4^2\right) \\ &= z_1^2 + z_2^2 + z_3^2 + z_4^2, \end{aligned} \tag{3}$$

by Step 1. The explicit expressions for the z_i's appear on the right-hand side of (1). These explicit expressions and the content of (2) allow us to conclude that $z_i \equiv 0 \pmod{m}$, $i = 1, 2, 3, 4$. (E.g., $z_2 = x_1y_2 - x_2y_1 + x_3y_4 - x_4y_3 \equiv x_1x_2 - x_2x_1 + x_3x_4 - x_4x_3 \equiv 0$.)

Dividing (3) by m^2 we obtain

$$np = \left(\frac{z_1}{m}\right)^2 + \left(\frac{z_2}{m}\right)^2 + \left(\frac{z_3}{m}\right)^2 + \left(\frac{z_4}{m}\right)^2,$$

with $0 < n < m$. This shows that m is not least if $m \ge 3$. Hence $m = 1$.

Steps 2 and 3 prove that every prime $p (p \ne 2)$ is the sum of four squares. But also $2 = 1^2 + 1^2 + 0^2 + 0^2$, and therefore every prime number is a sum of four squares. By Step 1 we conclude the proof.

2 WITT'S CANCELLATION LAW

The result we prove in this section is well known in the theory of symmetric bilinear forms. We use its matricial formulation, but the result is in actuality a geometric one.

Once and for all we make clear that *all vector spaces that occur are of finite dimension over the field Q of rational numbers*. [We thus do ourselves the favor of avoiding nightmarish descriptions of the structure of the underlying field, such as being of characteristic $\ne 2$, or having (or not having) all positive numbers squares, etc.]

2.1

Let V be a vector space. A *symmetric bilinear form* (or *form*, for short) on V is a function $(\,,\,) : V \times V \to Q$ satisfying

$$(x, y) = (y, x),$$

and

$$(ax + by, z) = a(x, z) + b(y, z).$$

[It follows from these assumptions that also

$$(z, ax + by) = a(z, x) + b(z, y).$$

If the form has the additional property that $x = 0$ whenever $(x, y) = 0$, for all $y \in V$, we call it an *inner product*.]

2.2

When we select a base in V [$u_1, u_2, \ldots, u_n$ (say)], the form yields a symmetric matrix A with (i, j)th entry $a_{ij} = (u_i, u_j)$. (For a fixed form on V, different bases yield different symmetric matrices.)

We call two symmetric matrices rationally *congruent* if they describe the same form in two different bases. Let the matrix of the form be A in one base, and B in another. How are A and B related? If P is a (nonsingular) transformation on V that carries the first base $(u_1, \ldots, u_n)$ into the second $(v_1, \ldots, v_n)$ (explicitly, $Pu_i = v_i$), then

$$B = PAP^t.$$

Indeed,

$$\begin{aligned} b_{ij} &= (v_i, v_j) = (\Sigma_k p_{ik} u_k, \Sigma_m p_{jm} u_m) \\ &= \Sigma_k \Sigma_m p_{ik} a_{km} p_{jm} = \Sigma_m (\Sigma_k p_{ik} a_{km}) p_{jm}, \end{aligned}$$

the last expression being the (i, j)th entry in the matrix PAP^t (by P^t we denote the transpose of the matrix P).

The writing $A \overset{c}{=} B$ intimates the fact that the symmetric matrices A and B (with rational entries) are rationally congruent. *Congruence is an equivalence relation*. [In one dimension, for example, two rational numbers are congruent if their quotient is a square (over Q). Thus 1 and 2 are not congruent, but 1 and 4 are.]

Observe that a symmetric matrix $A = (a_{ij})$ induces a form on V though the process of picking a base $u_1, \ldots, u_n$ and defining $(u_i, u_j) = a_{ij}$ (then extending the form by linearity to arbitrary linear combinations).

2.3

Given a form on V, we call vectors x and y *orthogonal* if $(x, y) = 0$. A vector x is said to be *isotropic* if $(x, x) = 0$, else we call it *nonisotropic*.

If the form is identically zero all vectors are isotropic, for example. On the other hand we can make the following statement:

$*$ *If the form is not identically zero, then nonisotropic vectors necessarily exist.*

For if $(x, x) = 0$ for all x, then $0 = (x + y, x + y) = (x, x) + (y, y) + 2(x, y) = 2(x, y)$, implying that $(x, y) = 0$ for all x, y, which contradicts the assumption.

For a vector u the subspace $\{y : (u, y) = 0\}$ is denoted by $\langle u \rangle^{\perp}$ and is called the *orthogonal complement* of u. Observe that:

$*$ *If $(u, u) \neq 0$, then $V = \langle u \rangle \oplus \langle u \rangle^{\perp}$ (and, in particular, the dimension of $\langle u \rangle^{\perp}$ is precisely one less than the dimension of V).*

Indeed, let n denote the dimension of V. It is clear that u does not belong to $\langle u \rangle^{\perp}$, since $(u, u) \neq 0$. Hence $\langle u \rangle^{\perp}$ is not the whole space and thus its dimension is at most $n - 1$. Evidently $\langle u \rangle \cap \langle u \rangle^{\perp} = 0$. It remains to prove that $\langle u \rangle$ and $\langle u \rangle^{\perp}$ span all of V. Let x be any vector in V. Then write

$$x = \frac{(x, u)}{(u, u)} u + \left(x - \frac{(x, u)}{(u, u)} u \right)$$

and observe that $x - (x, u)(u, u)^{-1} u$ is in $\langle u \rangle^{\perp}$. This shows that $V = \langle u \rangle \oplus \langle u \rangle^{\perp}$.

Furthermore,

$*$ *Any vector space equiped with a bilinear form has an orthogonal base.* [A base $u_1, \ldots, u_n$ is *orthogonal* if $(u_i, u_j) = 0$, for $i \neq j$.]

If the bilinear form is identically zero, any base is orthogonal. Else the bilinear form is not identically zero and thus a vector u_1 exists such that $(u_1, u_1) \neq 0$. Write the space V as follows:

$$V = \langle u_1 \rangle \oplus \langle u_1 \rangle^{\perp}.$$

The dimension of $\langle u_1 \rangle^{\perp}$ being one less than that of V we may assume by induction that $\langle u_1 \rangle^{\perp}$ has an orthogonal base. Let $u_2, \ldots, u_n$ be an orthogonal base for $\langle u_1 \rangle^{\perp}$; then u_1 together with $u_2, \ldots, u_n$ form an orthogonal base for V.

2.4

A bijective linear transformation T on V that preserves the bilinear form (in the sense that $(Tx, Ty) = (x, y)$, for all x and y in V) is called an *orthogonal transformation*.

An important example is the *reflection* in a nonisotropic vector x:

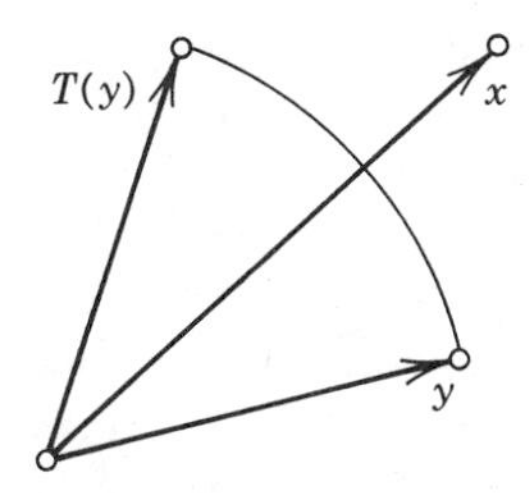

Algebraically the reflection T in x is defined by its (geometrically obvious) property:

$$T(y) + y = 2\frac{(y, x)}{(x, x)}x.$$

* *It is geometrically clear that* $T(x) = x$, *that* $T^2 =$ *identity* [i.e., $T(T(y)) = y$, for all y], *and, most importantly, an easy check shows that indeed* $(T(y), T(z)) = (y, z)$, *for all* y *and* z.

[To understand the geometry observe again the cannonical writing

$$y = \frac{(y, x)}{(x, x)}x + \left(y - \frac{(y, x)}{(x, x)}x\right),$$

with $y \to (y, x)(x, x)^{-1}x$ being the *orthogonal projection* of y on x. That is:

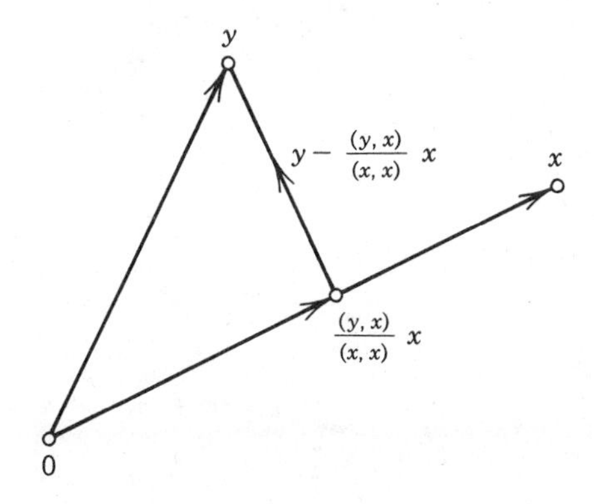

with

$$\left(\frac{(y, x)}{(x, x)}x, y - \frac{(y, x)}{(x, x)}x\right) = 0.]$$

To get to Witt's cancellation law one more step is necessary:

$*$ *If y and z are vectors in V that satisfy $(y, y) = (z, z) \neq 0$, then there exists an orthogonal transformation on V sending y into z.*

Indeed, if $y + z$ is nonisotropic, then the reflection in $y + z$ would do the trick (see below):

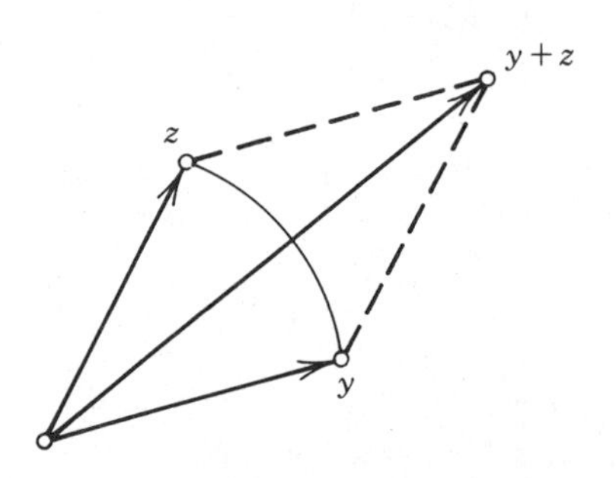

If $y + z$ is isotropic, then $y - z$ is not, because

$$(y - z, y - z) = 4(y, y) - (y + z, y + z) = 4(y, y) \neq 0.$$

We can therefore reflect in $y - z$ to send y into $-z$ and follow this by the orthogonal transformation that sends a vector into its negative to map y into z. The figure below captures the geometry:

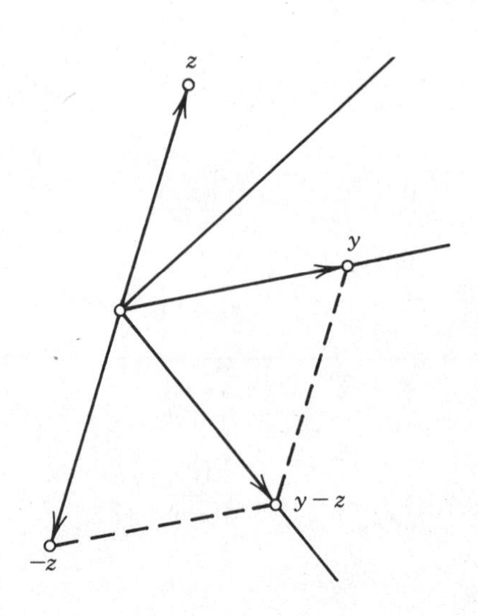

2.5

Given two symmetric matrices $A = (a_{ij})$ and $B = (b_{ij})$ let us remind ourselves what the notation $A \overset{c}{=} B$ stands for. We are in the presence of a bilinear form $(,)$ on a vector space V. Writing $A \overset{c}{=} B$ assumes the existence of two bases on V (say $u_1, \ldots, u_n$ and $v_1, \ldots, v_n$) such that $(u_i, u_j) = a_{ij}$, and $(v_i, v_j) = b_{ij}$.

This is in particular valid for *diagonal* matrices. Let us write $[a_1, \ldots, a_n]$ for the diagonal matrix with its ith diagonal entry equal to a_i. Then

$$[a_1, \ldots, a_n] \overset{c}{=} [b_1, \ldots, b_n]$$

assumes the existence of two (orthogonal) bases, $u_1, \ldots, u_n$ and $v_1, \ldots, v_n$ such that $(u_i, u_i) = a_i$ and $(v_i, v_i) = b_i$.

We now prove Witt's theorem.

Witt's Theorem. *If* $[a_1, \ldots, a_m, b_1, \ldots, b_n] \overset{c}{=} [a_1, \ldots, a_m, c_1, \ldots, c_n]$, *then* $[b_1, \ldots, b_n] \overset{c}{=} [c_1, \ldots, c_n]$.

We distinguish two cases: the former assumes that all the a_i's are nonzero, while in the latter case at least one a_i is zero.

Case 1. ($a_i \neq 0$, for $1 \leq i \leq m$). By iteration we can assume $m = 1$. To simplify notation write a for a_1. By assumption, two orthogonal bases of V exist (which we denote by $u, u_1, \ldots, u_n$ and $v, v_1, \ldots, v_n$) such that

$$(u, u) = a\ (\neq 0) \qquad (v, v) = a\ (\neq 0)$$
$$(u_i, u_i) = b_i \quad \text{and} \quad (v_i, v_i) = c_i.$$

We know (see Section 2.4) that we can find an orthogonal transformation T that carries u into v. Then T necessarily maps $\langle u \rangle^{\perp}$ into $\langle v \rangle^{\perp}$. Look at the space $\langle v \rangle^{\perp}$. In it we can easily identify two orthogonal bases: $Tu_1, \ldots, Tu_n$ and $v_1, \ldots, v_n$. The former yields $(Tu_i, Tu_i) = (u_i, u_i) = b_i$, the latter $(v_i, v_i) = c_i$, $1 \leq i \leq n$. Hence $[b_1, \ldots, b_n] \overset{c}{=} [c_1, \ldots, c_n]$.

Case 2 (Some a_i's are zero). The basic idea in this situation is to work modulo the radical. The radical R of V is the subspace

$$\{x : (x, y) = 0, \text{ for all } y \text{ in } V\}.$$

Observe firstly that the dimension of R equals the number of isotropic vectors in a (and consequently any) orthogonal base of V. This informs us that matrices $[a_1, \ldots, a_m, b_1, \ldots, b_n] \overset{c}{=} [a_1, \ldots, a_m, c_1, \ldots, c_n]$ have *the same* number of zero entries on the diagonal, the number being the dimension r (say) of R. Suppose k of the b_i's, and thus k of the c_i's, are zero. By an eventual relabeling we can assume $a_1, \ldots, a_{m-(r-k)} \neq 0$, $b_1, \ldots, b_{n-k} \neq 0$, and $c_1, \ldots, c_{n-k} \neq 0$.

Introduce a bilinear form on the quotient space V/R by defining $(\bar{x}, \bar{y})$ to be (x, y) (the form on V), with $\bar{z}$ signifying the coset $z + R$. The form so defined on R/V is well defined.

The $m + n - r$ nonzero images of the vectors of the first orthogonal base of V form an orthogonal base for V/R. Similarly for the second base. These two orthogonal bases of V/R allow us to write

$$[a_1, \ldots, a_{m-(r-k)}, b_1, \ldots, b_{n-k}]$$
$$\overset{c}{=} [a_1, \ldots, a_{m-(r-k)}, c_1, \ldots, c_{n-k}].$$

Case 1 allows us to conclude that $[b_1, \ldots, b_{n-k}] \overset{c}{=} [c_1, \ldots, c_{n-k}]$. In terms of matrices we can therefore write

$$C = PBP^t,$$

where $B = [b_1, \ldots, b_{n-k}]$, and $C = [c_1, \ldots, c_{n-k}]$. By way of the matrix $\begin{pmatrix} P & 0 \\ 0 & I \end{pmatrix}$, with I the identity matrix of dimension k, we obtain the congruence of $[b_1, \ldots, b_n]$ and $[c_1, \ldots, c_n]$. (Recall that $b_{n-k+1} = \cdots = b_n = 0$, and $c_{n-k+1} = \cdots = c_n = 0$.) This ends the proof.

Finally we deduce

Witt's Cancellation Law. *If* $\begin{pmatrix} A & 0 \\ 0 & B \end{pmatrix} \overset{c}{=} \begin{pmatrix} C & 0 \\ 0 & D \end{pmatrix}$ *and* $A \overset{c}{=} C$, *then* $B \overset{c}{=} D$. [All matrices are symmetric with rational entries.]

The information available [i.e., $\begin{pmatrix} A & 0 \\ 0 & B \end{pmatrix} \overset{c}{=} \begin{pmatrix} C & 0 \\ 0 & D \end{pmatrix}$, and $A \overset{c}{=} B$] suggests writing the vector space V as the orthogonal direct sum

$$V = U \oplus W.$$

The subspace U has dimension equal to that of A (or C), and W to that of B (or D). We change bases separately on U and W to obtain: $A \overset{c}{=} L$ and (since $C \overset{c}{=} A$) also $C \overset{c}{=} L$, with L diagonal. Also (on W) $B \overset{c}{=} M$ and $D \overset{c}{=} N$, with M and N diagonal. Now,

$$\begin{pmatrix} L & 0 \\ 0 & M \end{pmatrix} \overset{c}{=} \begin{pmatrix} A & 0 \\ 0 & B \end{pmatrix} \overset{c}{=} \begin{pmatrix} C & 0 \\ 0 & D \end{pmatrix} \overset{c}{=} \begin{pmatrix} L & 0 \\ 0 & N \end{pmatrix}$$

and Witt's results on diagonal matrices inform us that $M \overset{c}{=} N$. By transitivity $B \overset{c}{=} M \overset{c}{=} N \overset{c}{=} D$, and the proof is complete.

2.6 A Concluding Remark

The orthogonal transformations on V form a group under composition of functions, called the *orthogonal group*. Let us call two subspaces of V isometric if there exists a bijective linear transformation between them that preserves the form. Witt's results inform us that *the orthogonal group acts transitively on isometric subspaces* (of the same dimension), this last statement being, by and large, a geometric affair. We recommend [1] and [3] for further reading.

REFERENCES

[1] I. Kaplansky, *Linear Algebra and Geometry: A Second Course*, Allyn and Bacon, Boston, 1969.

[2] G. H. Hardy, *An Introduction to the Theory of Numbers* (4th ed.), Oxford Clarendon Press, London, 1960.

[3] E. Artin, *Geometric Algebra*, Interscience, New York, 1957.

Hints and Solutions to Exercises

About the missing answers:
Euripides was wont to say,
"Silence is an answer to a wise man."

Chapter 1

1. When rolling two fair dice the chance of rolling a sum of 9 is $4/6^2$; when rolling three, the chance is $25/6^3$ (a slightly greater chance).

2. Let P be a nonempty set. Fix, once and for all, a subset S of P of odd cardinality. Consider the bijection $A \to (A - S) \cup (S - A)$ on the subsets A of P. This bijection maps subsets of even cardinality into subsets of odd cardinality and vice versa.

3. 11; a solution for the general case is known (but is somewhat intricate).

4. $\binom{4}{2}\binom{5}{3}\binom{7}{1}\binom{4+5+7}{6}^{-1}$. Generalize.

5. We include 1 and 90 in all calculations: (a) The sum of three distinct integers is even if and only if one of them is even, or all of them are even. The answer thus is $45 \cdot 45 \cdot 44/1!2! + 45 \cdot 44 \cdot 43/3!$. Similar arguments are used to answer parts (b) and (c).

6. Let $I = \{1, 2, \ldots, m\}$, and $1 \le n \le m - 1$. Denote by I_r the set of r-subsets of I. Consider the map f from $I_n \times \{1, \ldots, n\}$ into $I_{n-1} \times \{1, \ldots, m - n + 1\}$ defined by

$$f : ((a_1, \ldots, a_n), j) \to ((a_1, \ldots, \hat{a}_j, \ldots, a_n), a_j - j + 1).$$

Then f is a well-defined bijection. (By writing $\hat{a}_j$ we indicate that a_j is omitted as an entry.)

8. (a) $\binom{13}{2}^4\binom{52}{8}^{-1}$.

(b) Think of the 52 cards being partitioned into 3 stacks: hearts, spades, and other. The answer is

$$\binom{52}{8}^{-1} \sum_{i=0}^{4} \binom{13}{i}\binom{13}{i}\binom{52-26}{8-2i}.$$

(c) One needs to carefully work out the possibilities.

9. We agree that there are 365 days in a year. Then the probability that no birthdates will match is

$$\frac{365}{365}\cdot\frac{364}{365}\cdot\frac{363}{365}\cdot\,\cdots\,\cdot\frac{365-n+1}{365} = p_n,$$

say. The answer to the initial question is $1 - p_n$.

10. The "types" of possible occurrences for visits are (0, 3, 3), (1, 2, 3), and (2, 2, 2). There are

$$3\frac{6!}{3!3!} + 3!\frac{6!}{1!2!3!} + \frac{6!}{2!2!2!} = 510$$

possible ways in all.

11. $\binom{20-8+1}{8}\binom{20}{8}^{-1}$.

13. A positive integer less than a prime p cannot have p in its prime factorization. We write $a \equiv b$ (modulo p) if the difference $a - b$ is divisible by p.

14. $\dfrac{n}{n-3}\dbinom{n-3}{3}$.

16–19. Several of these exercises can be verified by induction. Additional insights will be gained upon reading Section 3 of Chapter 2.

20. Let $c(n, r)$ be the number of such sequences. By deleting (and reinserting n) establish the recurrence

$$c(n,r) = \sum_{k=0}^{r} \binom{r}{k} c(n-1,k), \qquad 1 \le r \le n-1.$$

21. Think in terms of probabilities. Suppose there are n errors in the book. The chance that A spots one is a/n, the chance that B spots one is b/n. Since the reading is done independently, the fraction c/n of errors found by both is approximately equal to $(a/n)\cdot(b/n)$. Thus n is approximately ab/c, and we estimate to have $ab/c - (a+b-c) = (a-c)(b-c)/c$ errors remaining.

22. $\binom{m}{r}\binom{n}{r}r!$.

23. Note that each subset S of cardinality s may be written as a union of k subsets in $(2^k - 1)^s$ different ways.

24. $\binom{n}{[n/2]}$.

25. Each integer can be written in the form $2^p q$, with q odd.

26. Look at the $\binom{n+1}{r+1}$ subsets with $r + 1$ elements taken from the set $\{0, 1, \ldots, n\}$. Delete the smallest elements. Proceed.

27. Establish a bijection to permutations with exactly k cycles.

Chapter 2

Section 2

1. Coefficient of x^{14} in $(x + x^2 + x^3 + x^4 + x^5 + x^6)^4$.

2. Coefficient of x^n in $(1 - x - x^2 - x^3 - x^4 - x^5 - x^6)^{-1}$.

3. Coefficient of x^{12} in $(1 + x + x^2)^{15}(1 + x + x^4)$.

4. Coefficient of x^{20} in $(1 + x^2 + x^4 + x^6 + \cdots)(x + x^3 + x^5 + x^7 + \cdots)^2$.

6. Start out with $1 + x + x^2 + \cdots = (1 - x)^{-1}$ and appropriately adjust its derivatives to obtain:

(a) $x\dfrac{d}{dx}[x(1 - x)^{-2}]$.

(b) $\dfrac{x^3}{3!}\dfrac{d^3}{dx^3}[(1 - x)^{-1}]$.

7. $\binom{10}{2}\binom{8}{3}\binom{5}{5}$; $2\binom{5}{2}\binom{3}{3}\left[\binom{10}{2}\binom{8}{3}\binom{5}{5}\right]^{-1}$; $\binom{7}{1}\binom{6}{2}\binom{4}{4}\left[\binom{10}{2}\binom{8}{3}\binom{5}{5}\right]^{-1}$.

8. $\dfrac{11!}{5!2!2!1!1!}$.

Section 5

1. Establish bijective correspondences between the various subproblems, prove the recurrence, and read the c_n's out of the resulting generating function.

2. $A(x) = x(20x^3 - 16x^2 + x + 1)^{-1}$. Expand.

3. Induction will get you a long way, but generating functions will help too, for example, writing $(1 - x - x^2)^{-1} = \Sigma_n(x + x^2)^n$ yields (e) immediately, and so on.

4. $a_n = \binom{n}{4} + \binom{n}{2} + 1$.

Section 6

1. $$\begin{vmatrix} 2 & 0 & 0 & -1 \\ 0 & 2 & -1 & 0 \\ 0 & -1 & 3 & -1 \\ -1 & 0 & -1 & 3 \end{vmatrix} = 21.$$

2. In general, the sum of the determinants of the $\binom{n}{k}$ principal minors of $C(G)$ of dimension k equals the kth elementary symmetric sum of eigenvalues of $C(G)$—see Section 4.14 for details. When $k = n - 1$ this reduces to what the exercise requires.

3–5. In all these cases the eigenvalues of the Kirchhoff matrix can be computed without too much trouble. For example, $C(K_n) = nI - J$, with I the identity matrix and J the matrix with all entries 1. An orthonormal set of eigenvectors for J is clearly also a set of orthonormal eigenvectors for $C(K_n)$. Thus $C(K_n)$ has eigenvalues: n of multiplicity $n - 1$, and 0 of multiplicity 1 (associated to the eigenvector **1**). By the previous exercise K_n has $(1/n)\prod_{i=1}^{n-1} n = (1/n)n^{n-1} = n^{n-2}$ labeled spanning trees, a familiar expression. [Now that we discuss this, it may be appropriate to mention that the eigenvalues of $C(G)$ can be computed for any strongly regular graph G (cf. Section 4.12). We thus have a *general formula* for the number of (labeled) spanning trees of any strongly regular graph.]

6. We have $C(G) + C(\overline{G}) = C(K_n) = nI - J$. The Kirchhoff matrices $C(G)$ and $C(\overline{G})$ have therefore the same set of eigenvectors. Consequently their eigenvalues are related in the following simple way: $\mu_i(\overline{G}) = n - \mu_i(G)$, for $1 \le i \le n - 1$, and $\mu_0(\overline{G}) = \mu_0(G) = 0$.

7. A reference may help: see Cheng, Masaro, and Wong's article, "Do nearly balanced multigraphs have more spanning trees?" in *Journal of Graph Theory*, **8**, 342–345 (1985).

Chapter 3

Section 2

1. (a) We have $\omega^{mn} = e^{2\pi i n} = 1$; thus $0 = (\omega^n)^m - 1 = (\omega^n - 1)\sum_{k=0}^{m-1} kn$; when is $\omega^n = 1$?

(b)
$$\frac{1}{m}\sum_{j=0}^{m-1} F(x\omega^j)\omega^{-lj} = \frac{1}{m}\sum_{j=0}^{m-1}\Big(\sum_{k\ge 0} a_k x^k \omega^{jk}\Big)\omega^{-lj}$$
$$= \frac{1}{m}\sum_{k\ge 0} a_k x^k \sum_{j=0}^{m-1}\omega^{j(k-l)} = \frac{1}{m}\sum_{m\mid(k-l)} a_k x^k m.$$

2. (a) Let $m = 2$, $l = 1$, and $F(x) = (1 + x)^{2n+1} = \sum_{k=0}^{2n+1}\binom{2n+1}{k}x^k$ in the previous exercise.

(b) With $\omega = e^{2\pi i/3}$ we have

$$\sum_{k=0}^{n}\binom{3n}{3k} = \frac{1}{3}\sum_{j=0}^{2}(1 + \omega^j)^{3n} = \frac{1}{3}\left[2^{3n} + (-\omega^2)^{3n} + (-\omega)^{3n}\right].$$

(c) Use Exercise 1(b) and simplify.

3. Easy substitutions in the binomial inverse pair yield (a) and (b).

4. Abel's identities can help.

6. Use lattice paths to obtain the first identity, then invert to derive the second. The last is similar to the first.

Section 4

1. (a), (c), (e).

2. (a) -3.

(c) $(1,2)$.

(e) Does not exist.

(f) $(-1,0,1)$.

3. Using Lagrange's theorem we obtain

$$F(x) = \sum_{n\geq 1}\left(\frac{1}{n}\sum_{k=1}^{n}(-1)^k\binom{n}{k}\frac{k^{k-1}}{(k-1)!}\right)x^n.$$

5. $\ln(1+C) = \sum_{n\geq 1}\left(\frac{1}{n}\sum_{j=0}^{n-1}(-1)^j\binom{2n-2-j}{n-1-j}\right)x^n.$

6. $G(x, y) = \Sigma a_{mn}x^m y^n$, where

$$a_{mn} = (-1)^{n+m}\frac{1}{n}\binom{n}{m}\sum_{k\geq 1}(-1)^k\binom{n-m}{k}\binom{2k-2}{k-1}, \qquad n\geq 1,\ m\geq 0.$$

7. Use MacMahon's master theorem.

8. $(1-xy)^{-1} = 1 + \sum_{n\geq 1}(\frac{1}{n}\sum_{j=0}^{n-1}(j+1)y^{j+1}\frac{(-n)^{n-1-j}}{(n-1-j)!})(xe^x)^n.$

Section 5

1. We can write $x(d/dx)(S(x)-1) = nxS^2(x)$, which gives $S(x) = (1-nx)^{-1}$.

3. We seek the coefficient of x^n in $\Pi_{i=1}^k(x^{a_i} + x^{a_i+1} + \cdots + x^{b_i-1})$. Simplify.

5. Proceed as in Example 2 in Section 3.22.

7. Coefficient of $\mathbf{x}^2$ in $A(\mathbf{x})$, where $A(\mathbf{x})$ is as in Exercise 5.

8. Choose weights appropriately, then use the weight preserving bijection (3.29).

10. Let the weight of a planted plane tree with i nonroot vertices, j of which are of odd degrees, be (i, j). Use (3.29) to find the (bivariate) generating function F for such trees. The answer will be found in the coefficient of $x^{2n}y^{2m+1}$. [Note that F satisfies $F = x(y + F + yF^2 + F^3 + \cdots)$.]

11–12. Similar to 10.

13. Exercise 5 may be helpful.

Section 6

1. $1 + x + x^2$; there are seven two-dimensional subspaces.

2. $\langle(1,0)\rangle$, $\langle(0,1)\rangle$, $\langle(1,1)\rangle$, $\langle(-1,1)\rangle$.

4. $(q^n - 1)(q^n - q)\cdots(q^n - q^{k-1})$ and $(q^k - 1)(q^k - q)\cdots(q^k - q^{k-n-1})$, respectively.

5. Denote by c_{km}^p the number of nonnegative integral solutions to the system

$$x_0 + x_1 + x_2 + \cdots + x_p = k$$
$$x_1 + 2x_2 + \cdots + px_p = m.$$

By Section 8 of Chapter 2,

$$\prod_{l=0}^{p}(1 - tx^l) = \sum_{i,j} c_{ij}^p t^i x^j = \sum_i \Big(\sum_j c_{ij}^p x^j\Big) t^i$$
$$= \sum_i c_i^p(x) t^i = F_p(t),$$

where $c_i^p(x) = \sum_j c_{ij}^p x^j$. Note that $(1 - t)F_p(t) = (1 - tx^{p+1})F_p(tx)$. Equating the coefficients of t^i on both sides of this functional equation we obtain

$$c_i^p(x) - c_{i-1}^p(x) = x^i c_i^p(x) - x^{p+i} c_{i-1}^p(x).$$

Thus $c_i^p(x) = (x^{p+i} - 1)(x^i - 1)^{-1} c_{i-1}^p(x)$, and $c_0^p(x) = 1$. We iteratively obtain $c_i^p(x) = \prod_{l=1}^i (x^{p+l} - 1)/(x^l - 1)$. In particular, for $i = k$ and $p = n - k$ we obtain

$$c_k^{n-k}(x) = \sum_{l=1}^{k} \frac{x^{n-k+l} - 1}{x^l - 1} = \begin{bmatrix} n \\ k \end{bmatrix}(x).$$

Therefore c_{km}^{n-k} is the coefficient of x^m in $\begin{bmatrix} n \\ k \end{bmatrix}(x) = c_k^{n-k}(x) = \sum_j c_{kj}^{n-k} x^j$.

Chapter 4

Section 1

1. No, it cannot.

2. This is the simplest case of Turàn's theorem.

3. It may be easier to show that the complementary graph (a direct sum of complete graphs of the same size) has a maximal number of triangles among all regular graphs of the same degree. Count $|\{(y, \Delta) : y \in \Delta, \Delta$ a triangle$\}|$ two different ways. The result follows by complementation (with a little bit of work).

6–8. Generally straightforward (and time consuming).

9. Consider a complete graph on $R(p-1, q; 2) + R(p, q-1; 2)$ vertices with edges colored either red or blue. Let v be a vertex. Distinguish two cases: (a) There are $R(p, q-1; 2)$ red edges emanating out of v; (b) there are $R(p-1, q; 2)$ blue edges emanating out of v. One of these two cases must occur. Proceed.

10. By induction on $p + q$, and with the help of Exercise 9.

11. Let $a_1, a_2, \ldots, a_{n^2+1}$ be the sequence. Supposing that there is no increasing subsequence of length $n + 1$, we force the existence of a decreasing sequence of that length. For each a_k let i_k be the length of the longest increasing subsequence beginning with a_k. Since all $n^2 + 1$ of the i_k's are between 1 and n (inclusive), some length, say m, will be used $n + 1$ times at least. The corresponding $n + 1$ integers form the decreasing sequence sought.

12. Observe that $n(3) = 3$ and prove that $n(4) = 5$. Then show that if there are k points, no three of which are colinear, and such that all the quadrilaterals formed from them are convex, then the points are the vertices of a convex k-gon. (A set of points in the plane is called convex if together with two distinct points it contains the whole line segment joining the two points.) Proceed.

14. Consider the complete graph with vertices labeled $1, 2, \ldots, r_n$ [where $r_n = R(3, 3, \ldots, 3; 2)$, with n occurrences of 3]. Color the edges with colors $1, 2, \ldots, n$ such that edge uv is assigned color j if and only if $|u - v| \in S_j$. Use Ramsey's theorem.

Section 2

1. $(n, n-1-d, n-2d+c-2, n-2d+a)$.

3. Petersen's graph.

5. No.

7. See Delsarte, Goethals, and Seidel's article, Spherical codes and designs, *Geometriae Dedicata*, **6**, 363–388 (1977).

Section 3

1. $n^{-1}(d-\lambda_1)^{f_1}(d-\lambda_2)^{f_2}$, where λ_1, λ_2, f_1, and f_2 are explicitly written in terms of the parameters (n, d, a, c) of the strongly regular graph in (4.3).

3. $A^k x = \mu^k x$, thus by linearity $P(A)x = P(\mu)x$.

4. Pick a vertex out of each of the $n - k$ components of the forest and amalgamate them into one vertex: what results is a spanning tree in an amalgamated graph on $k + 1$ vertices. Relate this idea to the statement written at the very end of Section 4.17.

6–8. The answers to these problems are not known to the author.

Chapter 5

Section 3

1. Yes.

2. Yes.

3. The second.

4. Yes.

Section 5

1. The maximal flow has value 6.

4. 8, it seems.

Section 7

2. Yes.

5. Use Exercise 4.

Chapter 6

Section 4

3. (a) $P_G = \frac{1}{24}(x_1^{12} + 3x_2^6 + 6x_4^3 + 6x_1^2x_2^5 + 8x_3^4)$.

(b) $P_G = \frac{1}{24}(x_1^{12} + 3x_1^4x_2^4 + 6x_1^4x_4^2 + 6x_2^6 + 8x_3^4)$.

(c) They are not isomorphic.

(d) More patterns arise when coloring the 12 (inside and outside) faces.

(e) The answer is the coefficient of $y_1^2y_2^3y_3^7$ [multiplied by $n(n-1)(n-2)$] in

$$P_G\left(\sum_{i=1}^{n} y_i, \sum_{i=1}^{n} y_i^2, \ldots, \sum_{i=1}^{n} y_i^{12}\right),$$

where P_G is the respective cycle index calculated in parts (b) and (a), respectively.

4. The cube and the octahedron are "dual" (in the sense that the latter is obtained from the former by identifying the point in the middle of each face, and vice versa). It is evident, therefore, that the cycle indexes that arise are the same as those for the cube (when represented on faces, edges, and vertices, respectively).

6. Label the vertices of the Petersen graph as displayed in Section 7.31 (or as described in Section 4.11). This helps us show that the (automorphism) group in question is precisely A_5, the group of even permutations on five symbols. Go through Pólya's recipe. This is not a trivial exercise.

8. We encourage the student to first list the 18 pairs of $X \times Y$, then list the 36 group elements of $G \oplus H$. Proceed to write out the representation of each group element on $X \times Y$ (a somewhat laborious but instructive task).

Section 7

1–2. A step-by-step application of the recipe yields the answers sought.

3. This problem parallels that of enumerating nonisomorphic graphs presented in Section 6.26. In this case one needs to find the representation of S_v on ordered pairs of vertices. We discussed this in (7) of Section 4.

5. The patterns correspond to the orbits of the group H acting on the range R. In case of the symmetric group the action is transitive, thus only one pattern arises.

9–10. We direct the reader's attention to the *Remark* that appears at the very end of Section 1. Reference [5] in Chapter 6 contains these results.

Chapter 7

Section 3

2. A Hadamard matrix of order 2^n may be obtained by tensoring $\begin{pmatrix} 1 & 1 \\ -1 & 1 \end{pmatrix}$ with itself n times. One obtains a Hadamard matrix of order 12 by starting out with the Paley design (7.3) and reversing the process described in Section 7.3.

4. No, it cannot.

5. Fix a block α. Denote by B the set of blocks different from but not disjoint from α. Let n_i of these blocks intersect α in i points. For $j = 0, 1, 2$ count in two ways the cardinality of the set

$$\{(\sigma_j, \beta) : \sigma_j \subseteq \alpha, \sigma_j \subseteq \beta; \beta \in B\}.$$

Explicitly compute $\Sigma_i(i-x)^2 n_i$ and observe that it is nonnegative for all x.

7. Imitate the complex numbers by constructing finite analogs for them, such as $GF(p) + iGF(p)$, where $i \notin GF(p)$ and $i^2 = -1$; in other words p should be such that -1 is not a square in $GF(p)$. Consider the maps mentioned in the hint. What are the parameters of the resulting designs? The design in (7.3) is obtained by taking $p = 3$.

Section 6

1. In order to understand the proof of the BRC theorem it really helps to examine the case of a projective plane of order 6.

3–4. See [5, p. 11, 12, and 57] (reference in Chapter 7).

6. The group arises naturally as the group of nonsingular linear maps (or matrices) form a three-dimensional vector space over $GF(2)$ into itself (such matrices obviously map subspaces of dimension 2 into subspaces of dimension 2). By projectivizing, the two-dimensional (vector) subspaces become projective lines and the group in question may also be viewed as consisting of those permutations on the projective points that preserve the projective lines. It is doubly transitive on points. To show that it is simple compute the sizes of the conjugacy classes. A normal subgroup is a (disjoint) union of such classes. But no sum of sizes of conjugacy classes is a proper divisor of 168.

7. The group consists of linear maps and automorphisms of the field. See [7, Section 2.6] of Chapter 7.

Section 7

1. See Section 4.11.

7. The full automorphism group of the Petersen graph is S_5. This should help.

8. The full automorphism group has order 108.

9. No, it is not isomorphic to Desargue's configuration.

11. See Raghavarao, *Constructions and Combinatorial Problems in Design of Experiments*, Wiley, New York, 1971.

Chapter 8

Section 1

1. (a) $1/2, 1/\sqrt{3}$. (b) $1, \sqrt{2}$. (c) $2, \sqrt{2}$.

(d) The mean is infinite (you guessed 0, huh?). No sense in talking about variance.

(e) Look at $M(t) = E(e^{tY}) = \Sigma_y e^{ty}/2^y$. Observe that

$$\mu = \sum_y \frac{y}{2^y} = \sum_y \frac{d}{dt}\left(\frac{e^{ty}}{2^y}\right)\Bigg|_{t=0} = \frac{d}{dt}M(t)\Bigg|_{t=0}$$

$$= \frac{d}{dt}\left[(1 - 2^{-1}e^t)^{-1} - 1\right]\Bigg|_{t=0} = 2.$$

Differentiate twice (and adjust) to obtain the variance. (This trick is known as the moment generating function technique.)

2. $\int_{1/2}^{1}\int_0^{1/4} 4y_1y_2\,dy_1\,dy_2$; $\int_0^1\int_{y_1/2}^{1} 4y_1y_2\,dy_2\,dy_1$. [To compute the second integral draw out the region $\{(y_1, y_2): y_1 < 2y_2\}$ within the unit square.]

3. Equality is attained if and only if $c_1Y_1 + c_2Y_2 + c_3 = 0$ for some constants c_1, c_2, c_3.

4–6. Select suitable substitutions. These problems are fairly intricate, but not exceedingly intricate.

7. (a) OK. (b) Unusual.
(c) Extremely unusual. (d) Somewhat unusual.
The chance that an observation would be more than three standard deviations away from the mean is less than 1%.

8. (a) and (d) are extremely unusual; (c) is OK; (b) is unusual. The random variable $\chi^2(n)$ has mean n and variance $2n$.

Section 3

1. Main effects are 5 (for nickel) and 1 (for manganese). Interaction is -7. The high negative interaction suggests that presence of both alloys reduces the strength.

2. Seven. Sixteen. Two. Four. Eight.

3. Four.

7. Yes, it is.

Section 4

1. $\begin{matrix} 1 & 1 & 2 \\ 2 & 3 & 3 \end{matrix}$.

2. Any linear map from the image space to the original space would do, so long as it is the identity map when identifying the image and the kernel.

3. Differences $\alpha_i - \alpha_j$ are estimable if and only if the vectors $l_i - l_j$ (with l_m the vector with 1 in the mth coordinate and 0 elsewhere) are in the row span of C. This happens if and only if $l_1 - l_m$ form a basis for the row span of C, that is, if and only if C has rank exactly one less than the dimension.

4. B projects orthogonally onto the column space of A. The maps B and $I - B$ are orthogonal projections onto the column space of A and the orthogonal complement of that space, respectively.

5. The C matrices are just 3×3.

7. The graph $G(\delta)$ is obtained from $G(d)$ upon the addition of several edges or loops. The difference $C_\delta - C_d$ is simply equal to the information matrix of the set of edges and loops added; it is (as is the case with any information matrix) a nonnegative definite matrix. This difference is the zero matrix if and only if only loops were added.

8. The binary designs.

Section 5

1. 12.

2. Coordinatize over $GF(4)$. Sixty-four points.

4. 42?

Chapter 9

Section 3

1. Draw the Hasse diagrams. Compute the Möbius function in the form of an upper triangular matrix in each case; see the answer to Exercise 2, below. Verify Hall's results on a couple of diagrams.

2.

	(1)	⟨(12)⟩	⟨(34)⟩	⟨(12)(34)⟩	$Z_2 \oplus Z_2$
(1)	1	−1	−1	−1	2
⟨(12)⟩		1	0	0	−1
⟨(34)⟩			1	0	−1
⟨(12)(34)⟩				1	−1
$Z_2 \oplus Z_2$					1

4. A list of all subgroups (with inclusions) is necessary in each case. This is a long but instructive exercise. At least for case (b) the reader should try to give a complete solution.

7. Proceed as at the very end of Section 9.4.

Section 5

8. We have $g(S) = \Sigma_{T \leq S} f(T)$. By the Möbius inversion formula $f(S) = \Sigma_{T \leq S} g(T)\mu(T, S)$, where μ is the familiar Möbius function on the lattice of subspaces of V.

9. (a) For $m = 0$ or n the statement is clearly true. Fix therefore m, $0 < m < n$. Choose as "red" subsets (as mentioned in Exercise 8) all subgroups of order p^m. With the notation of Exercise 8 we clearly have $f(G) = 0$, and therefore

$$0 = f(G) = \sum_{T \leq G} g(T)\mu(T, G) = g(G) + \sum_{T < G} g(T)\mu(T, G).$$

Or

$$g(G) = -\sum_{T < G} g(T)\mu(T, G).$$

This formula allows us to work by induction. In addition $\mu(T, G) = 0$ (see Proposition 9.1) unless T is a major subgroup of G. Look at the last formula for $g(G)$ modulo p. By the inductive assumption $g(T) = 1$ (modulo p). But $\mu(T, G) = (-1)^k p^{\binom{k}{2}}$, where p^k is the index of the major subgroup T in G. How many major subgroups of index p^k in G are there? Work modulo p. The same ideas can be used on parts (b), (c), and (d).

12. Compute the number of automorphisms for $Z_2 \oplus Z_4 \oplus Z_4 \oplus Z_8$, and then for $Z_9 \oplus Z_{27}$ (using Delsarte's formula in each case). The answer is the product of the two numbers.

14. Derive the corresponding signatures and apply Delsarte's result.

15. What are the possible signatures?

17. The familiar expressions of the Eulerian function and the Gaussian polynomials will be obtained.

19. Yes.

20. Is the type $(1, 1, 1, \ldots, 1)$?

Index

Applied Probability and Statistics (Continued)

JUDGE, HILL, GRIFFITHS, LÜTKEPOHL and LEE • Introduction to the Theory and Practice of Econometrics
JUDGE, GRIFFITHS, HILL, LÜTKEPOHL and LEE • The Theory and Practice of Econometrics, *Second Edition*
KALBFLEISCH and PRENTICE • The Statistical Analysis of Failure Time Data
KISH • Statistical Design for Research
KISH • Survey Sampling
KUH, NEESE, and HOLLINGER • Structural Sensitivity in Econometric Models
KEENEY and RAIFFA • Decisions with Multiple Objectives
LAWLESS • Statistical Models and Methods for Lifetime Data
LEAMER • Specification Searches: Ad Hoc Inference with Nonexperimental Data
LEBART, MORINEAU, and WARWICK • Multivariate Descriptive Statistical Analysis: Correspondence Analysis and Related Techniques for Large Matrices
LINHART and ZUCCHINI • Model Selection
LITTLE and RUBIN • Statistical Analysis with Missing Data
McNEIL • Interactive Data Analysis
MAINDONALD • Statistical Computation
MANN, SCHAFER and SINGPURWALLA • Methods for Statistical Analysis of Reliability and Life Data
MARTZ and WALLER • Bayesian Reliability Analysis
MIKÉ and STANLEY • Statistics in Medical Research: Methods and Issues with Applications in Cancer Research
MILLER • Beyond ANOVA, Basics of Applied Statistics
MILLER • Survival Analysis
MILLER, EFRON, BROWN, and MOSES • Biostatistics Casebook
MONTGOMERY and PECK • Introduction to Linear Regression Analysis
NELSON • Applied Life Data Analysis
OSBORNE • Finite Algorithms in Optimization and Data Analysis
OTNES and ENOCHSON • Applied Time Series Analysis: Volume I, Basic Techniques
OTNES and ENOCHSON • Digital Time Series Analysis
PANKRATZ • Forecasting with Univariate Box-Jenkins Models: Concepts and Cases
PIELOU • Interpretation of Ecological Data: A Primer on Classification and Ordination
PLATEK, RAO, SARNDAL and SINGH • Small Area Statistics: An International Symposium
POLLOCK • The Algebra of Econometrics
PRENTER • Splines and Variational Methods
RAO and MITRA • Generalized Inverse of Matrices and Its Applications
RÉNYI • A Diary on Information Theory
RIPLEY • Spatial Statistics
RIPLEY • Stochastic Simulation
ROSS • Engineering Statistics
RUBIN • Multiple Imputation for Nonresponse in Surveys
RUBINSTEIN • Monte Carlo Optimization, Simulation, and Sensitivity of Queueing Networks
SCHUSS • Theory and Applications of Stochastic Differential Equations
SEAL • Survival Probabilities: The Goal of Risk Theory
SEARLE • Linear Models
SEARLE • Matrix Algebra Useful for Statistics
SPRINGER • The Algebra of Random Variables
STEUER • Multiple Criteria Optimization
STOYAN • Comparison Methods for Queues and Other Stochastic Models
TIJMS • Stochastic Modeling and Analysis: A Computational Approach
TITTERINGTON, SMITH, and MAKOV • Statistical Analysis of Finite Mixture Distributions
UPTON • The Analysis of Cross-Tabulated Data
UPTON and FINGLETON • Spatial Data Analysis by Example, Volume I: Point Pattern and Quantitative Data

(*continued from front*)